GENES, BRAIN FUNCTION, AND BEHAVIOR

GENES, BRAIN FUNCTION, AND BEHAVIOR

What Genes Do, How They Malfunction, and Ways to Repair Damage

DOUGLAS WAHLSTEN

Professor Emeritus, Department of Psychology, University of Alberta

ACADEMIC PRESS

An imprint of Elsevier

Academic Press is an imprint of Elsevier
125 London Wall, London EC2Y 5AS, United Kingdom
525 B Street, Suite 1650, San Diego, CA 92101, United States
50 Hampshire Street, 5th Floor, Cambridge, MA 02139, United States
The Boulevard, Langford Lane, Kidlington, Oxford OX5 1GB, United Kingdom

Notices
Knowledge and best practice in this field are constantly changing. As new research and experience broaden our understanding,
changes in research methods, professional practices, or medical treatment may become necessary.

Practitioners and researchers must always rely on their own experience and knowledge in evaluating and using any information,
methods, compounds, or experiments described herein. In using such information or methods they should be mindful of their
own safety and the safety of others, including parties for whom they have a professional responsibility.

To the fullest extent of the law, neither the Publisher nor the authors, contributors, or editors, assume any liability for any
injury and/or damage to persons or property as a matter of products liability, negligence or otherwise, or from any use or
operation of any methods, products, instructions, or ideas contained in the material herein.

Library of Congress Cataloging-in-Publication Data
A catalog record for this book is available from the Library of Congress

British Library Cataloguing-in-Publication Data
A catalogue record for this book is available from the British Library

ISBN 978-0-12-812832-9

For information on all Academic Press publications
visit our website at https://www.elsevier.com/books-and-journals

Working together
to grow libraries in
developing countries

www.elsevier.com • www.bookaid.org

Publisher: Nikki Levy
Acquisition Editor: Farra, Natalie
Editorial Project Manager: Bennett, Timothy
Production Project Manager: Vijayaraj Purushothaman
Cover Designer: Matthew Limbert

Typeset by SPi Global, India

Contents

Author Biography

Dr. Wahlsten received the BSc degree in physics from Alma College and the PhD in psychology and psychobiology from the University of California at Irvine. He then did postdoctoral study at the Institute for Behavioral Genetics at the University of Colorado before moving to the University of Waterloo and then the University of Alberta, where he is currently Professor Emeritus. He taught undergraduate and graduate courses in behavioral genetics from 1969 to 2010. He was cofounder and then served as President of the International Behavioural and Neural Genetics Society, and in 2006 received the Distinguished Scientist Award from that society. He was also cofounder of the journal *Genes, Brain and Behavior* and author of the treatise on *Mouse Behavioral Testing* published in 2010. He has published numerous articles concerning behavioral and neural genetics, ranging from learning and hereditary brain defects in mice to eugenic sterilization of children in Alberta as well as theoretical issues pertinent to human intelligence. He has also published many critical evaluations of statistical models that purport to separate influences of heredity and environment on human and animal behavior.

Preface

Genes are now in the news almost daily. Genes are huge molecules that are part of our DNA, and the total DNA in the nucleus of each cell in our body is the genome. Viewed at the molecular level, DNA is not so complicated. It consists of long chains of just four kinds of small chemicals that are abbreviated A, C, G, and T. The genome can be represented as a very long sequence of those four letters, and the unique orders of those bases give genes their properties. Inspired by that amazing DNA sequence, some writers have imbued the genes with magical abilities to make things happen. A more scientific approach examines what genes actually do and the roles they play in forming a brain, organizing behaviors, and processing information from the environment. It makes a clear distinction between the specific set of genes a person inherits from the parents, termed the *genotype*, and the properties of things we study and measure, termed *phenotypes*. Behaviors are phenotypes, as are features of the nervous system.

A landmark in the progress of modern science was announced on June 26, 2000, at the White House in Washington, DC, by US President Bill Clinton and Tony Blair, Prime Minister of Great Britain. They claimed that the full sequence of the human genome had been determined by two large, competing teams of scientists, and the press conference officially declared the race a tie. The press release on that day heralded the dawn of "a new era of molecular medicine" (Clinton, 2000). Later, leading scientists claimed that knowledge of the DNA would tell much about behavior too. James D. Watson, codiscoverer of the structure of DNA, told an interviewer that, if he were embarking on a research career after the DNA sequence was known, "I'd be working on something about connections between genes and behavior. You can find genes for behaviors" (Rennie, 2003). The National Institutes of Health (NIH) in the United States and other agencies began to fund large-scale studies using the new genome data to search for genes involved in important psychiatric and behavioral disorders.

Eighteen years after Clinton's proud announcement, much has been learned by knowing the DNA sequence intimately, but some of the promises of the genome projects have not been realized. The journey from knowledge of molecules to understanding of brain function and behavior proved to be long and steep with many pitfalls along the way. The official DNA sequence was determined for a very small number of anonymous blood donors. The original genome project took no account of the environments of the people who donated their DNA for research. Gradually, it became clear that a better understanding of brain function, mental disorders, and behaviors required an expanded scope of knowledge.

PRECISION MEDICINE AND CONSUMER GENETICS: THE PERSONAL GENOME

Consequently, a new project with a scope unimaginable in the year 2000 was begun on February 25, 2016, under the leadership of President Barack Obama and the NIH (Kaiser, 2016). The aim of the *Precision Medicine Initiative* is to collect DNA samples and determine the genome sequence of 1 million Americans of a wide range of ethnicities in order to "tailor medical treatments to individuals" and extend the potential benefits to "underserved groups." The project aims to collect extensive data on the environments of those who donate their DNA so that researchers can explore "the interplay among genetics, lifestyle factors, and health." The public interface of the program, named *All of Us*, seeks "better insights into the biological, environmental, and behavioral influences" on diseases that currently have no means of prevention or treatment.

In the United Kingdom, the *UK10K* project has been launched to sequence 10,000 genomes (Charolidi, 2015), and the National Health Service is expanding the effort to cover 100,000 genomes. The Canadian Cancer Research Alliance has launched the *Canadian Partnership for Tomorrow Project* to gather genomic data on 300,000 people, including tissue samples from a subset of 150,000, and results from that project are now becoming available (Fave et al., 2018). It is designed to assess lifestyle and health-related behaviors as well as environmental risk factors related to cancer. Entire genomes of many people are being made available in public access databases through the *Personal Genome Project* that includes volunteers from the United States, Canada, the United Kingdom, Austria, and China.

Much is already known and well understood about the genes, brain functions, and many behaviors, but research done since 2000 has uncovered a degree of complexity that was not anticipated. Consequently, on some days, it feels like the ocean of things not understood is expanding faster than the islands of good understanding. It is essential that we find a way to simplify this almost overwhelming complexity without oversimplifying. This book provides sufficient background knowledge to make sense of many of the new findings.

The range of people interested in genetics is expanding greatly with the advent of affordable personal genome testing (O'Connor, 2017). All one needs to do is spit into a tube and mail it to a company along with payment. The list of service providers maintained by the International Society of Genetic Genealogy is growing rapidly. Most recommend that a person take her results to a family physician for action or pay a fee online to consult a medical geneticist, but the consumer of the data is ultimately responsible for interpreting the report. Possibilities for misinterpretation are abundant. A recent analysis of the genomes of 56 adult volunteers who were initially thought to be healthy detected at least one faulty gene in 95% of the people and genetic defects with significant implications for the person's health in 14 people (Reuter et al., 2018). The authors pointed out possible benefits from using genome data "preemptively for precision medicine." Implications for the patients and the healthcare system are not trivial.

This book encourages the reader to approach genetic data and risk factors with caution. DNA technology may have good accuracy and even clinical value when a defect in one specific gene is detected, but for the many psychiatric and behavioral problems that are genetically complex and sensitive to many environmental influences, prediction of disease from results of a genome profile is generally poor (Bunnik, Schermer, & Janssens, 2011; Kalf et al., 2014).

CLEAR QUESTIONS, DIFFERENT ANSWERS, OR NO ANSWERS

When considering research that probes the frontiers of knowledge, we expect results of new studies to be somewhat ambiguous, conflicting, and perplexing. Clarity is a hard-won quality of mature fields of science. The study of genetics, brain function, and behavior is well beyond its infancy. Not yet mature, the field seems to be in an awkward adolescence, full of promise but embarrassed by the many things that remain unknown. We need to develop skills at seeing the larger picture and the trends, even when many of the finer details, such as the precise length of the genome or number of genes it encodes, are still in doubt.

The Clinton/Blair announcement conveyed to most people that the genome project was complete, but the completion of the project turned out to be an elusive goal because the finish line kept being moved. Maneuvers behind the scenes by competing teams of researchers, one of which was privately financed and hoped to profit from its work on the DNA sequence, injected urgency and uncertainty into deliberations about when to make major announcements (Wickelgren, 2002). In 2001, they agreed to publish simultaneously in February in the prominent journals *Science* (Venter et al., 2001) and *Nature* (International Human Genome Sequencing Consortium, 2001). The Venter group proposed that the human genome consisted of 2.905 billion chemical bases that embodied 26,383 genes but still had 116,442 gaps, while the IHGSC group claimed 2.847 billion bases encoding 30,000–35,000 genes with 102,068 gaps remaining in the sequence. Gradually, the gaps were filled. Researchers continued to publish "finished" sequence one chromosome at a time, completing the task in 2006 when chromosome 1 was said to be 99.4% finished (Gregory et al., 2006).

The official human genome sequence is a *reference sequence* not based on any one person's DNA but a composite of DNA from several individuals. Further investigation revealed that there is never going to be a definitive answer to the question about the length of the human genome. When investigators determined the full sequence for eight people not included in the original genome study, four of whom were from the Yoruba group in Nigeria and two from Asia (Kidd et al., 2008), the eight sequences involved a total of 724 newly discovered blocks of DNA inserted into the known sequence, while another 747 blocks in the reference sequence seemed to be deleted in at least one of the eight people. It became apparent that all population groups and indeed all individuals have several relatively large blocks of DNA sequence either deleted altogether or duplicated. A recent report of 56 whole genomes sequenced for the Personal Genome Project found that every individual had a unique set of DNA segment insertions that were not present in the reference genome or deletions that erased some of the reference sequence from their genomes (Reuter et al., 2018). It now appears that the lengths of the DNA molecules that comprise the genome are slightly different for everyone. There is no one "normal" genome.

Recent estimates of the numbers of confirmed human genes are 21,315 in the semiofficial *GeneCards* database and 20,719 in the *GENCODE* version 19 database, but these have been called into question by efforts to detect proteins in human tissue samples. Genes are thought to code for the structures of proteins (Chapter 2). A recent report cited more than 2000 so-called genes in the official catalogs that may not be genes at all because they never get translated into real proteins (Ezkurdia et al., 2014). A deeper look at the data indicates that the concept of the gene itself needs to be revised (Portin & Wilkins, 2017).

How serious are these ambiguities about the nature of genetic material? It depends very much on why someone wants to answer the questions. For those just beginning to explore the field of human genetics and its relations with the nervous system and behavior, a rough approximation of genome length and the number of genes is quite adequate. Public databases such as *Online Mendelian Inheritance in Man* (OMIM) provide convenient ways to find information about genes, but consumers of the information are advised to imbibe with caution. The March 18, 2018, version of OMIM update states, "OMIM is intended for use primarily by physicians and other professionals concerned with genetic disorders, by genetics researchers, and by advanced students in science and medicine. While the OMIM database is open to the public, users seeking information about a personal medical or genetic condition are urged to consult a qualified physician for diagnosis."

AUTHORITATIVE SOURCES

In the Internet age, there are many sources of information on genes and related topics that vary widely in quality and credibility, and all of them are readily accessible. Scientific knowledge relies on the authority of the sources.

The primary research literature in scientific journals uses anonymous peer view to control quality. Different teams of researchers often reach different conclusions. Ideally, a research project is judged by the quality of its methods, not the results themselves. This book makes use of the primary literature when citing authorities but does not seek to conduct a comprehensive review of the published literature on every topic. In many places, the primary literature is utilized as a source of *instructive examples* that aid understanding.

Original Articles. For those affiliated with a university, it may be convenient to obtain a free copy of a research report published in a scientific journal in the form of a digital PDF (public document format) file through a library that has access to the original article. A search of PubMed or Google Scholar can also be productive, but some journals request payment of a fee for downloading the article by the general public. *Open-access* articles require no payment.

The reader should be aware that some scientific journals actually charge a hefty fee to the author of an article in order to make it open access. The fee will be affordable only for large and well-funded research teams with generous sponsors. Consequently, a literature review limited to open-access articles will offer a biased glimpse of the full richness of scientific knowledge.

Secondary sources are reviews of the primary literature in which an author summarizes methods and findings of many articles. Those kinds of reviews are themselves subjected to anonymous peer review by a journal, and they are especially valuable when the primary articles present a wide range of findings.

An entire book by one author may lack authority, no matter how long it is. The authority of a book is no greater than the original sources it cites, and it can be far less if views of the original authorities are not respected. For example, a recent book on race (Chapter 19) was criticized by more than a hundred authors of original articles on genetics it cited, who complained their work had been misinterpreted.

The mass media can be valuable sources for calling attention to the latest findings and highlighting controversies, as well as exploring the political implications of massive research projects financed by governments, but it is important to consult the original sources of information that is presented in a media report.

Wiki. Wikipedia is not regarded as part of the peer-reviewed scientific research literature and is usually not cited in a scholarly article. Nevertheless, a wiki article offers a convenient entry point into the published literature and provides many references with links to the original sources.

Internet URLs. It is common practice to cite a database or other Internet resource using the URL (uniform resource locator), but over the years, this has not served us very well. URLs often change, as we all know from the ubiquitous message "page not found." Meanwhile, search engines are becoming ever more powerful. Consequently, there are many organizations and resources utilized in this book that are identified by their full names without a URL. Those names or sometimes a unique abbreviation can be inserted into a search box to find the latest version.

Technical terms. In many places in the following chapters, a term is defined or a concept explained the first time it is used. But readers have diverse backgrounds. When a term unfamiliar to a reader is encountered, its meaning can usually be found with a quick Internet search.

References

Bunnik, E. M., Schermer, M. H., & Janssens, A. C. (2011). Personal genome testing: test characteristics to clarify the discourse on ethical, legal and societal issues. *BMC Medical Ethics, 12,* 11.

Charolidi, N. (2015). First insights from UK genome-sharing project. *BioNews, 820.* https://www.bionews.org.uk/page_95202.

Clinton, W. J. (2000). *President Clinton announces the completion of the first survey of the entire human genome: Hails public and private efforts leading to this historic achievement.* The White House, Washington, D.C.: Office of the Press Secretary.

Ezkurdia, I., Juan, D., Rodriguez, J. M., Frankish, A., Diekhans, M., Harrow, J., ... Tress, M. L. (2014). Multiple evidence strands suggest that there may be as few as 19,000 human protein-coding genes. *Human Molecular Genetics, 23,* 5866–5878.

Fave, M. J., Lamaze, F. C., Soave, D., Hodgkinson, A., Gauvin, H., Bruat, V., ... Awadalla, P. (2018). Gene-by-environment interactions in urban populations modulate risk phenotypes. *Nature Communications, 9,* 827.

Gregory, S. G., Barlow, K. F., McLay, K. E., Kaul, R., Swarbreck, D., Dunham, A., ... Prigmore, E. (2006). The DNA sequence and biological annotation of human chromosome 1. *Nature, 441,* 315–321.

International Human Genome Sequencing Consortium. (2001). Initial sequencing and analysis of the human genome. *Nature, 409,* 860.

Kaiser, J. (2016). NIH's 1-million-volunteer precision medicine study announces first pilot projects. *Science.* https://www. sciencemag.org/news/2016/02/nih-s-1-million-volunteer-precision-medicine-study-announces-first-pilot-projects.

Kalf, R. R., Mihaescu, R., Kundu, S., de Knijff, P., Green, R. C., & Janssens, A. C. (2014). Variations in predicted risks in personal genome testing for common complex diseases. *Genetics in Medicine, 16,* 85–91.

Kidd, J. M., Cooper, G. M., Donahue, W. F., Hayden, H. S., Sampas, N., Graves, T., ... Eichler, E. E. (2008). Mapping and sequencing of structural variation from eight human genomes. *Nature, 453,* 56–64.

O'Connor, A. (2017). Personal genetic testing is here. Do we need it? *New York Times.* https://www.nytimes.com/2017/10/03/well/live/personal-genetic-testing-is-here-do-we-need-it.html.

Portin, P., & Wilkins, A. (2017). The evolving definition of the term "gene". *Genetics, 205,* 1353–1364.

Rennie, J. (2003). A conversation with James D. Watson. *Scientific American, 288,* 67–69.

Reuter, M. S., Walker, S., Thiruvahindrapuram, B., Whitney, J., Cohn, I., Sondheimer, N., ... Scherer, S. W. (2018). The personal genome project Canada: findings from whole genome sequences of the inaugural 56 participants. *Canadian Medical Association Journal, 190,* E126–E136.

Venter, J. C., Adams, M. D., Myers, E. W., Li, P. W., Mural, R. J., Sutton, G. G., ... Zhu, X. (2001). The sequence of the human genome. *Science, 291,* 1304–1351.

Wickelgren, I. (2002). *The gene masters: How a new breed of scientific entrepreneurs raced for the biggest prize in biology* (1st ed.). New York: Times Books/Henry Holt and Co.

Acknowledgments

Thanks are extended to Henry Klugh at Alma College who first introduced me to a rigorous approach to behavioral research and statistical data analysis, as well as Michael Cole, James L. McGaugh, and Richard Whalen at the University of California Irvine, and Gerald McClearn at the Institute for Behavioral Genetics in Boulder, who first inspired my interest in behavioral and neural genetics. Special appreciation is due to Jerry Hirsch, one of the founders of the field of behavioral genetics, who taught us much about the ethical aspects of research and writing in this discipline as well as technical matters. Work on this book began in 1982 when the author spent a sabbatical with Hirsch at the University of Illinois.

My own understanding of many theoretical and practical issues was deepened by extensive discussion over the years with my PhD students Patricia Wainwright, Barbara Bulman-Fleming, Melike Schalomon, and Dan Livy as well as colleagues George Michel, Wim Crusio, Marla Sokolowski, Cathy Rankin, and many others. Appreciation of sociological and social-demographic aspects of human behavioral genetics was heightened by interactions with Susan McDaniel.

It has been my good fortune to enjoy a professional collaboration and friendship with John Crabbe beginning in 1969 at the University of Colorado and continuing to the present time. The work done with Crabbe reached a level of behavioral test sophistication and understanding of environmental aspects of psychological data in behavioral genetics that could never have been attained on my own.

While at the University of Alberta I learned much from Sandra Anderson, Jon Faulds, and Rob Wilson about the real-world application of eugenic ideology in a government-sponsored sterilization program. Special thanks are due to Leilani Muir, Judy Lytton, and Glen Sinclair who shared with me intimate details about their lives at the Provincial Training School for Mental Defectives in Alberta, their experience of eugenic sterilization, and their struggles to lead fuller lives after they returned to civilization.

More recently, work on this book was encouraged and supported in many important ways by my dear friend Caffyn Kelley. The entire manuscript was subjected to a careful reading by Barbara Bulman-Fleming who detected glitches and gaffes that needed to be expunged. Finally, preparation of the chapters for publication required the expertise of my editor Timothy Bennett and production assistants Vijayaraj Purushothaman and Ashwathi Aravindakshan. Their contributions are greatly appreciated.

1

Levels and Explanations

Understanding the relations between genes, the brain, and human behavior is a very challenging task because the three things exist at different levels (Table 1.1). Behavior and thinking are done by an individual, a whole person. The science of psychology teaches us how to define and measure mental functions and behaviors. Genes, on the other hand, are molecules contained in the nucleus of almost all cells in the body (except red blood cells). A gene, part of a DNA molecule, codes for the structure of another kind of molecule, a protein, which is a long chain of smaller amino acid molecules. The study of genes and the inheritance of DNA molecules is the domain of genetics. The genes transmitted from parents to offspring constitute the person's *genotype*, whereas the characteristics that are measured constitute the *phenotypes*, and phenotypes can exist at several levels.

Once we have a fairly good idea of how the gene works and what its protein product does, we can then begin to understand how it might be involved in brain function and behaviors. Hopefully, we will also gain some good ideas about how to ameliorate the effects of a defective gene by making adjustments to the environment or devising new medical treatments. This book explains how genes are related to thought and behavior by exploring examples in which the role of a specific gene is quite well understood. The examples show how a gene can influence behaviors, even though it does not code specifically for those behaviors. The concept is not an easy one to grasp. Real examples can guide us to a better understanding of difficult concepts.

LEVELS AND SCIENTIFIC DISCIPLINES

The concept of integrative levels has been applied in different ways by different fields of science, as described in a historical review by Kleineberg (2017). Here, the concept is applied to things that differ in size and the kinds of connections within and between the levels. An entity at one level is made up of several smaller things at the next lower level that are connected and work together. Small molecules such as water or ethanol are composed of two or more kinds of atoms connected by chemical bonds (Fig. 1.1). Macromolecules such as proteins are made up of long chains of smaller molecules known as amino acids, each of which is made of hydrogen, carbon, oxygen, nitrogen, and sometimes sulfur. Organelles are built from several kinds of macromolecules, while an entire cell such as a neuron in the brain contains many types of organelles. Several kinds of cells unite to form a specific tissue, such as the quadriceps muscle in the leg, and an assemblage of different tissues forms an organ—a liver or a brain. Combining the brain with other essential organs such as the heart, lungs, eyes, and hands, we arrive at an entire organism, the individual person, who in turn is part of a social group. The brain itself does not express behavior; it does not move. The brain itself cannot think; it needs connections with sense organs and a long tutoring in language to engage in thought and express ideas to other members of a social group. The whole person thinks and speaks. Only as parts of a social group do people have anything to say.

Because the levels of reality are so different, specialized disciplines have arisen to study and explain them. Physics studies atoms such as carbon and oxygen, but it cannot tell us much about the properties of macromolecules such as genes or proteins. To be a good physicist, a scholar does not need to know anything at all about genes. Likewise, a geneticist will not be able to explain inheritance by doing an in-depth investigation of the nitrogen atom or the electron. There are many commonalities across levels in the fundamental methods of doing a scientific study and analyzing the data with mathematics, and a few broad generalizations about nature can also be made. Nevertheless, the large bulk of knowledge of a specific level is encapsulated in the texts and journals of a specific discipline.

A few hybrid disciplines span several levels. Neuroscience is a prime example. Molecular neuroscience examines how small neurotransmitter molecules such

1

TABLE 1.1 Levels of Explanation and Scientific Disciplines

Level	Examples	Specialty
Geographic cluster	Village, province, nation	Political science
Social organization	Family, club, orchestra, hospital	Sociology
Individual	Fruit fly, mouse, dog, human being	Psychology
Organ	Eye, brain, ovary, arm	Physiology
Tissue	Retina, gums, bicep muscle	Physiology
Cell	Neuron, leucocyte, sperm	Cell biology
Organelle	Synapse, myelin, mitochondria	Neurobiology
Macromolecule	DNA (gene), protein, omega-3 fatty acid	Biochemistry, genetics
Molecule	Water, carbon dioxide, ethanol, lysine	Chemistry
Atom	Carbon, hydrogen, oxygen, iron, silver	Physics

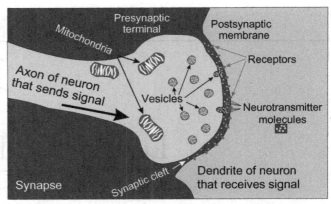

FIG. 1.2 Diagram of a synapse, a specialized organelle that conveys signals from one nerve cell to another using small neurotransmitter molecules that are synthesized in one cell, stored in small vesicles and then released into the gap or cleft between the two cells, where they stimulate receptors on the next neuron. A synapse is so small that about 1000 of them placed side by side would amount to just 1 mm. Mechanisms of synaptic function are described in Chapter 4.

as dopamine are synthesized by large macromolecules called enzymes and stored in an organelle, a synapse that connects two neurons (Fig. 1.2). The enzymes needed to make dopamine arrive in the synapse, and the transmitter molecules that they synthesize are then stored in small vesicles. Things happen in the person's environment that can stimulate the release of dopamine from the vesicles and deliver a pulse of chemical energy that influences the next nerve cell in a circuit. If enough neurons get involved in the action, this can change the individual's behavior. Thus, a student of neuroscience needs to study sciences at different levels from molecule to mind in order to achieve a good understanding of how the entire nervous system works to regulate behavior.

LEVELS AND SIZE

The sizes of parts that are involved at each level increase greatly from atom to macromolecule, to cell, to organ, to the whole person. The metric system provides a convenient way to express sizes in powers of 10 (Table 1.2). A superb illustration of the scale of size is provided by the web page maintained by the University of Utah that can be accessed using the search term "cell size and scale." A diagram (Fig. 1.3) shows the range of things that can be seen with the unaided eye, a light microscope, and an electron microscope. Living organisms can be relatively small, such as a minute HIV virus (125 nm) or the familiar *Escherichia coli* bacterium ($1.5 \times 4 \mu m$) that populates our digestive tract. A bacterium is a single cell and is much smaller than just one nerve cell in a human brain that can grow to have a cell body with a diameter of 40 μm or microns (Herndon,

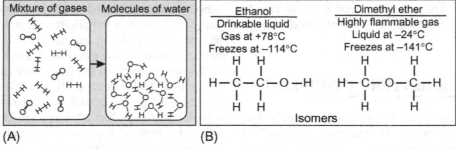

FIG. 1.1 Atoms of carbon, hydrogen, and oxygen are shown as the letters C, H, and O, respectively. Chemical bonds are shown as small lines, and a double bond is two lines. Hydrogen can form just one bond, oxygen two, and carbon four. (A) When each atom of H or O is bound to another of the same kind, the two kinds form a mixture of gases. When a spark triggers oxidation of the hydrogen, the result from the same atoms arranged differently is liquid water. (B) The isomers ethyl alcohol and dimethyl ether both have the chemical formula C_2H_6O, but the properties of the molecules differ greatly when the oxygen is in a different position.

TABLE 1.2 The Metric System and Powers of 10

Prefix	Multiplier	Power of 10	Examples
Giga	1,000,000,000	10^9	Gigabyte
Mega	1,000,000	10^6	Megaton
Kilo	1000	10^3	Kilometer
One	1	1	Gram
Centi	0.01	10^{-2}	Centimeter
Milli	0.001	10^{-3}	Millisecond
Micro	0.000001	10^{-6}	Microgram
Nano	0.000000001	10^{-9}	Nanometer

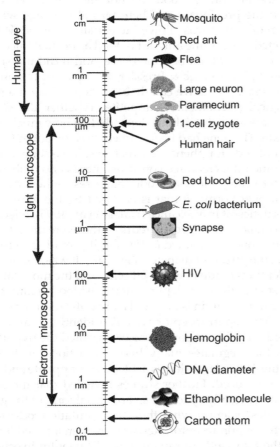

FIG. 1.3 Sizes of things, ranging from small animals that can be seen with the unaided eye to a single atom. The smallest object that most people can see without a microscope is in the range 0.1–0.2 mm. One micrometer (μm) or micron is 0.000001 m or 0.001 mm. One nanometer (nm) is one millionth of a millimeter or 0.000001 mm. A very large protein such as hemoglobin, when folded into a ball, is about 7 nm in diameter, which means that about 140 of them side by side would amount to 1 μm and 140,000 of them side by side would span 1 mm.

1963) and branches extending outward >250 μm (0.25 mm).

As we proceed above the level of cells, most things can be seen by the unaided eye, and it thus is easier to know

their forms and functions. An entire blue whale can reach 180 Mt ($1 t = 1000$ kg), more than the combined mass of 7 million 25 g mice. Nevertheless, the nervous systems of whale, mouse, and other mammals are remarkably similar. Size by itself does not tell us much about the individual parts or the internal organization of an organism.

Gravity provides a good example of levels and size. Isaac Newton made the fundamental discovery that every speck of matter in the universe attracts every other speck. The amount of matter in a body is indicated by its mass in grams or kilograms. Newton showed that the force of gravitational attraction between a person and the earth is determined by the mass of a person (m_1) multiplied by the mass of the earth (m_2) and divided by the square of the distance of the person from the center of the earth (d). We experience the force of gravitational attraction toward the center of the earth as body weight and measure it with a bathroom scale. In the era of space travel, astronauts who visited the moon testified that the force of gravity there is far less than on earth because the moon is so much less massive. Gravitation applies equally well to a simple rain drop falling to earth and immense galaxies of stars involving vast distances. For gravitational attraction, it makes no difference how the atoms that comprise one body are joined together; the important thing is simply how many atoms of which elements it contains. The scientific explanation of gravity involves things that transpire at the atomic level. The formula for computing the force of gravity is $F = C(m_1 m_2 / d^2)$, where the number C is the gravitational constant.

PROPERTIES AND CONNECTIONS

When we study a group of objects at one level, we can see how the properties of a group often depend on how the parts are connected with each other. There are situations in which the same parts are connected differently, which results in very different properties. Several of these are quite familiar. Consider the atoms hydrogen and oxygen (Fig. 1.1A). Each can exist as a colorless, odorless gas wherein pairs of atoms of one kind are loosely bound to each other and free to float in a container. They are so small that gravity has little effect, and the gas particles fill the container from top to bottom. When a spark is sent through the mixture of gases, this triggers oxidation with a loud pop and flash. Suddenly, the hydrogen and oxygen atoms recombine into a familiar molecule: water or H_2O. It is a clear liquid, and gravity draws most of it to the bottom of the container. The water will freeze into solid ice at 0°C, whereas the mixture of gases will not freeze at all at any temperature people are likely to experience. Thus, although the constituent parts—the atoms—are the same in the mixture of gases and the pool of liquid water, the properties of the mixture of gases and the water molecule are radically different.

Whether the water actually exists as liquid, solid, or gaseous water vapor depends on the temperature of the molecules' environment. The freezing point of a molecule in turn depends on the number and kinds of its atoms as well as how they are interconnected. Water freezes at 0°C, whereas ethanol freezes at −114°C. Thus, the properties of a particular kind of substance depend on both its internal structure formed from interconnected atoms and its external environment. It makes no sense at all to say that its existence as a gas, liquid, or solid depends more on the internal than the external factor. Both must be fully taken into account in order to understand the properties of the whole.

Many kinds of molecules are *isomers* in which the same constituent atoms are interconnected in different ways. Fig. 1.1B shows ethanol, a liquid at room temperature, and dimethyl ether, a highly flammable gas. Both have the chemical composition C_2H_6O. Only the location of the oxygen atom differs, but this changes the properties of the two molecules dramatically.

LEVELS AND GENES

Here, we are mainly interested in medical disorders and complex behaviors, not basic physics. Nevertheless, some of the same principles seem to be involved, albeit at different levels. Consider diabetes, a common malady. Blood sugar levels become very high. Untreated, the high blood sugar contributes to obesity, the loss of vision, kidney damage, poor circulation in the feet, and many other symptoms. Type 1 diabetes results from insufficient insulin production from the type B cells of the islet of Langerhans that are distributed throughout the pancreas, and it is treated with insulin injections. Insulin stimulates the absorption of glucose from the blood into the muscle by activating a small structure on the surface of the muscle fiber, the insulin receptor. Type 2 diabetics have enough insulin, but the receptor does not respond well to it, so there is insulin insensitivity or resistance, and blood sugar levels rise to unhealthy levels when enough glucose cannot get into the muscle fibers.

Mice sometimes are diabetic and obese too, and they show almost the same array of symptoms we term diabetes in humans. In 1950, a new genetic mutation was discovered in lab mice that led to extreme overeating and obesity, and it was named the *obese* gene (Ingalls, Dickie, & Snell, 1950), abbreviated *ob*. A mouse had to have two copies of the mutation and have genotype *ob/ob* in order to show the extreme weight gain, the obesity phenotype. In the 1950s in mouse genetics, it was customary to name a gene for an abnormal phenotype, the disease that appeared when it was mutated. The normal

form of the gene was symbolized +. A few years later, another gene was discovered that also resulted in obesity and rapid onset of diabetes, and it was named the *diabetes* gene, abbreviated *db* (Hummel, Dickie, & Coleman, 1966). It turned out that the insulin receptor was normal in both kinds of mice. The problems in the mutant mice originated in unknown molecules.

For the *diabetic* mice (Fig. 1.4A), one of the phenotypes was behavioral: overeating. The diabetic mice (*db/db*) ate far more than their normal siblings and became obese. So, researchers employed a psychological method called *pair-feeding*. A normal mouse (+/+) was housed by itself in a cage, and its daily food intake was measured. The next day a *db/db* mouse housed in an adjacent cage was given only the amount of food the normal mouse had consumed the previous day. There was also a cage with a *db/db* mouse that was allowed to eat all it wanted. When restricted to the same amount of food as normal mice, the *db/db* mouse did not become obese, and it did not show elevated glucose (Lee & Bressler, 1981). Thus, the physiological symptoms we call diabetes depended on a psychological phenomenon—appetite. Whether the *db/db* mice were actually diabetic depended on how much food they ate. The gene had been named *diabetes*, but in fact, it does not code for phenotypic diabetes. If we simply limit how much the creatures consume, there is no diabetes.

After years of investigation, other researchers discovered a new hormone that is encoded by the *obese* gene and is deficient in *ob/ob* mice (Halaas et al., 1995). Injecting it into the *ob/ob* mice greatly reduced the overeating. Later, the hormone was named *leptin*, and it provided a crucial part of the puzzle of diabetes (Fig. 1.4B). It was discovered that leptin is made in white fat cells, and the more fat has been stored, the more leptin enters the bloodstream. Like all hormones, leptin is detected by specific receptor molecules. The leptin receptor was also identified, and it is located on cells in a part of the brain called the hypothalamus that regulates many bodily functions, including appetite. When leptin levels in blood rise, appetite and eating are reduced. Further studies revealed that the *obese* mutation actually codes for an abnormal form of the gene that codes for leptin, while the *diabetic* mutation codes for a defective leptin receptor. Researchers then injected leptin into the two kinds of mutant mice. The leptin injections had no effect on the *diabetic* mice because their defective leptin receptor could not detect it (Schwartz, Seeley, Campfield, Burn, & Baskin, 1996), but they greatly reduced diabetic symptoms in the *obese* mice because the injected leptin reduced their appetites for food.

After the functions of the two genes were known, they were renamed. In mice, the one that coded for the leptin hormone was officially designated the leptin gene (*Lep*), whereas the other became the leptin receptor gene (*Lepr*). The normal leptin gene was symbolized Lep^+, and the mutation became Lep^{ob}. The normal gene

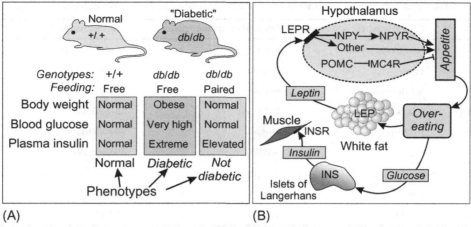

(A) (B)

FIG. 1.4 (A) When allowed to feed freely, mice with two copies of a mutant gene named "diabetic" (*db*) became obese and showed symptoms of severe diabetes, whereas mice with two normal copies of the gene (+/+) remained lean and healthy. When the (*db/db*) mice were restricted to only the amount of food eaten daily by their lean siblings (pair-fed), they did not become obese or show symptoms of diabetes, even though they carried abnormal forms of the gene (Lee & Bressler, 1981). Data shown are relative to +/+ baseline values. (B) Further study revealed that appetite, obesity, and diabetes are regulated by a complex system involving several genes in both humans and mice acting in several organs at several different levels, including behavior. The gene formerly known as "obese" codes for the hormone leptin and is now named LEP, while the old gene named "diabetes" codes for the leptin receptor (LEPR). Several genes are part of the circuit in the hypothalamus that includes neuropeptide Y (NPY) and its receptor (NPYR). Connections that involve activation are shown by *arrows*, whereas those inhibiting the downstream process end in a *bar*. Normally, obesity is prevented by a feedback loop whereby excess eating grows more white fat that in turn synthesizes more leptin that then stimulates the leptin receptor and turns down the appetite control.

encoding the receptor became *Lepr*[+], and the mutation became *Lepr*[db]. The genotypes of what were formerly known as the *obese* and *diabetes* mice were in all subsequent research symbolized as *Lep*[ob]/*Lep*[ob] and *Lepr*[db]/*Lepr*[db], respectively. The genes were also found in humans and named using capital letters (LEP and LEPR; see Fig. 1.4B). Thus, the modern practice in naming genes is to name them for what they normally do at the molecular level, not for a phenotypic disease that sometimes appears under certain conditions.

This example demonstrates the importance of knowing how things work at several levels in order to comprehend the origins of disease symptoms and learn how to treat them. Although genetic mutations were the original sources of the problems, they act via a complex, multilevel system that involves a psychological process—appetite—and the animals' nutritional environment. A behavioral method, pair-feeding, helped to understand the origins of the obesity and diabetes. The genes are macromolecules, but the disease we call diabetes is not simply a molecular issue and cannot always be treated by administering some specific molecule. The disease is a multilevel phenotype.

ANIMAL MODELS

Conducting experiments with lab animals is far easier than doing so with humans because their environments and breeding can be strictly controlled and many things can be done that would be impossible with humans for both practical and ethical reasons. Research in neuroscience relies heavily on the study of animal models. Four of the most commonly studied species in the genetic analysis of the nervous system and behavior are listed in Table 1.3. We hope the findings for animals will give us a pretty good idea of what is taking place in humans as well, but caution is warranted when comparing species. How well results are likely to generalize across species depends strongly on the level at which we are making the comparisons.

The tiny nematode worm about 1 mm long that is abundant in garden soil has only 302 nerve cells, yet it has about as many different genes as humans. A large fraction of nematode genes are counterparts of human genes that have very similar chemical structures and must have been inherited from a common ancestor long ago. About half of proteins that occur in those worms are also found in humans, sometimes with minor variations in structures, and about 40% of genes known to be important for human diseases have counterparts in the worms (Corsi, Wightman, & Chalfie, 2015). Many of the large molecules that are so important for human brain function also occur in nematodes (Hobert, 2013), but the synapses that connect one nerve cell to another are different from ours, and the overall structure of the two nervous systems and the kinds of nerve cells they contain are very different indeed. Nematodes show a number of simple reflexes that are quite easy to study, and they are also capable of learning and remembering things that happen in their

TABLE 1.3 Features of Four Species Commonly Studied in the Field of Genes, the Brain, and Behavior[a]

Scientific name	*Caenorhabditis elegans*	*Drosophila melanogaster*	*Mus musculus domesticus*	*Homo sapiens*
Common name	Nematode worm	Fruit fly	House mouse	Human
Adult size	1 mm	2.5 mm	15–25 g	150–180 cm
Reproduction	Self-fertilizing	Sexual	Sexual	Sexual
Age at first reproduction	3 days	7–20 days	46–62 days	14–20 years
Life span or expectancy	18–20 days	30 days	600–800 days	60–85 years
Number of neurons	302	340,000	100 million	100 billion
Kinds of chromosomes	6	4	21	23
Genome (million bases)	100.3	175	2803.6	3257.3
Protein-encoding genes	20,444	17,717	23,202	<20,000

[a] *Sources of information (will change with future updates of databases): Nematode worm: search terms WormBook and WormBase; Fruit fly: search term FlyBase; Mouse: search terms Mouse Genome Informatics and MGI 6.11 Introduction to Mouse Genetics; Human: search terms Genetics Home Reference and Human Genome Assembly GRCh38.p12.*

environments (Ardiel & Rankin, 2010). Nevertheless, at the level of behavior, there is a wide divergence between worms and people. The nematodes eat bacteria living on the same broth as the worm. They can move by undulating in S-shaped curves, but they have no legs or arms for propulsion. Neither do they have eyes or visual organs. The life span is short, and social interactions are rudimentary. Thus, they are remarkably similar to humans at the level of macromolecules but diverge at higher levels.

The ubiquitous fruit fly has a much larger nervous system than nematodes and well-developed vision plus remarkable capacities for flight. The adults engage in a kind of courtship during mating. Both the larvae and adults are capable of learning and memory. The flies share thousands of genes in common with humans that are implicated in some kind of disease (Hu et al., 2011). Nevertheless, the structure of their nervous system and behavioral repertoire are substantially unlike humans.

The humble house mouse is a mammal with much greater similarity to us. Its genome is almost as large as humans, and it has just as many genes (Table 1.3). Mice have 17,098 genes that are very similar to human genes, and there are more than 1300 human diseases for which there is a fairly good mouse model of the disease process (Mouse Genome Informatics). Young mice are born hairless with their eyes closed, and they rely on care by the mother for several weeks. The nervous system of mice has many regions that are essentially the same as humans in their locations and connections with other parts of the brain, including the cerebellum, hypothalamus, hippocampus, and corpus callosum (Chapter 4). The eye and retina are very similar, but color vision is more highly developed in humans. The human cerebral cortex is folded extensively and contains many more nerve cells relative to the rest of the brain, while the mouse cortex is smooth and relatively smaller, but the layers and connections are strikingly similar in the two species. A wide variety of tests of complex behaviors are available for mice (Crawley, 2007; Wahlsten, 2011). Mice do not have our kind of language, but there is evidence that they emit patterns of sounds in the ultrasonic range, especially during courtship, that resemble songs (Arriaga, Zhou, & Jarvis, 2012; Chabout, Sarkar, Dunson, & Jarvis, 2015).

As a general pattern, similarities of nervous system anatomy and behavioral functions are greater in animals that are more closely related to humans because of more recent evolution from a common ancestor. All species of animals have DNA that encodes genetic information, and the mechanisms by which that information is processed are the same throughout the animal kingdom. The relations between DNA and proteins are also essentially the same. Thus, we expect experiments that are done at the molecular level will yield similar results for a wide range of species, while studies of complex behaviors are likely to reveal many things that are unique to a particular species.

Whatever the level of a research project, results always need to be confirmed in humans. An animal model of a human function is not a substitute for investigation of the human condition. It needs to be verified. Nevertheless, many kinds of exploratory studies are far easier and more efficient to do with animal models. Landmark studies of genes and the basic processes of brain development and nervous system function were virtually all done with experiments on animals and then later confirmed in work with humans.

THE EXPERIENCE OF PAIN

It is interesting to inquire whether animals experience pain the way we do. This is an important question from the standpoints of ethical treatment of animals and the use of animal models in research on pain. Without a doubt, pain has important cognitive aspects in people that involve language, and these higher-level thought processes are just about impossible to model with mice or rats. At the same time, the way pain is processed in the nervous system is remarkably similar for all mammals, and many of the drugs used to treat pain in people are also effective for aiding mice. Receptors in the brain that detect pain signals are encoded by genes that are very similar, and the manner of synthesizing chemicals in the brain that signal pain is virtually the same as well. The nerve endings that sense damage to peripheral tissue are very similar, as are the pain pathways via nerves from the periphery into the brain (Chapter 4).

Nevertheless, the assessment of pain in people often involves verbal communication with medical personnel who make judgments of pain intensity based upon how their patients rate their own pain levels. When children are not able to rate the pain on many dimensions, they may be rated by the Wong-Baker FACES pain rating scale that asks a child to point to a symbol on a card showing smiling and frowning faces (Wong & Baker, 1988). There are six faces ranging from a score of 0 (no hurt at all) to 10 (hurts worst of all). Several versions of this scale are returned by a Google Images search, one of which is sanctioned by the Wong-Baker FACES Foundation.

Mice can obviously not be rated in this way, but a clever adaptation of pained faces has been used to construct a rating scale for mice that looks closely at *their* faces. Investigators with extensive experience studying pain in mice had noticed that animals tended to grimace during the period following a surgical operation. They devised the mouse grimace scale in which video images are collected of the face and then rated with regard to the eyes, nose, cheek, ear, and whiskers (Langford et al., 2010). Each feature is scored as 0 = grimace not present, 1 = moderately visible, and 2 = severe, and the five numbers are added to obtain an index ranging from 0 to 10. The scale proved to be very effective in rating postoperative pain and assessing the effects of analgesic drugs (Matsumiya et al., 2012). The grimace scale has a major advantage over the scales used in clinical practice with people because the ratings of mice from video images are done "blind" by technicians who do not know what treatments the mice had received, thereby avoiding possible biases in making judgments.

A number of psychological and social factors have major impacts on the human experience of pain. People who experience greater social support from family, friends, and the helping professions report less severe pain from cancer and during childbirth, and they require lower doses of analgesic drugs to achieve acceptable levels of pain control. The "placebo effect" that alleviates pain when the substance given to patients actually has no active ingredient also highlights the role of psychological processes (Kaptchuk & Miller, 2015). Thus, for anyone interested in the origins and amelioration of pain, understanding the phenomenon at several levels is essential.

EXPERTISE AND APPLICATIONS

There is far too much knowledge embodied in the disciplines in Table 1.1 for one person to understand them all in depth, but deep knowledge, real expertise, is not necessary for many people, including readers of this book. We need enough knowledge to answer the main questions we ask about genes, brains, and behaviors. This is sometimes referred to as reading knowledge. One can read an essay, a news article, or a book written for nonspecialists and grasp the general ideas of how things work. Perhaps, one might even be able to discern when an article in a newspaper contains a serious error and then write a letter to the editor. But this depth of knowledge would not be enough to apply it in biomedical or psychological practice. It would not enable a person to make a formal diagnosis of a behavioral disorder or prescribe a drug to treat it, even though the person might have a pretty good idea of how a drug is supposed to function in the brain. A practitioner must also know about making fine distinctions among several possible diagnoses that involve similar symptoms, the merits of alternative therapies, a diverse array of possible side effects of a drug (Chapter 17), and the realities of family dynamics and other social factors. If and when the time comes for we ourselves to be examined and possibly treated for some kind of biochemical defect or a mental disorder, many of us want it to be done by genuine experts.

It is possible to read this chapter without knowing just what a gene is. The term is used several times here, and simply knowing that it is a very big molecule that codes for the structure of a protein is enough to get the gist of these paragraphs. To go further with the story of genes, the brain, and behavior, we must gain a deeper understanding of genes.

HIGHLIGHTS

- Things exist at different levels, ranging from atoms to molecules, to cells, to organs, then upward to an entire organism living in a society.
- Each level is studied by a specialized field of science using terms and methods that are appropriate for that level.
- The properties of something existing at a specific level depend on the natures of its parts and the pattern of interconnections among them.
- A gene is a portion of a large molecule (DNA) and functions at the molecular level.

- A person's genotype is the set of genes transmitted by the parents, whereas phenotypes at other levels are characteristics that develop and can be measured.
- A gene is usually named for what it normally does at the molecular level, not a disease that may occur when there is a defect in the gene. The actual occurrence of disease often depends not only on the genotype but also on features of other levels, including the individual's environment.
- Different species of animals are very similar at the molecular level but often differ greatly at the level of behavior. Lab animals provide good models of things that happen at the molecular level, and at that level, they are very similar to humans.
- To understand complex behavioral phenomena such as hunger or pain, it is important to study them at different levels ranging from the molecular to behavioral and to consider the social context of behavior.

References

Ardiel, E. L., & Rankin, C. H. (2010). An elegant mind: learning and memory in *Caenorhabditis elegans*. *Learning and Memory, 17,* 191–201.

Arriaga, G., Zhou, E. P., & Jarvis, E. D. (2012). Of mice, birds, and men: the mouse ultrasonic song system has some features similar to humans and song-learning birds. *PLoS One, 7,* e46610.

Chabout, J., Sarkar, A., Dunson, D. B., & Jarvis, E. D. (2015). Male mice song syntax depends on social contexts and influences female preferences. *Frontiers in Behavioral Neuroscience, 9,* 76.

Corsi, A. K., Wightman, B., & Chalfie, M. (2015). A transparent window into biology: a primer on *Caenorhabditis elegans*. *Genetics, 200,* 387–407.

Crawley, J. N. (2007). *What's wrong with my mouse?: Behavioral phenotyping of transgenic and knockout mice* (2nd ed.). New York: Wiley.

Halaas, J. L., Gajiwala, K. S., Maffei, M., Cohen, S. L., Chait, B. T., Rabinowitz, D., et al. (1995). Weight-reducing effects of the plasma protein encoded by the obese gene. *Science, 269,* 543–546.

Herndon, R. M. (1963). The fine structure of the Purkinje cell. *Journal of Cell Biology, 18,* 167–180.

Hobert, O. (2013). The neuronal genome of *Caenorhabditis elegans*. In vol. 13. *WormBook: The online review of* C. elegans *biology* (pp. 1–106). Pasadena, CA: WormBook.

Hu, Y., Flockhart, I., Vinayagam, A., Bergwitz, C., Berger, B., Perrimon, N., & Mohr, S. E. (2011). An integrative approach to ortholog prediction for disease-focused and other functional studies. *BMC Bioinformatics, 12,* 357.

Hummel, K. P., Dickie, M. M., & Coleman, D. L. (1966). Diabetes, a new mutation in the mouse. *Science, 153,* 1127–1128.

Ingalls, A. M., Dickie, M. M., & Snell, G. D. (1950). Obese, a new mutation in the house mouse. *Journal of Heredity, 41,* 317–318.

Kaptchuk, T. J., & Miller, F. G. (2015). Placebo effects in medicine. *New England Journal of Medicine, 373,* 8–9.

Kleineberg, M. (2017). Integrative levels. *Knowledge Organization, 44,* 349–379.

Langford, D. J., Bailey, A. L., Chanda, M. L., Clarke, S. E., Drummond, T. E., Echols, S., et al. (2010). Coding of facial expressions of pain in the laboratory mouse. *Nature Methods, 7,* 447–449.

Lee, S. M., & Bressler, R. (1981). Prevention of diabetic nephropathy by diet control in the *db/db* mouse. *Diabetes, 30,* 106–111.

Matsumiya, L. C., Sorge, R. E., Sotocinal, S. G., Tabaka, J. M., Wieskopf, J. S., Zaloum, A., et al. (2012). Using the Mouse Grimace Scale to reevaluate the efficacy of postoperative analgesics in laboratory mice. *Journal of the American Association for Laboratory Animal Science, 51,* 42–49.

Schwartz, M. W., Seeley, R. J., Campfield, L. A., Burn, P., & Baskin, D. G. (1996). Identification of targets of leptin action in rat hypothalamus. *Journal of Clinical Investigation, 98,* 1101.

Wahlsten, D. (2011). *Mouse behavioral testing.* London, UK: Elsevier.

Wong, D. L., & Baker, C. M. (1988). Pain in children: comparison of assessment scales. *Pediatric Nursing, 14,* 9–17.

2

Genes

DNA, RNA AND PROTEIN

A gene is a large molecule that resides in the nucleus of cells throughout the body. It is part of a deoxyribonucleic acid (DNA) molecule that occurs in a chromosome. The DNA molecule is a long chain of four kinds of small chemicals known as nucleotide bases. The bases are adenine, cytosine, guanine, and thymine, abbreviated ACGT (Fig. 2.1). Each gene has a unique sequence of the four bases that are joined by chemical bonds into an extraordinarily long molecule. The DNA actually consists of two strands wrapped around each other in a double-helix configuration that makes the DNA very stable. Wherever there is an A base on one strand, it is paired with a T base on the other, and a C is always paired with a G on the other strand. The extraordinary stability of DNA is illustrated by a recent discovery of a 700,000-year-old fossil horse in permafrost in a Yukon gold mine that yielded good enough DNA to compare its sequence with living horses (Chung, 2013; Orlando et al., 2013).

When geneticists describe the structure of a gene, they just list a long series of the four letters because the description would occupy far too much space on a page and be incomprehensible to us if the actual bases were shown as entire molecules. The machinery of a cell can make sense of the chemicals, but we the readers are better able to sense the nature of a gene from the rows of letters. The four letters are symbols of a biological language, just as the symbols A to Z are letters in the English language that can be arranged into different kinds of words and sentences. The 27 DNA bases that code specifically for the structure of the hormone oxytocin, for example, are TGCTACATCCAGAACTGCCCCCTGGGA. Adjacent bases in the string inform the cell of where the gene starts and stops. When a gene has many bases, it will be shown on the printed page as several lines of ACGT symbols, but in reality, the bases are all in one long line that is folded over and over to make it more compact. Line breaks on a page are just conveniences for us to fit everything into the width of one printed page.

The complete human genome is distributed over 23 distinct kinds of chromosomes. Actually, there are 46 chromosomes in each cell nucleus because one copy of each is obtained from the mother and another comes from the father. Twenty-two of the distinct kinds of chromosomes are termed *autosomes* and are the same in males and females. The *sex chromosomes* are of two kinds, the large X and the much smaller Y. Only males have a Y chromosome from the father and one X from the mother, whereas females have two Xs, one from each parent. Chromosomes are large enough that they can be stained with a chemical and then seen with a light microscope, but the DNA, genes, and nucleotide bases are too small to be seen that way. The chromosomes are numbered 1–22 according to the size they have in a microscopic image, plus the large X and diminutive Y (Fig. 2.2). The locations of lightly and darkly stained bands on each chromosome are very consistent across different people and have been given official numbers on the smaller (p) and larger (q) arms of each chromosome. It is now possible to localize each unique gene in relation to a specific band on the chromosome where it occurs.

The sequence of the human genome distributed over the 23 chromosomes entails more than 3 billion nucleotide bases. In 2015, one source gave the size as 3,228,894,042 nucleotide bases (see Genome Reference Consortium). Now, anyone can examine and even download the most recent version of the sequence at official websites. All that DNA is contained in the nucleus of just one cell. It is difficult for us to comprehend how so many bases could be folded and packed into such a small volume. Three billion is an immense number. If one had a book containing 200,000 words and each word has an average of about five letters long plus space between words, that would amount to 1.2 million letters a to z and A to Z plus spaces and punctuation in the entire book. To hold 3 billion of the four letters A, C, G, and T, one would need 2500 books of that size. If the 23 chromosomes of one cell were unfolded and laid end to end, they would span more than 2 m, but the DNA strand is so

Genes, Brain Function, and Behavior
https://doi.org/10.1016/B978-0-12-812832-9.00002-6

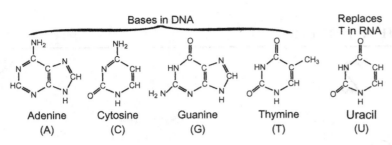

FIG. 2.1 Four nucleotide bases found in DNA. In RNA, T is replaced by U. Certain features of a molecule are not shown in order to make the diagram more compact. The symbols H_2 and H_3 indicate that two or three hydrogen atoms are bonded to nitrogen or carbon, but the bonds are not shown as lines. Where two letters NH or CH are adjacent, the chemical bond between them is not shown.

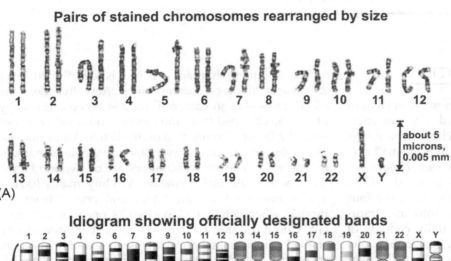

FIG. 2.2 (A) Staining human chromosomes with the quinacrine or the Giemsa stain reveals dark bands that are rich in the nucleotides A and T (Schreck & Disteche, 2001). (B) Idiograms from Ensembl database showing standard band designations adopted by the International System for Cytogenetic Nomenclature (O'Connor, 2008). *A: Reprinted from Jorde, L. B., Carey, J. C., & Bamshad, M. J. (2016). Medical genetics (5th ed.). Philadelphia, PA: Elsevier Health Sciences, Fig. 6-1, Copyright 2016, with permission from Elsevier.*

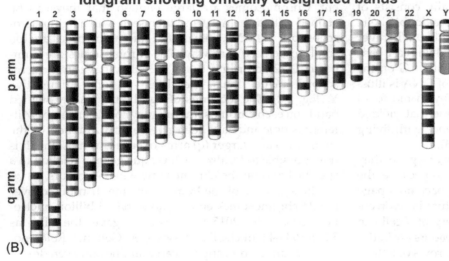

fine (2×10^{-6} mm) that all those bases can be folded to fit inside a cell nucleus.

The DNA of just one cell embodies thousands of genes. The first published estimates from the Human Genome Project in 2001 set the range for the number of human genes as 30,000–38,000, but a recent study concluded there might be only about 19,000 human genes that have well-defined functions (Ezkurdia et al., 2014). "Only" is perhaps not a good word. Any way one views the estimates of gene numbers, there are surely an immense number of different genes in the nucleus of every cell in our bodies. The number of nucleotide bases in the DNA of each chromosome is proportionate to its physical size (Fig. 2.2), whereas the numbers of genes contained in a chromosome range widely from 45 on the Y chromosome and 225 on chromosome 21 to more than 1000 genes on chromosomes 1, 2, 3, 6, 11, 12, 17, and 19. Genes themselves can differ greatly in length, which makes the numbers of genes on a chromosome only loosely related to its physical size.

At the present time, the reference sequence of the full human genome in DNA is quite well established, and only minor adjustments to the databases are now being made from year to year. How the full DNA sequence is divided into genes and just how many functional genes exist are far from certain. How the actions of all those genes are regulated in a cell is also the subject of much current research. Extracting meaningful knowledge from that collection of over 3 billion letters, AGTC remains a daunting challenge.

When cells divide, beginning with the fertilized egg (the zygote), the DNA is copied faithfully to the daughter cells. Thus, the DNA is almost exactly the same in every cell in the body. As a cell matures, portions of its DNA are *transcribed* into an intermediate molecule termed ribonucleic acid (RNA) that leaves the nucleus, migrates to the cytoplasm of the cell, and attaches to a small organelle known as a *ribosome*, where the code in the RNA is *translated* into a protein molecule, a long chain of amino acids. The specific set of genes that is transcribed and then translated is different in different kinds of cells. The proteins that a cell synthesizes confer many of its unique characteristics.

RNA also consists of four bases, where the thymine in DNA is replaced by uracil, so that RNA is a long sequence of the bases ACGU. The RNA system involves an amazing facility known as the genetic code. Each short sequence of just three bases (a triplet or *codon*) in RNA codes for an amino acid molecule that is added to a chain of other amino acids, resulting in a protein molecule. The complete set of triplets and their amino acids are shown in Table 2.1.

What does a gene do? As a first approximation, we can say that it codes for the structure of a protein molecule, the unique sequence of amino acids that comprise the specific protein found in humans. DNA and proteins are macromolecules, and one macromolecule codes for the structure of another. Genes do their work at the macromolecular level. Synthesis of a protein by assembling amino acids into a chain is the opposite of the familiar process of digestion whereby protein from the diet is broken down into small amino acid molecules by enzymes such as pepsin in the stomach. Protein macromolecules from plant and animal tissues in our food are first disassembled in the digestive tract, absorbed into the bloodstream and cells, and then reassembled into proteins that become parts of humans that ate the food.

In mammals, a gene consists of several blocks of bases termed "exons" that code for portions of the protein molecule, separated by long blocks of code termed "introns" that are not translated into protein. RNA from the several exons gets spliced together and then passes into the cytoplasm where translation occurs. One big surprise from the full genome sequence was the realization that the exons coding for protein structures comprise <2% of the total genome, whereas introns account for about 98% of human DNA. The functions of all that DNA in introns are not well understood. Some portions of introns play important roles in regulating gene action, but much remains to be learned about the vast majority of the DNA landscape.

Some genes and their RNA transcripts are relatively short. Consider the oxytocin (OXT) gene (Fig. 2.3) that is located in the p13 band on chromosome 20. The entire gene consists of 898 bases, 378 of which comprise three exons, and its exons code for a protein with 125 amino acids. After it is translated, the protein is cleaved into two smaller chains that have different functions. One of

TABLE 2.1 The Genetic Code: RNA Codons Specifying Amino Acids in Protein

Codons	Amino acid (abbreviation)	Codons	Amino acid (abbreviation)
UUU, UUC	Phenylalanine (Phe, F)	AUU, AUC, AUA	Isoleucine (Ile, I)
UUA, UUG	Leucine (Leu, L)	AUG	Methionine (Met, M) START
UCU, UCC, UCA, UCG	Serine (Ser, S)	ACU, ACC, ACA, ACG	Threonine (Thr, T)
UAU, UAC	Tyrosine (Tyr, Y)	AAU, AAC	Asparagine (Asn, N)
UAA, UAG	STOP	AAA, AAG	Lysine (Lys, K)
UGU, UGC	Cysteine (Cys, C)	AGU, AGC	Serine (Ser, S)
UGA	STOP	AGA, AGG	Arginine (Arg, R)
UGG	Tryptophan (Trp, W)	GUU, GUC, GUA, GUG	Valine (Val, V)
CUU, CUC, CUA, CUG	Leucine (Leu, L)	GCU, GCC, GCA, GCG	Alanine (Ala, A)
CCU, CCC, CCA, CCG	Proline (Pro, P)	GAU, GAC	Aspartate (Asp, D)
CAU, CAC	Histidine (His, H)	GAA, GAG	Glutamate (Glu, E)
CAA, CAG	Glutamine (Gln, Q)	GGU, GGC, GGA, GGG	Glycine (Gly, G)
CGU, CGC, CGA, CGG	Arginine (Arg, R)		

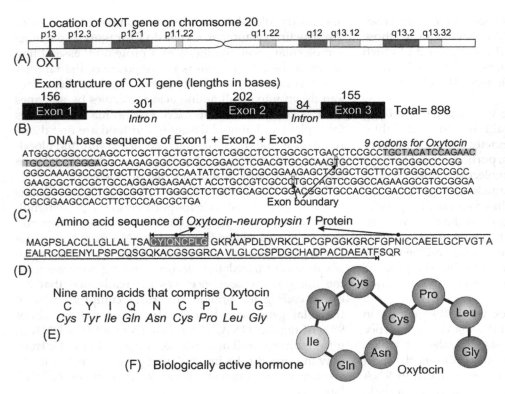

FIG. 2.3 (A) The OXT gene located on chromosome 20 consists of (B) three exons separated by noncoding introns. (C) The 898 nucleotide bases of the exons code for (D) the amino acid sequence of the oxytocin-neurophysin 1 protein. That longer protein is cleaved into smaller subunits, one of which (E) is the peptide hormone oxytocin. The remnants of the longer protein are broken apart and recycled. (F) The active form of oxytocin depends on a bond forming between the two cysteine amino acids so that it folds into the correct 3-dimensional shape. *Sources: NCBI report for CCDS13044.1, Ensembl report for OXT-001 ENST00000217386, NCBI RNA and amino acid sequence NM_000915.3, OMIM.*

these is the hormone oxytocin that has just nine amino acids, located from positions 20 to 28 in the longer chain, whereas another is the hormone neurophysin 1 that extends from positions 32 to 125. (When a chain consists of fewer than 20 amino acids joined by peptide bonds, it is often termed a polypeptide rather than a protein.)

Other genes are immense and consist of many exons whose products can be spliced together in several different ways. TTN on chromosome 2, for example, is a monster gene with 304,814 bases that codes for a protein named titin or connectin that has 34,350 amino acids (see GeneCards). It not only is involved in the contraction of muscle, especially in the heart, but also appears in cells in the brain. It consists of 363 exons that can be spliced together in dozens of ways that differ in the stiffness they confer on a muscle fiber (Granzier & Labeit, 2004). Thus, the one gene codes for the structures of many related but different protein molecules. The largest known gene is the 2.2 million bases of DNA of the DMD gene on the X chromosome that codes for the 3685 amino acids of the dystrophin protein (see GeneCards). Much of the DMD gene is introns.

It is now apparent that in most instances, a single gene codes for the structure of more than one kind of protein. The transcription, translation, and splicing or cleaving of gene products are regulated by molecules beyond the confines of the specific gene itself (Chapter 3). A gene is not an automaton; it is just one part of a dynamic, living system. It performs its tasks at the molecular level and is controlled by other kinds of molecules that are part of the system.

GENE NAMES AND GENE FUNCTIONS

Genes themselves have no names in their own world. They are just chains of small molecules that scientists term A, C, G, and T. All those fine abbreviations for genes and chemicals are human creations, convenient ways to summarize the variety that exists at the molecular level. The names given to the molecules often correspond to what they do at the chemical level. In other instances, when the gene function is not understood, the name is assigned by the scientist who first discovered the gene. This is a privilege of discovery.

When the role of a gene is fairly well understood, it is usually named for the kind of chemical reaction in which it is involved. By convention, human genes are given formal abbreviations consisting of capital letters and numbers that are convenient for finding information about a gene in a database. Just the name of the gene gives a biochemist a very good idea of what the gene does. Chemical reactions usually proceed in discrete steps, whereby one kind of molecule is converted into another kind and that product is in turn converted into yet another kind. In some situations, a step is facilitated by an enzyme that binds briefly to a specific molecule and adds or subtracts another small molecule to or from it. An enzyme is a large protein molecule, and a gene codes for its structure. The enzyme itself is not consumed in the reaction, and it can mediate the same step several times before it decays or is broken apart.

Consider the neurotransmitter acetylcholine (Ach), a small molecule that stimulates the contraction of a muscle fiber when it is released from the terminal of a nerve attached to the muscle (Fig. 2.4). No gene codes for the structure of Ach. Instead, it is synthesized in the nerve cell by transferring a small acetyl group to a molecule of choline obtained from blood. Accordingly, the enzyme mediating this reaction is called choline acetyltransferase. It is a very large protein molecule consisting of 748 amino acids, and its structure is specified in the choline acetyltransferase (CHAT) gene on chromosome 10. When Ach is released from the nerve terminal, it is detected by a cluster of large molecules on the muscle fiber. Because these are highly specific in shape and respond almost exclusively to the presence of acetylcholine, they

are termed cholinergic receptors. Many kinds of these are found in various combinations on muscle fibers and in the brain (Chapter 4). Each is encoded by a separate gene (the cholinergic receptor (CHRN) family). When the nerve stimulation abates, the Ach neurotransmitter molecules are broken down by the enzyme acetylcholinesterase (ACHE) gene into smaller molecules that do not stimulate the cholinergic receptors, and muscle contraction ceases. ACHE removes an ester group from Ach, so it is called an esterase. Enzymes usually are given names ending in the suffix "-ase" when the smaller molecules on which they operate are known, and the gene name corresponds to the enzyme for which it codes, not to the end product of the reaction.

Another example is provided by the synthesis of the brown melanin pigment in skin and hair cells that protects the body from harmful ultraviolet rays from the sun. The amino acid molecule tyrosine is absorbed from the blood into a melanocyte cell and converted into melanin via several discrete steps (see Fig. 2.5). Two separate steps are catalyzed by the enzyme tyrosinase. The TYR gene on human chromosome 11 codes for the structure of the enzyme tyrosinase that has 529 amino acids.

FIG. 2.4 Synthesis of the neurotransmitter acetylcholine (Ach) is mediated by the enzyme choline acetyltransferase encoded in the CHAT gene (chromosome 10, band q11.23), and breakdown is mediated by acetylcholinesterase encoded in the ACHE gene (chromosome 7, band 22.1). When released on muscle fibers where there are receptors to detect it, Ach causes contraction of the muscle fiber. Diagram is simplified by the omission of certain atoms and bonds. *Source: GeneCards, OMIM.*

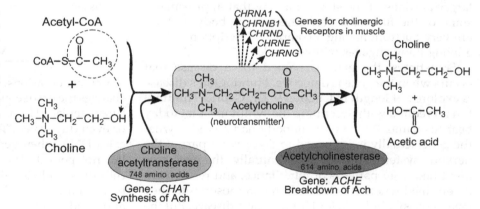

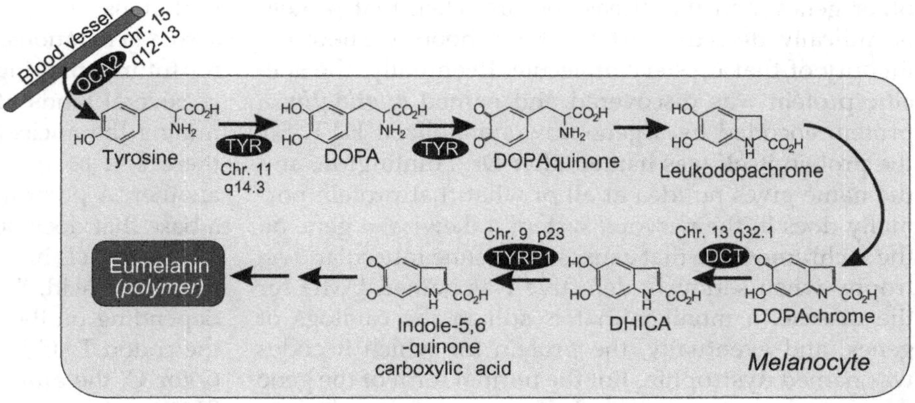

FIG. 2.5 Synthesis of the dark-brown pigment eumelanin in melanocyte cells in the skin begins when the amino acid tyrosine is absorbed into the cell from the bloodstream. That step is mediated by the oculocutaneous albinism type 2 (OCA2) gene. If the step is blocked, there will be no melanin, and the person will be albino. The next two steps on the chain of events are mediated by the tyrosinase enzyme encoded by the TYR gene, and several further steps are mediated by other enzymes, including dopachrome tautomerase (DCT) and tyrosinase-related protein 1 (TYRP1). If the TYR gene has a serious defect, there will be little or no melanin synthesized, resulting in another kind of albinism phenotype (Table 2.2). *Source: GeneCards, OMIM.*

Tyrosinase is folded into a specific shape that enables it to bind briefly to a tyrosine molecule and attach a hydroxyl (OH) group to it, generating a DOPA molecule. It can also bind briefly to DOPA and remove two hydrogen atoms, which yields DOPAquinone. Thus, the same gene can have more than one biochemical function. Looking at the chain of events, it is evident that the TYR gene does not code for melanin. It mediates two steps in a longer series. At the very end of the series, several identical melanin molecules are joined end to end (polymerized) to form the pigment eumelanin. The diagram shows some but not all of the other genes involved in the process.

The reader is advised to study Figs. 2.4 and 2.5 carefully because they show the kinds of things that real genes actually do. A gene often does not code for the end product of a reaction; it codes for an enzyme that catalyzes just one step along the way. Several genes must work together with other molecules as a team to make the whole process do the right thing. The diagrams show only portions of all the processes involved in making muscles contract and the skin darken. Although a living system does all of these things at the same time in one grand symphony, we can only comprehend smaller parts of the whole. Gradually, we try to assemble larger and larger portions of the story in a way that approximates some of the functions performed in the body, but we are very far from having a complete description of how a living thing engages with its world.

Sometimes, the gene is named for a disease that occurs when the gene is defective. Huntington's disease is explored at length in Chapter 9. Dr. George Huntington is credited with describing the symptoms that today bear his name. At the time, nobody had any idea what the gene actually is supposed to do or what part of the nervous system is involved. Eventually, the gene was identified from patterns of inheritance, and researchers determined where it is located on chromosome 4. They gave it the symbol HD, for Huntington's disease. But the gene does not code for a specific disease. The normal form of the gene codes for a protein molecule, just like other genes, and the disease occurs when that protein is radically defective. At the time, nobody knew the identity of that mystery molecule. Eventually, the specific protein was discovered and named *huntingtin*, a protein encoded by a gene now symbolized HTT. So, the protein itself was named after Dr. Huntington, and the name gives no idea at all of what that protein normally does in the nervous system. Likewise, a gene on the X chromosome that causes Duchenne muscular dystrophy when seriously defective was named DMD for the disease, a moniker that is still in the catalogs of genes, and eventually, the protein for which it codes was named dystrophin. But the normal form of the gene does not code for any kind of muscle dystrophy. As described in Chapter 1 in the section about diabetic mice

and further in later chapters, a defective gene does not automatically code for a disease. It codes for a specific protein that, when seriously defective, may result in disease under certain conditions.

Experts in molecular genetics are well aware of the distinction between the name for a gene and its normal function, but journalists and other nonspecialists are often misled into thinking that the gene actually codes for the end product, the disease. They equate genetic coding with unalterable fate. It is a widespread mistaken belief that is difficult to dispel.

The gene in the nucleus knows only its offspring, RNA, and chemicals that instruct it to synthesize particular kinds of RNA, whereas the RNA that moves out of the nucleus knows its offspring, protein (Chapter 3). The connection between DNA in the nucleus and the protein for which it codes is indirect. The gene itself is imprisoned in the cell nucleus and usually has no direct contact with the protein for which it codes or the end products of a long and complex reaction. It is a source of vital information but cannot "know" what it is supposed to be doing for the organism. Neither does it code "for" a specific phenotype. It is just a big molecule interacting with nearby molecules. How the actions of a gene are regulated is discussed in the next chapter.

MUTATION

On rare occasions, some kind of accident happens that changes the base sequence in the DNA, so that one kind of base is replaced by another, maybe eliminated altogether, or even duplicated. The molecular machinery that replicates DNA when generating a new cell is very reliable but not perfect. The mistakes we term mutations. The cells of an individual consist of those that comprise the body (somatic cells) but are not passed to the next generation and germ cells (egg and sperm) that generate offspring. Somatic mutations usually have little effect and are not transmitted to the next generation. Germ cell mutations are quite rare, but they can be transmitted across generations, and they can give rise to big troubles for future offspring.

Several kinds of mutation can be detected by determining the nucleotide base sequence of a gene. Rarely, there is a *point mutation* that substitutes one base for another. A point mutation may be silent if it substitutes a base that does not change the amino acid encoded by the triplet (Table 2.1). If the substitution does change the amino acid, the effect may be very subtle or drastic, depending on the nature of the change. For example, if the codon TAC is changed to TAG by a substitution of G for C, the amino acid tyrosine will be replaced by a STOP codon, and only part of the protein will be translated. This most likely would render it completely ineffective, sometimes with tragic consequences. A *deletion* removes one or more bases, while an *insertion* adds one or more bases into the gene sequence. Removing or inserting just one base will shift the reading of all subsequent codons over by one space, which in turn will drastically affect the amino acid sequence.

Mutations can be caused by radiation or toxic chemicals termed mutagens. In most cases, the effect seems to be rare, random, and unpredictable. Irradiating the gonads (ovary and testis) will generate mutations, and the stronger the radiation, the more mutations there will be. Adding the chemical *N*-ethyl-*N*-nitrosourea (ENU) to the diet can generate numerous mutations at very low doses in mice. Those mutations are so common that in a single mouse, there will be about one mutation in a block of 1 million bases and a noteworthy mutation in 1 of every 700 genes, meaning that on average, about 28 new gene mutations in the egg or sperm will be generated from a single dose of ENU given to one mouse. This method of generating genetic diversity is used only in lab research on animal genetics.

Large numbers of mutations have been detected in almost every human gene that has been studied thoroughly. They generally afflict very few people, especially if the mutation is very harmful. In most cases, it is not known when or why the mutation occurred in the first place. Often, the mutation is detected because a person shows some kind of disease or peculiar symptoms that draw the attention of a medical doctor and result in a referral to a medical geneticist.

Effects of a mutation are sometimes obvious and startling. Mutations that affect melanin synthesis can cause a person to lack pigmentation, such that the skin and hair are white and eyes are pink. This condition is known as albinism, the oculocutaneous form because both the eyes (oculo) and skin (cutaneous) are affected. If an albino child is born to parents with normal pigmentation, this will be noticed. The most common form of this disorder is called type 1, abbreviated OCA1. In 1990, a research team reported gene sequences for 30 albino individuals from two families and their relatives with normal pigmentation (Giebel, Strunk, King, Hanifin, & Spritz, 1990). Six of the albinos had a mutation in the TYR gene that substituted a thymine base for cytosine at position 242 of the gene sequence, which changed the CCT codon to CTT, resulting in a substitution of leucine for proline at amino acid 81 (Fig. 2.6A). The mutation is termed P81L. The tyrosinase enzyme was rendered ineffective, so that the affected people could not produce melanin pigment at all. The next year, another team reported an albino individual with an extra thymine base inserted into the TYR gene after base 1467 in exon 5 (Fig. 2.6B), which resulted in a shift of all downstream codons and amino acids and then a STOP codon that truncated the enzyme entirely (Chintamaneni et al., 1991). The mutation is

The P81L point mutation in exon 1 of TYR gene

FIG. 2.6 (A) Base sequence of exon 1 of the TYR gene showing a substitution of T for C at position 81 that yields an abnormal form of tyrosinase with leucine instead of proline (Giebel et al., 1990). People with two copies of the mutation are albino. (B) An insertion of the base T at position 1467 in exon 5 radically alters the tyrosinase enzyme (Chintamaneni, Halaban, Kobayashi, Witkop, & Kwon, 1991), resulting in albinism. *Gene and amino acid sequences from Ensembl and CCDS 8284.1 from NCBI.*

termed 1467insT. No melanin could be made at all in that person. Since those initial reports, more than 100 other mutations have been identified in the TYR gene that result in OCA1 (Chapter 7).

Most mutations that disable an enzyme are quite uncommon in the human population because they generate problems that make it difficult for the affected person to live or have children. People with albinism, for example, have a number of serious defects of the eye and vision, and they are prone to several kinds of cancer. In some societies, they are even the targets of vicious discrimination, even murder (Bever, 2015). The net result is that the frequency of OCA1 in several populations in Europe and Africa is about 1 per 28,000 births. The frequency represents a balance between the rate at which new mutations occur and the rate at which existing gene defects fail to be reproduced in the offspring of afflicted people who inherit two copies of a mutation.

To determine which of the many possible mutations indeed causes trouble for a new patient who shows albinism, a medical geneticist would need to gather information about that person's DNA sequence and look for something out of the ordinary in the TYR gene. For example, detailed genetic analysis of the DNA sequence of 62 individuals with oculocutaneous albinism (OCA) in Denmark revealed that 26% had a mutation in TYR, as expected, but several showed a defect in another gene related to melanin synthesis and 15% showed no defect in any gene currently known to cause OCA (Grønskov et al., 2009). Clearly, there is much more to the story of albinism than just the TYR gene.

For many mutations, the defect in the DNA base sequence is passed from parent to offspring. Because the germ cell, either egg or sperm, carries the mutant gene and all cells in the body of the offspring are subsequently derived from the zygote formed by the union of egg and sperm, all cells in the body of the offspring will also possess the mutation. In rare instances where a new mutation occurs in one of the eggs or sperm in a parent, the body cells of both parents are normal genetically, but the child has a mutation. That defect is then termed a de novo mutation.

With the advent of large-scale, rapid sequencing of the entire genome of an individual, it is now possible to pinpoint new mutations by comparing whole genomes of parents and their children. A recent study done in Iceland examined 78 mother-father-child trios by whole-genome sequencing (Kong et al., 2012). Among all the genes of 219 individuals, 4933 de novo mutations were detected in the child, about 63 per trio, and 73 of the 4933 occurred in exons, the portion of DNA that codes for parts of protein. The estimate of the mutation rate was 1.2×10^{-8} per nucleotide base per generation. The number 10^{-8} is small indeed, about one chance in 100 million for any one nucleotide base, but there are more than 3 billion bases in one genome, so the probability is high that any person, including the author of this book and the reader, will carry several de novo germ line mutations. Whether they have any noteworthy effect depends on *where* they occur in the genome.

The Icelandic study also confirmed something noted in many other large-scale research projects: There were many more mutations from the father than the mother, and mutations were considerably more common in offspring sired by older fathers. The mutation rate in fathers rose steadily with age and was twice as high in 40-year-old fathers than 20-year-olds. The reasons for this pattern are not entirely clear. It could be that the longer a man has lived, the more exposure he likely has had to radiation and mutagenic chemicals in the environment, but it is possible that the molecular mechanisms for replicating the DNA while making new sperm become fragile with age. Clearly, future studies of mutations induced by adverse environments must exercise great care to control for or take account of father's age.

GENOTYPE AND PHENOTYPE

Older sources often termed the gene that gives rise to albinism the albino gene, but this is misleading in two ways. First, the TYR gene does not code for the pigmentation defect we call albinism. It codes for the enzyme tyrosinase. Some mutant forms of TYR result in albinism; others do not. Second, several more genes have been found that can result in OCA. For example, OCA type 2 is related to a defect in a gene that codes for another protein that normally transports tyrosine from the blood into the cell (melanocyte) that manufactures melanin (Fig. 2.5). In cases where the defect is severe, tyrosine cannot get into the melanocyte from the blood, and melanin cannot be made, resulting in a physical appearance of the skin and eyes very similar to classical type 1 albinism that is related to a TYR mutation (Table 2.2). The critical distinction between OCA types 1 and 2 is that there is normal tyrosinase activity in type 2, so that OCA2 is now termed tyrosinase-positive oculocutaneous albinism. It is a kind of albinism that is *not* caused by a mutation in the TYR gene. The exact nature of the chemical defect in OCA2 is only vaguely understood, and the gene itself has been named OCA2, which is also the name of the clinical disorder, not the protein involved. This can be confusing. Because melanin is manufactured via a long chain of chemical steps mediated by different proteins or enzymes, there are several kinds of genetic mutations that can stop the process short of its goal. When it stops, the result is called albinism. Why it stops needs to be investigated in greater depth.

These examples highlight the critical distinction between *genotype* and *phenotype*. Genotype refers to the set of genes, the DNA that the individual inherits, whereas the phenotype is a measurable characteristic such as skin color, visual acuity, or intelligence. Oculocutaneous albinism (OCA) is a phenotype, and it can be caused by several different genotypes. Likewise, a mutation of the TYR gene that causes OCA1 gives rise to several abnormal phenotypes, especially those involving vision (Table 2.2). Variants of the OCA2 gene are involved in geographic differences in human skin color (Chapter 19). The science

TABLE 2.2 OMIM Clinical Phenotypes for Oculocutaneous Albinism OCA1A and OCA2

OCA type 1A (TYR mutation #203100)	OCA type 2 (OCA2 mutation #203200)
SAME CHARACTERISTICS	
Eye—decreased acuity	Eye—decreased acuity
Eye—severe nearsightedness	Eye—severe nearsightedness
Eye—misrouted optic nerve	Eye—misrouted optic nerve
Eye—nystagmus, strabismus (crossed eyes)	Eye—nystagmus, strabismus (crossed eyes)
Skin—white at birth	Skin—white at birth
Skin—no tanning	Skin—no tanning
DIFFERENT CHARACTERISTICS	
Eye—no iris or retinal pigment	Eye—decreased iris and retinal pigment
Skin—no pigment	Skin—highly variable, freckles from sunlight
Hair—white	Hair—white to blonde or red
Tyrosinase activity—none	Tyrosinase activity—normal
Population frequency—1/28,000	Population frequency—1/10,000

of genetics seeks to understand how variation in phenotypes is related to genotypes. One lesson that has been learned well is that there is no one-to-one correspondence between genotype and phenotype. Even for relatively stable traits such as skin or eye color, the situation is seen to be rather complex when one takes a closer look.

In the early years of the twentieth century, many biologists studied simple traits such as color of garden peas or severe defects of motor coordination in mice, and they referred to genes using terms describing the characteristics they hoped to explain. Gene names often corresponded to an observable trait, such as yellow versus green seeds that had been studied by Mendel or "obese" and "diabetes" in mice (Chapter 1). Genetic factors were thought to code for the observable trait itself. Bateson (1913) used the term "unit character" to describe what was inherited, and he emphasized that "characters behave as units." He noted that sometimes, a character might deviate from the typical form "due to the disturbing effects of many small causes not of genetic but presumably of environmental origin." The deviations presumably were "due to interference that is external or environmental in the wide sense." The choice of wording is important. It implies that the hereditary unit codes for a specific outcome of development, the characteristic itself, and environments can only disturb or interfere in minor ways with the internally predetermined result.

An alternative view was championed by Johannsen in a landmark 1911 essay (Johannsen, 1911). He contrasted two fundamentally different views of heredity: (1) the transmission conception that the personal qualities or characteristics of ancestors are transmitted to offspring through heredity and (2) the genotype conception that genes in the gametes are transmitted from ancestor to offspring but their personal qualities are not transmitted. For Johannsen, the personal qualities of individuals are reactions of the genotype to conditions encountered during development. He proposed that "A 'genotype' is the sum total of all the 'genes' in a gamete or zygote," whereas "phenotypes" are characteristics of the individual that may be observed by direct inspection or measuring. He lamented that "over and over, we find in the current literature this confusion of genotypes and phenotypes." He urged strongly that "the talk of 'genes for any particular character' ought to be omitted, even in cases where no danger of confusion seems to exist."

More than 100 years ago, most experts in genetics adopted Johannsen's terms and definitions, even though the chemical nature of genes was not yet known. Decades after Johannsen, when the fundamental discoveries about the structure of DNA and the genetic code were made, it became clear why the genotype conception of heredity is indeed correct. Unfortunately, many outside the genetics world today still have not assimilated this lesson. Virtually every day, we encounter in the mass media and casual conversation the antiquated view that a gene codes for some specific trait or phenotype.

There cannot be a gene that codes *for* a rather simple trait such as skin pigmentation or complex traits such as autism, schizophrenia, or depression. Instead, there are genes that code for proteins involved in the development and functioning of the skin or nervous system, and mutations in some of these might influence skin color or mental development. The distinction between the two ways of thinking and talking about genes is not merely semantic. It involves fundamentally different notions of what genes do. Do genes code for behavioral phenotypes, or do they code for molecules that are expressed in specific kinds of cells? The consensus among biological scientists favors the latter overwhelmingly. The former is a relic of a previous century.

SOURCES OF INFORMATION

Vast troves of information about genes and genetics are now readily available to anyone with an Internet connection. Governments of several countries and large foundations have devoted billions of dollars to collecting and disseminating data about genes. Several authoritative websites are listed in Table 2.3. They have been created by leading scientists and reviewed extensively by many other scientists. Some contain up-to-date details about intricate features of the genome and are useful mainly to medical geneticists, while others provide educational resources for the general public. The reader is invited to browse and search the databases for information on diseases or human characteristics that may be of special interest.

HIGHLIGHTS

- A gene is a macromolecule, a long chain of four chemical bases (ACGT) that comprise DNA. Portions of the base sequence termed exons are transcribed into RNA in the nucleus.
- RNA moves out of the nucleus and is translated into a protein, a chain of amino acids. The protein may then act as an enzyme to facilitate chemical reactions, a hormone, or a part of the structure of the cell. Thus, genes code for the structures of protein molecules.
- The human genome contains more than 3 billion bases, and about 20,000 genes code for proteins.
- A gene sequence can be altered by rare mutations that can be transmitted to offspring and can alter the structure of the protein.
- The official name of a gene should reflect what it does, but many gene names are misleading.

TABLE 2.3 Internet Sources of Information About Genes and Proteins[a]

Type	Search term	Sources
Comprehensive	GQuery NCBI	National Center for Biotechnology Information, NIH
Gene sequence	Ensembl	European Bioinformatics Institute Wellcome Trust Sanger Institute
Gene sequence	GenBank	GenBank database at National Institutes of Health (NIH)
Gene sequence	NCBI GRC	Genome Reference Consortium, NIH
Genes	OMIM	Online Mendelian Inheritance in Man, NIH
Genes	GeneCards	GeneCards database, Weizmann Institute of Science, Israel
Genes	NCBI gene	NCBI gene database, queries in many formats
Proteins	UniProt	European Bioinformatics Institute, Swiss Institute of Bioinformatics, Protein Information Resource
Proteins	CCDS	Consensus Coding Sequence Project, NIH
Educational	GHR NIH	Genetics Home Reference, *Handbook of Genetics*, NIH
Educational	NSW genetics	Centre for Genetics Education; fact sheets; NSW, Australia
Educational	NHS genetics	National Genetics and Genomics Education Centre, the United Kingdom

[a] *Type denotes emphasis of the database. Most give some data about both genes and proteins. Search term or organization name can be inserted into search window in Internet browsers to find the current Internet address for the site. All terms gave correct results in 2018.*

- The genotype consists of genes that are transmitted across generations, whereas the phenotype is something we can observe or measure but is not transmitted directly to offspring. It develops during the lifetime of the individual.
- Genes do not code for any measurable characteristic of the individual.

References

Bateson, W. (1913). *Mendel's principles of heredity*. Cambridge: Cambridge University Press.

Bever, L. (2015). Where albino body parts fetch big money, albinos still get butchered. *The Washington Post*. March 13.

Chintamaneni, C. D., Halaban, R., Kobayashi, Y., Witkop, C. J., & Kwon, B. S. (1991). A single base insertion in the putative transmembrane domain of the tyrosinase gene as a cause for tyrosinase-negative oculocutaneous albinism. *Proceedings of the National Academy of Sciences of the United States of America*, 88, 5272–5276.

Chung, E. (2013). Ancient Yukon horse yields oldest genome ever. *CBC News*. July 27.

Ezkurdia, I., Juan, D., Rodriguez, J. M., Frankish, A., Diekhans, M., Harrow, J., et al. (2014). Multiple evidence strands suggest that there may be as few as 19,000 human protein-coding genes. *Human Molecular Genetics*, 23, 5866–5878.

Giebel, L. B., Strunk, K. M., King, R. A., Hanifin, J. M., & Spritz, R. A. (1990). A frequent tyrosinase gene mutation in classic, tyrosinase-negative (type IA) oculocutaneous albinism. *Proceedings of the National Academy of Sciences of the United States of America*, 87, 3255–3258.

Granzier, H. L., & Labeit, S. (2004). The giant protein titin: a major player in myocardial mechanics, signaling, and disease. *Circulation Research*, 94, 284–295.

Grønskov, K., Ek, J., Sand, A., Scheller, R., Bygum, A., Brixen, K., et al. (2009). Birth prevalence and mutation spectrum in Danish patients with autosomal recessive albinism. *Investigative Ophthalmology and Visual Science*, 50, 1058–1064.

Johannsen, W. (1911). The genotype conception of heredity. *The American Naturalist*, 45, 129–159.

Kong, A., Frigge, M. L., Masson, G., Besenbacher, S., Sulem, P., Magnusson, G., et al. (2012). Rate of de novo mutations and the importance of father's age to disease risk. *Nature*, 488, 471–475.

O'Connor, C. (2008). Chromosome mapping: idiograms. *Nature Education*, 1, 107.

Orlando, L., Ginolhac, A., Zhang, G., Froese, D., Albrechtsen, A., Stiller, M., et al. (2013). Recalibrating Equus evolution using the genome sequence of an early Middle Pleistocene horse. *Nature*, 499, 74–78.

Schreck, R. R., & Disteche, C. M. (2001). Chromosome banding techniques. *Current Protocols in Human Genetics*, 4, 2.

Further Reading

Jorde, L. B., Carey, J. C., & Bamshad, M. J. (2016). *Medical genetics* (5th ed.). Philadelphia, PA: Elsevier Health Sciences.

3

Gene Expression

Although almost every cell in the body has a complete set of genes, only a fraction of those genes is expressed as RNA and protein in any particular cell or kind of tissue. Gene expression is exceedingly complex, and even today, more than 10 years after the complete human genome sequence was determined in 2006, much remains to be learned. A presentation of the New Genetics by the National Institute of General Medical Sciences described the details of gene expression as "murky." Fortunately, the fine details of mechanisms of gene expression are not critically important for understanding later chapters.

INSIDE AND OUTSIDE THE NUCLEUS

In order to read the information in a specific gene and transcribe it into RNA, the strands of the DNA double helix must be unwound in the vicinity of that gene. There is a promoter region at the beginning of a gene that provides a site to which the enzyme RNA polymerase can attach and start the process of *transcription* (Fig. 3.1A). The promoter region does not become part of the RNA that is *translated* into protein. There are also *enhancer* sequences along the DNA that serve as binding sites for regulatory proteins that aid in preparing the DNA so that the RNA polymerase can gain access to just one DNA strand. These enhancer sites can be located within an intron, so the intron can play an important role in gene expression, even though it does not code for protein structure. They can also occur in the long stretches of DNA located between two genes. Thus, the high proportion of human DNA that is not in exons is not all "junk." The RNA polymerase moves down the DNA strand and generates a long RNA molecule, sometimes termed pre-RNA, that contains bases corresponding to the promoter and introns as well as exons. Other enzymes then clip out and discard the promoter and intron sequences. Finally, the exon-derived sequences are spliced together into one complete molecule of messenger RNA (mRNA) that exits the nucleus.

The mRNA migrates to an organelle in the cytoplasm termed the endoplasmic reticulum, where there are many thousands of ribosomes ready for action. A ribosome attaches to an mRNA molecule and begins to move along it (Fig. 3.1B). In the fluid surrounding the ribosome are small molecules of transfer RNA (tRNA), each of which has three nucleotide bases (each base being A, C, G, or U) at one end and a specific amino acid molecule at the other end. The genetic code (Table 2.2) describes which amino acid will occur on each possible sequence of the three bases, the codon. If a specific tRNA has three bases that complement the bases on the mRNA currently located at the ribosome, the tRNA binds to the mRNA just long enough for another enzyme, peptidyl transferase, to join the amino acid to a growing chain of amino acids by forming a peptide bond. Thus, the chain of amino acids grows longer and longer until the ribosome encounters a STOP codon, which stops the process and releases the finished protein, which is then transported to other places in the cell or even the bloodstream.

The intricate details of these processes may seem to have little relevance for understanding behavior, but there is an important lesson embedded in them. Transcription and translation take time, and those times can be estimated. Generating a new pre-RNA molecule from a gene in DNA proceeds at about 25 nucleotide bases per second, the speed at which RNA polymerase moves along the DNA strand (Heyn et al., 2014). Time is also needed to remove the introns and splice together the exon-derived mRNA segments, but this can take place while transcription proceeds. In mammals, the initial transcription includes the introns, and introns comprise the large majority of bases in most genes. The human dystrophin gene (DMD) has 2,241,933 bases, more than 99% of which are introns. At a transcriptional speed of 25 bases per second, it would take 89,677 s to complete the job or 25 h. Some sources place the time at closer to 16 h, but in any way one views it, transcribing that gene requires a large portion of a day. Translating the mRNA into protein at the ribosome also takes time; it proceeds at about six amino acids per second in mice (Ingolia, Lareau, &

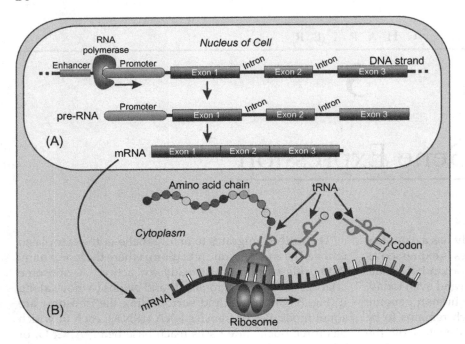

FIG. 3.1　(A) Transcription of DNA sequence into RNA sequence involves several steps that occur inside the cell nucleus. Only the finished messenger mRNA is exported to the cell cytoplasm. (B) Translation of the base sequence in mRNA into a protein occurs at the ribosome via transfer tRNA that has a triplet of bases (codon) at one end and a specific amino acid at the other end (Table 2.2) that allows the chain of amino acids to grow step by step.

Weissman, 2011). The dystrophin protein consists of 3685 amino acids, which would require about 614s or 10 min to assemble from the mRNA. All of these times are probably underestimates because other chemical changes are involved.

Table 3.1 shows estimated times to generate a complete protein, beginning from the onset of transcription of a gene for several genes known to be involved in behavior. Three genes shown there are critically important for contraction of skeletal muscle fibers (Fig. 2.4). The gigantic titin protein is responsible for the elasticity of those fibers. Complex motor actions such as running a race or playing a Chopin sonata can be initiated within 0.1s of a signal to start, and performance of the sequence of motions is coordinated with millisecond precision. Many runners could finish a 5K race, and a pianist could

complete a Chopin piano sonata before a large protein derived from a freshly transcribed gene is ready for work. Everything must be in place in the muscles and nervous system before a new behavior sequence begins. There must already be muscle fibers and nerve cells connected in a circuit, and there must be energy stores and vesicles full of neurotransmitters (Fig. 1.2) ready for action. Of course, genes are intimately involved in the formation of neural and muscle structures, but they do not contain a code for the sequence of movements embodied in a behavior. Gene expression is much too slow to keep pace with intricate and sprightly behaviors. Genetically encoded information is important for constructing the parts of the system, but the coding of a complex sequence of movements is done at higher, more nimble levels.

TABLE 3.1　Times to Transcribe DNA Into mRNA and Translate mRNA Into Protein

Gene symbol	Protein name	Bases in gene	Time (min)	Amino acids in protein	Time (min)	Total time (min)
CHAT	Choline acetyltransferase	56,010	37	748	2.1	39
ACHE	Acetylcholinesterase	6981	4.6	614	1.7	6
CHRNA1	Cholinergic receptor A1	16,881	11	482	1.3	11
OXT	Oxytocin-neurophysin	898	0.6	125	0.3	1
TYR	Tyrosinase	118,308	79	529	1.5	80
TTN	Titin	304,814	203	34,350	95	298

GeneCards. Rates: 25 bases per second for transcription, 6 amino acids per second for translation.

GENE EXPRESSION

There are two ways to detect the expression of a specific gene in a cell or tissue. One detects the presence of the mRNA sequence that corresponds to the gene, whereas the other detects the protein itself using antibodies that recognize it specifically in a sea of thousands of other proteins. Antibodies generated to a specific protein will bind to it when applied in a solution to a thin section of a brain on a glass slide, and their locations can then be detected with a light microscope. As shown in Fig. 3.2 for the orexin-1 receptor protein in a rat brain, the method reveals the location of a protein in great anatomical detail, showing specific cells in many brain regions that express the protein and even the places within a cell where it is located. Orexin-1 receptor protein is present in moderate to extensive amounts in 56 different regions throughout the rat brain, including several portions of the hypothalamus involved in hunger and feeding (Hervieu, Cluderay, Harrison, Roberts, & Leslie, 2001). The gene is also implicated in processes that control the wake-sleep cycle. The exquisite detail revealed by antibodies is readily seen only when the investigators examine just one or two kinds of protein at a time. Doing this for each of thousands of genes expressed in just the brain would be an exhausting task.

New technologies are now able to detect mRNA made from thousands of different genes in a single assay. Knowing the base sequence of a given gene, a computer working with a robot generates a synthetic chain of just a few nucleotides (an *oligonucleotide*), perhaps 25 or 30 bases long in the same order as in the gene, and affixes many copies

of it to a small spot on a glass slide. If the mRNA from the gene that contains the complement of that oligonucleotide sequence comes into contact with it in a solution, it will bind to it and form a dense cluster that can be seen with a microscope. The machine creates a *gene expression array* that has thousands of dots of different oligonucleotides, each corresponding to a coding (exon) sequence from a specific gene. When a piece of tissue is homogenized and then applied to the slide, certain dots will have many mRNA strands adhering to them. Those dots reveal which genes are expressed abundantly as mRNA in that piece of tissue. Homogenizing the tissue obliterates fine anatomical details, and it combines the mRNA from genes that are expressed in different parts of the tissue.

The gene expression array can reveal how many of the thousands of human genes are expressed in the brain, for example. This gives some idea of the biochemical complexity of the brain. The answer is mind-boggling: about 84% of all human genes are expressed somewhere in the adult brain (Hawrylycz et al., 2012). The same methods can be used to generate profiles of which genes are expressed at highest levels in which parts of the brain, as will be discussed further in Chapter 4.

Gene expression is not always a static, stable characteristic of a tissue. Living things change all the time, and a gene expression array can provide snapshots at different times to show the dynamic nature of genes. It would not be surprising, for example, to find that some genes are more intensely expressed during daylight than at night or vice versa. The methods used to show this are not readily applied with people. Independent samples of a large block of tissue or an entire organ need to be taken every hour or two and analyzed with the gene expression arrays (Hughes et al., 2007). This is easy enough to arrange with mice. A recent study examined changes in mRNA profiles in the liver of lab mice (Hughes et al., 2009). First, the animals were maintained in a lab with 12 h light and 12 h dark each day, and they had free access to food at all times. Mice are nocturnal and do most of their eating and exploration of the environment in the dark phase. Investigators wanted to learn not only which genes change with light and dark but also which ones become "entrained" so that they will continue to show the 12:12 rhythm even when the environment is dark at all times. This revealed a mousy kind of "jet lag" arising from an internal clock. The researchers switched the lab to constant darkness and then tested different mice every hour for 48 h. A sophisticated statistical analysis of gene expression detected those genes that continued to show a 24 h cycle of activity even in constant darkness (Fig. 3.3). They found 3667 such genes, a staggering number.

Another physiological cycle shows similarly dramatic changes in genes: the estrus or reproductive cycle. In mice, the female comes into estrus and is receptive to mating every 4–5 days, and estrus is followed by proestrus

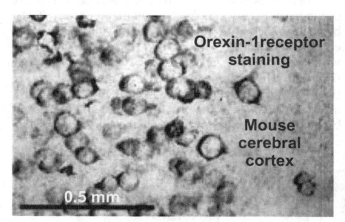

FIG. 3.2 Neurons that do not produce the orexin-1 protein are not stained by the antibody. The protein is concentrated around the perimeter of neurons that do express the orexin-1 receptor. The antibody does not appear in the cell nucleus. *Adapted from* Neuroscience, Vol. 103, *Hervieu, G. J., Cluderay, J. E., Harrison, D. C., Roberts, J. C., & Leslie, R. A. Gene expression and protein distribution of the orexin-1 receptor in the rat brain and spinal cord, pp. 777-797, copyright 2001, with permission of Elsevier. Scale bar shows 0.5 mm. Orexin-1 is one of two peptide hormones that are encoded by the HCRT (hypocretin) gene, and there are two genes for its receptors (HCRTR1, HCRTR2).*

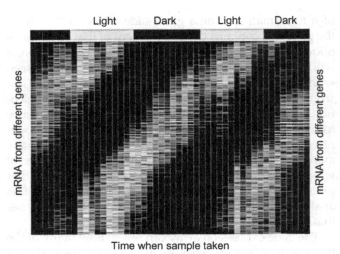

FIG. 3.3 Results over 2 days for genes that show peak expression levels at different times in the light-dark cycle. Each fine horizontal band represents a different gene; brightness indicates intensity of gene expression above background levels (Hughes et al., 2009; Fig. 5A).

when the lining of the uterus sloughs off and mating does not happen. Investigators removed the uterus from females that were in either estrus or proestrus, extracted the mRNA, and analyzed it with gene expression arrays (Yip, Suvorov, Connerney, Lodato, & Waxman, 2013). They found 2428 genes that differed between the two phases, 1127 with higher activity during estrus and 1301 higher during proestrus. Some genes changed their expression levels by only twofold, but there were 56 for which expression in one phase was 10 times higher than the other. Remodeling of the uterine tissue was extensive, and it depended on the combined activities of numerous genes involved in cell division, growth of the lining of the uterus, and preparations for receiving and nurturing a fertilized egg, should one appear.

Clearly, gene expression is a dynamic process involving extraordinarily many genes in a particular tissue. Gene activity is strongly influenced by internal, physiological processes such as the estrus cycle and external environmental influences such as lighting and feeding.

At the same time, many genes in a tissue do not change their expression levels appreciably between night and day or with phase of the estrus cycle. No general statement can be made about whether gene expression tends to be stable or fluctuate over time. It all depends on which genes in which cells and what kinds of environmental changes we examine. Any suggestion that genes are generally not subject to influences from the environment is nonsensical and counterfactual. Changes in gene activity in response to environmental change are critically important for the ability of an organism to adapt to new challenges. Given what is known about how the nervous system works (Chapter 4), we can assert confidently that any change in the environment that is powerful enough to alter behavior in a major way must also change gene expression, and changes in gene expression always accompany a change in behavior.

SUPPRESSION OF GENE EXPRESSION

Some genes are very active early in the development of an embryo but then become silent and are not expressed in the adult. Others are expressed in a particular kind of tissue, but in other kinds of tissue, the gene may be shut off altogether. Researchers studying cancer cells made an important discovery about ways this can happen. There are sites along the DNA where small molecules can become attached, and a gene whose promoter region has many such marks will often be silenced. Removing those marks can reawaken a gene and enable the renewed synthesis of its mRNA and protein. The chemical nature of the mark is very simple. A small methyl group (CH_3) becomes attached to a specific carbon atom (#5) in the cytosine nucleotide base of the DNA, almost always when the cytosine is adjacent to a glycine (G) base (Fig. 3.4). The enzyme DNA methyltransferase type 3 (DNMT3A or B) mediates this event. When many sites are methylated along the promoter region in the DNA, it becomes very difficult or even impossible to uncoil the

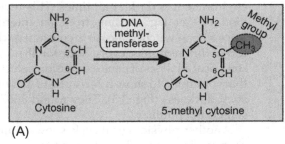

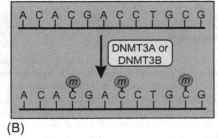

FIG. 3.4 (A) The enzyme DNA methyltransferase can attach a methyl group to the carbon in cytosine at position 5. (B) Methyl groups, a form of epigenetic mark, are usually attached to a cytosine that is adjacent to a guanine (G) nucleotide. The DNA methyltransferase enzymes from genes DNMT3A or B mediate the change.

DNA, and the RNA polymerase cannot move along the DNA strand.

Recent surveys of the human and mouse genomes have revealed that methylation is a common state of affairs for many genes (Kundaje et al., 2015). About 70% of all adult CG nucleotide pairs in DNA of mice and humans are methylated (Pelizzola & Ecker, 2011). When methylation is frequent in the promoter region of a gene, its expression is repressed. Two features of genome methylation have generated great interest among scientists and even the general public. First, changes in an animal's early environment can alter the pattern of DNA methylation, and the changes can persist for a lifetime, because the methylation state can be passed from one cell to another through mitosis or cell division during growth. Second, there is evidence that the methylation state of some genes can be passed from a mother or father to the offspring. What is clearly an environmental influence on the parent's biochemistry can become part of the offspring's gene expression apparatus. A chemical mark imposed on the DNA without changing the nucleotide sequence of the DNA itself is called an *epigenetic* change. The study of epigenetics is a very active frontier of research today.

EARLY EXPERIENCE AND LASTING CHANGES

It is well established that many kinds of experience early in the life of a mammal can exert lasting effects on the behavior of the adult. There must be some kind of chemical and/or anatomical change that is involved. A well-studied example in rats was one of the first to implicate changes in methylation in behavioral effects of early experience. Newborn rats and mice are virtually helpless and rely on care by the mother for the first 3 weeks after birth. Not only does the mother suckle the pup, but also she cleans it with her tongue and grooms it. Licking of the pup's anus and genitals is essential in order to stimulate early urination and defecation, and the extensive touching is important in many ways for future health. It was noted that in lab rats, some mothers show large amounts of anogenital licking and grooming (LG) of pups after birth, while others are relatively low in LG. The difference in behavior appeared in the next generation as well. Female pups reared by high LG mothers tended to show more LG behavior toward their own offspring. This could arise from a genetic influence on behavior, in which genes that encourage high LG are passed from parent to offspring, but it might also be transmitted via an environment or an epigenetic change.

Fostering of pups was used to distinguish those effects (Champagne et al., 2006). Mothers were observed with their first litters and classified as either high or low licking and grooming (LG). Then, they were remated and allowed to give birth to another litter. Within 12 h of birth, four pups were removed briefly from the cage and marked to identify them. In the *unfostered* condition, those pups were returned to their birth mother for rearing. The *infostered* control entailed pups being switched between two high LG and two low LG mothers, in order to assess whether the fostering itself had an effect. In the *cross fostered* condition, pups born to a high LG mother were transferred to a low LG mother and vice versa. All pups were weaned at 22 days after birth and housed two rats per cage until 90 days old, at which time their brains were removed and analyzed chemically in a region at the base of the brain called the hypothalamus. Measures were made of the amount of mRNA for the estrogen receptor type α (ERα) derived from the rat *Esr1* gene, a receptor known to be an important regulator of maternal behavior.

Results indicated that the amount of ERα mRNA depended strongly on the mother who reared the pups after birth (Fig. 3.5A), not the genotype of the birth mother. Analysis done in separate samples of rats found that there was considerably more methylation at 13 CG sites in the ERα promoter region of the DNA in the low LG females (Fig. 3.5B). It appeared that the amount of licking and grooming received by a pup after birth affected the methylation of the ERα gene promoter region, which in turn influenced the amount of the ERα receptor expressed in the hypothalamus.

Many studies have confirmed that stressful early experiences can exert long-lasting changes on brain function involving changes in DNA methylation (Bagot & Meaney, 2010; Moore, 2015; Provencal & Binder, 2015). A thorough review of studies with rats and mice noted numerous examples in which methylation was altered by forced swimming, restraint in a small box, long periods away from the mother, social isolation, repeated social defeats, exposure to a predator, and chronic moderate pain (Stankiewicz, Swiergiel, & Lisowski, 2013). Suggestive evidence exists that similar lasting effects occur in human brains as a result of child abuse (McGown, Kerber, Fujii, & Goldberg, 2009) and mental distress leading to suicide (Keller et al., 2010). Now that we know how widespread the methylation sites are in the brain (70% of CG sites in adult humans), there is good reason to expect that enduring changes in DNA methylation can result from many kinds of early experience.

As researchers delve more deeply into epigenetics, it has become apparent that the situation is far more complicated than just methylation of promoter regions (Daxinger & Whitelaw, 2012). Many other molecules are involved in the process of gene expression, and many studies have indicated that epigenetic marks on DNA "have diverse and seemingly conflicting influences on transcription" (Meaney & Ferguson-Smith, 2010). Many drugs are now being tested for their abilities to alter

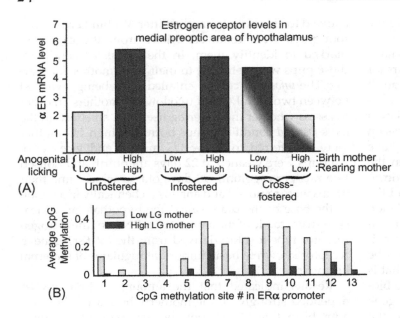

(A)

(B)

FIG. 3.5 (A) Levels of estrogen receptor type α in the hypothalamus of rats having mothers low or high in anogenital licking or cross fostered shortly after birth onto the opposite type of mother. (B) Methylation of sites in the gene promoter region in relation to maternal anogenital licking of pups. High methylation impedes gene expression. *Source: Champagne FA, et al; Maternal Care Associated with Methylation of the Estrogen Receptor-α1b Promoter and Estrogen Receptor-α Expression in the Medial Preoptic Area of Female Offspring, Endocrinology 2006; 147 (6): 2909–2915, doi:10.1210/en.2005-1119. Reproduced by permission of Oxford University Press on behalf of the Endocrine Society.*

methylation and other chemically based changes in gene expression. Much remains to be learned in this domain, and we are not yet able to alter methylation deliberately in order to achieve superior results of development.

LEARNING AND MEMORY

Memory is a relatively long-lasting change in the brain, enduring for months in mice or years in humans. Many biochemical changes in the brain accompany learning and memory (Kandel, 2012), and changes in methylation can now be added to that list (Levenson & Sweatt, 2005). Enzymes from genes that mediate methylation (DNMT1 and 3A) are present in adult nerve cells, and drugs that interfere with methylation also impair

memory formation and recall in several kinds of tasks (Yu, Baek, & Kaang, 2011). The situation is not entirely clear because certain kinds of experience lead to removal of methylation marks, and no specific enzymes have yet been found that work this way. It is virtually certain that learning and memory involve several kinds of changes in addition to methylation.

A recent study uncovered a remarkable degree of specificity in certain kinds of epigenetic changes during learning (Dias & Ressler, 2014). Mice were exposed to one of two kinds of odor molecules, acetophenone or propanol (Fig. 3.6). A mouse gene named *Olfr151* codes for an odor receptor protein expressed in the lining of the nasal cavity (Chapter 4) that detects acetophenone, while *Olfr6* detects propanol. After exposing a mouse to one of the odors, the mouse was given a brief electric shock. On a test given

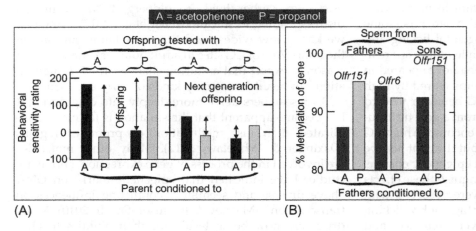

(A)

(B)

FIG. 3.6 (A) Father rat was conditioned by exposing him to an odor (A = acetophenone and P = propanol) that was followed by shock. Offspring were tested with either A or P but not shocked. Arrows show size of effect of training. A smaller but significant effect was evident in the second generation offspring. (B) There was higher methylation of the *Olfr151* gene that codes for a receptor that detects acetophenone in the fathers that were conditioned to A rather than P and the sons that were never exposed to A or P. *Reprinted by permission from Dias, B. G., & Ressler, K. J., Parental olfactory experience influences behavior and neural structure in subsequent generations.* Nature Neuroscience, 17, Copyright 2014.

later, mice showed altered behavior in the presence of the odor that had been paired with shock, whereas reactions to the other odor were unchanged. Additionally, the promoter of the *Olfr151* gene was demethylated in mice shocked in the presence of acetophenone but not when shocked during propanol exposure. Reviewing these and other recent experiments, Szyf (2014) observed that unraveling all the links between methylation and brain function in memory "is bound to be a formidable challenge." Nonetheless, it is evident that such links exist. Gene expression plays an important role in learning and memory.

TRANSGENERATIONAL EFFECTS OF ENVIRONMENT

It is not at all surprising that gene expression is changed by a wide range of environments. The storage of information in a nervous system over long periods of time requires that something like this must be happening. Discoveries about DNA methylation were generally met with applause because the phenomena showed how long-lasting changes at the molecular level could occur in the brain. Before long, those observations were confirmed in many labs, and a new discipline emerged— behavioral epigenetics.

In several of the early studies (Champagne, 2008; Francis, Diorio, Liu, & Meaney, 1999) and many times thereafter, researchers were surprised to find that methylation marks also showed up in the offspring and even the grand-offspring of those who had been exposed to vivid experiences early in life, even though the descendants never had those experiences (Babenko, Kovalchuk, & Metz, 2015; Jiang et al., 2008; McCarrey, 2014; Sharma, 2015; Wei, Schatten, & Sun, 2015). Dias and Ressler (2014) found that the sensitivity to acetophenone and the effects on methylation of the *Olfr151* gene were also found in the offspring that had never been exposed to either odor but whose father had been exposed to acetophenone and shocked (Fig. 3.6B) (Dias & Ressler, 2014).

A dogma has been embraced by many biologists that all epigenetic marks on the DNA are erased during the formation of sperm and egg, so that the single-celled zygote begins life as a blank slate epigenetically. It was believed that environmental effects on the mother can affect her offspring only via her uterine and early postnatal environment, not via changes associated with her DNA. Indeed, Johannsen's genotype conception of heredity (Johannsen, 1911) insisted that experience could not change the gene itself, but epigenetic marks that are transmitted to the next generation challenge the old dichotomy of genes and the environment in a big way (Crews, Gillette, Miller-Crews, Gore, & Skinner, 2014; Lane, Robker, & Robertson, 2014).

GENES, ENVIRONMENT, OTHER

Discoveries about the molecular biology of genes and epigenetics are now calling into question some widely accepted definitions of certain terms. What a gene is (Chapter 2) is not so simple to discern (Portin & Wilkins, 2017). The exon portions that code for protein have all been tabulated for many species, and where those sequences in the DNA start and stop is clear enough. The promoter sequence upstream of the first exon (Fig. 3.1) can usually be defined too, but there is great uncertainty about the locations of all the enhancer and silencer regions that can act to enable or block expression of a gene. Should those sequences be considered part of the gene? After all, they are made from DNA. Even the most powerful computer working on the sequence of ACGTs has trouble detecting all those control regions, many of which are dispersed across large introns and in the spaces between genes (*intergenic* regions). The cellular machinery itself knows how to make sense of all this and rarely makes a mistake, so there must be some kind of biological code for gene expression. Perhaps, this is one of those situations in which the more we know, the less we understand.

RNA Genes

Just as the human genome sequence was taking shape in the official databases, it became evident that some bona fide genes are present in DNA and expressed in RNA but are not then translated into protein. Termed "noncoding RNA genes" (Eddy, 2001), they are biologically active but are "effectively invisible" to methods designed to detect protein-encoding genes. Some of them are very short DNA sequences that encode transfer RNAs or ribosomal RNA (Fig. 3.1), but others are much longer and reside in the large regions between protein-encoding genes. A systematic search of mammalian genomes uncovered evidence for about 1600 such long noncoding RNAs (Guttman et al., 2009), and more recent evidence points to a much larger number. It appears that many RNA genes are important for the expression of the protein-encoding genes and are involved in epigenetic processes of methylation that are known to regulate expression of many genes (Skinner, 2015). They are also implicated in transgenerational transmission of methylation states (Frías-Lasserre & Villagra, 2017; Vogt, 2017). The full scope of the functions of RNA genes, especially their relevance to human disease states, is not yet known, but it is likely to be large because of the sheer bulk of DNA

(>98%) that exists outside of the narrowly delimited exons that code for protein.

Beyond Dichotomies

Should we regard epigenetic marks on the DNA as genetic or environmental? As shown in Fig. 3.4A, the methyl group is joined to the cytosine molecule by a chemical bond. It is an epigenetic mark because it is imposed upon (epi) the DNA without changing the sequence of nucleotide bases in the DNA. Many of the influences that can alter DNA methylation are clearly environmental. If we consider only the ways in which methylation can be altered during the lifetime of one individual, methylation looks much like a kind of memory stored in molecular form. It is the result of the environment and does not appear to be genetic in the sense of an altered DNA sequence. Or does it?

The debate has intensified with new observations of epigenetic changes that are transmitted to future generations via the germ cells. Transgenerational epigenetic effects act much like conventional genetic effects, at least in the next generation. Eventually, they may fade away if the original stimulus that caused them does not recur (Burggren, 2015), but when comparing parents and offspring, epigenetic effects can mimic simple genetic effects. Shall we say that the environment can be inherited, or should it be said that the environment can outright change a gene? Should those methylation marks be regarded as environment hitch-hiking on the genome, or are they an integral part of the genome, once they have traversed the chasm between generations?

Nature obeys its own laws and does not draw boundaries to fit our categories. What if a provocateur was to propose that DNA actually consists of five, not just four kinds of nucleotide bases. There are the usual ACGTs (Fig. 2.1) plus methylated cytosine (Fig. 3.4) that we might symbolize as C^m, the fifth base, so that DNA is a long chain of the bases ACGTC^m. Without any doubt, methylated cytosine is a kind of nucleotide molecule. The methyl group CH_3 is not merely pasted over it; it is an integral part of the molecule joined by a bona fide chemical bond. It happens that the transition from a bare C to methylated C^m and back is dynamic and responsive to environmental conditions. It is not generally accepted in biology that there is a fifth kind of base (C^m). Sometime in the remote history of molecular biology, some theorists decided that the parts of a gene must be very stable, *not* subject to environmental perturbation. They insisted on a rigid dichotomy between gene and environment: an epigenetic mark must be viewed as a stroke of chalk on a blackboard, something easily erased and not really part of the underlying message, something that is in essence part of the environment. But

"genetic" in the form of a purely 4-base DNA was defined by scientists before the molecular bases of epigenetic changes were discovered. Ideology came first. Then came the new facts.

Rather than chasing our tails while trying to define and count genes, it may be helpful to define certain phenomena more clearly. Skinner, Manikkam, and Guerrero-Bosagna (2010) regard epigenetics as "molecular processes around the DNA that regulate genome activity independent of DNA sequence." Their term *around the DNA* clearly includes methylation marks and thereby avoids any need to classify methylation as either something genetic with five kinds of nucleotide bases or something environmental. It is a *response* to the environment without itself being part of the environment. It is a third category of causation. Likewise, Skinner (2015) recognizes "epigenetic transgenerational inheritance" or transmission as something that occurs via the germ line without direct environmental exposure, as when a parent is exposed to a significant environmental stressor that later has an impact on offspring who never experienced the stressor. It is a third source of transgenerational similarity. Thus, one way to address the ambiguities is to use a third category, epigenetic, that is neither genetic nor environmental. There is nothing in biology or psychology that requires all things to be divided into two piles.

Environment itself is easier to define. Environment consists of all those things outside an organism that impinge on it. That is, an environment in the way we use it when discussing genetics and behavior is defined in relation to the individual organism. Experience is a subset of factors in the environment that are detected via sense organs. What is the organism is usually clear enough because it is enclosed in some kind of skin or shell, so environment exists outside of that covering.

So often when we try to dichotomize, there are situations that frustrate the effort. What shall we make of the simple matter of feeding and ingestion of food? When the meal is still on the dinner table, it is obviously environment. After the meal has been digested and absorbed into the bloodstream, it is obviously part of the organism. But, what should we say about a freshly consumed meal that is in the stomach but has not yet been digested and absorbed? Is it still environment, or is it now part of the organism? Here, we need to employ an *operational definition* that defines a thing by the operations we use to measure it. For example, the body weight of a mouse or a person is defined by a weighing device. You get on the bathroom scale, and it shows your body weight. Even if you just had breakfast, that is your weight, defined operationally. Perhaps, this seems like a small point, but there are other more weighty questions for which fine distinctions can have large implications.

HEREDITY

Until now, the use of the term "heredity" has been carefully avoided, for good reason. There is wide disagreement about the proper meaning of the word. Some insist that heredity consists only of genes, things encoded in the DNA sequence of the germ cells, and they consider these as the basis for effects of "nature" as opposed to environmental "nurture." Others regard all things that are transmitted from parent to offspring and that generate resemblance between traits of offspring and the parents to be part of heredity. The two perspectives take us back to the discussions at the time of Johannsen (1911; see Chapter 2). Are traits passed from one generation to the next, or are genes transmitted while traits or phenotypes develop anew each generation? The genotype conception of heredity, which draws a clear line of demarcation between genotype and phenotype, has been accepted by experts for decades but is now being challenged by epigenetics.

Wealth and other components of social status tend to be conveyed across generations, but they are widely understood to be environmental effects that arise from the structure and laws of a society, not an individual's molecules. For some other things conveyed across generations, the boundary between heredity and the environment is not quite so clear. As discussed at length by Moore (2015) and Crews et al. (2014), the new epigenetics has occasioned a rethinking of many things that transpire between generations. Epigenetic marks can be transmitted across generations. They act like part of heredity, but their origins are different than those of genes inscribed in DNA millions of year ago.

A GUT FEELING ABOUT HEREDITY

One factor that is now receiving great attention in science and the media is the flora of bacteria that populate several parts of our bodies. The human "microbiome" includes more than 1000 species of bacteria that themselves possess 100 times as many genes as the human genome (Clemente et al., 2015; Guinane & Cotter, 2013). The NIH Human Microbiome Project in the United States is sampling bacteria from the nose and sinuses, the mouth, the skin, the stomach, the intestines and colon, and the vagina and urinary tract. The genomes of the bacteria studied in large samples of normal, healthy individuals now provide baselines against which disease states can be compared (see also the European Metagenomics of Human Intestinal Tract). Together, those bacteria constitute a diverse community of organisms, a community that includes the human host. Many of those bacteria are very important for human health, helping to digest complex sugars and carbohydrates, synthesize certain vitamins, condition the immune system, and protect against harmful infectious species (Fouhy, Ross, Fitzgerald, Stanton, & Cotter, 2012; Guinane & Cotter, 2013; Latuga, Stuebe, & Seed, 2014). They are mutually symbiotic with us and evolved in close association with our species over many thousands of human generations. We require them in order to perform many functions, and they need us to provide a suitable environment in which to prosper. The microbiome is inside and upon us. It is an integral part of the individual organism. Yet, it can be seen as part of an organism's environment.

The gut microbiome participates in a two-way communication with the brain and forms what has been termed the "gut-brain axis." Experimental changes in gut microbiome of mice in the lab have substantial effects on several kinds of behaviors (Diaz Heijtz et al., 2011), and evidence indicates similar things happen with humans (Carabotti, Scirocco, Maselli, & Severi, 2015). At the same time, social stress in animals can alter the gut microbiome within just a few hours (Galley et al., 2014). The brain via the sympathetic nervous system (Chapter 4) directs many of the activities of the gut that are critical for digestion, and there is evidence that neural signals influence the workings of the microbiome directly through receptors for neurotransmitter molecules that are located on many bacterial cell surfaces in the gut (Mayer, Knight, Mazmanian, Cryan, & Tillisch, 2014).

Those microorganisms are transmitted from the mammalian parent to offspring, usually via the mother. Not only the species but also specific strains of bacteria are transmitted from mother to child (Latuga et al., 2014). In mammals, it is generally not possible for an entire cell such as a bacterium to cross via the placenta into the fetus. Consequently, the infant at birth usually has no bacteria in its digestive tract or dwelling in the mouth, nasal passages, or pores in the skin. In mice and many other mammals, the newborn passes through the mother's vagina during birth, it lives in a nest provided by the mother, and she licks the newborn all over its body, including the genitals and anus. This intimate contact with the mother provides abundant opportunity for bacteria living on and in her body to find a hospitable home in her offspring. Similar routes of transmission exist in humans, although the cultural practices differ somewhat from mice. Environmental factors that can influence the population of bacteria in the human newborn include natural vs caesarian surgical delivery, breast vs bottle-feeding, time of weaning, maternal use of antibiotics, and the hospital environment itself (Fouhy et al., 2012; Latuga et al., 2014). Mode of delivery of the infant at term is especially important because natural birth conveys in a very direct manner the mother's microbiome to her child, whereas an infant delivered by caesarian section accumulates its microbiome from a variety of environmental sources, including the hospital environment that often entails harmful strains of *Clostridium* and other pathogens (Fouhy et al., 2012). Once it has matured, the microbiome can be altered by the kind of diet one consumes (Wu et al., 2011).

Is the microbiome part of heredity because it is transmitted from the mother, or is it strictly environmental? The terms heredity, hereditary, and heritable foster confusion. It is proposed that heredity, hereditary, heritable, and heritability be used sparingly in scientific discourse. In the next few chapters, care will be taken to avoid them altogether. Little or nothing is lost because of this choice. Any characteristic of the brain or behavior needs to be studied carefully in order to learn precisely what kinds of transmission are giving rise to resemblance of parent and offspring. If bacteria are passed from mother to child, the phenomenon can be well described as environmental transmission across generations. If, on the other hand, one wishes to search for effects of heredity by computing statistical correlations between phenotypes of mother and offspring (Chapters 14 and 15), differences in the microbiome could have a major impact on those correlations.

A more profound consequence of these distinctions arises if we now contemplate the meaning of an individual. The person is actually a composite of two things: an organism with the same DNA in all its cells and a rich community of microorganisms that colonize the gut and virtually all surfaces of the body exposed to the environment. A proper balance between the two is essential for good health and normal behavior. Conceptually, we can differentiate between the organism proper and its intimately associated microbiome that, strictly speaking, is part of our environment, but in almost everything we do in daily life, we combine the two into one individual. The human being is an animal bearing a rich coat of microorganisms.

DEFINITIONS USED IN THIS BOOK

- *Gene*: A linear sequence of nucleotide bases in DNA that codes for the structure of a protein and enables its expression.
- *Genotype*: The set of all kinds of genes possessed by an individual.
- *Genome*: The entire DNA sequence of an individual, more than 3 billion bases in humans that includes long introns that do not code for protein.
- *Gene expression*: Synthesis of mRNA and protein molecules on the basis of the DNA sequence in a gene.
- *RNA gene*: A portion of DNA that is transcribed into RNA in which the RNA itself is biologically active but is not translated into protein.

- *Epigenetic*: Chemical changes in molecules that are part of a gene or closely associated with it and that influence the expression of the gene.
- *Mutation*: An abrupt change in the DNA sequence of a gene. A change in a body cell that is not a germ cell is a somatic mutation, whereas a change that is transmitted in a germ cell (sperm or egg) is a germ line mutation.
- *Environment*: Things outside an individual organism that impinge on it.
- *Experience*: A subset of features of an environment that act via the individual's sense organs.
- *Phenotype*: A characteristic of an organism that develops after conception and is measured.
- *Genetic transmission*: Passing of genes from parent to offspring.
- *Epigenetic transmission*: Passing of epigenetic changes in the DNA or molecules closely associated with the DNA from parent to offspring.
- *Environmental transmission*: Passing of features of the parents' environment to offspring.

HIGHLIGHTS

- Each cell contains a complete set of genes (DNA) in its nucleus, but only those that are transcribed into RNA molecules can influence other processes in the cell and beyond.
- A gene expression array can detect RNA transcribed from thousands of genes in one cell or a piece of tissue.
- More than 15,000 different genes are expressed in the brain.
- Gene expression from DNA to RNA to protein requires several minutes or even hours and is a much slower process than dynamic changes in behavior.
- In RNA genes, the RNA itself is biologically active without being translated into protein.
- Environmental factors can turn expression of a gene on or off. Which aspects of the environment are most effective in doing this depend on the specific gene.
- Durable changes in gene expression are generated by adding a methyl (CH_3) group to bases in DNA of a gene. These *epigenetic* changes suppress expression of a gene.
- Behavior of a mother toward her offspring or severe environmental stressors can induce lasting changes in methylation that affect offspring behavior later in life.
- Many other kinds of changes in behavior, such as learning and memory, are mediated by epigenetic changes in DNA.
- Epigenetic marks on the DNA induced by experience can be transmitted from parent to offspring so that the offspring show the methylation marks without having the experience.

- There are three kinds of transmission from parent to offspring: genetic, environmental, and epigenetic.
- Bacteria populating the digestive tract and other body surfaces constitute the microbiome and can be passed from mother to offspring via environmental transmission.

References

Babenko, O., Kovalchuk, I., & Metz, G. A. (2015). Stress-induced perinatal and transgenerational epigenetic programming of brain development and mental health. *Neuroscience and Biobehavioral Reviews, 48,* 70–91.

Bagot, R. C., & Meaney, M. J. (2010). Epigenetics and the biological basis of gene x environment interactions. *Journal of the American Academy of Child and Adolescent Psychiatry, 49,* 752–771.

Burggren, W. W. (2015). Dynamics of epigenetic phenomena: intergenerational and intragenerational phenotype 'washout'. *Journal of Experimental Biology, 218,* 80–87.

Carabotti, M., Scirocco, A., Maselli, M. A., & Severi, C. (2015). The gut-brain axis: interactions between enteric microbiota, central and enteric nervous systems. *Annals of Gastroenterology, 28,* 203–209.

Champagne, F. A. (2008). Epigenetic mechanisms and the transgenerational effects of maternal care. *Frontiers in Neuroendocrinology, 29,* 386–397.

Champagne, F. A., Weaver, I. C., Diorio, J., Dymov, S., Szyf, M., & Meaney, M. J. (2006). Maternal care associated with methylation of the estrogen receptor-alpha1b promoter and estrogen receptor-alpha expression in the medial preoptic area of female offspring. *Endocrinology, 147,* 2909–2915.

Clemente, J. C., Pehrsson, E. C., Blaser, M. J., Sandhu, K., Gao, Z., Wang, B., ... Dominguez-Bello, M. G. (2015). The microbiome of uncontacted Amerindians. *Science Advances, 1,* e1500183.

Crews, D., Gillette, R., Miller-Crews, I., Gore, A. C., & Skinner, M. K. (2014). Nature, nurture and epigenetics. *Molecular and Cellular Endocrinology, 398,* 42–52.

Daxinger, L., & Whitelaw, E. (2012). Understanding transgenerational epigenetic inheritance via the gametes in mammals. *Nature Reviews Genetics, 13,* 153–162.

Dias, B. G., & Ressler, K. J. (2014). Parental olfactory experience influences behavior and neural structure in subsequent generations. *Nature Neuroscience, 17,* 89–96.

Diaz Heijtz, R., Wang, S., Anuar, F., Qian, Y., Bjorkholm, B., Samuelsson, A., & Pettersson, S. (2011). Normal gut microbiota modulates brain development and behavior. *Proceedings of the National Academy of Sciences of the United States of America, 108,* 3047–3052.

Eddy, S. R. (2001). Non–coding RNA genes and the modern RNA world. *Nature Reviews Genetics, 2,* 919–929.

Fouhy, F., Ross, R. P., Fitzgerald, G. F., Stanton, C., & Cotter, P. D. (2012). Composition of the early intestinal microbiota: knowledge, knowledge gaps and the use of high-throughput sequencing to address these gaps. *Gut Microbes, 3,* 203–220.

Francis, D., Diorio, J., Liu, D., & Meaney, M. J. (1999). Nongenomic transmission across generations of maternal behavior and stress responses in the rat. *Science, 286,* 1155–1158.

Frías-Lasserre, D., & Villagra, C. A. (2017). The importance of ncRNAs as epigenetic mechanisms in phenotypic variation and organic evolution. *Frontiers in Microbiology, 8,* 2483.

Galley, J. D., Nelson, M. C., Yu, Z., Dowd, S. E., Walter, J., Kumar, P. S., ... Bailey, M. T. (2014). Exposure to a social stressor disrupts the community structure of the colonic mucosa-associated microbiota. *BMC Microbiology, 14,* 189.

Guinane, C. M., & Cotter, P. D. (2013). Role of the gut microbiota in health and chronic gastrointestinal disease: understanding a hidden metabolic organ. *Therapeutic Advances in Gastroenterology, 6,* 295–308.

Guttman, M., Amit, I., Garber, M., French, C., Lin, M. F., Feldser, D., ... Cassady, J. P. (2009). Chromatin signature reveals over a thousand highly conserved large non-coding RNAs in mammals. *Nature, 458,* 223–227.

Hawrylycz, M. J., Lein, E. S., Guillozet-Bongaarts, A. L., Shen, E. H., Ng, L., Miller, J. A., ... Jones, A. R. (2012). An anatomically comprehensive atlas of the adult human brain transcriptome. *Nature, 489,* 391–399.

Hervieu, G. J., Cluderay, J. E., Harrison, D. C., Roberts, J. C., & Leslie, R. A. (2001). Gene expression and protein distribution of the orexin-1 receptor in the rat brain and spinal cord. *Neuroscience, 103,* 777–797.

Heyn, H., Sayols, S., Moutinho, C., Vidal, E., Sanchez-Mut, J. V., Stefansson, O. A., ... Esteller, M. (2014). Linkage of DNA methylation quantitative trait loci to human cancer risk. *Cell Reports, 7,* 331–338.

Hughes, M., Deharo, L., Pulivarthy, S. R., Gu, J., Hayes, K., Panda, S., & Hogenesch, J. B. (2007). High-resolution time course analysis of gene expression from pituitary. *Cold Spring Harbor Symposia on Quantitative Biology, 72,* 381–386.

Hughes, M., DiTacchio, L., Hayes, K. R., Vollmers, C., Pulivarthy, S., Baggs, J. E., ... Hogenesch, J. B. (2009). Harmonics of circadian gene transcription in mammals. *PLoS Genetics, 5,* e1000442.

Ingolia, N. T., Lareau, L. F., & Weissman, J. S. (2011). Ribosome profiling of mouse embryonic stem cells reveals the complexity and dynamics of mammalian proteomes. *Cell, 147,* 789–802.

Jiang, Y., Langley, B., Lubin, F. D., Renthal, W., Wood, M. A., Yasui, D. H., ... Beckel-Mitchener, A. C. (2008). Epigenetics in the nervous system. *Journal of Neuroscience, 28,* 11753–11759.

Johannsen, W. (1911). The genotype conception of heredity. *The American Naturalist, 45,* 129–159.

Kandel, E. R. (2012). The molecular biology of memory: cAMP, PKA, CRE, CREB-1, CREB-2, and CPEB. *Molecular Brain, 5,* 14.

Keller, S., Sarchiapone, M., Zarrilli, F., Videtic, A., Ferraro, A., Carli, V., ... Chiariotti, L. (2010). Increased BDNF promoter methylation in the Wernicke area of suicide subjects. *Archives of General Psychiatry, 67,* 258–267.

Kundaje, A., Meuleman, W., Ernst, J., Bilenky, M., Yen, A., Heravi-Moussavi, A., ... Roadmap Epigenomics, C. (2015). Integrative analysis of 111 reference human epigenomes. *Nature, 518,* 317–330.

Lane, M., Robker, R. L., & Robertson, S. A. (2014). Parenting from before conception. *Science, 345,* 756–760.

Latuga, M. S., Stuebe, A., & Seed, P. C. (2014). A review of the source and function of microbiota in breast milk. *Seminars in Reproductive Medicine, 32,* 68–73.

Levenson, J. M., & Sweatt, J. D. (2005). Epigenetic mechanisms in memory formation. *Nature Reviews Neuroscience, 6,* 108–118.

Mayer, E. A., Knight, R., Mazmanian, S. K., Cryan, J. F., & Tillisch, K. (2014). Gut microbes and the brain: paradigm shift in neuroscience. *Journal of Neuroscience, 34,* 15490–15496.

McCarrey, J. R. (2014). Distinctions between transgenerational and non-transgenerational epimutations. *Molecular and Cellular Endocrinology, 398,* 13–23.

McGown, A. J., Kerber, W. D., Fujii, H., & Goldberg, D. P. (2009). Catalytic reactivity of a meso-N-substituted corrole and evidence for a high-valent iron-oxo species. *Journal of the American Chemical Society, 131,* 8040–8048.

Meaney, M. J., & Ferguson-Smith, A. C. (2010). Epigenetic regulation of the neural transcriptome: the meaning of the marks. *Nature Neuroscience, 13,* 1313–1318.

Moore, D. S. (2015). *The developing genome: An introduction to behavioral epigenetics.* New York: Oxford University Press.

Pelizzola, M., & Ecker, J. R. (2011). The DNA methylome. *FEBS Letters, 585,* 1994–2000.

Portin, P., & Wilkins, A. (2017). The evolving definition of the term "gene". *Genetics, 205,* 1353–1364.

Provencal, N., & Binder, E. B. (2015). The effects of early life stress on the epigenome: from the womb to adulthood and even before. *Experimental Neurology, 268,* 10–20.

Sharma, A. (2015). Systems genomics analysis centered on epigenetic inheritance supports development of a unified theory of biology. *Journal of Experimental Biology, 218,* 3368–3373.

Skinner, M. K. (2015). Environmental epigenetics and a unified theory of the molecular aspects of evolution: a neo-lamarckian concept that facilitates neo-darwinian evolution. *Genome Biology and Evolution, 7,* 1296–1302.

Skinner, M. K., Manikkam, M., & Guerrero-Bosagna, C. (2010). Epigenetic transgenerational actions of environmental factors in disease etiology. *Trends in Endocrinology and Metabolism, 21,* 214–222.

Stankiewicz, A. M., Swiergiel, A. H., & Lisowski, P. (2013). Epigenetics of stress adaptations in the brain. *Brain Research Bulletin, 98,* 76–92.

Szyf, M. (2014). Lamarck revisited: epigenetic inheritance of ancestral odor fear conditioning. *Nature Neuroscience, 17,* 2–4.

Vogt, G. (2017). Facilitation of environmental adaptation and evolution by epigenetic phenotype variation: insights from clonal, invasive, polyploid, and domesticated animals. *Environmental Epigenetics, 3,* 1–17.

Wei, Y., Schatten, H., & Sun, Q. Y. (2015). Environmental epigenetic inheritance through gametes and implications for human reproduction. *Human Reproduction Update, 21,* 194–208.

Wu, G. D., Chen, J., Hoffmann, C., Bittinger, K., Chen, Y. Y., Keilbaugh, S. A., … Lewis, J. D. (2011). Linking long-term dietary patterns with gut microbial enterotypes. *Science, 334,* 105–108.

Yip, K. S., Suvorov, A., Connerney, J., Lodato, N. J., & Waxman, D. J. (2013). Changes in mouse uterine transcriptome in estrus and proestrus. *Biology of Reproduction, 89,* 13.

Yu, N. K., Baek, S. H., & Kaang, B. K. (2011). DNA methylation-mediated control of learning and memory. *Molecular Brain, 4,* 5.

4

The Nervous System

The nervous system is the grand organizer of behavior, taking in vast quantities of sensory data about the external world and the internal state of the organism and synchronizing an intricate dance of muscle contractions and glandular secretions. It extends its nerve fibers to virtually every part of the body in order to gather data and stimulate actions. It includes the central nervous system, consisting of the brain and spinal cord, which control voluntary movements of skeletal muscles, and the peripheral or autonomic nervous system, a parallel system consisting of clusters of nerve cells termed ganglia chained together outside the spinal cord, which regulate activities of most internal organs, glands, and smooth muscles (Fig. 4.1).

The nervous system can be studied and understood at different levels (Chapter 1). The molecular level involves thousands of different genes, proteins, and other substances. Many molecular parts join together to form organelles such as a synapse or myelin sheath around an axon, and an entire cell is assembled from a large number and variety of organelles. The neurons are then organized into intricate circuits that send and receive signals to and from other nerve cells, and large assemblies of neurons form expanses of tissue that organize particular kinds of functions. Put all those tissues together, and we have an entire organ, a brain that is part of the larger nervous system. Behavior emerges from the nervous system. The brain itself is not capable of performing any behavior.

A major objective of this book is to explain how genes are involved in normal and abnormal behaviors. This cannot be done without exploring the nervous system, because no gene can act directly to influence any behavior. However, most genes have well-defined relationships with some specific part of the nervous system. Extensive use is made here of databases of genes (GeneCards, OMIM) that indicate which of the many thousands of genes are especially important for specific brain functions.

GROSS ANATOMY AND ART

Scientific study of the nervous system began to advance rapidly during the Renaissance in Europe. Renaissance painters were keenly interested in accurate depiction of the world. They discovered the principles of perspective and applied them in works of art. They studied human anatomy in order to portray the three-dimensional structures of persons and things in two dimensions on a wall or canvas. Curiosity about the body resulted in bold dissections that revealed intricate details of skeletons and muscles, and facts laid bare by dissections challenged the authority of Galen and other ancients whose opinions held sway over physicians for more than 1000 years. Dissections were often performed in public on someone who had died or been executed shortly before the dissection performance began. The tissue was not preserved in any way (Jones, 2012), and drawings of the details of a body had to be done rapidly. Thus, the study of gross anatomy began in the early 16th century. Calling it gross means that early anatomists examined large parts visible to the unaided eye, as opposed to anatomy of small details that thrived after invention of the microscope. It was also gross in 21st-century parlance because the bodies under the knife quickly decayed, which destroyed the finer anatomical details and emitted foul odors.

Foremost artists of that period who dissected human bodies included da Vinci, Michelangelo, Raphael, and Dürer (Gross, 1997). Da Vinci (1452–1519) was the first to portray intricate details of human nerves and the brain with considerable accuracy. In 1490, he made diagrams of the ventricles of the brain and connections of the eye with the brain, and later, he sketched the topography of the optic chiasm, cranial nerves, and the vagus nerve. Many of his colleagues viewed his masterful drawings, but they were not published in multiple copies for more than 200 years. The importance of art for anatomy is well summarized by da Vinci's own words: "Dispel from your mind the thought that an understanding of the human body in every aspect of its structure can be given in words; for the more thoroughly you describe, the more you will confuse the mind of the reader ... it is therefore necessary to draw as well as describe ... I advise you not to trouble with words unless you are speaking to a blind man" (cited in Gross, 1997, p. 353).

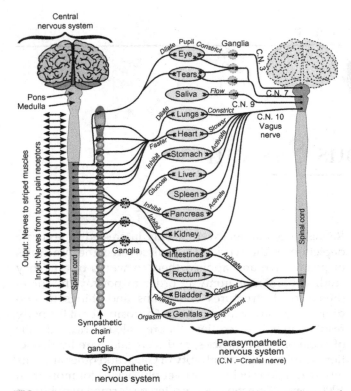

FIG. 4.1 Major parts of the central and peripheral (sympathetic and parasympathetic) nervous systems, showing some of the internal organs that they innervate. The central nervous system connects with muscles that can move the skeleton consciously and precisely, whereas the sympathetic and parasympathetic often do their work without awareness or conscious control.

TABLE 4.1 Anatomical Terms Describing Parts of the Brain and Latin Meanings

Latin	Meaning	Latin	Meaning
Amygdala	Almond-shaped	Locus coeruleus	Place that is sky blue
Caudate	Tail	Mammillary	Resembles breasts
Cerebellum	Little cerebrum	Medulla	Middle
Cingulum	Girdle or belt	Parietal	Wall
Claustrum	Cloister, locked place	Pineal	Pine cone shaped
Corpus callosum	Tough body	Pons	Bridge
Dendron; Greek	Tree (dendrite)	Striatum	Striped
Hippocampus	Seahorse	Substantia nigra	Black substance
Hypothalamus	Under chamber	Thalamus	Chamber
Inferior colliculus	Lower hill	Vagus	Vague

The first authoritative publication of fine details of human anatomy was *De Humani Corporis Fabrica* (The Fabric of the Human Body) by Vesalius in 1543. His drawings revealed a cerebral cortex that had many gyri (plural of gyrus) separated by fissures or sulci (plural of sulcus). His masterwork was published in Latin with notes in the margin in Greek (Scatliff & Johnston, 2014). Scholarly work and medical education at the time was conducted in Latin. Anatomical dissections of human brains were described by Willis in his *Cerebri Anatome* in 1664. There is evidence that Willis and others in this period sought to preserve the brain tissue with alcohol mixed with vinegar, so that the brain became "soused and pickled."

History helps to understand terms in use today to describe the brain. There is a fundamental distinction between gray matter and white matter. White matter includes regions that have many large nerve fibers covered with sheaths of myelin, a substance rich in lipids, whereas gray matter involves regions where there are many nerve cell bodies and relatively few long fibers covered with myelin. But living tissue is not actually gray. It just looks more or less gray in pickled brains. In living tissue, the zones with many nerve cells are tan, and their many small blood vessels are red, so the living tissue appears pink. The so-called white matter has fewer blood vessels and therefore is light pink. Only when the blood is pumped out of the organism shortly after death do the areas rich in myelin actually appear white. Replacing the blood by a solution to arrest decay and "fix" the tissue became common practice in anatomy labs after the time of Willis. Many people today think of gray matter as the outer layer of the cerebral cortex where they believe thinking occurs. In fact, gray matter is present in many zones of the pickled nervous system, including neuron-rich parts near the central canal of the spinal cord.

Parts of the brain that drew the attention of early anatomists were relatively large. The observers had little idea of what a particular part did, so it was often described merely by its shape. The descriptions utilized Latin terms, and many of them have been bequeathed on modern neuroscience. Table 4.1 provides translations of some of the terms now in common use. A section through the hippocampus reminded some anatomists of the curled tail of a seahorse. The pineal gland seemed to be shaped like a small pine cone, and the amygdala had a shape like an almond. The corpus callosum was made of sturdier stuff than the soft and mushy cerebral cortex.

PARTS OF THE BRAIN

Today's medical students, many of them future surgeons, spend months or even years dissecting bodies and learning human neuroanatomy before touching a knife to a living person. For our purposes, a few diagrams will suffice. Fig. 4.1 sketches the broad scheme of the human nervous system, divided into three main parts. The central nervous system (CNS), consisting of much of the brain and spinal cord, controls the rapid and intricate movements of numerous finely striped muscles attached to the skeleton, and it collects information from the skin, the joints, and the muscles themselves, as well as sense organs such as eyes, ears, and nose. Many of the movements of the skeletal muscles are under conscious control by the cerebral cortex at the top of the brain, and their movements are said to be voluntary. The autonomic nervous system (ANS), on the other hand, controls the much slower actions of smooth muscles and glands associated with internal organs, actions that are often performed reflexively without awareness and are therefore said to be involuntary. The ANS has two divisions, termed sympathetic and parasympathetic, which usually exert opposite effects. Activation of the sympathetic circuits generally prepare the individual for action by accelerating the heart, dilating bronchi in the lungs to gather more oxygen, increasing blood glucose level, and shutting down digestion. It also stimulates the release of the hormone adrenalin into the bloodstream. Parasympathetic stimulation, on the other hand, slows the heart, reduces glucose and adrenalin levels, and boosts digestive activity.

The CNS and both components of the ANS all rely on activities of the brain and spinal cord, but they are controlled by separate areas of the brain and spinal cord, although there are numerous connections between CNS and ANS. Many parts of the body, including the sense organs, depend on the well-coordinated actions of both CNS and ANS. The eye, for example, tracks objects using striped or striated muscles linked to pattern data from rods and cones in the retina that reach the cerebral cortex via the thalamus. The amount of light admitted into the eye is controlled by the pupil that dilates when the influence of the sympathetic nervous system prevails and constricts under the influence of the parasympathetic nervous system. Pupil size is of course related to the brightness of the scene, but it also depends on the emotional state of the individual.

The brain itself can be divided into dozens of regions with different kinds of tissue having different patterns of inputs and outputs. Several of these are shown in Fig. 4.2. Information about the surface of the body and positions of the limbs enters the brain through the spinal cord, and information from the senses of vision, hearing, smell, and taste enter the brain via several cranial (pertaining to the cranium or head) nerves. From the spinal cord, information moves upward through several parts of the hindbrain and midbrain to the thalamus, a major relay nucleus that routes the data to the cerebral cortex. The cerebral cortex has several major lobes and numerous

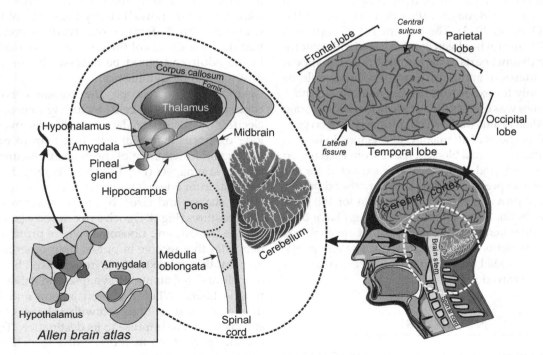

FIG. 4.2 Major parts of the human brain, with many names derived from Latin, as indicated in Table 4.1. The view of the cerebral cortex with its lobes represents the outer surface of the brain, whereas structures shown in the inset are at or near the middle of the brain. Smaller regions within the hypothalamus and amygdala are represented in the Allen Brain Atlas.

smaller fissures, and it is a prominent feature covering most of the outer surface of the human brain. The kinds of functions that are represented in each major region of the cerebral cortex are fairly well known. From the cerebral cortex, elaborate circuits descend to areas such as the pons that regulate motor actions and other regions such as the hypothalamus and amygdala that are critically important for the emotions. The hypothalamus sends signals over short distances to the pituitary gland, and that gland generates numerous hormones and releases them into the bloodstream to regulate the metabolic activities and growth processes of many internal organs. The cerebellum at the base of the brain has many folds and fissures that allow a very large number of cells to be packed into a relatively small volume of tissue. It plays an important role in the coordination of motor movements. The corpus callosum, a large bundle of nerve fibers coated with myelin, transmits vast quantities of information between the right and left halves of the cerebral cortex, so that the entire brain functions as one unified whole, rather than two independent half brains.

As the study of neuroanatomy progressed and more scientists from many countries became involved, there was a proliferation of names used for what seemed to be the same brain parts. The advent of large-scale databases on gene expression in the brain made it essential that a common set of names, abbreviations, and anatomical definitions be adopted so that data could be combined from many labs around the world. The NeuroNames project achieved this for a collection of over 16,000 anatomical names in eight languages (Bowden et al., 2011; Bowden & Dubach, 2003). The current list of approved names (see BrainInfo) has been condensed to 850 anatomical terms for distinct parts of the human brain that appear in standard atlases of anatomy, most of which would be meaningful only for experts on human neuroanatomy.

NeuroNames was used as the authority for naming brain parts in the Allen Brain Atlas, which compiled data on gene expression in portions of all major brain regions in human and mouse brains. A small block of tissue from each region was homogenized, and its mRNA was extracted for the analysis of gene expression (Chapter 3). Another database (see BrainMap) provides rapid access to data for 168 parts of the human brain. Some regions of the brain have been subdivided into several smaller portions that are analyzed separately for patterns of gene expression. The amygdala, for example, is divided into six groups of nerve cells, four of which are portrayed in Fig. 4.2.

NEURONS

The invention of a compound microscope by Galileo in 1609 was exploited by Hooke (1665) in his influential work *Micrographia* to show how some living things are made of discrete cells. It took more than 200 years to confirm that the nervous system itself is not an amorphous mass of soft gray and firmer white matter; it is made of small cells, just like other parts of the body. Anatomists devised ways to slice a preserved brain very thinly and stain the tissue with different kinds of chemicals that revealed the cell body and its nucleus in brain cells, but they could not determine whether cells of the nervous system were joined in a huge latticework of tubes or were distinct entities that only made contacts with each other.

The answer to this question emerged in 1873 when Camillo Golgi, after much experimentation, found a way to stain an entire nerve cell. First, he placed a small block of preserved tissue into a solution of potassium dichromate for 2 days and then immersed it in a solution of silver nitrate for another 2 days. Only a portion of the cells in the tissue block was stained, and this made it possible to perceive details of only one cell. The method was applied to many kinds of brain tissue by the anatomist Santiago Ramon y Cajal, who later described his delight at viewing the "unexpected spectacle" of fully elaborated brain cells (quote from DeFelipe, 2010). The complete cell consisted of a body or soma with many dendrites extending outward and an axon exiting the soma and growing long distances (Fig. 4.3). In 1888, Cajal described the existence of the large numbers of small spines protruding from the dendrites. He saw that the extraordinary complexity of nerve cells could serve as a basis for higher mental processes. He wrote eloquently of "the mysterious butterflies of the soul, the beating of whose wings may someday (who knows?) clarify the secret of mental life" (DeFelipe, 2010). Further observations convinced him that the finer details of the neurons were malleable even in the adult and could be shaped "by means of well-directed mental gymnastics."

Under the microscope, the human nervous system consists of billions of cells, including neurons that send electric/chemical signals to other cells over relatively long distances and glial cells that help to maintain the health and functioning of the neurons. Neurons communicate via synapses (Fig. 1.2) and generally do not transfer material from one cell to the next. They have a wide variety of shapes and sizes, but most neurons share a few basic features (Fig. 4.4) such as a cell body (soma) containing the nucleus and ribosomes where proteins are made, dendrites that receive impulses from other neurons, and an axon that can extend far from the cell body and then branch and terminate in synapses on other neurons or muscle fibers. When the axon is long and has a large diameter, it is usually coated with a sheath of myelin that speeds the nerve impulse to its destination. The myelin is actually part of the membrane of a special kind of glial cell termed an oligodendrocyte.

The membrane of a neuron in living tissue is electrically polarized so that the inside is slightly more negative

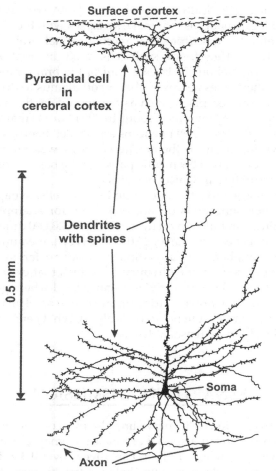

FIG. 4.3 Drawings of a neuron by Cajal, adapted by the author. Dendrites of a pyramidal neuron in the cerebral cortex are covered with thousands of small spines where synapses occur. DeFelipe (2010) gives the original source of the drawing as Figure 689 in Cajal (1899–1904) *Textura del Sistema Nervioso del Hombre y de los Vertebrados*.

than the outside. When a sufficient number of incoming signals activate the dendrites enough to exceed a threshold, the cell fires an impulse down the axon. The impulse involves a few millisecond when the voltage on the membrane becomes positive, followed by a brief refractory period when it cannot fire again. When the axon is coated with myelin, the nerve impulse can travel along the axon at speeds up to 100 m/s, which is about 10 cm in 1/1000th of a second. Input stimulation to the dendrites that does not exceed the threshold is not transmitted to the synapses at the end of the axon. Thus, nerve conduction is all or none within any short period of time, but the rate of firing of a neuron each second can range widely. It might not fire at all for a minute or more, or it might fire 10 times in 1 s, depending on activity at its dendrites. Some neurons send signals to the next cell in a circuit that actually inhibit its firing.

COMPLEXITY

The stunning complexity of the human brain can be sensed from two facts—the number of neurons and the number of synapses on just one neuron. Counts of the numbers of neurons in different brain regions (Table 4.2) reveal that densities of neurons in the cerebral cortex and cerebellum differ because the cerebellum has an extraordinarily large number of very small neurons that interconnect the massive Purkinje neurons. Even in the cerebral cortex where there are many relatively large neurons, there is an astounding number and density of neurons, roughly 14,000 of them on average in just one tiny cube of tissue 1 mm × 1 mm × 1 mm.

Spines on neurons usually involve one or more synapses. The number of spines with synapses on some kinds of

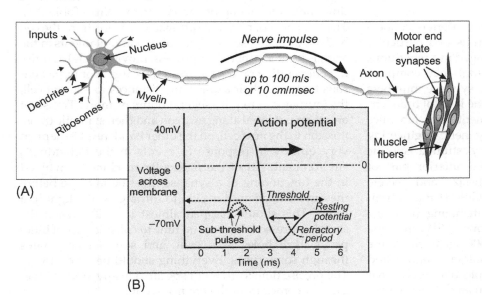

FIG. 4.4 Motor neuron and graph showing voltage of the action potential as it races down an axon. (A) The nerve cell body in the spinal cord receives inputs along its dendrites. It manufactures proteins at the ribosomes in the cytoplasm, and the proteins are then transported very slowly to other parts of the cell. (B) When sufficiently strong and frequent excitatory inputs occur at the dendrites, the neuron fires an action potential down the axon coated in myelin, which speeds the journey to synapses on muscle fibers.

TABLE 4.2 Numbers of Neurons in Human Brain[a]

	Total brain	Cerebral cortex	Cerebellum
Mass (g)	1508	1233 (82%)	154 (10%)
Volume (cc)	1400	1142	143
Neurons	86 billion	16 billion	69 billion
Neurons/mm^3	61 thousand	14 thousand	482 thousand

[a] *Volume obtained using density of 1.08 g/cc. One cc is one cubic centimeter or 1 mL (milliliter). A tiny cube of tissue that is 1 mm on each side has a volume of 1 mm^3. Cube size is a good indicator of difference in average neuron size in different parts of the brain. Herculano-Houzel, 2009*

neurons is mind boggling. Careful counts have found that there are sometimes more than 100,000 spines on just one rat Purkinje cell (Takacs & Hamori, 1994) and there are probably even more in humans. A single cortical pyramidal cell can have more than 10,000 spines. Small neurons of course harbor fewer synapses. If there are perhaps 2000 synaptic connections on average for each of the 86 billion neurons, there will be 172,000,000,000,000 synapses in the human brain. This figure can be compared with the familiar laptop computer. A machine might feature 10 GB of random access memory (RAM), a volatile kind of storage that is erased or sent to long-term storage when the power is shut down. A gigabyte is 1 GB of data or one billion bytes, and a word having one byte of data contains eight binary bits, each being 0 or 1. Thus, 10 GB of computer RAM can process 80,000,000,000 bits of information. If a single synapse can be either on or off at one point in time, like a digital computer that uses only 0s or 1s, the approximate equivalent processing power of just one human brain would require the combined activity of (172,000)/80 or about 2000 digital laptop computers. These figures are solely for comparison between the brain and computer. The human brain does not work like a digital computer.

Another source of complexity is the pattern of connections among the parts. If many neurons in a particular cluster and their connections are arranged in a repetitious way that confers a high amount of redundancy, complexity might be somewhat less than synapse numbers suggest. A glimpse of complexity at the level of brain circuitry was provided by an experiment with lab mice that used an ingenious method of genetic engineering (Boehm, Zou, & Buck, 2005). They studied specific kinds of neurons found only in the anterior part of the hypothalamus, cells that synthesize and secrete gonadotrophin-releasing hormone (GnRH) that stimulates other nerve cells to release luteinizing hormone (LH) and follicle-stimulating hormone (FSH) into the bloodstream. There are only about 800 such cells in the mouse brain, a minute fraction of the total number, and they were tagged with a special molecular probe that allowed researchers to detect GnRH using a microscope.

They also genetically engineered the DNA so that they could locate neurons that make synapses with the GnRH cells and other neurons that receive axonal projections from them. They found such cells in 53 different areas of the brain, about 26 in each half of the brain, ranging from other zones nearby in the hypothalamus to diverse and unexpected places such as the amygdala, thalamus, and pons. They identified about 10,000 neurons that send axons to the 800 GnRH neurons and 40,000 neurons that receive inputs from them. The circuitry was far more complex than had been anticipated on the basis of previous research with GnRH neurons.

At first glance, the large numbers of cells, synapses, and interconnections do not bode well for attempts to explain brain function on the basis of just 20,000 different kinds of genes. Of course, 20,000 is a very large number, but the numbers of neurons and synapses are far greater. An entire neuron and circuitry are things that exist at a much higher level than the molecular level where genes reside. Moving downward to the molecular level, matters are very complex but not impossibly so. What genes do at that level can be reasonably clear.

SYNAPTIC TRANSMISSION

Transmission of the nerve impulse from one neuron to the next occurs at a special organelle, a synapse (simplified version in Fig. 1.2), and involves several kinds of protein molecules, each encoded by a specific gene. Each protein is synthesized at ribosomes in the neuron cell body and then transported down the axon to the synapse where it performs a specific task. As discussed in Chapter 3, the generation of a new protein molecule from a gene and its mRNA intermediary takes considerable time, hours for some of the larger proteins (Table 3.1). The transport of newly synthesized proteins from the cell body to the synapse via microtubules and filaments in the axon is itself a rather slow process, moving at 5–100 mm/day, depending on the nature of the cargo. For large neurons that send their axons long distances to a target cell, the journey can require several hours or more. Transit of an action potential along an axon and then synaptic transmission to the next cell, on the other hand, need to happen very quickly in response to events in the individual's environment. All of the molecular machinery involved in the functioning of a synapse must be in place before the time for action. It is akin to turning on the light in a room. This should happen almost instantly when the switch is pressed. There is no time to consult a set of blueprints, drill holes in a wall, and start feeding wires through some tubes. Everything should be ready to go. The precise time to go is not specified in any kind of blueprint; it comes from the environment.

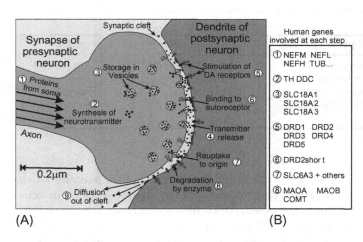

FIG. 4.5 Parts of a synapse that use dopamine (DA) as the neurotransmitter, showing genes that encode proteins with important roles in the process. The enzymes that synthesize DA are produced in the cell body and then transported to the synapse via neurofilaments that are encoded in genes. Another family of genes transports the DA into storage vesicles. DA released at the synaptic cleft either stimulates receptors on the next cell or autoreceptors that return it to storage vesicles. DA remaining in the cleft is broken down by other enzymes or diffuses away. *Source: GeneCards.*

The synapse is constructed so that a rapid response to nerve impulses can occur (Fig. 4.5). Transmitter molecules are synthesized at the synapse with the aid of specific enzymes and then stored in minute synaptic vesicles until they are needed. There are special protein molecules called transporters that help to transport a specific kind of neurotransmitter molecule into a vesicle. Once they are released and enter the synaptic cleft, the transmitter molecules can make contact with a receptor molecule, which will then stimulate the postsynaptic neuron. Generally speaking, the higher the concentration of transmitter in the cleft, the more intense will be the stimulation of the postsynaptic dendrite. It is important that transmitter levels be kept within certain limits in the cleft. In many kinds of neurons, there are special receptor molecules termed autoreceptors on the presynaptic part of the synapse that protrude into the cleft and detect the transmitter molecules. When an autoreceptor is stimulated, it can activate another kind of protein, a reuptake transporter on the presynaptic side that transports excess transmitter molecules back into the synapse for storage and reuse in a vesicle. Neurotransmitter levels in the cleft can also be reduced by a specific enzyme that breaks down or degrades the molecule into smaller parts or converts it into an inactive molecule that can then diffuse away from the synapse. Thus, the transmission of neural impulses at a synapse is a dynamic affair involving a balance between the production and release of transmitter molecules into the synaptic cleft and other processes that remove transmitter from the cleft.

Each step in the process of synaptic transmission is mediated by one or more proteins encoded in specific genes. The main steps are shown in Fig. 4.5 for an example involving the neurotransmitter dopamine (DA). The gene symbols (Chapter 2), shown here in capital letters, are the official abbreviations in the 2018 version of Gene Cards. This scheme is itself a simplification, and many more genes are involved in the process. Nevertheless, it does convey an impression of the intricate nature of things happening in just one kind of synapse:

1. Transport down axon involves neurofilaments (NEFM, NEFL, NEF) and neurotubules (more than 20 kinds of Tubulin (TUB) genes).
2. Synthesis involves tyrosine hydroxylase (TH) that acts on tyrosine from blood to make DOPA and then dopa decarboxylase (DDC) that converts it into dopamine transmitter (DA).
3. Three solute carrier proteins transport it into the vesicle (SLC18A1, 2, 3).
4. Transmitter release involves several genes not shown here.
5. Detection of DA in the cleft can involve five kinds of dopamine receptors (DRD1 to DRD5).
6. Activation by DA involves the short form of type 2 autoreceptor (DRD2short).
7. Reuptake of DA involves several kinds of solute carrier family 6A proteins (SLC6A3 and others)
8. Degradation involves monoamine oxidase A or B (MAOA and MAOB) or inactivation by catechol-O-methyl transferase (COMT).

The proper functioning of a dopamine synapse thus requires the coordinated actions of a substantial number of genes but not an impossibly large number. The specific genes involved, the proteins for which they code, and the role of each in the process of transmission are reasonably well understood.

Neurotransmitters

The neurotransmitters acetylcholine (Fig. 2.4) and dopamine are only two of a much larger population of chemicals that mediate neural activity. A more complete list is provided in Table 4.3. Virtually, all of these are found in

TABLE 4.3   Classes of Neurotransmitter Molecules in Relation to the Type of Receptor With the Number of Known Receptor Genes Shown in Brackets [ ]

Excitatory ion channels	Inhibitory ion channels	G-protein coupled	Unconventional	Peptides
Acetylcholine [16]	GABA [20]	Serotonin [10]	Anandamide [1]	>200 [1 each]
Glutamate [30]	Glycine [5]	Dopamine [5]	2-AG [1]	Enkephalin
		Norepinephrine [9]	Nitric oxide	Endorphins
		Acetylcholine [5]		Oxytocin ACTH
				Insulin Prolactin

GeneCards and OMIM.

other mammals and are not unique to humans. The release of certain kinds of transmitters excites the postsynaptic neuron, whereas others inhibit it. Some involve receptors that act very quickly to open small channels or pores in the postsynaptic membrane that allow ions to pass, whereas others act more slowly via a kind of protein termed a G-protein that is attached to the membrane. The first three categories in the table are considered "classical" neurotransmitters that are stored in vesicles and released to cross the cleft and stimulate receptors on the adjacent neuron, as diagrammed in Fig. 4.5. Each involves a specific set of genes, most of which are unique to the specific transmitter, and the identity of those genes has been established.

One class of transmitters discovered relatively recently came as a genuine surprise to neurochemists. They were exploring the brain to find possible receptors that responded to an ingested substance, Δ^9-tetrahydrocannabinol (THC), the principal psychoactive ingredient of marijuana. It happened that there were indeed receptors with a high affinity for THC, and they were widely distributed throughout the brain. But why would specific receptors detect a chemical so different from all known neurotransmitters? THC is a lipid, and it cannot be stored in vesicles. After working on this puzzle for several years, researchers concluded that the brain actually manufactures its own supply of a class of compounds they decided to name endocannabinoids. Endo signifies endogenous. Cannabinoids signify that they resemble chemicals made by the plant *Cannabis sativa*. In a rather delicious irony, it happens that even the most rigidly antidrug zealot manufactures his or her own supply of cannabinoids every day for free. Further investigation uncovered a specific receptor in the brain and the gene that encodes it (CNR1). Then, the identity of one of the transmitters itself was confirmed. What to name it? This was the prerogative of those who discovered it. They called it "anandamide," derived from the Sanskrit term for joy or bliss. A closely related molecule with similar functions carries the technical moniker 2-arachidonoylglycerol (2-AG), which has not yet been converted into Sanskrit.

The cannabinoids are regarded as "unconventional" neurotransmitters, an irony not lost on many neuroscientists. They seem to violate almost every rule followed so regularly by the conventional varieties. They are synthesized on demand via a complex array of enzymes and are not stored in the brain. Furthermore, they actually work in reverse, being synthesized by enzymes located in dendrites of the postsynaptic neuron and crossing the cleft to stimulate CNR1 receptors on the presynaptic side. Thus, they are regarded as "retrograde" transmitters. They occur in synapses that use either the excitatory glutamate transmitter or the inhibitory GABA transmitter. High levels of anandamide transmitter are able to suppress the release of either glutamate or GABA, so they are considered modulators of synaptic transmission. They therefore violate the once-sacrosanct rule that each neuron utilizes one and only one kind of transmitter to convey messages to the next neuron in a chain. Glutamate and GABA transmitters are abundant and widely distributed throughout the brain, and the endocannabinoids involved in those kinds of synapses have a wide range of possible effects on numerous brain functions.

Another class of transmitters involves neuropeptides, short chains of amino acids such as oxytocin (Chapter 1) that mediate a great diversity of brain functions. These share a common feature, in that one gene usually encodes several peptide transmitters. The mRNA is translated into a long protein termed a prepropeptide, and that chain is then cleaved into several smaller peptides that serve specific biological functions. Special enzymes expressed in a cell dictate that specific peptide will be produced there in abundance. For example, the proopiomelanocortin gene (POMC) specifies a long protein with 267 amino acids that can be cleaved in various ways to obtain the peptide transmitters β-endorphin, MET-enkephalin, β- and γ-lipotropins, adrenocorticotropic hormone (ACTH), and three forms of melanocyte-stimulating hormone (α-, β-, and γ-MSH). More than 90 genes are known that yield one or more neuroactive peptide transmitters. The peptide can be stored in a vesicle and released to stimulate a specific receptor. An important feature of many peptides is that they are synthesized and released from a synapse that also utilizes one of the classical transmitters such as acetylcholine, serotonin, dopamine, or GABA. Furthermore, some neurons utilize more than one kind of peptide transmitter, and vesicles with large amounts of peptides can be located in several places in a neuron in addition to the synapse. Thus, peptides break many old rules. The errant cannabinoid is not the only offender.

The question of what a transmitter such as acetylcholine or dopamine does, what its function is, has no easy answer. The DA transmitter molecule itself is rather small and relatively simple, being derived from the amino acid tyrosine. The receptor, on the other hand, is a very large protein, and there are several kinds of DA receptors, each encoded by a specific gene. Dopamine can excite five different types of receptors (the DRD family of genes). A postsynaptic neuron usually expresses only one of these types. The D1 and D5 types (DRD1 and DRD5 genes) specifying 446 and 477 amino acids (aa), respectively, are very similar, but D1 is expressed widely throughout the brain, while D5 is found mainly in the limbic lobe and is much more sensitive to DA than the D1 type. Both excite the postsynaptic neuron, while D2, D3, and D4 usually inhibit it. The D2 type (DRD2gene, 443aa) is abundant in the striatum and is involved in motor coordination as well as reward, learning, and memory. The translated protein can be cleaved into two forms, the long chain D2long that serves as a classical neurotransmitter receptor and the short chain D2short that serves as a DA autoreceptor that can inhibit DA release. The D3 type (DRD3 gene, 400aa) has a distribution and function similar to D2. The D4 type (DRD4 gene, 467aa) is abundant in the frontal lobe of the cortex, the midbrain, and the amygdala. The functions of DA thus involve what form of receptor is present on the postsynaptic neuron and the part of the brain where the receptor is expressed.

The situation is vastly more complicated for the versatile transmitter acetylcholine (ACh; Fig. 2.4). It is the sole transmitter that causes contraction of skeletal muscles when released at a specialized synapse on a muscle fiber known as the motor end plate (Fig. 4.4), whereas it tends to inhibit and slow the beating of the heart. It is widely utilized in the central nervous system and abundant in the various ganglia of the autonomic nervous system (Fig. 4.1). It also mediates effects on several internal organs via the vagus nerve and causes the adrenal gland to release adrenalin. It is important for conveying nerve impulses to the cerebral cortex and hippocampus.

There are two broad classes of ACh receptors, nicotinic and muscarinic, that are preferentially stimulated by the drugs nicotine and muscarine. The nicotinic receptors form ion channels, whereas muscarinic receptors are similar to the slower acting G-protein receptors (Table 4.3). The ion channels are formed from five different proteins arranged in a pentagon-like structure (pentamer) that can admit sodium and calcium ions down the opening at its center when a receptor protein binds ACh. There are five families of nicotinic receptors (α, β, δ, ϵ, and γ). Ten genes code for different forms of the α protein (CHRNA1 to 10), one of which is part of the motor end plate (CHRNA1), and four genes code for the β form (CHRNB1 to 4), one of which is used at the motor end plate (CHRNB1). The receptors forming the pentamer on muscle fibers usually have two α1 subunits, one β1, and one each of δ and ϵ forms (CHRND and CHRNE). In the central and autonomic nervous systems, on the other hand, the subunits α2–10 and β2–4 can join in a dizzying number of combinations, perhaps thousands of different receptor configurations having different properties and localized in functionally different kinds of neurons. Thus, not only is any general statement about the function of ACh itself devoid of meaning, but also most of the CHRN receptor subtypes in the central and autonomic nervous systems subserve many different functions. The functional properties of the pentamer depend on the specific combination of receptor subtypes and the place in the brain where they are located. Genetic diversity of a particular kind of transmitter receptor prevails for all but the endocannabinoids and peptides (Table 4.3). This diversity enables relatively few genes to join forces to generate a rich variety of types of neurons. It defies all attempts to describe the genetic basis of brain functions in just a few words or sentences.

GENE EXPRESSION IN BRAIN

The information contained in a DNA molecule is transcribed into mRNA in the cell nucleus and then translated into protein in the cell body before being transported to places in the cell where the protein has a role (Chapter 3). The genes expressed in a particular kind of tissue help to understand its functions. A comprehensive database of gene expression has now been compiled for the mouse and human brains, the Allen Brain Atlas. The levels of expression of more than 17,000 genes have been documented for more than 168 regions of the human brain. This rich depository of data is now available for mining by experts adept at statistical analysis, but potential users are cautioned that it "can be overwhelming for neuroscientists" (French & Paus, 2015). Those outside of neuroscience will find it very challenging to explore what is really a frontier of brain science. As shown in Chapter 3, levels of expression of many genes change dramatically with the time of day, diet, and even the emotional state of the individual. The Allen Brain Atlas does not reveal anything about which genes are most strongly influenced by common features of a fluctuating environment. This could be one reason why expression profiles for many genes differ markedly among the brains currently in the atlas (French & Paus, 2015).

The atlas provides strong evidence that most genes pertinent to brain-specific functions "are expressed in multiple regions and nonuniformly within regions," which suggests that a particular kind of gene and its protein product typically subserve many different functions in different brain regions (Hawrylycz et al., 2012). For example, several genes relevant to dopamine synthesis, storage, release, receptor function, and degradation were expressed at high levels in the striatum, amygdala,

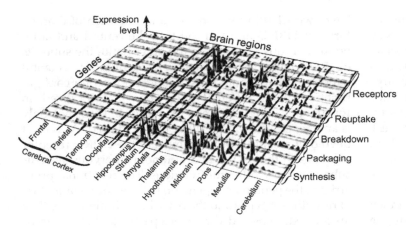

FIG. 4.6 Intensity of expression in mRNA of genes involved in different aspects of dopamine function in several brain regions. Genes related to the dopamine system are active in all major regions of the brain, although activity is far higher in some regions than others. Within a region, there are major differences in dopamine-system activity. *Reprinted by permission from Hawrylycz, M. J., Lein, E. S., Guillozet-Bongaarts, A. L., Shen, E. H., Ng, L., Miller, J. A., ... Jones, A. R., An anatomically comprehensive atlas of the adult human brain transcriptome, Nature, 489, 391–399, Copyright 2012.*

thalamus, hypothalamus, midbrain, pons, and medulla (Fig. 4.6). Those brain regions are important parts of systems that regulate motor actions, emotions, motivation and reward, and several sensory functions. Moderate but significant levels of DA-related gene expression were evident in some part of almost every other brain region. Levels of dopamine-related gene expression were relatively low and uniform across the cerebral cortex, in contrast with the CNR1 cannabinoid receptor gene that was expressed at significant levels in some but not all cortical zones (French & Paus, 2015). These findings demonstrate that there is no single, simply stated function of dopamine in the brain. It subserves many functions in various parts of the nervous system.

Another finding of considerable importance is that expression patterns for some genes differ markedly between human brains and brains of several other mammals, including the ubiquitous lab mouse (Hawrylycz et al., 2012; Zhang et al., 2016). The similarity of the genes themselves is remarkable for humans and mice (Chapter 2), whereas species differences in things that influence the expression of genetic information present formidable challenges to the utility of animal models for the discovery and validation of new psychiatric drugs (Markou, Chiamulera, Geyer, Tricklebank, & Steckler, 2009).

SENSORY INPUT

Information about the environment enters the nervous system via sense organs that detect stimuli with extraordinary sensitivity and specificity. Genes play central roles in sensory systems, and the function of many of those genes is plainly evident. Sensory modalities such as smell, taste, and vision rely on receptors that are encoded by genes. Not all sensory systems are presented here. Instead, focus is on those where genetic aspects have been well investigated.

Smell or Olfaction

Odors are detected by special nerve cells in the membrane lining the nasal cavity. One end of each neuron divides into many fine processes where odor receptors are located. The olfactory receptors are similar to those for G-protein-coupled neurotransmitters (Table 4.4), except that the chemicals most effective in activating them are located in the environment. A gene codes for the structure of a single kind of odor receptor (OR). In humans, 390 distinct kinds of OR genes that are part of the OR gene family have been identified (Olender, Lancet, & Nebert, 2008), and a recent count with Gene-Cards expands the total to 402. Mice and many other mammals rely strongly on odor cues, and mice have more than 1300 kinds of OR genes. Generally speaking, each kind of OR is sensitive to more than one chemical in the environment, and each kind of odor molecule activates more than one kind of receptor. It is the *combination* of OR activated by a chemical that informs the brain about what is present in the environment. Systematic experiments with different blends of chemicals have demonstrated that the human sense of smell can discriminate millions of distinctly different odors (Bushdid, Magnasco, Vosshall, & Keller, 2014).

The vast number of different OR was discovered only recently, after the human genome was sequenced, on the basis of similarity of DNA sequences in genes scattered widely across every chromosome except number 20 and the Y, which carry no OR genes. Structures of the OR-encoded proteins comprising the receptors were deduced from the DNA sequences. Further examination of the genome revealed a very surprising fact: in addition to the 390 protein-encoding OR genes, the human genome contains about 465 OR-type *pseudogenes* that look very much like an ordinary gene but have become disabled by changes in gene structure that make it impossible for them to be expressed as protein. The remnants of our evolutionary past are effectively silenced, yet they are

TABLE 4.4 Sensory System Genes Coding for Receptors and Transduction

Sense	Gene names or families	Protein-encoding genes	Pseudogenes
Odor	OR1A__ to OR56B__	390–402	461–465
Pheromones	VN1R1, 2, 3, 4, 5	5	105
Pheromones	VN2R family	0	18
Taste (sweet and savory)	TAS1R1, 2, 3	3	0
Taste (bitter)	TAS2R__	32	8
Taste (salty and sour)	ENAC	?	
Taste (cold and hot)	TRPM8, TRPV1	?	
Vision (rods; gray) 120 million/eye	RHO; GNAT1; PDE6A, B, D, G	8	0
Vision (cones; color) 7 million/eye	OPN1LW, MW, SW; GNAT2; PDE6C, H	6	0
Vision (light-dark)	OPN4	1	0

GeneCards, OMIM.

still part of the genome and are passed from parent to offspring, just like bona fide genes. They can generate uncertainty in the official counts of bone fide genes.

Olfactory neurons send their axons directly into an extension of the brain, the olfactory bulb, that is located above the nasal cavity. There, the axons synapse onto a dense cluster of cells termed a *glomerulus* that in turn sends its output deeper into the brain. Each glomerulus receives inputs only from sensory neurons that express the same OR gene. The synapses of olfactory neurons on cells in a glomerulus utilize the neurotransmitter glutamate, whereas further steps in sensory processing in the glomerulus involve the transmitters GABA and dopamine. Parts that reside deeper into the brain detect the specific kind of chemical by noting the combination of receptors that are activated.

When neuroscientists first studied the expression of the OR genes, they expected them to be localized in nasal tissue, but a few kinds of OR were found in the testes where they help to guide sperm cells on their long journey via a kind of chemical sense. OR are also located in several internal organs, including the liver, heart, lung, and even the muscle and skin, where they can promote healing of wounds in the presence of certain odors in the environment (Busse et al., 2014).

Pheromones

In many species of vertebrates, there is also a highly evolved sense for detecting air-borne *pheromones* that are important in social signaling and reproductive function. The V1R and V2R gene families code for about 165 and 60 different kinds of receptors, respectively, in mice. These are expressed primarily in a specialized structure termed the vomeronasal organ that is adjacent to the olfactory bulb and sends axons to the amygdala. There has been great interest in a possible role of pheromones in human social affiliation, romantic attraction, and related behaviors. Genetic data now suggest that our remote ancestors may have had such sensory capacities, but now, almost all in this domain have been disabled or lost in us. There is no convincing evidence that the vomeronasal organ even exists in adult humans. The human VN1R gene family has 110 entries in the GeneCards database, but only three of them are conventional protein-encoding genes, while 105 are clearly pseudogenes that once played a role but now are silenced, and two others are evolving into pseudogenes. The VN2R gene family in humans has 18 different kinds, all of which are now pseudogenes.

Ligands

It is remarkable that so much is now known about olfactory receptor genes and the receptors themselves, but the specific *ligands*, chemicals that activate specific receptors, are almost entirely unknown in humans. Even in lab mice, the function of more than 90% of OR gene-encoded receptors is unknown (Peterlin, Firestein, & Rogers, 2014), and there is no guarantee that a ligand shown to be effective for a specific receptor in mice will also play the same role in humans. Chemical assays *in vitro* (performed in a glass dish) have suggested ligands for many human OR-type genes, but the findings need to be confirmed in living persons. A recent study found evidence that variation in the human OR10G4 receptor is related to individual sensitivity to the chemical guaiacol (Mainland et al., 2014), so it seems that a single mutation can influence perception of a specific chemical. Because a single chemical usually activates more than one kind of receptor, we expect that a mutation in a single receptor gene will not expunge all sensitivity to a specific chemical. At the present time, knowing the structure of a particular form of an OR gene does not tell us what chemical(s) can be sensed in the environment. How those genes work in a general way is well enough understood, but at the level of molecules, where the structure of an OR gene and its associated receptor are known exactly, its specific function cannot yet be discerned.

Taste

The sense of taste detects specific classes of chemicals in the environment via taste buds or papillae on the

TABLE 4.5 TAS2R Receptors Activated Out of 25 and Effective Ligands Out of 104[a]

Number of ligands detected by a specific number of receptors							
Number of receptors	0	1	2	3	4	5	>5
Number of ligands	20	31	20	10	4	7	8
Number of receptors able to detect a specific number of ligands							
Number of ligands detected	0	1	2	4	6–10	11–20	>20
Number of receptors	6	2	3	1	5	4	4

[a] *Adapted from Meyerhof, W., Batram, C., Kuhn, C., Brockhoff, A., Chudoba, E., Bufe, B., … Behrens, M. (2010). The molecular receptive ranges of human TAS2R bitter taste receptors. Chemical Senses, 35, 157–170 by permission of Oxford University Press.*

surface of the tongue and in several other organs. There are several families of taste receptor genes in humans (Table 4.5). T1R consists of three genes that work together to detect two broad classes of tastes. Two kinds of genes are expressed in a single receptor to make a *heterodimer*, dimer meaning two parts and hetero meaning different kinds. For example, TAS1R1 + TAS1R3 detects a wide range of amino acids in food, including glutamate that functions elsewhere as a neurotransmitter. This sense is often described as savory or umami. Many kinds of chemicals can activate that dimer. TAS1R2 + TAS1R3 detects a broad range of sugars and similar sweet compounds. TAS2R is a large family of genes for receptors that detect what we usually experience as bitter taste. Bitter taste is associated with a large number of compounds in the environment that have widely divergent structures, and it is also a characteristic of many synthetic compounds. It is believed that bitter taste served as a flag of caution to our ancestors that prevented them from eating dangerous foods, but we now know that people can learn to enjoy many kinds of bitter tastes in moderation. The question of what specific chemical (ligand) is detected by which receptor has no simple answer. By working with receptor proteins *in vitro*, researchers have been able to screen many chemicals for the activation of known receptors. A recent study assessed 104 bitter tastes for their abilities to activate 25 different human TAS2R taste receptors (Meyerhof et al., 2010). Results summarized in Table 4.5 indicate that only two receptors were activated by just one ligand, whereas six others were not activated by any of the 104 test compounds. There were 31 chemicals that activated just one kind of receptor, but most receptors responded to several kinds of chemicals. Eight different receptors responded to more than 10 different compounds. As occurs for odor sense, any one chemical in the environment appears to be identified in most cases by the *combination* of receptors that it activates. That is how a relatively small number of genetically unique taste receptors can detect such a vast number of distinctly different environmental chemicals.

Other classes of taste have been less well investigated, and the specific receptors for certain ones remain unclear, for example, salty and sour. A sodium channel in mice seems to mediate the sense of saltiness, but this has not yet been verified for humans. The TRPM8 receptor confers a sense of cool temperature and also responds to chemicals such as menthol that are perceived as having a cool taste, whereas TRPV1 confers a sense of heat and is especially sensitive to capsaicin, an active ingredient in hot peppers.

Whereas the taste buds on the tongue and in the mouth detect chemicals as they are being eaten, chemosensory cells in several internal organs detect the chemicals that have actually been consumed and are undergoing digestion (Welcome, Mastorakis, & Pereverzev, 2015). What we perceive as taste relies on signals from taste receptors on the tongue that transmit signals to the brain in the pyriform cortex. Taste receptors in the internal organs that are sensitive to sugars, amino acids, and especially glutamate can stimulate the reflexive release of hormones that promote digestion or stimulate the vagus nerve that transmits signals to the nucleus of the solitary tract in the brain stem, which then sends signals back to the organs via the vagus nerve to regulate digestion (San Gabriel, 2015). Chemosensory cells that send signals via the vagus nerve utilize serotonin as the neurotransmitter, whereas the messages coming from the vagus to several internal organs involve acetylcholine as a transmitter. Vagal activation promotes the flow of saliva, production of insulin and other pancreatic secretions, and release of gastric acid into the stomach. It also provokes the release of serotonin that causes contractions or peristalsis of smooth muscles that move food along the digestive tract. The gene-derived taste receptors active in the internal organs are the same as many of those present in the taste buds on the tongue, but the pathways of information into and out of the central nervous system are quite different.

Vision is a highly developed sense in humans. Chemically and genetically, there are several steps in the transduction of light energy via receptors in rods and cones of the retina (Fig. 4.7). The receptor in rods is a combination of two molecules, one a relatively small one called *retinal* that is derived from Vitamin A and the other a large protein known as *rhodopsin* that is encoded by the RHO gene. The combination is very stable in the dark, but retinal is exquisitely sensitive to light. Under conditions of almost complete darkness, a rod neuron can detect just one photon of light energy and change the membrane potential of the rod enough to alter the output of the cell. The photon of light is absorbed by retinal, and this changes its shape. The new form cannot fit so neatly into the rhodopsin, and it soon departs, leaving rhodopsin unstable and vibrating. This prompts a cascade of molecular events (Fig. 4.7) involving proteins derived from GNAT (G protein, alpha transducing) and PDE6 (phosphodiesterase

Signal transduction in rod cell in retina

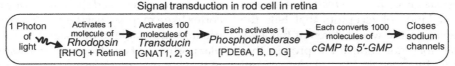

FIG. 4.7 One photon of light striking retinal bound to rhodopsin is amplified into the conversion of 100,000 molecules of cGMP to 5'-GMP, which closes sodium channels in the rod cell. Gene abbreviations from GeneCards for several steps in the processes are shown.

type 6) genes that greatly amplify the signal, to the extent that just one photon of light can prompt the conversion of 100,000 molecules of cyclic GMP (cGMP) to GMP, which in turn can close sodium channels and alter the amount of glutamate transmitter acting on glutamate receptors on the next neuron in the circuit. The rods do not fire action potentials. Instead, their outputs of neurotransmitter are graded and reflect the intensity of light impinging on the retina.

Molecules activated by the photon of light need to be deactivated in order to detect the next photon that arrives. Several other genes are important for the termination of activation. Retinal is restored to its original state by a cycle involving at least six different genes, whereas rhodopsin itself if deactivated by three or four genes and PDE is shut down by a cluster of proteins from three genes. Meanwhile, the resting level of cyclic GMP is restored by several forms of GUCA1 and GUCY genes.

The principal lesson is that, just like the examples for acetylcholine and melanin (Figs. 2.4 and 2.5), chemical change occurs via a series of small steps, each mediated by a specific gene or a combination of two or three proteins encoded by different genes. Only one molecule shown in Fig. 4.7 is specifically sensitive to light—retinal. No gene codes for retinal itself. Instead, several genes are needed to generate it and regenerate it. Vision based on rods requires proper functioning of all of the genes named in the figure, plus a large number of others that are not shown or discussed here.

Rods are numerous and widely distributed across the retina, except for the central point of vision, the fovea, where receptors are mainly cones. Rods are responsible for peripheral vision, especially the detection of motion of objects away from the central focus of vision, and are almost entirely responsible for vision in low light. Retinal plus rhodopsin is most sensitive to photons with wavelengths in the green portion of the color spectrum but relatively insensitive to red light. Consequently, visibility of red objects in low light conditions is greatly reduced. Rod-based vision cannot perceive color. The signals sent to the brain from rods will be virtually the same for photons having wavelengths of 440 and 540 nanometers (nm) when the two light sources are equally intense. Likewise, rods cannot tell the difference between relatively weak green light and much more intense blue light. Most of us experience the familiar fact that the world at night consists of shades of gray. That is the signature of rod-based vision.

Near the center of the field of vision are cone cells, each of which expresses only one of three kinds of opsin molecules that sense particular ranges of wavelengths of light. The three opsins are much less sensitive to low light levels than rhodopsin and function best in bright daylight. The short-wavelength form (encoded by the OPN1SW gene) not only is maximally sensitive to blue light but also responds, albeit weakly, to green light (Table 4.6). Medium-wavelength opsin (OPN1MW) is most sensitive to yellow-green light but has a wide range of sensitivity, as does the long-wavelength form (OPN1LW), which is most sensitive to yellow light. It is not accurate to term these kinds of rods blue, green, and red receptors. Color is perceived by the brain only by comparing the intensities of signals coming from all three kinds of cones in a small patch of the retina. Blue light of 420 nm wavelength will produce a much stronger signal from SW rods than MW or LW rods. Green light at 470 nm will produce some activity from SW rods, more from LW, and higher amounts from MW, whereas yellow light at 550 nm will produce no activity from SW rods and roughly equally amounts from MW and LW. What we perceive as red light weakly activates LW cones but has almost no effect on MW cones.

Thus, as was noted for the senses of odor and taste, a gene-derived receptor molecule does not code in a direct way for the response to a specific kind of stimulus from

TABLE 4.6 Sensitivity of Opsin Photoreceptors to Light of Different Wavelengths (nm, nanometers)

Gene	Name	Location	Peak (nm)	Range (nm)	Response to light > 600 nm
RHO	Rhodopsin	Rods	490 (green)	400–475	None
OPN1SW	Short-wavelength opsin	"Blue" cones	415 (blue)	375–475	None
OPN1MW	Medium-wavelength opsin	"Green" cones	530 (green)	400–600	Weak
OPN1LW	Long-wavelength opsin	"Red" cones	560 (yellow)	400–650	Moderate

the environment. Instead, it is the *combination* of signals from several different kinds of receptors that tells the brain what is happening in the world. Even at the level of molecules, a gene does not code for a protein that detects a unique or highly specific kind of signal from the environment. We rely on the nervous system to interpret the relative strength of signals coming from multiple kinds or receptors. The biochemical functions of genes in the signal transduction pathway in the retina are fairly well understood. Knowing what each of those genes does, we can then say something about what they do *not* do: A gene does not code for a molecule that detects a unique kind of signal that impinges on a sensory receptor cell. Instead, it enables the receptor to detect a certain range or variety of signals. Then, a network of different kinds of neurons analyzes and interprets the signals from several receptors.

Outputs from the rods and cones are initially processed in the retina by several other kinds of cells before being sent to the brain via the optic nerve. Near the center of the field of vision, one rod or one cone generally sends its output to just one bipolar cell, which then relays the signal directly to a ganglion cell, which fires an action potential along a myelinated axon into the brain. Away from the center, most ganglion cells receive inputs from several receptor neurons, which enable the system to detect edges in the visual field where there is a strong contrast between light and dark zones. Some ganglion cells are wired to detect an "on" signal when a receptor suddenly detects a photon, whereas others are tuned to sense "off" signals when the flow of photons to a particular spot ceases. About 2% of the ganglion cells are themselves sensitive to light, owing to their expression of the OPN4 gene, which produces a form of opsin protein called melanopsin. These cells send information about the overall brightness of a scene but do not inform the brain about patterns or rapid movements. They help to synchronize the day-night rhythm of activity via connections with the hypothalamus, and they also regulate the size of the pupil opening in the eye.

Touch

The senses of touch and pain apparently involve relatively few distinct kinds of receptor molecules compared with odor and taste. Although showing little chemical diversity, they express great spatial diversity, being widely distributed across the entire surface of the body and in many internal organs too. The sensory nerve endings near the skin surface conduct signals to relay neurons in the dorsal root ganglia and then into the spinal cord where they course upward to relay neurons in the thalamus that terminate on sensory neurons in the primary somatosensory region of the cerebral cortex. Distinct regions of the body surface are represented spatially in primary somatosensory cortex like a map, so that the cortex represents *where* a touch occurs and what kind of touch it is. The map is a very regular portrayal of touch over the body surface. The sequence of patches of somatosensory cortex proceed from deep in the fissure between the hemispheres to patches on the outer surface of the cortex in the following order: the genitals, toes, foot, knee, leg, hip, trunk, neck, head, arm, elbow, forearm, hand, fingers, thumb, eye, nose, face, lips, teeth, gums, tongue, and pharynx.

An object that exerts pressure on the skin causes a small indentation that stimulates mechanoreceptors in the cell membrane, which convert mechanical motion into action potentials. There are five kind of small organelles that respond to touch. Three of them fire rapidly at the start of a stimulus but then become silent during sustained pressure (Meissner and Pacinian organelles, hair bulb), whereas two send action potentials along their axons as long as pressure is applied, gradually reducing their frequencies with sustained touch (Merkel disk, Ruffini organelle). The Pacinian corpuscle is exquisitely sensitive to very weak vibrations of high frequency but has relatively low acuity or spatial resolution, whereas the Merkel disk can sense objects against the skin that are less than 1 mm wide, such as the small bumps on a braille card used by the blind for reading.

Given the advanced state of knowledge of molecular genetics, as summarized in Chapters 2 and 3 as well as the present chapter, it is noteworthy that so little is known about the genes involved in the sense of touch. Several authors have remarked about this deficiency (Reed-Geaghan & Maricich, 2011; Tan & Katsanis, 2009). The sparse knowledge of human touch receptors contrasts strongly with the comprehensive data on touch sensors in the nematode worm and fruit flies (Lumpkin, Marshall, & Nelson, 2010; Reed-Geaghan & Maricich, 2011).

Nociception

The skin and many internal organs and tissues contain a vast number of free nerve endings that detect tissue damage or stimulation so strong that there is a risk of tissue damage (Woolf & Ma, 2007). The nerve endings lack specialized organelles such as the Merkel disk, but they are populated with a considerable number of different receptors that warn the brain of hazards and harm. The receptors at the periphery are not pain receptors. They are termed *nociceptors*. They provide ways of sensing harmful stimuli in the environment, whereas pain is a perception by more central regions of the brain. The nervous system uses different receptors and circuits to sense touch and nociception.

Table 4.7 lists several nociceptors that are encoded by specific genes. One surprising omission from the list is a

TABLE 4.7 Genes That Code for Nociceptors[a]

Symbol	Nociceptor functions
TRPV1	Activated above 42°C; detects capsaicin and other noxious chemicals
TRPV2	Activated above 52°C
TRPV3	Activated from 22°C to 40°C
TRPA1	Detects mustard oil, cinnamon, garlic, tear gas; responds to cold
TRPM8	Responds to cold below 22°C; detects menthol
HRH1	Itch, inflammation arising from histamine release
ASIC1	Activated by acidic cell environment; acid sensing ion channel
ASIC3	Detects inflammation that results in acidic cell environment
P2RX3	Detects ATP spilling from fractured cells
BDKRB1, 2	Detects release of bradykinin from tissue damage

[a] *Source: GeneCards, OMIM.*

receptor for mechanical stimulation, such as puncture by a sharp needle that fires a fast signal into the central nervous system that tells much about the place and intensity of the damage. Investigators have documented the rapid transit of action potentials along axons from the periphery when there is focal damage, but the nature of the receptor for that kind of damage is not yet known, despite many studies.

Nociceptors that detect hot objects and noxious chemicals applied to the skin, as well as tearing of cell walls, are better understood. At least five kinds of receptors in the TRP family provide input to the brain about temperatures above or below the healthy range—temperatures that might even kill a cell if no action is taken to escape an extreme environment. They also detect several chemicals that create a sensation of extreme heat (e.g., capsaicin in chili peppers) or a cool sensation (menthol). Inflammation of skin and underlying tissue releases histamine that generates a sensation of itching and creates an acidic cell environment. When cells are torn open, the highly active ATP molecule is spilled into its locale, and a powerful peptide termed bradykinin is released that can generate intense pain.

Signals travel down the axon from the nociceptors, enter the spinal cord, and stimulate the release of glutamate transmitter, which in turn activates neurons that relay action potentials to the medulla and then the thalamus, traveling along paths that are close to but distinct from those for nonpainful touch (Dubin & Patapoutian, 2010). Just as happens for the sense of touch, the signals reach well-defined zones in the primary somatosensory cortex that inform the brain of *where* the damage is occurring as well as its modality and intensity.

Pain

We all have experienced pains of different kinds and probably have taken drugs to treat them. Pain is a perception that is organized in the cerebral cortex by combining data from the nociceptors with facts about emotions, stress levels, and even memories about a certain kind of experience. The experience of pain is usually accompanied by behaviors that express discomfort and attempts to escape the noxious situation. The degree of pain can be judged with a rating scale, especially one that is based on the expression of the face (Chapter 1).

Long before the dawn of modern neuroscience, it was widely appreciated that certain chemicals from plants can relieve pain. Acetylsalicylate from willow shrubs is very effective at reducing tissue inflammation after injury. It is now recognized that synthetic versions, such as aspirin and ibuprofen, block the action of the cyclooxygenase enzyme that generates inflammation. Those drugs are widely used because of their mild side effects. In situations of more intense pain, smoke from the opium poppy or a more potent extract known as heroin has proved helpful. Opium and heroin have many other effects, including a psychological sense of euphoria and less pleasant things such as constipation. Chemical analysis revealed that the active ingredient in opium is a compound named morphine, a drug that can be synthesized and then administered in purified form as a tablet or injection with a needle. Morphine is widely used to treat pain, but this imposes a severe penalty on the patient because people often come to require higher and higher doses to achieve the same level of pain relief and subsequently become addicted to the drug.

Research on the biological effects of morphine uncovered specific receptors in the spinal cord, medulla, cerebral cortex, and elsewhere in the body that detect the presence of opioid chemicals. There are three genes in humans that encode slightly different forms of opioid receptors (OPRD1, OPRK1, and OPRM1), one of which (OPRM1) also detects the presence in the blood of morphine, heroin, fentanyl, and methadone. It was surprising to find receptors in the nervous system that are so highly sensitive to a chemical from a plant that many people never encounter. Further study of this system revealed that there are endogenous peptide neurotransmitters that act on what scientists named opioid receptors and create effects much like morphine, but they are localized in certain pathways in the nervous system. Two of these are beta-endorphin and enkephalin. It happens that there is a parallel pathway of axons that descend from the cerebral cortex to the medulla and spinal cord where they

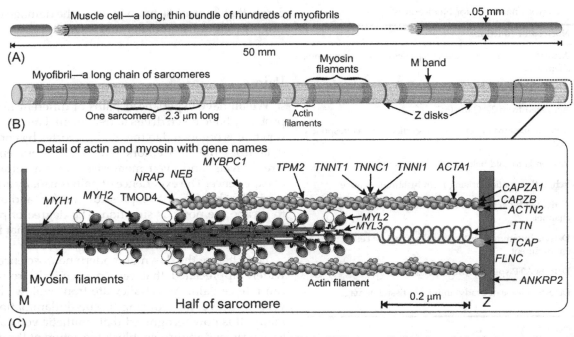

FIG. 4.8 (A) One muscle cell is a very long and thin bundle of many myofibrils. (B) Each myofibril is a long chain of very complex sarcomeres, each with filaments of actin and myosin proteins. (C) Many kinds of protein, each encoded by a gene, comprise the intricate structure of the sarcomere that can generate mechanical motion. Gene names are given in Table 4.8. White heads on the myosin filaments show head location in a different position that moves the structure by a small amount relative to actin. *Sources: Cooper, G. M. (2000). The cell: A molecular approach. Sunderland, MA: Sinauer Associates; Liu, D., Sartor, M. A., Nader, G. A., Gutmann, L., Treutelaar, M. K., Pistilli, E. E., …Gordon, P. M. (2010). Skeletal muscle gene expression in response to resistance exercise: sex specific regulation. BMC Genomics, 11, 659.*

form synapses on nerve cells that in turn have synapses on the neurons that send fibers from nociceptors upward to the cortex (Al-Hasani & Bruchas, 2011). The receptors on these regulatory neurons detect opioids, whereas the transmitter released from the synapse on the upward-coursing neurons is GABA, an inhibitory substance. Thus, activation of the regulatory neuron by beta-endorphin can block the transmission of signals upward to the brain that normally would cause pain. A similar blockage of signals from the nociceptors to the brain occurs when morphine-like drugs are given; there is no perception of pain, even though the conditions at the periphery may activate the nociceptors intensely. The production and release of beta-endorphin and enkephalin can be influenced by many psychological processes such as pleasure seeking, reward, and stress and cognitive processes such as expectation and belief. The placebo effect that often confers pain relief presumably arises from these complex circuits.

MOTOR OUTPUT

Most of what we regard as behavior is mechanical motion of the parts of our bodies. This is as true for lifting a heavy weight or swimming as it is for playing a musical instrument or checking little boxes in a multiple-choice exam. Muscles attached by tendons to bones in the skeleton provide the force for most kinds of movements we can observe directly. Those skeletal muscles have a similar structure, no matter where they are positioned in the body, and their contractions are initiated by the release of the neurotransmitter acetylcholine from the motor end plate of motor neurons in the spinal cord (Fig. 4.4, Fig. 2.4). The muscle cell is very long and thin (Fig. 4.8), and it contains hundreds of myofibrils that are able to contract on command from the central nervous system (Cooper, 2000). Each myofibril is itself a long chain of more than 10,000 organelles termed *sarcomeres*. The sarcomere is an intricate and elegant structure built from dozens of different proteins, each encoded by a different gene. Whereas the rod in the retina can convert photons of light into nerve impulses via several discrete steps (Fig. 4.7), the sarcomere can convert nerve impulses into mechanical motion. Names of several genes important for this work are provided in Table 4.8.

At the core of the sarcomere is a long molecule of the giant protein titin (TTN) that joins the middle zone (M band) and end disk (Z) and gives the structure its elasticity or springiness. Titin can be compressed when the sarcomere is activated by acetylcholine, and it can stretch when the activation ceases. The contraction or elongation of the sarcomere occurs when two kinds of filaments slide past each other. One kind consists of many filaments of

TABLE 4.8 Genes Involved in Sarcomere Function in Fast- and Slow-Twitch Muscle Fibers[a]

Name (function)	Fast twitch	Slow twitch	Both, other
Actin alpha 1 (basic subunit of actin filament)			ACTA1
Actinin alpha 2 (binds end of actin chain to Z disk)			ACTN2
Ankyrin repeat domain 2 (senses muscle fiber stretch)	ANKRP1	ANKRP2	
Tropomyosin 2 (binds to actin filament)	TPM3	TPM2	
Troponin type C1 (stops inhibitory action of T1 form)	TNNC2 TNNI2	TNNC1 TNNI1	
Troponin type I1 (stops inhibitory action of T1 form) Troponin type T1 (inhibits actin-myosin activation)	TNNT3	TNNT1	
Nebulin (binds to and stabilizes actin filament)			NEB
Nebulin-related anchoring protein (anchor to myofibril)			NRAP
Filamin B, beta (anchors actin to membranes)			FLNB
Filamin C, gamma (cross-links actin to Z disk)			FLNC
Capping protein on actin in muscles at Z disk; alpha 1 Capping protein on actin in muscles at Z disk; beta			CAPZA1 CAPZB
Tropomodulin 4 (blocks elongation of actin filament)			TMOD4
Titin (huge protein that gives springiness to myofibril)			TTN
Titin cap (binds to titin-Z disk domain)			TCAP
Myosin heavy chain (large filament with neck and head)	MYH13	MYH4	MLH1,2
Myosin light chain, regulatory form	MYL1, F	MYL3	
Myosin-binding protein C (crossbridge actin and myosin)	MYBPC2	MYBPC1	

[a] Source: GeneCards, OMIM.

myosin, each of which is made from six large molecules (a hexamer). Two molecules of heavy-chain myosin (MYH) wrap around each other to create a long tail and form an oval-shaped head that protrudes from a flexible hinge region. Each head region is stabilized and activated via two smaller molecules termed light chain myosin (MYL). When activated, the head region can move in a small arc, and this motion repeated across many myosin hexamers can move the entire bundle of myosin along adjacent chains of filaments made from many molecules of the protein actin (ACTA1) that are stabilized by a chain of nebulin proteins (NEB). Myosin does not bind strongly to the actin, because this would prevent motion. The myosin and actin chains are kept closely aligned by myosin-binding protein (MYBPC1) that nonetheless allows for some movement of the two chains relative to each other. Along the surface of the actin filament is the protein tropomyosin (Tpm) that participates in the transfer of chemical energy from the actin chain to myosin when it is activated. Normally, this action is inhibited by a cluster of three troponin proteins (TNN) at each end of the tropomyosin. TNNT has binding sites on TPM, whereas TNNI inhibits activation and TNNC abolishes this inhibitory activity. Release of ACh from the motor endplate reduces the inhibitory effect of TNNI on TPM and thereby contributes to the activation of actin and myosin, resulting in movement in the flexible head of the MYH chain and movement of the myosin bundle relative to the actin chain. Other proteins mentioned have roles in anchoring molecules to the Z disk and capping the chain of actin molecules.

Skeletal muscle cells are present in two configurations—slow twitch and fast twitch—that are related to the specific kinds of genes that are expressed in the myosin and actin components. Slow-twitch sarcomeres are well adapted for muscle contractions of long duration and moderate force that do not fatigue quickly, and their onset of contraction is relatively slow. Fast-twitch sarcomeres are quick to respond to stimulation from the motor endplate with large force but cannot maintain contractions for very long. Although many of the protein constituents of the fast and slow cells are the same, several involve different forms of proteins coded by different genes.

The function of the totality of a sarcomere is to contract a myofibril, which, when repeated across thousands of myofibrils, shortens a muscle. Each gene plays a discrete role in the larger process, and its protein product engages in interactions with other molecules that are adjacent to it. Thus, muscle contraction is a property of a large system of interconnected organelles made up of molecules. Hundreds of proteins comprise a sarcomere and are essential components of normal muscle contraction and relaxation. Fig. 4.8 shows some of the better known structural components, but there are others that play important roles, including many whose contributions are not yet well understood. And there are many others involved in glucose and lactate metabolism that provide energy via the bloodstream to drive muscle contraction.

A further glimpse of the complexity of muscle can be obtained by measuring expression of genes in muscle tissue, especially those that are substantially altered by exercise (Liu et al., 2010). Aerobic exercise training over

6 weeks on a bicycle in a lab altered the mRNA expression levels in muscle biopsy samples taken before and after training for more than 400 different genes, many of which are involved in muscle structure and contraction (Timmons et al., 2005). A recent study of cycling that used a nonexercised control group vs 10 weeks of training found that more than 1000 genes changed expression levels, depending on the kind of exercise (Vissing & Schjerling, 2014). Another experiment involved humans in a lab who underwent 3 months of intensive training for just one leg. Biopsies were then compared for the two legs (Lindholm et al., 2014), and gene expression in mRNA and epigenetic methylation were assessed. Exercise changed methylation levels substantially for more than 800 genes, and mRNA levels were also altered for about 800 genes, many of which were related to muscle contraction and the formation of new blood vessels. There was a general pattern whereby gene expression increased for genes whose enhancer regions showed reduced levels of methylation, a relationship explained in Chapter 3.

FUNCTIONS AND SYSTEMS

How we can see and move can be well described in terms of genes that are expressed at the level of organelles such as the synapse and sarcomere. *What* we see and what kinds of actions we perform cannot be described at the level of organelles and instead are regulated at higher levels involving circuits of nerve fibers and large numbers of neurons in different brain regions. The higher level functions do not correspond in any simple way to particular kinds of neurotransmitters or other small molecules. Those functions emerge from complex systems of interconnected neurons. Of course, those neurons express and depend on hundreds or even thousands of different genes, but knowing only those arrays of genes tells little about the functions.

Knowledge of brain function, especially what parts of the brain are involved with which functions, accumulated from experiences with stroke, gunshot wounds to the head, industrial and automobile accidents, cancer surgery, and other invasive procedures. That is how people first learned that the right side of the brain controls many aspects of things that happen on the left side of the body. Experimental neuroscience later approached these issues more systematically with surgical procedures performed on rats, dogs, cats, and monkeys and electric recordings from specific parts of the brain enabled circuits involved in vision and motor actions to be traced. Functional magnetic resonance imaging (fMRI) extended this work to the human brain without the need for inflicting major damage on the volunteer. The body of knowledge compiled by neuropsychology grew rapidly, but this took place in parallel with genetic studies of the nervous system. It was learned that functionally distinct regions of the cerebral cortex, for example, express remarkably similar arrays of neurotransmitters and genes involved in neuronal functioning. At the molecular and organelle levels, there is little to distinguish most regions of the cortex. There are fissures and sulci that provide handy landmarks to delineate functional modules (Fig. 4.2) without telling much at all about the actual functions of each place. Where the regions obtain their inputs and send their outputs differ greatly and the secrets to those patterns reside in the dynamics of brain development (Chapter 5). Later chapters will introduce further details about the nervous system as needed, such as Chapter 17 on schizophrenia where the basics of psychopharmacology are presented.

HIGHLIGHTS

- Neurons receive inputs from other neurons via synapses on the dendrites, and a large neuron can receive inputs from more than 100,000 other nerve cells. All this information is integrated, and the neuron sends an output signal via just one axon that exits the cell. The axon can then branch and send the signal to dozens of other neurons.
- An electric impulse can travel down the axon at speeds up to 100 m/s when the axon is coated in a sheath of myelin.
- When the impulse reaches a synapse where neurotransmitter molecules are stored in vesicles, the contents of the vesicle are poured into the synaptic cleft or gap where the transmitter molecules then contact receptor molecules on the next neuron in the circuit.
- There are more than 10 kinds of neurotransmitters employed in the nervous system plus more than 100 small peptides that can act as transmitters. Certain of these excite the next cell in the series, while others inhibit its firing. Any one neuron usually employs just one kind of transmitter, although its activity can be affected by local modulators.
- More than a dozen genes are expressed in just one synapse that are specific to the kind of transmitter, and each performs an essential function such as synthesis, storage and release of the transmitter, reuptake of excess transmitter in the cleft, detection of a transmitter by a receptor, and breakdown and recycling of the transmitter.
- A specific kind of transmitter molecule is used in many regions of the nervous system to perform a variety of different physiological functions. No transmitter is dedicated to just one kind of behavioral function such as muscle contraction, reward, or fear. The function of any one neuron depends on both the transmitter it uses and its interconnections with other cells.

- Sensory receptors can detect far more kinds of stimuli than there are kinds of receptors. The identity of a stimulus such as a specific smell, taste, or light is established by the *combination* of different receptors it activates.
- Some sensory receptors such as those involved in olfaction and taste are also expressed in internal organs where they help to organize digestion and promote healing after injury.
- Transduction of a light stimulus in rods of the retina occurs via a cascade of five gene-mediated steps that amplify the signal from just one photon of light 100,000 times, and more than 20 other genes are involved in restoring the molecules to the dark state, ready for action when the next photon arrives.
- Color vision arises from cones, each of which expresses one of three kinds of gene-encoded opsin molecules. Each kind of opsin has a wide spectrum of sensitivity but a different location of peak sensitivity. Color is perceived by the brain by comparing intensities of signals from the three kinds of cone cells.
- More than 20 kinds of gene-encoded nociceptors detect potentially harmful or damaging stimuli to the body, but pain itself is judged by the brain as a quality that includes cognitive aspects and memory. Relief from pain is often provided by gene-encoded opioid receptors that can prevent signals from the nociceptors from reaching the brain.
- Muscle contraction is implemented by long chains of organelles termed sarcomeres that involve the actions of more than 30 different gene-encoded parts. Proper function of sarcomeres is essential for any kind of behavior, but what specific behavior occurs depends on the pattern of signals organized by the brain and sent to the periphery.
- When we zoom in to the level of organelles, it is possible to perceive where specific genes are expressed and what they do. When we zoom out to the level of a small brain region containing several kinds of cells, gene expression arrays can detect hundreds of genes that are expressed therein but cannot readily show what any one gene is doing or why.

References

Al-Hasani, R., & Bruchas, M. R. (2011). Molecular mechanisms of opioid receptor-dependent signaling and behavior. *Anesthesiology, 115,* 1363–1381.

Boehm, U., Zou, Z., & Buck, L. B. (2005). Feedback loops link odor and pheromone signaling with reproduction. *Cell, 123,* 683–695.

Bowden, D. M., & Dubach, M. F. (2003). NeuroNames 2002. *Neuroinformatics, 1,* 43–59.

Bowden, D. M., Johnson, G. A., Zaborsky, L., Green, W. D., Moore, E., Badea, A., ... Bookstein, F. L. (2011). A symmetrical Waxholm canonical mouse brain for NeuroMaps. *Journal of Neuroscience Methods, 195,* 170–175.

Bushdid, C., Magnasco, M. O., Vosshall, L. B., & Keller, A. (2014). Humans can discriminate more than 1 trillion olfactory stimuli. *Science, 343,* 1370–1372.

Busse, D., Kudella, P., Grüning, N.-M., Gisselmann, G., Ständer, S., Luger, T., ... Gkogkolou, P. (2014). A synthetic sandalwood odorant induces wound-healing processes in human keratinocytes via the olfactory receptor OR2AT4. *Journal of Investigative Dermatology, 134,* 2823–2832.

Cooper, G. M. (2000). *The cell: A molecular approach.* Sunderland, MA: Sinauer Associates.

DeFelipe, J. (2010). *Cajal's butterflies of the soul: Science and art.* New York: Oxford University Press.

Dubin, A. E., & Patapoutian, A. (2010). Nociceptors: the sensors of the pain pathway. *Journal of Clinical Investigation, 120,* 3760–3772.

French, L., & Paus, T. (2015). A FreeSurfer view of the cortical transcriptome generated from the Allen Human Brain Atlas. *Frontiers in Neuroscience, 9,* 323.

Gross, C. G. (1997). Leonardo da Vinci on the brain and eye. *The Neuroscientist, 3,* 347–355.

Hawrylycz, M. J., Lein, E. S., Guillozet-Bongaarts, A. L., Shen, E. H., Ng, L., Miller, J. A., ... Jones, A. R. (2012). An anatomically comprehensive atlas of the adult human brain transcriptome. *Nature, 489,* 391–399.

Herculano-Houzel, S. (2009). The human brain in numbers: a linearly scaled-up primate brain. *Frontiers in Human Neuroscience, 3,* 31.

Hooke, R. (1665). *Micrographia: Or some physiological descriptions of minute bodies made by magnifying glasses. With observations and inquiries thereupon.* London: J. Martyn and J. Allestry.

Jones, R. (2012). Leonardo da Vinci: anatomist. *British Journal of General Practice, 62,* 319.

Lindholm, M. E., Marabita, F., Gomez-Cabrero, D., Rundqvist, H., Ekström, T. J., Tegnér, J., & Sundberg, C. J. (2014). An integrative analysis reveals coordinated reprogramming of the epigenome and the transcriptome in human skeletal muscle after training. *Epigenetics, 9,* 1557–1569.

Liu, D., Sartor, M. A., Nader, G. A., Gutmann, L., Treutelaar, M. K., Pistilli, E. E., ... Gordon, P. M. (2010). Skeletal muscle gene expression in response to resistance exercise: sex specific regulation. *BMC Genomics, 11,* 659.

Lumpkin, E. A., Marshall, K. L., & Nelson, A. M. (2010). The cell biology of touch. *Journal of Cell Biology, 191,* 237–248.

Mainland, J. D., Keller, A., Li, Y. R., Zhou, T., Trimmer, C., Snyder, L. L., ... Matsunami, H. (2014). The missense of smell: functional variability in the human odorant receptor repertoire. *Nature Neuroscience, 17,* 114–120.

Markou, A., Chiamulera, C., Geyer, M. A., Tricklebank, M., & Steckler, T. (2009). Removing obstacles in neuroscience drug discovery: the future path for animal models. *Neuropsychopharmacology, 34,* 74–89.

Meyerhof, W., Batram, C., Kuhn, C., Brockhoff, A., Chudoba, E., Bufe, B., ... Behrens, M. (2010). The molecular receptive ranges of human TAS2R bitter taste receptors. *Chemical Senses, 35,* 157–170.

Olender, T., Lancet, D., & Nebert, D. W. (2008). Update on the olfactory receptor (OR) gene superfamily. *Human Genomics, 3,* 87–97.

Peterlin, Z., Firestein, S., & Rogers, M. E. (2014). The state of the art of odorant receptor deorphanization: a report from the orphanage. *Journal of General Physiology, 143,* 527–542.

Reed-Geaghan, E. G., & Maricich, S. M. (2011). Peripheral somatosensation: a touch of genetics. *Current Opinion in Genetics and Development, 21,* 240–248.

San Gabriel, A. M. (2015). Taste receptors in the gastrointestinal system. *Flavour, 4,* 14.

Scatliff, J. H., & Johnston, S. (2014). Andreas Vesalius and Thomas Willis: their anatomic brain illustrations and illustrators. *American Journal of Neuroradiology, 35,* 19–22.

Takacs, J., & Hamori, J. (1994). Developmental dynamics of Purkinje cells and dendritic spines in rat cerebellar cortex. *Journal of Neuroscience Research, 38,* 515–530.

Tan, P. L., & Katsanis, N. (2009). Thermosensory and mechanosensory perception in human genetic disease. *Human Molecular Genetics, 18,* R146–R155.

Timmons, J. A., Larsson, O., Jansson, E., Fischer, H., Gustafsson, T., Greenhaff, P. L., … Sundberg, C. J. (2005). Human muscle gene expression responses to endurance training provide a novel perspective on Duchenne muscular dystrophy. *FASEB Journal, 19,* 750–760.

Vissing, K., & Schjerling, P. (2014). Simplified data access on human skeletal muscle transcriptome responses to differentiated exercise. *Scientific Data, 1,* 140041.

Welcome, M. O., Mastorakis, N. E., & Pereverzev, V. A. (2015). Sweet taste receptor signaling network: possible implication for cognitive functioning. *Neurology Research International, 2015,* 606479.

Woolf, C. J., & Ma, Q. (2007). Nociceptors—noxious stimulus detectors. *Neuron, 55,* 353–364.

Zhang, J., Gao, G., Begum, G., Wang, J., Khanna, A. R., Shmukler, B. E., … Kahle, K. T. (2016). Functional kinomics establishes a critical node of volume-sensitive cation-Cl−cotransporter regulation in the mammalian brain. *Scientific Reports, 6,* 35986.

5

Development

The marvelous complexity of the nervous system sketched in Chapter 4 describes neurons, synapses, transmitter molecules, and receptors in adults. Those billions of distinctly different cells arose from just one large cell, the zygote, which is the beginning of a new individual. All the features of a mature nervous system are generated through the process of development. They emerge anew in each individual from primordial structures that do not resemble a nervous system at all. The study of anatomy can tell us when nerve cells or synapses first appear, but it cannot explain what causes each advance in the embryo to occur. Explanation requires experimentation.

Experimentation with developing human embryos is usually considered unethical. Thus, experimental analysis of development relies almost entirely on the study of animal models (Table 1.3). The very earliest human embryos just before and shortly after fertilization are being studied in clinics that perform *in vitro* fertilization (Chang et al., 2015), and some early embryos in some jurisdictions are made available for research (Zernicka-Goetz, 2016). Nevertheless, most of what we know about early nervous system development comes from work with fruit flies, frogs, flies, mice, and rats. The dynamics of early development of most vertebrates are remarkably similar, as are the genes they inherit. We can generalize most findings to human embryos while remaining alert for situations that are unique to our species.

Embryologists and geneticists have tried to use the same names for genes that have the same origins in different species. Naming a gene is the privilege of the person who discovered it, which can give rise to some peculiar monikers, e.g., *Sonic hedgehog* discussed in this chapter. The name is first assigned for the species wherein the gene was originally discovered and then adapted for use in other species. This can be puzzling to many readers, as when a human gene involved in axon guidance is named *Wingless Integration Factor* (WNT). It was originally discovered in a fruit fly with no wings. In nonhuman animals, the gene symbol usually is given in lower-case italics except for capitalization of the first letter, whereas for humans, the gene name is all capital letters and numbers. The protein for which the gene codes may be expressed using the same letters as the gene name but in lower case without the italics, or it may be described with complete words without italics. The reader is forewarned that this technical language can be confusing. In this book, the results of a specific study with a particular species will usually be described using gene names appropriate for that species, whereas patterns that are portrayed as having broad generality are described using terms appropriate for humans.

THE EARLY EMBRYO

Rapid Transformations

A progenitor cell in the ovary that has two copies of each chromosome (diploid set) undergoes meiotic cell division to generate an *ovum* that has just one copy of each chromosome (haploid set). The ovum then leaves the ovary and begins a journey down the fallopian tube, where it may encounter haploid sperm cells. If the timing and place are right, one sperm may penetrate the ovum and trigger the process of fertilization and cell division of the resulting *zygote* by mitosis. The cells inside the *zona pellucida* sheath become more numerous but smaller, and they soon form a compact ball of cells, the *morula*. With the next round of cell divisions, something changes in a qualitative way. Suddenly, some of the new cells end up inside the morula and are no longer in contact with the zona pellucida, while those on the outer portion of the embryo touch the zona pellucida. The cells in the center become the *inner cell mass*, which in turn becomes the embryo proper, the part that develops into a complete organism, while the outer cells become the *trophoblast* that gives rise to the placenta and the chorion that contains and nurtures the embryo. The inner cell mass and trophoblast together comprise the *blastocyst*. The amnion, chorion, and fetal side of the placenta have the same genotype as the embryo proper.

Gene Expression

The two kinds of germ cells themselves are grossly unequal. The sperm is very small and simple, not much larger than a bacterial cell, and its head contains little more than chromosomes. The midpiece is the motor with several mitochondria that provide energy to move the tail that propels the sperm forward. At fertilization, the midpiece and mitochondria from the father degenerate after their tasks are complete. Mitochondria from the father do not become part of the new embryo. The ovum is the largest single cell known in humans with a diameter of about $100 \mu m$, and its volume is about 8000 times greater than the sperm. The ovum contains the same amount of chromosomal DNA as the sperm, but its cytoplasm is vastly greater in volume and it contains roughly 500,000 mitochondria that possess their own DNA.

Clear evidence of the expression of genes in the mRNA of healthy sperm has been reported (Bonache, Mata, Ramos, Bassas, & Larriba, 2012), but RNA of many expressed genes is present only in fragments that cannot serve as a template for making proteins (Casas & Vavouri, 2014). Many more genes are involved in the formation of sperm in the testes (Bonache et al., 2012), but most are no longer expressed when the sperm enters the female. Large numbers of sperm genes carry epigenetic marks that may influence later gene expression. Thousands of genes in sperm are methylated (Casas & Vavouri, 2014; Pacheco et al., 2011), which suggests the paternal genes are at least temporarily silenced (Chapter 3).

The situation in ova and the early embryo is much different. A recent study applied enhanced-gene-expression techniques that allowed investigators to determine which genes are expressed in a single cell (Shaw, Sneddon, Zeef, Kimber, & Brison, 2013). Three each of ova, 4-cell embryos, and blastocysts were analyzed. There were 9745 genes expressed in the ovum from the mother but only 338 in the 4-cell embryo. In the 4-cell embryos, the stores of maternal RNA were almost exhausted. Then, there was a shift to a new phase when both maternal and paternal genes were expressed, and 5133 genes produced RNA in the blastocyst. Despite the biochemical complexity of events in the embryo, the cells remained anatomically simple and essentially identical until the inner cell mass appeared.

Extreme Phenotypic Plasticity

Before the embryo implants into the uterus, it is floating in a rich liquid environment and can be removed from the mother and grown in a lab for a few days in a solution with appropriate nutrients. A variety of experiments can be performed during this time to explore alternative pathways of development and test the fate and potential of those cells. *Fate* denotes what kinds of cells and tissue will normally arise from a particular cell in the early embryo, whereas *potential* denotes the varieties of tissues the cell *could* become under different conditions. Three kinds of experiments done with lab mice and rats demonstrate the remarkable plasticity of early embryo development (Fig. 5.1).

Twins occur when one zygote or early embryo splits into two separate balls of cells, a fairly common event in humans. Those two embryos will necessarily have the same genotype. Being from one zygote, they are termed monozygotic (MZ) twins, as opposed to two ova fertilized by different sperm that result in dizygotic (DZ) twins conceived at the same time. If a mouse embryo is removed from the uterus, a fine thread carefully tightened around its middle can separate it into two embryos, and when transferred to a recipient female at the correct stage of pregnancy, those embryos can mature into two healthy mice (Fig. 5.1A). Clearly, the fates of the cells in the morula have not yet been specified or differentiated. Otherwise, we might find that one MZ twin has only a right foot while the other gets the left foot. Instead, both individuals develop healthy left and right limbs. The material present in one ovum and one sperm can give rise to just one brain or to two separate brains in different individuals, depending on the local conditions in the maternal environment shortly after fertilization.

One can become two, and two can also become one. If the zona pellucida is gently pulled away from two 8-cell embryos and they are pressed together, their cells can then intermingle and form one large morula that grows into a giant blastocyst. When transferred to a receptive female mouse uterus, that embryo becomes a healthy *chimera* (Fig. 5.1B). A chimera is just one individual consisting of two populations of genetically distinct cells (Mintz & Silvers, 1967). Each cell in the inner cell mass of the blastocyst divides many times by mitosis, giving rise to a large number of cells in the adult animal. Some phenotypes are then seen to arise strictly according to genes contained in the ancestral cell. Pigmentation is a good example. If one parent of the chimera is albino because of a mutation that prevents the formation of melanin (Fig. 2.5) and the other parent is darkly pigmented, skin cells in the adult that receive the albino mutation will lack pigment, regardless of the genotype or phenotype of neighboring cells. The result is an animal with patches of albino fur and patches in the retina that also lack pigment. Many other features of the animal appear to be quite normal, despite the random jumble of cells from which they arose. Even though the blastocyst has twice the usual number of cells, the adult typically is normal in size, and many traits are close to the average of phenotypes of the parents. The blastocyst may be a mosaic of two genotypes, but the adult is usually not a mosaic of phenotypes, except for traits that are unaffected by interactions among different cells.

There is little harm done if one of the cells of the 8-cell embryo is removed altogether (Fig. 5.1C), and then, the

FIG. 5.1 Plasticity of development in mouse embryos. In each case, the manipulated embryo(s) is returned to the uterus of a surrogate female who is at the proper stage of pregnancy to receive and nurture the embryos. (A) Separating a ball of 16 cells into two balls of cells gives rise to monozygotic twins. (B) Fusing two embryos from different genetic strains gives rise to a chimera, a single individual composed of two genetically distinct populations of cells. If the parent strains have different coat colors, the chimera has patches of fur of different colors. (C) Removing one cell for genetic testing does little harm to the embryo, and it can grow to become a normal adult.

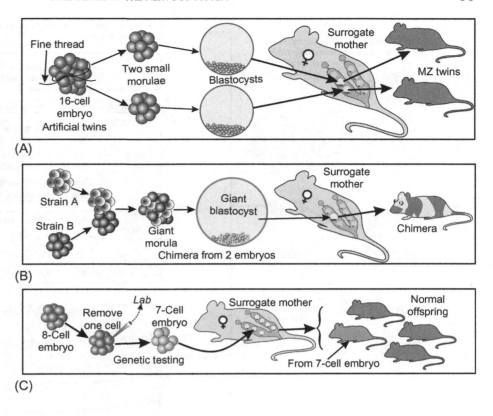

7-cell embryo is transferred to a surrogate female. The resulting offspring is generally quite normal, no matter which of the cells is removed. This phenomenon is important for preimplantation genetic testing in humans, whereby the genotypes are determined for several embryos from the same woman fertilized by a partner or a donor, and only the ones lacking a harmful mutation are returned to the uterus of the mother. It is now possible to scan the entire genome from just one cell of the embryo in a search for harmful mutations (Kumar et al., 2015).

After the inner cell mass is formed and the blastocyst implants into the uterus, cells multiply and begin to differentiate so that they become specialized for life as certain kinds of tissues. The potential of any one cell in the embryo to become other kinds of cells gradually becomes restricted. After dividing, new cells usually migrate to new locations in the embryo, sometimes moving a considerable distance. To determine cell lineage and fate, there must be a way to label a cell and then see where its labeled offspring are later found in the embryo. One effective method for tracing cell lineages is to inject a tracer molecule into a cell at a specific location in the embryo and then examine the tissue hours or days later to see where its offspring cells went. The tracer molecule should be quite large, so that it cannot leak out of the cell or its offspring. It should be biologically inert and not interact with the host cell in a way that might influence development. It should also not be easily digested by

the host enzymes. It was discovered that a large peroxidase enzyme from the horseradish plant has all these qualities. It can be passed unchanged to several generations of new cells. When small quantities of HRP (horseradish peroxidase) are injected into a cell, a stain that detects HRP can then determine which other cells in an embryo are descended from that one cell (Fig. 5.2A). This and similar methods have been used to map the *fate* of many cells in amphibian and mouse embryos. The branching diagram in Fig. 5.2B shows which tissues throughout the body are usually derived from which parts of the early embryo. It is somewhat simplified, in that most entire organs of the mature individual arise from several kinds of ancestral cells in the embryo. Nevertheless, it illustrates the stunning phenotypic plasticity of those few seemingly simple cells in the early embryo and the spectacular power of development to generate new forms.

EMERGENCE OF THE NERVOUS SYSTEM

After implantation into the lining of the uterus, cells continue to divide by mitosis; the embryo grows rapidly and becomes notably more complex. Cells migrate and form sheets or layers that bend and form tubes. The inner cell mass becomes differentiated into three layers of cells that are the forerunners of all the major organ

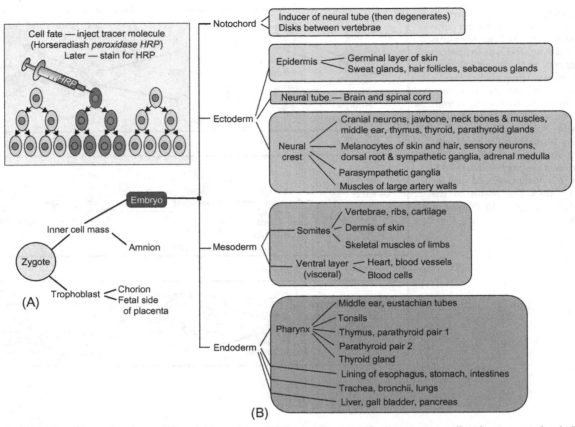

FIG. 5.2   Developmental fates of various cell lines in the embryo. (A) Fate is determined by injecting one cell with a tracer molecule (HRP in this example) and then later staining tissues to determine which of them received the HRP. (B) Ectoderm cells that can give rise to all parts of the nervous system are sometimes designated as neural stem cells.

systems—*ectoderm* at the top, *mesoderm* in the middle, and *endoderm* at the bottom layer with the *notochord* just below the portion of the ectoderm that will become the neural tube (Fig. 5.3). Those tissues are almost impossible to manipulate in mammals in which the fragile embryo is embedded deep inside the mother's womb, but in frogs, newts, and fish that develop in pond water, the outer layer of the ectoderm is amenable to fine surgical operations in the lab. It was noted that a streak forms in the ectoderm and there is a small pore where eventually the head will form. Researchers discovered that a small piece of tissue near the pore (termed the "organizer") could be dissected and transplanted to a different location in the ectoderm, where it induced the formation of a second head and nervous system in the same embryo (Spemann & Mangold, 1924). The landmark discovery inspired many other studies of tissue grafting in embryos of several species (Elliott, 2016; Harland, 2008), and it was firmly established that many important steps in embryo development entail interactions between different kinds of cells that induce changes in each other. Proximity to the notochord, for example, is essential for inducing the ectoderm to transform into the neural tube.

Once the neural tube has closed, the numbers of cells in the nascent nervous system begin to increase rapidly. To increase cell number, the cell must divide by mitosis. This event can be seen with a light microscope powerful enough to note the condensed chromosomes that are apparent just before the cells separate. Cell division in the neural tube takes place at the surface of the ventricle that contains a rich supply of nutrients (Fig. 5.4A). After the cell divides, the daughter cells then migrate along radial glial cells toward the upper surface of the neural tube that is covered by a membrane known as the *pia mater* (Fig. 5.4B). When the cell arrives at the zone near the pia, it begins to form an elaborated nerve cell with dendrites and an axon (Fig. 5.4C). Generally speaking, after a cell has begun to mature into a neuron having connections with other cells, it can no longer divide or generate new neurons. Its migration has ended, and further growth of the cell entails greater volume and complexity and more interconnections with other neurons. Each wave of neurons that migrates outward from the ventricle forms a layer of neurons, resulting in a cerebral cortex with seven distinct layers in adult humans. The outermost layer forms last. Each successive wave of

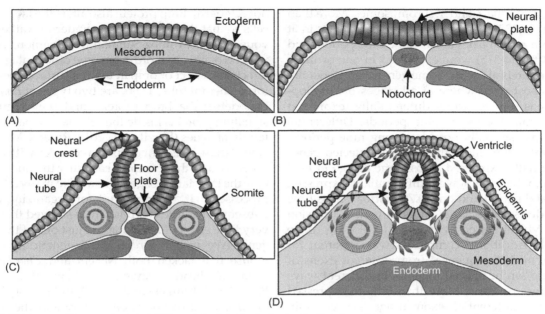

FIG. 5.3 (A) The inner cell mass grows and differentiates into three sheets of cells—the ectoderm, mesoderm, and endoderm—whose fates are shown in Fig. 5.2. (B) When a notochord forms from mesoderm, it induces changes in the adjacent ectoderm that will form a central nervous system. (C) As cells multiply at different rates in the ectoderm, it folds inward to form a tube. (D) When the neural tube closes, cells from the neural crest are freed to migrate widely and form the peripheral nervous system. The notochord will soon disappear, its mission accomplished.

FIG. 5.4 (A) After the neural tube forms, cells proliferate through mitosis at the ventricle. (B) Some cells become radial glia that serve as guides for neurons that migrate to outer layers. (C) After reaching their destinations, immature neurons form dendrites to receive inputs and send out axons with growth cones in search of suitable targets.

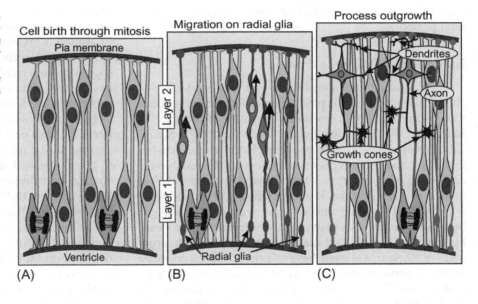

migrating neurons must pass through the forest of cells already on-site in earlier layers.

Early Gene Expression

The great diversity of cell types in the adult is derived from just one cell, the zygote. All descendants of the zygote have essentially the same genotype or genome encoded in the DNA. What makes the cells different is the arrays of genes that are expressed in them, which in turn are related to the cells' locations, their neighbors, and chemicals or other stimuli from the environment that bathe them. Cell and tissue types are phenotypes that develop, and their relations with genotype are exceedingly complex. The degree of complexity can be seen in a recent study of gene expression in 16 different brain regions in 57 postmortem human embryos, fetuses, infants, and adults using the technique of RNA expression arrays (Kang et al., 2011). The arrays were capable of revealing the expression of 17,565 different genes, including some on the X and Y chromosomes. More than 86% of them (15,132) were seen at significant levels in at least one brain region from at least one time period, and more than 81% (14,375) were expressed in one or more regions of the cerebral cortex. Some genes were generally expressed almost everywhere at all stages of development, whereas others were differentially expressed mainly in certain tissues or time periods. Differential expression was especially notable among time periods. In the cerebral cortex, more than 57% of detected genes were differentially expressed across time or age in fetal development, whereas 9% were differentially expressed in the postnatal period and only 0.7% varied across time periods in adults. Thus, large shifts in gene expression levels were typically seen in rapidly developing tissue prior to birth. Additional complexity was apparent for genes whose mRNA sequences from different exons of the same gene can be spliced together in different ways to form distinct proteins. More than 90% of expressed genes showed differential exon usage across brain regions; alternative splicing was especially pronounced across fetal time periods, when 83% of expressed genes showed usage of different exons to form unique proteins in rapidly developing fetuses. Thus, the pattern whereby one gene gives rise to more than one kind of protein was evident for the vast majority of genes active during the fetal period.

Induction of Nervous System

Early studies with amphibian embryos showed that chemicals from a small patch of ectoderm could induce the formation of a nervous system (Spemann & Mangold, 1924). A quest began to identify those chemicals. Pieces of frog ectoderm were placed in glass dishes in the lab, and various chemical extracts of the patch were applied to fresh embryos. One of them prompted the formation of a neural tube. It was purified and proved to be a protein. The gene encoding it was named *Noggin* because it seemed to provoke the formation of a head. Further studies revealed that *Noggin* is not the sole organizer of a nervous system. Several other proteins encoded by other genes were found to have similar effects (e.g., *Chordin and Gremlin*; Brazil, Church, Surae, Godson, & Martin, 2015). Elaborate experiments revealed that neither Noggin nor Chordin proteins directly activates genes in the cell nucleus. Instead, they inhibit the action of another protein that has a central role in the chain of causation— bone morphogenetic protein (from *Bmp* gene in animals and *BMP* gene in humans). Noggin fits neatly around the Bmp protein and binds to it, so that Bmp cannot activate its receptor (Groppe, Greenwald, Wiater, & Rodriguez-Leon, 2002; Walsh, Godson, Brazil, & Martin, 2010). The Bmp protein was first discovered in studies of bone formation, something that involves mesoderm,

not ectoderm. Bmp protein also alters the fate of ectoderm cells so that they develop into epidermis rather than nervous system (Munoz-Sanjuan & Brivanlou, 2002). Only when Noggin or Chordin or some other antagonist counteracts the effect of Bmp at the cell surface does the tissue form a neural tube. There are two types of *Bmp* receptors that detect the Bmp protein, and they regulate other signaling proteins inside the cell that are part of the *Smad* family of genes (Brazil et al., 2015; Rider & Mulloy, 2010). One of those, *Smad-4*, can form a complex with other Smad proteins, and that complex can enter the nucleus and initiate the translation of genes related to many other kinds of processes that are important for organizing epidermis. Subsequent studies with lab mice showed that there was very little effect of knocking out just one of the inhibitory genes *Noggin* or *Chordin*, but a genetically engineered mouse that lacked both *Noggin* and *Chordin* failed to form a head (Anderson, Lawrence, Stottmann, Bachiller, & Klingensmith, 2002). Either *Noggin, Chordin,* or *Gremlin* in combination with one of the other genes can block the action of Bmp, a phenomenon termed *functional redundancy*.

No single molecule qualifies as the organizer or inducer of a nervous system. Instead, there is an interconnected network of molecules that is the basis for organization. The result of this elaborate network is that a nervous system forms when enough Noggin, Chordin, or Gremlin is present, whereas in their absence, the tissue becomes epidermis. Bone morphogenetic protein and Noggin protein have important effects in many other tissues in the embryo and the adult. Bmp of course is involved in the formation of bones and cartilage as well as the nervous system. Furthermore, it has widespread effects in other organs. Noggin that acts on Bmp is also involved in the formation of mesodermal tissues including the skeleton, kidney, and heart as well as pituitary, prostate, and thymus glands (Rider & Mulloy, 2010). In human adults, BMP and its antagonists are important contributors to diseases of the bone, lung, liver, and kidney, and they are implicated in several kinds of cancers (Walsh et al., 2010). Thus, the gene-encoded proteins that are so important for the very first steps in the formation of the nervous system in the embryo from ectoderm are equally important for the origins of several mesodermal tissues in embryos and diseases in adults. *Bmp, Noggin,* and other genes do not code for a nervous system as such. Acting in the proper place at the proper time, they *enable* the embryo to embark on a long journey that results in a nervous system. They are necessary parts but not a sufficient explanation of early brain development. Each of the proteins is an integral part of the molecular system, and none has any organizational priority in the embryo. Induction of the neural tube is more complex and involves many more molecules than are mentioned here (Stern, 2005).

An important feature of early embryonic development is that chemical signals typically extend only very short distances from the cell of origin to the target cell with its receptors. In the case of induction of the neural tube, the target is on an adjacent cell or only one or two cells away. Further development of the neural tube from ectoderm occurs during its close contact with the notochord (Fig. 5.3). A critical chemical signal responsible for this transition has been identified. It is a protein that is synthesized in the notochord and then diffuses outward along the spaces separating the first few cells in the neural tube. The name of the gene is sonic hedgehog (*Shh*; SHH in humans). This name warrants some explanation (see Box 5.1).

The SHH gene initiates a chain of events that transforms the primitive cells in the neural tube into specific kinds of neurons (Choudhry et al., 2014). To achieve this, nature devised something elaborate, a puzzle with many pieces, some of which have not yet been identified. This molecular system has been studied intensively because SHH and associated proteins are known to play important roles in several common kinds of cancer that claim many lives every year (Chen, Gao, & Luo, 2013). The same gene that is so important in organizing the nervous system in embryos can create grave damage when reactivated in an adult.

Many neurotransmitters are synthesized with the aid of enzymes, whereas SHH codes for a protein that is the signal itself. Table 5.1 summarizes the genes that are involved in each step of the process in humans. All of the steps were initially identified in fruit flies and mice, and equivalent genes were then identified in humans. The long Shh protein must have a way to escape from the notochord cell and then diffuse among neural tube cells. This happens spontaneously when the protein separates into two parts and the shorter adds a cholesterol molecule at one end. Then, another gene adds a palmitate molecule to the other end to form the M-Shh-N molecule, and a third gene releases the M-Shh-N molecule to begin diffusion. Upon reaching a neural tube cell, the Shh molecule is moved close to its receptor with the aid of another protein and then binds to a large protein from the patched 1 (PTCH1) gene. In the absence of Shh, the Ptch1 receptor inhibits the adjacent Smoh protein and a complex of other regulatory molecules (from *SUFU, KIF7, PRKACA* genes) prevents the activation of the Gli1 protein derived from the *GLI1* gene. When Shh binds to the Ptch1 receptor, on the other hand, Smoh becomes active and releases Gli1 from the grip of the other molecules, so that it becomes activated, enters the nucleus, and transcribes several target genes that transform the neural tube cells into new kinds of neurons.

Shh diffuses among cells and reaches some that are 8–10 cells away from the notochord. The concentration of Shh is much higher nearest the notochord and then declines in cells closer to the top of the neural tube (Fig. 5.5). Several of the target genes that are sensitive to the Shh signal respond best to specific concentrations of Shh, and the different genes are thereby activated in different zones of the neural tube, giving rise to differentiation of the structure. Other cells secrete two other proteins (from *BMP4* and *WNT1* genes) that establish a concentration gradient from top to bottom of the neural tube and activate another set of genes that cause further differentiation in the neural tube. The two concentration gradients thereby combine to divide the neural tube into functionally distinct groupings of different neuron types in the dorsal-ventral dimension. There is no spatial mapping executed by any one gene on its own. Functional differentiation depends on

---

BOX 5.1

Gene Names

Sonic hedgehog (*Shh*) has nothing to do with sound or hedgehogs. The gene was first detected in a large study of mutations induced in fruit flies (Nüsslein-Volhard & Wieschaus, 1980). The larvae had a peculiar pattern of bumps all over the surface that resembled, to the investigators, spines on a charming little rodent, the hedgehog, so they named the gene "hedgehog" (*Hh*). Thirteen years later, three genes were identified in mice that closely resembled the fly *Hh* gene (Riddle, Johnson, Laufer, & Tabin, 1993). What to name them? Riddle, a postdoctoral fellow working in the lab of Tabin at Harvard University, had a 6-year-old daughter who owned a Sega comic book that featured a character named "Sonic the Hedgehog," and this was adopted because the Tabin lab "had a reputation for being loud and boisterous" (Keen & Tabin, 2004). Before the Riddle et al. scientific paper was published, Sega began to promote a new video game titled Sonic the Hedgehog, the second version of which is still available as a game online. The mouse and human gene names survived, despite a critical commentary in the New York Times (A Gene Named Sonic, 1994) and reservations expressed by some scientists.

TABLE 5.1　Sonic Hedgehog and Genes Involved in Its Expression in Neural Tube (GeneCards and OMIM)

#	Gene	Chrom. [amino acids]	Gene name	Function
1	SHH	7 [462]	Sonic hedgehog	Notochord cells synthesize Shh protein with 462 aa. The protein is spontaneously cleaved into two fragments; shorter one becomes the signal; cholesterol is added to it to form Shh-N
2	HHAT	1 [493]	Hedgehog acetyltransferase	Adds palmitic acid to other end of Shh-N to form a multimer M-Shh-N; tethered to the notochord cell
3	DISP1	1 [1524]	Dispatched 1	Releases M-Shh-N for diffusion away from notochord
4	HHIP	4 [700]	Hedgehog interacting protein	Binds M-Shh-N protein at target cell, which removes Shh from the primary signal stream; modulates signal
5	PTCH1	9 [1447]	Patched 1	Primary receptor for M-Shh-N signal in cell membrane
6	SMOH	7 [794]	Smoothened	Inhibited or suppressed by PTCH1 when there is no Shh signal; activated by PTCH1 when it binds M-Shh-N, causes cascade of effects
7	GLI1	12 [1106]	Glioma-associated oncogene 1	Transcription factor that promotes readout of several genes when SMOH is activated; inhibited by Sufu protein from SUFU gene
8	SUFU	10 [484]	Suppressor of fused	Suppressor or negative regulator that binds to GLI1; regulation of SUFU is complex
9	PRK-ACA	19 [351]	Protein kinase cAMP-activated catalytic subunit α	Cofactor involved in inhibition and activation of GLI1 via Smoh
10	KIF7	15 [1343]	Kinesin family 7	Cofactor involved in inhibition and activation of GLI1 via Smoh

interactions among several genes and diffusion of their protein products, all of which need to be taken into account to achieve even a first-order approximation to the complexity of the system of molecules and cells in just a small region of the neural tube.

HOX and the Head-Tail Axis

It is evident that the genes involved in the SHH signaling pathway are distributed widely across several chromosomes (Table 5.1), just as what occurs with genes involved in the synthesis of melanin and acetylcholine. There is no clustering of genes into a small region of one chromosome, even when they are part of a common functional system. On the other hand, establishment of the anterior (head) to posterior (tail) axis in the embryo nervous system relies on clusters of closely spaced genes on a chromosome. This was originally discovered in fruit flies in which mutations generated little monsters with two pairs of wings or a leg where there should be an antenna. Typical body segment patterns could be radically disrupted by genetic mutations. It was learned that the anatomical order of types of body segments corresponds closely with the order on a chromosome of the *homeodomain* or *Hox* genes. There are eight of them in flies, and they all share a common DNA sequence of 180 base pairs that generates highly similar sequences of 60 amino acids in the proteins. The complete proteins encoded by the *Hox* genes are generally hundreds of amino acids in length, but there is usually a shorter sequence within each gene that seems to regulate how it is expressed. The *Hox* genes regulate the expression of other genes that then generate cells with characteristics typical of a particular body segment. One such fly gene has been found to regulate expression of more than 80 other genes.

Detailed studies of chick and mouse embryos have shown that it is not a simple matter of each *Hox* gene coding for a single segment of the brain stem or spinal cord. Regional identity is specified by *combinations* of *Hox* genes (Mark, Rijli, & Chambon, 1997). This has been demonstrated by studies of gene expression and of genetically engineered mutations or knockouts that disable a specific *Hox* gene or pair of genes in mice. For example, clusters of cells in the brain stem that give rise to different cranial nerves are specified by different combinations of *Hox* genes.

HOX genes are also important for the formation of mesodermal tissues such as the skeleton and digits of the paw or hand (Mark et al., 1997). The more posterior HOX genes are involved in the formation of the uterus and ureter in females and testes-related tissues and vas deferens in males. The type 12 and 13 genes contribute to the formation of digestive tract and smooth muscles of the rectal sphincter. Thus, HOX genes participate in differentiation of all three major zones of the early embryo—ectoderm, mesoderm, and endoderm. However, they do not contribute to the formation of brain structures at a higher level than the hindbrain. Mutations in HOX genes often lead to malformation of hands and face as well as other features of the skeleton, but they sometimes cause abnormalities of many organ systems (Quinonez & Innis, 2014). Changes in HOX gene

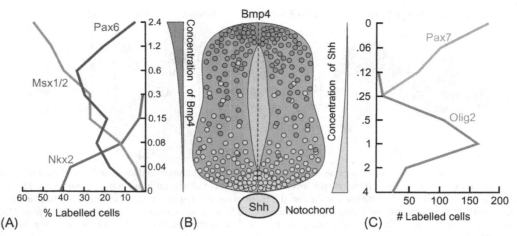

FIG. 5.5 Further differentiation of the neural tube along its dorsal-ventral axis (B) depends on concentration gradients of the signaling molecules Bmp4 and Shh. Many genes are regulated by the relative concentrations. For example, (C) Pax7 is abundant where Bmp4 is high and Shh is low, while (A) Nkx2 is high where Shh is high and Bmp4 is low. *(A) Adapted from Mizutani et al. (2006). Fig. 4A. (C) Reprinted by permission from Dessaud, E., Yang, L. L., Hill, K., Cox, B., Ulloa, F., Ribeiro, A., … Briscoe, J., Fig. 1E, Copyright 2007. Interpretation of the sonic hedgehog morphogen gradient by a temporal adaptation mechanism. Nature, 450.*

expression are believed to play a key role in many kinds of cancer (Bhatlekar, Fields, & Boman, 2014).

HOX genes contribute to the early formation of the more posterior portions of the nervous system, and they act by regulating the expression levels of many other genes that are important for a particular segment of the head-tail axis. Consequently, they are often regarded as "critical master regulatory" factors (Bhatlekar et al., 2014). But detailed examination of nervous system origins reveals that HOX genes are not themselves the master switches or organizers. Instead, they are controlled by other genes that regulate the formation of the head-tail axis. Retinoic acid that is generated in early mesoderm cells establishes a gradient of concentration and diffuses from highest levels in the hindbrain region to lower levels at the tail end of the embryo. This process involves several steps mediated by other genes. Opposing concentration gradients of retinoic acid and fibroblast growth factor (FGF gene) establish the head-to-tail axis, and the various HOX genes then respond to specific concentrations of those proteins (Lippmann et al., 2015).

Retinoic acid (RA) has the ability to regulate the expression levels of many other genes in addition to the HOX group. A specific sequence of bases has been identified in which RA can bind to the DNA of a gene, and genes with this kind of RA response element are likely to have their expression levels regulated by the concentration of RA. Rhinn and Dolle (2012) documented more than 65 genes that are sensitive to RA levels, including six *Hox* genes in mice. Several other genes are important for establishing the head-to-tail axis via concentration gradients, and the HOX genes respond to those gradients. Those signaling molecules are thought to have a crucial role "in the onset of patterning in the neural tube" (Deschamps & van Nes, 2005). The concentration of retinoic acid regulates many genes, but RA expression itself is in turn the product of

an array of protein-encoding genes that are active earlier in the embryo.

Several other genes play an important part in dividing the brain into distinct zones in regions where HOX genes are not expressed. Among these are the PAX and EMX gene families, some of whose members have important roles in dividing the brain anatomically and functionally. Two of these, *Pax6* and *Emx2*, have been studied extensively in mouse embryos. As shown in Fig. 5.6, *Emx2* is expressed at high levels in the rear part of the cerebral cortex, tapering to low levels in the front part, whereas *Pax6* shows an opposite pattern. Certain areas where *Pax6* expression is high and *Emx2* is low become motor cortex, whereas visual cortex arises where *Emx2* is high and *Pax6* is low. Experiments with mice in which one of the two genes has been disabled reveal how boundaries between areas are not defined directly by information encoded in just one gene; instead, function depends on relations between expressed genes (Muzio & Mallamaci, 2003). If *Emx2* is deleted, the area of cortex devoted to motor control is much greater, and the visual area is greatly reduced, whereas the opposite pattern occurs when *Pax6* is deleted. Only when both *Emx2* and *Pax6* are deleted in the same mouse is cortical anatomy grossly and fatally changed.

MAKING CONNECTIONS

Axon Outgrowth and Path Finding

After the newly generated neurons have moved to the region of cortex where they will reside, they send out dendrites and axons that will set up connections with other cells, some nearby and others at remote sites. The actual connections are synapses that use specific

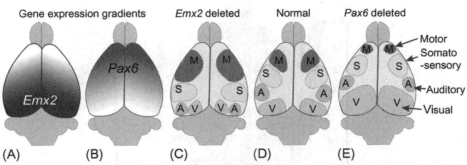

FIG. 5.6 Locations of expression of *Emx2* and *Pax6* genes in mouse brains in relation to functional areas of the cerebral cortex. (A) Emx2 protein is most abundant in the posterior portion of the cerebral cortex, while (B) Pax6 is most abundant in the anterior portion. (C) Sizes of cortex that are devoted to four different functions in mice lacking the *Emx2* gene. (D) Sizes of cortical regions in intact animals. (E) Sizes in mice lacking the *Pax6* gene. Size of a specific functional area is reduced when a gene involved with its normal expression is deleted. *Adapted from Muzio and Mallamaci, Emx1, emx2 and pax6 in specification, regionalization and arealization of the cerebral cortex. Cerebral Cortex, 13, 641–647, Fig. 1AB, copyright 2003, by permission of Oxford University Press.*

neurotransmitters to convey information between cells (Chapter 4). Genes play a major role in guiding axons and forming synapses, but genes do not embody a wiring diagram or a blueprint. Things progress one step at a time until a chemical on a specific kind of cell signals to the axon that "this is the place" to stop and form a synapse. When the immature neuron sends an axon on its journey, the growth cone at the tip of the axon finds no information about its destination. Instead, it encounters local molecules that attract or repel it. Only when it gets close to a suitable target do the axon tip and the target dendrite mutually embrace each other.

The axon finds its way among a labyrinth of other cells, fibers, and molecules by means of an active growth cone at its tip (Fig. 5.4). The structure of the growth cones involves many of the same kinds of proteins that are involved in building muscle—actins and myosins. Microtubules from the cell body (soma) extend down the axon

to the growth cone and deliver the molecules needed to build new structures. The filopodia at the growth cone tip form and retract rapidly, testing the local environment and finding a path forward. The filopodia are covered with receptors that respond to specific kinds of chemical ligands or guidance molecules along the path. Activation of a receptor by a ligand can result in attraction of the growing axon toward the source of the ligand or repulsion away from it, depending upon the specific kinds of molecules. When a repulsion cue is encountered, the filopodia in that zone begin to degenerate, while those opposite that zone thrive, and these changes gradually turn the cone away from the negative signal.

Several kinds of ligands are abundant during nervous system formation in the embryo (Table 5.2). Some of them (group 3 netrins and semaphorins) are secreted from another cell and diffuse outward, forming a concentration gradient as it happens with the SHH and BMP genes

TABLE 5.2 Growth Cone Guidance Genes (from OMIM, GeneCards, and HGNC)

Ligand	Genes	Response mediated by receptor		Other phenotypes
		Attraction	**Repulsion**	
Netrin	NTN1,3–5, NTNG1,2	DCC	UNC5A,B, C,D	Mammary gland, ovary, testes, lung, pancreas, muscle, blood vessels; several kinds of cancer in adults
Ephrin A	EFNA1–5	EFHA1–10	–	Blood vessel formation, intestines; breast, colon cancer
Ephrin B	EFNB1–3	EPHB1–6	–	Reciprocal activation
SLIT	SLIT1–3	–	ROBO1,2,3,4	Interacts with netrins and DCC; several adult cancers
Semaphorin	SEMA3A–7A	PLXA1–D1	NRP1,2	Immune system, blood vessels, bone formation, scar tissue; leukemia
Sonic hedgehog	SHH	BOC	BMPs	Skeleton, hand and limb formation; tumors of brain, prostate, lung, digestive tract
WNT	WNT1,3,3A,5A	RYK, FZD8	–	Formation of bone, heart, muscle, skin; breast and prostate cancer, diabetes

in the neural tube (Fig. 5.5). For an axon, such a gradient serves to attract it toward the source of the secretion. Other ligands are embedded in a cell membrane and can interact only with adjacent cells (e.g., ephrins). Each is a protein encoded by a gene, and most ligands can be present in several forms that work most effectively with specific receptors. Most receptors expressed on a growth cone exist in several forms that comprise a gene family. This diversity allows for a large number of specific ligand-receptor combinations. Usually, there are several kinds of genes active, often from more than one group in Table 5.2.

The semaphorin group of signals is active in many parts of the early nervous system and several other tissues including blood vessels (Yoshida, 2012). The SEMA3 group codes for a protein that is secreted from a cell, whereas other members of the SEMA group are attached to a cell membrane and influence only growth cones that come into contact with that cell. Sema proteins activate various members of the plexin (PLX) group of receptors and in some cases neuropilin (NRP1 or 2) receptors. Specific PLX and NRP receptors respond to signals from different semaphorins (Neufeld & Kessler, 2008), each encoded by a different gene.

A similar kind of analysis has been conducted with several other ligand-receptor combinations. One general conclusion is that any one kind of ligand typically stimulates more than one kind of receptor protein, and any one receptor protein is commonly activated by more than one specific kind of ligand. A functional role of a specific ligand is often detected in other kinds of tissue, such as neuropilins, which are also involved in the development of blood vessels throughout the body of the embryo (Pellet-Many, Frankel, Jia, & Zachary, 2008). Most of the signal and receptor molecules are reactivated in various cancers in the adult. It thus seems fair to conclude that the fantastic array of signals and receptors involved in axon guidance are not specifically devoted to nervous system development. The embryo utilizes many of the same molecules to do different things in different tissues.

Synapse Formation

Chemical cues that signal to the growth cone when it has reached its destination are not well known to researchers, but they are clear enough to the axons themselves. When the growth cone of an axon contacts and bypasses vast numbers of cells during its journey, it can quickly detect cells with which a functional synapse can be formed. Prior to the arrival of the axon's growth cone at its destination, the dendrite of another neuron does not show any clear sign of where a synapse will form. The exact location is determined only when the growth cone makes contact with the membrane of the dendrite. Just before contact with the dendrite, rapid

extension of the axon stops, and filopodia retract. When the axon makes contact with a suitable dendrite, the two membranes are bound together by cadherin proteins (CADH2), and the dendrite sends chemical signals to the presynaptic part of the axon via WNT genes, a family of 19 different signaling proteins that generate complex changes in the tip of the axon and its neuron (Dickins & Salinas, 2013). Microtubules in the axon conduct chemical messengers both to and from the cell nucleus, and a variety of newly synthesized proteins are shipped downstream to the axon tip. Reciprocal changes (*reciprocal induction*) are induced in the dendrite so that it begins to accumulate proteins that will build the postsynaptic portion of a synapse. These include proteins from the postsynaptic density (PSD) genes, especially PSD95. The nascent synaptic membranes are held in place by neuroligin (NLGN1) expressed on the dendritic side that binds to neurexins (NRXN) on the presynaptic surface. Packets of several proteins organized in a vesicle are sent down the axon from the neuron soma, and they form parts of the active zone where the release of neurotransmitters from vesicles will occur (Chapter 4). Release of neurotransmitter at the synaptic cleft and electric activity of the cell commence while the synapse is getting organized, and this vital activity contributes to the final wiring pattern of the brain.

Although very complex at first reading, these descriptions of genes involved in brain development are actually minimalist. The full picture involves thousands of genes differentially expressed in a multitude of distinct kinds of tissues and cells, and the roles of many of them remain largely unknown. It appears that each anatomical area, including each functionally unique zone of the cerebral cortex, has a unique signature of genes that are expressed therein (Santiago & Bashaw, 2014). Identity of a region is never specified by just one special gene. Instead, a region is defined by the *combination* of different genes that are expressed, combinations that are manifest in the myriad connections it forms with other regions.

PLASTICITY OF DEVELOPMENT

Cell Death and Axon Pruning

If the precise wiring of the brain is not strictly encoded in genetic information, how can we explain it? One of the extraordinary secrets to its success is failure. The embryo brain generates far more neurons than are found in the adult brain. Many of them perish at a young age, and those that survive and thrive are typically the ones that form good connections with other cells. This was discovered when neuroscientists sliced the spinal cord of a chick embryo into many thin sections and stained them so that individual nerve cells could be viewed and counted. This painstaking and tedious work revealed that large numbers of neurons die and degenerate not long after they are born. The portion of the spinal cord near the tail end of the embryo contains motor neurons that form connections with muscles in the leg. The primordium of the leg is a small limb bud that appears several days before the chick hatches. Counting neurons revealed that their numbers peaked at about 10 days after the limb bud first appeared and then declined precipitously for the next 4 weeks, to the extent that more than half of the neurons perished (Hughes, 1961). The situation stabilized after hatching.

Further investigation showed that all those dead and dying cells were no accident. If the limb bud was removed surgically in the early embryo, the chick later showed a striking deficit of motor neurons in the spinal cord just on the side that sent axons to the zone of the missing limb muscles. Evidently, the motor neurons that could not locate a good target on a muscle degenerated. To strengthen this conclusion, researchers then removed the limb bud from one embryo and grafted it onto another embryo of the same age at a spot right next to its own limb bud. When the graft operation was successful, something remarkable occurred: there were far more motor neurons surviving in the spinal cord on the side where neurons could connect with either of the two limbs (Fig. 5.7A). Evidently, many of the motor neurons that would ordinarily have perished were then able to find a muscle cell and form a healthy connection, and this rescued them (Hollyday & Hamburger, 1976). The name for this phenomenon is *functional validation*. Those cells that form viable connections will survive, and those that do

not will perish. By generating a surplus of neurons and then culling them to preserve the most successful ones, the embryo ensures that all of the viable muscle fibers will indeed connect with a motor neuron. Things do not need to be wired correctly from the outset.

A similar fate awaits many of the axons themselves and even the synapses. Most axons form several branches when they near the target zone, and many of them eventually degenerate because they do not form active connections with a target neuron. The adroit removal of entire axons or some of their branches is termed neural *pruning*. This has been demonstrated by careful counting of axons in bundles of nerve fibers at different ages. Fig. 5.7B provides an example from the optic nerve of cats (Williams, Bastiani, Lia, & Chalupa, 1986). The counts revealed clearly that numbers of axons in the optic nerve peak at about 6 weeks after conception, 2 weeks before birth, and then plummet in a spectacular way, going from a total of about 600,000 to less than 200,000 axons over a 6-week period. Selective pruning favored the largest and fastest axons, the ones that had established strong connections with the brain.

Synapses also undergo a period of substantial decline in most brain regions as the less effective connections are pruned. A major study of human cerebral cortex examined one small region with the aid of a powerful microscope (Huttenlocher & Dabholkar, 1997). Postmortem samples of normal brains were obtained for a wide range of ages from shortly before birth to more than 70 years. Both synapse numbers and densities peaked at about 8 months after birth and then declined steeply for the next 6 years, stabilizing after the age of 10 years (Fig. 5.7C).

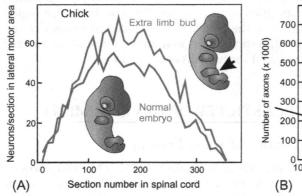 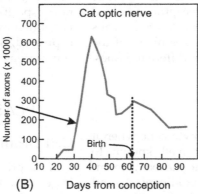 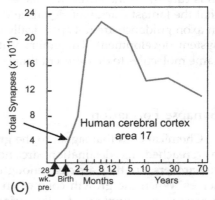

FIG. 5.7 Sculpting of the nervous system by death of cells, axons, and synapses. (A) When an extra limb bud is grafted onto a chick embryo, many more neurons populate the lateral motor area of the spinal cord. (B) The numbers and densities of axons in the cat optic nerve increase greatly until about 6 weeks after conception and then decline rapidly as axons are lost. Density declines further as many axons form thick myelin sheaths. (C) Synapses in the human cerebral cortex increase rapidly until about 6 months after birth and then decline greatly. *(A) Adapted from Hollyday and Hamburger, Reduction of the naturally occurring motor neuron loss by enlargement of the periphery.* Journal of Comparative Neurology, 170, *311–320, Fig. 4, Copyright 1976, with permission from John Wiley & Sons; (B) Adapted from Williams et al., Growth cones, dying axons, and developmental fluctuations in the fiber population of the cat's optic nerve.* Journal of Comparative Neurology, 246, *32–69, Fig. 1, Copyright 1986, with permission from John Wiley & Sons; (C) Reprinted from Huttenlocher, P. R., Synaptic density in human frontal cortex—developmental changes and effects of aging,* Brain Research, *163, 195–205, Fig. 3, Copyright 1979, with permission from Elsevier.*

Synaptic Competition and Refinement

The losses of neurons, axons, and synapses occur when some form better connections than others. The weaker ones are more likely to perish. Hence, the term "competition" is often used to describe this process. It is clear that the early connections between two parts of the nervous system are usually diffuse and relatively widespread, whereas they become more precise and refined through selective processes that depend on synaptic function. There is abundant evidence that synapse elimination in mammals reflects patterns of postnatal motor activity or behavior (Brown, Jansen, & Van Essen, 1976; Buffelli, Busetto, Bidoia, Favero, & Cangiano, 2004; Jansen & Fladby, 1990). In a very real way, behavior therefore has an important role in shaping the nervous system itself.

Early Visual Experience and Critical Periods

One of the best-understood sensory systems is vision. The remarkable precision of retinal ganglion cell mapping onto visual cortex has been studied in great detail. Electric recordings from awake animals with the aid of fine microelectrodes lowered into the cortex have enabled researchers to assess the response of single neurons to a wide range of visual stimuli. One consistent finding is that many cells respond equally well to stimuli presented to either left or right eye; these are termed binocular or bilateral cells. Others respond mainly to stimuli presented to the eye on the opposite side (contralateral) or to the eye on the same side of the head (ipsilateral) but not both. An ocular dominance score ranging from 1 to 7 expresses the extent to which a single cell responds best to stimuli presented to the opposite- or same-side eye. It was found that when kittens are reared with both eyes open, cells of visual cortex of adult cats most often respond to stimuli from either eye. Researchers then examined the consequences of closing the eyelid of one eye for 2.5 months beginning shortly after birth. Effects were profound (Wiesel & Hubel, 1963); there were no binocular cells at all, and even the response of ipsilateral cells was considerably reduced. The eye itself seemed to be perfectly healthy. It was the connections of the eye with visual cortex that depended so strongly on visual experience. Closing one eye after 2 months of normal vision had much less effect, and closing the eye of an adult cat after 3 months had no effect at all. Thus, there was a *critical period* for the organizing effect of visual experience that began shortly after birth and extended for several months. Another demonstration of this point was achieved by inducing squint or strabismus by severing a few small muscles that control eye movement, so that one eye always looked off to one side. In that situation, the information sent from the two eyes to visual cortex was never the same, and as a result, there was an almost total loss of bilateral responses, whereas connections with one eye or the other were well formed (Hubel & Wiesel, 1965). A similar result occurred if the kitten was raised with the left and right eyes being blocked by a patch on alternate days for several weeks; there was a striking absence of binocular responding cells. Thus, the influence of early visual experience on the detailed wiring and synaptic activity of neurons in visual cortex was of utmost importance for achieving normal visual function.

It is apparent from these and many other experiments that patterns of gene expression are very important for regional differentiation of the cerebral cortex and most other parts of the nervous system. Furthermore, patterns of gene expression at one phase of development are themselves consequences of differences in spatial expression of other genes at an earlier stage. A final phase of nervous system development ensues in which experience that shapes many connections has a major impact on perception, motor coordination, and other behavioral processes. Experience plays a role in strengthening the most effective connections and pruning away the weaker ones.

HIGHLIGHTS

- Most of what we know about nervous system development in the embryo comes from experiments done with fruit flies, frogs, chicks, and mice. Genes and the dynamics of early development are remarkably similar in mice and humans.
- A very small number of cells in the inner cell mass of the 8-day-old embryo gives rise to thousands of kinds of cells in the adult that express radically different phenotypes, even though they have identical genotypes. Development involves an elaborate series of changes in gene expression.
- One layer of cells, the ectoderm, gives rise to the neural tube through induction by the notochord that later disappears. There is no single gene that codes for this important step. Instead, there is a network of genes that initiates the formation of the nervous system.
- Most gene-derived molecules that are involved in early nervous system development are also active in forming many other tissues such as the bone, heart, kidney, and many glands.
- Differentiation of the neural tube along its dorsal-ventral axis depends on two opposite concentration gradients of signaling molecules. What genes are turned on in a particular location depends on the relative concentrations of the different signals.
- The head-tail axis of the embryo is established through relative concentrations of several other kinds of signals, and these are able to control the expression levels of dozens of genes that combine to generate a

variety of structures from head to tail. Many of the genes involved in head-tail differentiation of the nervous system are also involved in the formation of the skeleton, limbs, and reproductive organs.

- Concentration gradients of still other signaling molecules are important for dividing the cerebral cortex into functionally different regions. Thus, regional differentiation of the cortex depends on combinations of different signals.
- Neurons send growing axons in search of suitable places to make connections. Guidance is provided by growth cones at the tip of the axon that are either repelled or attracted by gene-encoded protein molecules they encounter along the journey. There are dozens of combinations of pathway signaling molecules and their receptors on the growth cones. This remarkable specificity allows an axon to grow through a forest of different kinds of cells until if finds a good place to stop and form a synapse.
- The embryonic nervous system generates far more neurons, axons, and synapses than are found in the adult. Many of them perish when they fail to form viable connections (pruning). This dynamic process makes it possible to form a multitude of connections without needing to have a code to specify the location of each one.
- The activity of the sense organs at the periphery can sculpt the patterns of synaptic connections with sensory areas of the cerebral cortex. There is usually a critical period soon after birth when sensory activity has the greatest impact on synaptic connections with the cortex.
- There is no genetic blueprint for brain structure. No gene specifies where a neuron should make a connection. Brain organization arises from interacting networks of gene-encoded proteins and many aspects of early sensory experience that generate patterns of connections with the cerebral cortex.

References

A Gene Named Sonic. (1994). New York times. January 11 Retrieved from http://www.nytimes.com/1994/01/11/science/a-gene-named-sonic.html

Anderson, R. M., Lawrence, A. R., Stottmann, R. W., Bachiller, D., & Klingensmith, J. (2002). Chordin and noggin promote organizing centers of forebrain development in the mouse. *Development, 129,* 4975–4987.

Bhatlekar, S., Fields, J. Z., & Boman, B. M. (2014). HOX genes and their role in the development of human cancers. *Journal of Molecular Medicine (Berlin, Germany), 92,* 811–823.

Bonache, S., Mata, A., Ramos, M. D., Bassas, L., & Larriba, S. (2012). Sperm gene expression profile is related to pregnancy rate after insemination and is predictive of low fecundity in normozoospermic men. *Human Reproduction, 27,* 1556–1567.

Brazil, D. P., Church, R. H., Surae, S., Godson, C., & Martin, F. (2015). BMP signalling: agony and antagony in the family. *Trends in Cell Biology, 25,* 249–264.

Brown, M. C., Jansen, J. K., & Van Essen, D. (1976). Polyneuronal innervation of skeletal muscle in new-born rats and its elimination during maturation. *The Journal of Physiology, 261,* 387–422.

Buffelli, M., Busetto, G., Bidoia, C., Favero, M., & Cangiano, A. (2004). Activity-dependent synaptic competition at mammalian neuromuscular junctions. *Physiology, 19,* 85–91.

Casas, E., & Vavouri, T. (2014). Sperm epigenomics: challenges and opportunities. *Frontiers in Genetics, 5,* 330.

Chang, Y., Li, J., Chen, Y., Wei, L., Yang, X., Shi, Y., & Liang, X. (2015). Autologous platelet-rich plasma promotes endometrial growth and improves pregnancy outcome during in vitro fertilization. *International Journal of Clinical and Experimental Medicine, 8,* 1286–1290.

Chen, Q., Gao, G., & Luo, S. (2013). Hedgehog signaling pathway and ovarian cancer. *Chinese Journal of Cancer Research, 25,* 346–353.

Choudhry, Z., Rikani, A. A., Choudhry, A. M., Tariq, S., Zakaria, F., Asghar, M. W., … Mobassarah, N. J. (2014). Sonic hedgehog signalling pathway: a complex network. *Annals of Neurosciences, 21,* 28–31.

Deschamps, J., & van Nes, J. (2005). Developmental regulation of the Hox genes during axial morphogenesis in the mouse. *Development, 132,* 2931–2942.

Dickins, E. M., & Salinas, P. C. (2013). Wnts in action: from synapse formation to synaptic maintenance. *Frontiers in Cellular Neuroscience, 7,* 162.

Elliott, E. (2016). *Women in science: Hilde Mangold and the embryonic organizer.* JAX Blog October 20 Retrieved from https://www.jax.org/news-and-insights/jax-blog/2016/october/women-in-science-hilde-mangold.

Groppe, J., Greenwald, J., Wiater, E., & Rodriguez-Leon, J. (2002). Structural basis of BMP signalling inhibition by the cystine knot protein Noggin. *Nature, 420,* 636.

Harland, R. (2008). Induction into the Hall of Fame: tracing the lineage of Spemann's organizer. *Development, 135,* 3321–3323.

Hollyday, M., & Hamburger, V. (1976). Reduction of the naturally occurring motor neuron loss by enlargement of the periphery. *Journal of Comparative Neurology, 170,* 311–320.

Hubel, D. H., & Wiesel, T. N. (1965). Binocular interaction in striate cortex of kittens reared with artificial squint. *Journal of Neurophysiology, 28,* 1041–1059.

Hughes, A. (1961). Cell degeneration in the larval ventral horn of Xenopus laevis (Daudin). *Journal of Embryology and Experimental Morphology, 9,* 269–284.

Huttenlocher, P. R., & Dabholkar, A. S. (1997). Regional differences in synaptogenesis in human cerebral cortex. *Journal of Comparative Neurology, 387,* 167–178.

Jansen, J., & Fladby, T. (1990). The perinatal reorganization of the innervation of skeletal muscle in mammals. *Progress in Neurobiology, 34,* 39–90.

Kang, H. J., Kawasawa, Y. I., Cheng, F., Zhu, Y., Xu, X., Li, M., … Sestan, N. (2011). Spatio-temporal transcriptome of the human brain. *Nature, 478,* 483–489.

Keen, A., & Tabin, C. (2004). Cliff Tabin: super sonic an interview. *The Murmur Weekly,* 1–5.

Kumar, A., Ryan, A., Kitzman, J. O., Wemmer, N., Snyder, M. W., Sigurjonsson, S., … Lewis, A. P. (2015). Whole genome prediction for preimplantation genetic diagnosis. *Genome Medicine, 7,* 35.

Lippmann, E. S., Williams, C. E., Ruhl, D. A., Estevez-Silva, M. C., Chapman, E. R., Coon, J. J., & Ashton, R. S. (2015). Deterministic HOX patterning in human pluripotent stem cell-derived neuroectoderm. *Stem Cell Reports, 4,* 632–644.

Mark, M., Rijli, F. M., & Chambon, P. (1997). Homeobox genes in embryogenesis and pathogenesis. *Pediatric Research, 42,* 421.

Mintz, B., & Silvers, W. K. (1967). "Intrinsic" immunological tolerance in allophenic mice. *Science, 158,* 1484–1487.

Munoz-Sanjuan, I., & Brivanlou, A. H. (2002). Neural induction, the default model and embryonic stem cells. *Nature Reviews Neuroscience, 3,* 271–280.

Muzio, L., & Mallamaci, A. (2003). Emx1, emx2 and pax6 in specification, regionalization and arealization of the cerebral cortex. *Cerebral Cortex, 13,* 641–647.

Neufeld, G., & Kessler, O. (2008). The semaphorins: versatile regulators of tumour progression and tumour angiogenesis. *Nature Reviews: Cancer, 8,* 632–645.

Nüsslein-Volhard, C., & Wieschaus, E. (1980). Mutations affecting segment number and polarity in Drosophila. *Nature, 287,* 795–801.

Pacheco, S. E., Houseman, E. A., Christensen, B. C., Marsit, C. J., Kelsey, K. T., Sigman, M., & Boekelheide, K. (2011). Integrative DNA methylation and gene expression analyses identify DNA packaging and epigenetic regulatory genes associated with low motility sperm. *PloS One, 6,* e20280.

Pellet-Many, C., Frankel, P., Jia, H., & Zachary, I. (2008). Neuropilins: structure, function and role in disease. *Biochemical Journal, 411,* 211–226.

Quinonez, S. C., & Innis, J. W. (2014). Human HOX gene disorders. *Molecular Genetics and Metabolism, 111,* 4–15.

Rhinn, M., & Dolle, P. (2012). Retinoic acid signalling during development. *Development, 139,* 843–858.

Riddle, R. D., Johnson, R. L., Laufer, E., & Tabin, C. (1993). Sonic hedgehog mediates the polarizing activity of the ZPA. *Cell, 75,* 1401–1416.

Rider, C. C., & Mulloy, B. (2010). Bone morphogenetic protein and growth differentiation factor cytokine families and their protein antagonists. *Biochemical Journal, 429,* 1–12.

Santiago, C., & Bashaw, G. J. (2014). Transcription factors and effectors that regulate neuronal morphology. *Development, 141,* 4667–4680.

Shaw, L., Sneddon, S. F., Zeef, L., Kimber, S. J., & Brison, D. R. (2013). Global gene expression profiling of individual human oocytes and embryos demonstrates heterogeneity in early development. *PloS One, 8,* e64192.

Spemann, H., & Mangold, H. (1924). Über induktion von embryonalanlagen durch implantation artfremder organisatoren. *Archiv für Mikroskopische Anatomie und Entwicklungsmechanik, 100,* 599–638.

Stern, C. D. (2005). Neural induction: old problem, new findings, yet more questions. *Development, 132,* 2007–2021.

Walsh, D. W., Godson, C., Brazil, D. P., & Martin, F. (2010). Extracellular BMP-antagonist regulation in development and disease: tied up in knots. *Trends in Cell Biology, 20,* 244–256.

Wiesel, T. N., & Hubel, D. H. (1963). Single-cell responses in striate cortex of kittens deprived of vision in one eye. *Journal of Neurophysiology, 26,* 1003–1017.

Williams, R. W., Bastiani, M. J., Lia, B., & Chalupa, L. M. (1986). Growth cones, dying axons, and developmental fluctuations in the fiber population of the cat's optic nerve. *Journal of Comparative Neurology, 246,* 32–69.

Yoshida, Y. (2012). Semaphorin signaling in vertebrate neural circuit assembly. *Frontiers in Molecular Neuroscience, 5,* 71.

Zernicka-Goetz, M. (2016). Andrzej K. Tarkowski 1933-2016. *Nature Cell Biology, 18,* 1261.

6

Behavior

Behavior is the way we interact with the world. Eating, running, singing, weeping, and even reading involve motion of an individual with respect to the surroundings. Behavior is a phenotype. There is no purely genetic kind of behavior or any behavior attributable solely to environment. Behavior is a product of a nervous system that processes sensory data from the environment, and that nervous system is integrated with the skeletal, circulatory, digestive, and other systems to generate and regulate behaviors. The activities of that nervous system involve thousands of expressed genes, but the nervous system itself does not express behavior. Brain activity can be detected in the form of electric impulses or bright spots on a computer screen during a brain scan, but those phenotypes are not behaviors as such. Thought is an internal process that takes place in the brain but cannot be detected directly by an external observer. The outward manifestations of thought in the form of speech, writing, or typing are behaviors. Devious thoughts expressed in the form of imaginative literature may earn praise and prizes, whereas actually performing those misbehaviors may lead to a medical diagnosis or worse. Epileptic seizures, dangerous driving, hateful or disorganized speech, suicide, and excessive food or drug consumption are significant social problems that involve behavior.

Specialized methods for testing and measuring human behavior are primarily the domain of psychology. The American Psychological Association, Psych Central, Assessment Psychology, and other organizations provide copious details about tests via their websites. The Educational Testing Service maintains an online list of more than 25,000 tests that have been used at one time or another to assess human qualities. Psychological tests often seek to quantify differences among people who are regarded as more or less normal, whereas some tests are expressly designed to identify kinds or degrees of behavior that are exceptional—outside the normal range of variation. Psychiatric diagnosis, on the other hand, aims to discriminate between behaviors that are merely different and those that are outright pathological and warrant treatment. When discussing behavior, the principal distinctions between things

that pertain to psychology as opposed to psychiatry are professional qualifications and legal competencies. Psychiatrists, who are interested mainly in abnormal behaviors, have medical training and an MD degree and can prescribe drugs, whereas a psychologist, primarily interested in variations among normal people, may have a PhD degree plus expertise in statistical analysis and may recommend changes in lifestyle and government policies. A psychiatrist will usually have detailed knowledge of the physiology and chemistry of the nervous system, whereas a psychologist may have little idea of what is taking place in the brain but will know how to design and administer a test of behavior and be aware of the many things that can influence the interpretation of test results.

KINDS OF TESTS AND MEASURES

Only a few of the many available tests of behavior and psychological constructs are mentioned here. The ones highlighted and discussed at length are those that commonly occur in the research literature on genetics and behavior. Several kinds of tests are examined in greater depth in Chapters 14–17 on complex traits.

The Quest for Natural Behavior

The quest for "natural" behavior seeks to measure a behavior without changing it. This is possible when the observer is concealed and the individual being measured is not aware of being observed. Naturalists studying bird courtship or parental care of the young by mammals strive to blend into the background or watch from a great distance, or they may rely on a camera concealed in a shrub. Films of behavior made in natural settings appear at face value to provide intimate looks at nature free from human influence. Unfortunately, some of those marvelous scenes that have captivated audiences were not so natural. The award-winning 1958 Disney film *White Wilderness* claimed to show lemmings during group

Genes, Brain Function, and Behavior
https://doi.org/10.1016/B978-0-12-812832-9.00006-3

migration committing suicide by jumping from a cliff to the Arctic Ocean. It turns out the scenes were brazen deceptions staged in the Province of Alberta using lemmings purchased from the Inuit people of the northern regions and shipped south to a studio where they were forced to run on a turntable covered with snow and then herded over a cliff into the Bow River near Calgary (Cruel Camera, 1982; Woodford, 2003). The overt behaviors shown on the film were real enough, but the context was far from natural for those hapless creatures.

Unobtrusive Observations

Unobtrusive observations of humans who do not realize they are being assessed by scientists can run afoul of privacy rules and ethical standards for research, and they are rarely reported in the published literature that is cited in this book. It is important that someone who is part of a scientific study of behavior be informed of the nature of the procedures and agree to participate. The measured behavior is then a response not only to items on a test but also to a formal test situation that includes the involvement of a researcher. People are likely to behave and answer questions somewhat differently when they know they are being watched and judged. Measurement itself can change behavior.

Physical Measures

Physical measures can be applied to many kinds of behaviors, especially those assessed in a laboratory full of instruments. Some tests can be scored by the time in seconds required to make a choice between two options, the accuracy in millimeters of a person touching a spot on a computer screen, the primary frequency of the voice in hertz, the percentage of fats in the diet when someone is allowed to choose options in a cafeteria, and the distance traveled by one foot when pacing the floor in a waiting room. The measures themselves can be well defined in this way, but their meanings depend on psychological features of the test such as subtle aspects of the stimuli or variations in the instructions. The measures can be well expressed in metric system units, but what the instruments actually tell us about the mind or nervous system needs to be interpreted with due regard for the context of the measurement.

Self-report

Self-report is used in many kinds of rating scales, such as the facial grimace scale of children's pain (Chapter 1). For some kinds of behavioral processes, such as detection of odors, self-report is the only available source of knowledge about what another person perceives. Whether a unique chemical odor can be detected at all is well expressed by a simple Yes-No choice or a preference test.

R D S H K
O Z C N V
H O V N C
D Z H K R
D Z C N S
O V K H R
C H V R N

FIG. 6.1 Simplified version of the ETDRS visual acuity test. Raw score is the total number of letters the person can read correctly, beginning at the top left. This is then converted into an index of visual acuity using tables in a manual.

The threshold for detecting an odor can then be found by exposing a person to successively lower concentrations until a dilution is presented where the Yes-No or bottle A vs bottle B choice does not exceed chance. Whether someone finds the odor pleasant or noxious can be assessed with another rating scale from +3 (very pleasant), to 0 (neither good nor bad), to −3 (noxious and offensive).

A self-report test familiar to almost all of us is visual acuity using rows of letters (Fig. 6.1). The person reads aloud the letters shown on a printed chart or a screen at a standard distance from the eye and continues until the print is too small to see. This form of self-report compares the person's response with a known standard. If the response is correct, the person must have seen the letter. Near the visual threshold, the person expresses doubts or may incorrectly name a letter that is quite similar to the one on the screen, and only a fraction of the five letters in a row will be named correctly. The ETDRS chart was devised by the Early Treatment Diabetic Retinopathy Study group to measure the extent of visual impairment in diabetics and has become widely used in clinical practice (Bailey & Lovie-Kitchin, 2013). Each row has five letters of the same height. Knowing that height and the distance of the person from the screen, the angle subtended by the strokes in the letter can be computed to determine the minimal angle of resolution for the smallest detail that can be seen. Clinicians then take the logarithm of this value and report acuity as the log of minimal angle of resolution (LogMAR). For this test, the *raw score* is simply the number of letters the person can see, and the LogMAR is the *transformed score* that indicates visual acuity but makes sense only to an expert.

The thoughtful design of the ETDRS test corrects for several kinds of biases and sources of error. Because it is often used to monitor deterioration in vision arising from diabetes, the person is likely to be tested more than once over a period of months. If the same chart were used each time, the person might recall some of the letters from a previous test. For this reason, there are several forms of the test with different letters in the various rows. When any one letter is presented in different rows, it is always in a different order in nearby rows. The letters are chosen so as to be equally easy to identify, which excludes those that have just one stroke (I), are markedly asymmetrical (F, T, and Y), or have more than two horizontal or vertical strokes (B, E, and W).

A good behavioral test rates the behavior but does not attempt to diagnose the cause of any deficit. The test score is a phenotype, not an explanation. Different people can suffer similar symptoms for different reasons. This is clearly true for the ETDRS vision test. Someone might exhibit a loss of visual acuity because of gradually worsening myopia that can be corrected with new eye glasses, a worsening cataract that might require surgical replacement of the lens, chronically high blood sugar that calls for the treatment of the diabetes, or a loss of myelin from multiple sclerosis.

Self-ratings for Preliminary Screening

Self-ratings for preliminary screening commonly allow a wide range of scores, whereas clinical judgments by psychotherapists often categorize or dichotomize. The Beck Depression Inventory (Table 6.1) provides a well-known example that highlights the two approaches. The original test (Beck, Ward, Mendelson, Mock, & Erbaugh, 1961) presented 21 items, each of which can be rated from 0 (no problem) to 3 (serious symptoms). The total score can range from a placid 0 to severely disturbed at 63 points. The current version (BDI-2) is structured the same way, but some items have been revised to use more current terms (Beck et al., 1996). Psychiatrists sometimes regard anyone in the general population with a score less than 21 as more or less normal and those with 21 or more as depressed (Polgar, 2003). The cutoff at a score of 21 is arbitrary and not universally embraced by professionals. The online test notes that it is often self-administered but should primarily be administered by an experienced clinician. The test itself cannot distinguish between someone suffering from a long-term negative self-evaluation and another who has recently experienced a grievous loss of a loved one. A case history explored in depth during a clinical interview is essential before any formal diagnosis is made.

A few studies have been done with the BDI given to a representative sample of an entire population. Lasa, Ayuso-Mateos, Vazquez-Barquero, Diez-Manrique, and Dowrick (2000) obtained BDI ratings for 1250 people living in the city of Santander, Spain, and then, those with higher BDI scores were interviewed by a psychiatrist to assess depression. About one-third of the sample reported a score of 0, 80% scored less than 5, and almost everyone scored less than 12. Even in samples from a UK population with a much higher frequency of mild depressive symptoms (Veerman, Dowrick, Ayuso-Mateos, Dunn, & Barendregt, 2009), the large majority of people scored less than 10 (Fig. 6.2). Because the BDI yields such

TABLE 6.1 Item Names in Beck Depression Inventory-II[a]

1. Sadness
2. Pessimism
3. Past failure
4. Loss of pleasure
5. Guilty feelings
6. Punishment feelings
7. Self-dislike
8. Self-criticalness
9. Suicidal thoughts or wishes
10. Crying
11. Agitation
12. Loss of interest
13. Indecisiveness
14. Worthlessness
15. Loss of energy
16. Changes in sleeping pattern
17. Irritability
18. Changes in appetite
19. Concentration difficulty
20. Tiredness or fatigue
21. Loss of interest in sex

[a] *Concise titles for each item are taken from Table 1.1 in Beck, Steer, and Brown (1996), whereas the actual test presents whole sentences describing feelings. Each item is rated from 0 (no problem) to 3 (most depressed). A brief guide to the interpretation of scores is provided by Polgar (2003), and extensive experience with its use is described by Beck, Steer, and Carbin (1988).*

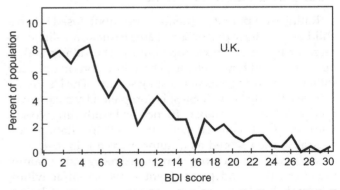

FIG. 6.2 Percentage of 239 women in a sample from the United Kingdom showing specific Beck Depression Inventory scores. The scores exist on a continuum, and there is no sharp boundary between those not at all depressed and those who might be rated as clinically depressed. *Adapted from Veerman et al., Population prevalence of depression and mean beck depression inventory score. The British Journal of Psychiatry, 195, 516–519, Copyright 2009, reproduced with permission from Cambridge University Press.*

a narrow range of scores at the low end of the scale for a nonclinical population, it is not useful for measuring commonplace and subtle variations in mood in the general population. It can be useful, however, as an inexpensive and quick-screening device to detect people who may need professional help.

The cutoff score that best divides normal from nonnormal mood ratings is not so obvious. A large review of dozens of studies of the BDI in countries around the world found a wide range of cutoff scores applied in different situations (Wang & Gorenstein, 2013). The challenge arises from the fact that mood exists on a continuum and fluctuates over days, weeks, and months. This was apparent in a study of 137 students who sought help from a university counseling center in the United States (Sprinkle et al., 2002). Each person completed the BDI rating form and was then evaluated for depression according to clinical diagnostic criteria. There were 87 students who qualified as clinically depressed by psychiatric criteria. If the cutoff was set at 9, all 87 had a BDI score of 9 or higher and were classified as "true positives." The problem, however, is that 28 of the 50 nondepressed students also scored 9 or higher on the BDI and constituted "false positives." Such a high false-positive rate shows that a cutoff score of 9 is much too low. Every step up for the cutoff score led to not only fewer false positives but also more false negatives in which genuine clinical depression was not detected by the BDI score. A cutoff of 17 BDI units resulted in a false-positive rate of 18% at the cost of 18% real cases of clinical depression that would not be detected by the BDI. This suggests that a diagnosis of clinically significant depression should never be based on the BDI alone.

Rating by a Parent or Friend

Rating by a parent or friend is commonly used when a child or an elderly person is not able to make a valid judgment of his or her own capabilities. The ratings reflect opinions about how well the person functions in daily life rather than during a formal testing session. The Dementia Severity Rating Scale (DSRS) is completed by a caregiver who is a family member, friend, or health-care worker with frequent contact with the person. Ratings on 12 scales, each scored as a number from 0 to 3, 4, 5, or 6, are made for memory, language, recognizing family members, time and place orientations, social activities, personal cleanliness, eating, and other features (Clark & Ewbank, 1996; Moelter et al., 2012). Points are summed over the 12 scales, and the total score is used to classify a person as showing mild, moderate, or severe dementia. Within each scale, the descriptors are arranged in order of severity, but there is no assurance that the difference between, for example, a normal rating of 0 and the least

severe rating of "sometimes confused" (score 1) is worth the same amount, 1 unit, as the difference between "usually confused" (score 3) and "almost always confused" (score 4). Questions of what the individual ratings mean are submerged by summing the scores over the several scales. The principal concern is what use will be made of the rating of dementia. Will it serve to decide who should be living in a nursing home vs their own home? This is a matter of great importance, but the quality of the psychological questionnaire is just one factor in a complex decision that involves available family supports and wealth.

Rating by a Trained Observer

Rating by a trained observer is often employed in judging dementia and behavioral disorders. Some of the available tests have undergone extensive refinement and improvement over the years and are now available in standard formats that are intended for use by those with formal training in giving and interpreting the specific test. Version 2 of the Dementia Rating Scale (DRS-2), for example, consists of 32 stimulus cards, standard scoring booklets, and a manual intended for professionals in psychological testing. Skill of the rater plays a much more important role in these kinds of tests than in the ETDRS test of visual acuity or heart-rate measures of fitness. To a large extent, this difficulty arises from the multifaceted and often ambiguous nature of human cognition. For dementia, we cannot rely on a self-rated score of mental function because one common feature of midstage dementia is denial that there is a problem. The major advantage of ratings on a standard scale by trained observers is that people receiving similar scores should indeed show similar degrees of impairment, something less likely for ratings by family members who may have widely differing amounts of education and experience with rating scales and strong personal interests in the portrayal of their relative. The tests do not reveal the sources of the dementia, but they can be valuable for evaluating the rate of decline in competence.

Multiple-Choice Items

Multiple-choice items limit the possibilities and usually have one choice that is correct. The Raven's Progressive Matrices test asks the person to choose which of six options best fits into the blank patch. The test booklet presents 36 such puzzles, and the score is the number out of 36 that are answered correctly. The test is designed for use with children who do not know how to read, although it can be applied with any child. The Army Alpha exam (Bregman, 1925) was used to select soldiers in the US Army for training as officers. Five of the eight subtests involved true-false or multiple-choice items that limited the options. The raw score on the test was the total number of items answered correctly in a limited period of time, much like a school exam. The Wechsler-Bellevue test of adult intelligence presented pictures where something was missing or wrong and asked the person to identify it; only one answer was correct in a sea of wrong answers. These different kinds of intelligence tests are discussed at length in Chapter 15.

Diagnosis by a Professional

Diagnosis by a professional is generally done on the basis of a clinical interview with the person and review of the information in the person's medical file. Symptoms and case history are compared with standard criteria for assigning a person to a diagnostic category such as depression, alcoholism, or autism. Two sets of criteria are in general use: the Diagnostic and Statistical Manual (DSM) by the American Psychiatric Association (2013) and the International Classification of Diseases (ICD) by the World Health Organization. These systems employ *differential diagnosis* in which a person must fit the description for one disorder and not fit all of the criteria for a similar disorder. Because there are usually several criteria that must be met to proclaim a specific disorder, borderline cases are possible, and clinicians may not always agree on which category best fits a particular person.

BEHAVIORS VS PSYCHOLOGICAL CONSTRUCTS

Psychology has many concepts and tests, but measurement remains a challenge. Just what is measured by a psychological test? Although most tests in common use rate the overt behaviors of a person, the details of the behavior such as checking a box or saying "true" are themselves trivial. The interesting thing is the totality of responses to questions. The goal of many tests is to assess or measure some psychological construct that depends on patterns of mental activity that cannot be observed directly. Intelligence, anxiety, and memory are elusive constructs that are not easily defined in a way that is universally accepted among professional psychologists. Consequently, there are several tests of ostensibly the same construct in common use. If the observed scores on different tests of the same construct are highly correlated with each other, this is evidence that they are indicators of the same construct. But which of them is correct or the most correct? The subject is considered at greater depth in texts on psychological testing.

Physical sciences have long-established definitions and standards for key features such as mass, length, time, and temperature. A standard for a mass of 1 kg, the international prototype kilogram or *Le Grand K*, was adopted by physicists in 1889 and is embodied in a standard cylinder of platinum that is stored securely at the International Bureau of Weights and Measures in Sèvres, France. When *Le Grand K* was removed from its airtight storage vault in 1946, it was found to have lost $30\,\mu g$ of mass, and at its next official test in 1991, it was down by $50\,\mu g$. Fifty millionths of a gram or 50 billionths of a kilogram after 100 years is not much, but a good standard should not change at all. Physicists and engineers take their standards very seriously, and they are now working to redefine the kilogram so that it will not shrink (Folger, 2017).

A key challenge for psychology is defining its constructs in a manner that is widely accepted by professionals and clearly specified so that tests can be designed to measure them. The precision of a test can be increased by using more items and evaluating it with more people, but a good standard for the accuracy of many measurements remains elusive in psychology. Constructs such as intelligence and anxiety can change in an entire population over a period of years. Behavior arises from dynamic processes where nothing seems to be truly fixed by nature. Psychologists long ago decided to sidestep this difficulty by converting many of their measures to scales that rate people relative to each other. The hallmark of psychological measurement is a focus on differences among people (Box 6.1).

BOX 6.1

Accuracy and precision are important features of all tests. Accuracy denotes how close a measure of behavior is to a known standard. This is an important issue for a Global Positioning System (GPS) watch, because some models in some situations consistently overestimate the total distance traveled along a smooth path. The measure of elevation gain is often shown as substantial when someone is running on a perfectly flat track. Precision denotes the number of meaningful digits or decimal places in the measure. For the LogMAR chart with five letters per row (Fig. 6.1), the difference between rows is 0.1 LogMAR unit, and there are five steps in a row, so precision, the smallest step between two raw scores, is 0.1/5 or 0.02 LogMAR unit. Generally speaking, the more items there are on a test, the greater will be its precision and the better its capacity to make fine discriminations between different people.

GOALS OF TESTING AND APPLICATIONS OF TEST SCORES

Psychiatrists, being medically trained and licensed, are looking for signs of an abnormality of thought or behavior—a disorder that warrants treatment. Measures of blood sugar or blood pressure can range widely without cause for alarm because there is a well-defined normal range of variation among people who are not ill. Many psychological tests also yield results that are considered to be in a normal range. Psychiatrists are interested in extreme manifestations of thought and behavior that cause difficulties in daily life, and they strive to dichotomize people into those with a bona fide disorder and those who are normal. A patient usually presents herself to someone with an MD degree because there is some kind of trouble such as insomnia or infertility. So, the first task of the physician is usually to establish a medical diagnosis, and the second phase is to begin a quest for a treatment or cure. If it happens that the patient's symptoms are actually quite common among apparently normal people, there may be no need to prescribe treatment at all. Almost everyone experiences a sleepless night now and then. Many women miss one menstrual period from time to time. A physician offering reassurance can be comforting.

Psychologists, on the other hand, are keenly interested in differences among people within the normal range of variation or perhaps the average test scores of large groups of people. They frequently measure and then debate the meaning of cognitive gender differences, for example. They strive to discern a role for heredity in differences among people who would all be judged mentally normal by a psychiatrist. They search for evidence concerning what age is best for starting school or whether experiences of a fetus before birth can influence later behavior. Consequently, psychologists have devised an elaborate and sophisticated technology for designing, giving, and interpreting behavioral and mental tests. The results of many tests are often reduced to a concise number such as the IQ score that tells an expert how well the person scored relative to others of the same age. Numerical scores have an advantage in that they can reveal markedly excellent performances just as well as deficits. They need not get mired in debate about just how good or bad someone must be to qualify as "atypical."

STANDARDIZED PSYCHOLOGICAL TESTS

How to devise and score a psychological test is well established (Kaplan & Saccuzzo, 2017) and has been used to formulate many of the tests in common use today. There are four major steps in this process:

Choice of Test Items

Choice of test items is perhaps the most challenging aspect. In order for a test to have high *validity*, all of the items must indeed reflect the psychological construct that the test purports to measure. If it is a test of anxiety, for example, people known to have high anxiety on the basis of clinical interviews should have high scores on a test item, whereas those with low levels of anxiety should have low scores. Because anxiety is a complex trait with many gradations in severity, it cannot be well represented by just one or two items. Instead, a variety of test items are devised to assess different aspects of anxiety. Each item on the test should correlate with a clinical indicator of anxiety and with other test items too.

Test methodology was originally devised to construct intelligence tests. Intelligence (Chapter 15) is regarded as a general kind of mental ability, something that is not limited to a specific skill such as memory or numerical computation. The test is usually divided into several subtests that tap different components of intelligence, and then, an average of scores on all subtests is calculated. Several brands of intelligence tests are commonly used in clinical practice, education, and personnel selection. Some of them underwent a long period of preliminary assessment of potential test items, and researchers explored different ways to administer the test, whereas others were pressed into service urgently, such as the Army Alpha test used in the United States during World War I to select men for officer training. Although test items vary widely among the different intelligence tests, calculations are done in essentially the same way for many tests to convert results into the familiar IQ score.

Standardization

Standardization of a psychological test is done with respect to a specific population of people, perhaps the population of an entire country. A *population* consists of everyone in the defined group, whereas a *sample* is a smaller fraction of people from the population. Measuring an entire population is done in a national census that strives to collect data on everyone. A census generally does not include items to measure psychological constructs. Instead, a psychological test is given to a sample of the population that ideally will have characteristics similar to the entire population. To achieve this, care is taken to avoid bias in choosing people to be in the sample. In an *unbiased sample*, every person in the population has the same chance of being chosen to provide data. One way to do this is to choose the sample randomly by using random numbers generated by a computer. Random choice avoids bias, but unless the sample is very large, it could result in a sample that has a higher proportion of males or well-educated people, for example, than the population. A good solution to this problem is to divide the population into a number of subgroups or *strata* differing by gender, age, education, and geographic region (urban, rural, state or province, etc.). Then, people within strata are chosen randomly. The result, known as a *stratified random sample* or a *representative random sample*, avoids many kinds of bias and usually provides a good estimate of qualities of the entire population. This method is used for the better psychological tests that are intended to identify people who have noteworthy mental problems or disabilities and those who should receive special privileges such as admission to a college or university.

Statistical Transformation

Statistical transformation is the next step when the investigator wants to *convert the raw test score to a standard score*. It will be familiar to most students studying psychology in a university, but the details will be obscure and arcane to almost everyone else. Some compare it with the process of making sausage. The end product may be quite pleasing to many, but the experience may be less so if one knows how that sausage is made. The goal of this phase is to estimate the average raw score of all people in the sample who take the test and then find a convenient indicator of how much people differ from the average score (Fig. 6.3). This chore is usually done with a computer. It is common practice in psychological statistics to represent the raw score of a person by X. This has nothing to do with the X chromosome. It is adapted from algebra wherein an unknown quantity is often represented by X. Then, the average or *mean score* (M) of the number of people who took the test (N) is found by summing all X scores and dividing by N. Next, the deviation of each person's raw score X from the mean is determined. Some deviations will be positive and some negative. If the deviations are simply added together, their sum will be 0. A more useful statistic is obtained by squaring each deviation and then finding the average squared deviation of X

from M, a quantity known as the *variance*. Variance will not illuminate things for most of us because of the squaring, but taking the square root of variance gets us to something quite useful: the *standard deviation* (S) of X. For a distribution of test scores that is approximately bell-shaped, most scores occur within two standard deviations of the mean (M). The Z score is the number of standard deviations that a person's X score is from the mean (M). $Z = 0$ indicates the score was right at the group mean, whereas $Z = 2$ is greater than almost all other people taking the test. When the distribution of scores is bell-shaped, the percentile rank can be determined from the Z score using special statistical tables or a computer program. *Percentile rank* is the percent of people who score at or below that X or Z score. For $Z = 2$, percentile rank is 98%, and only 2% of all people will have a higher score.

Generate the IQ Score

To generate the IQ score from the Z score is the last step and is easy. One must simply decide what average score and standard deviation are desired for the transformed score that we agree to call *IQ, the intelligence quotient*. Yes, the values of IQ are entirely arbitrary. They tell nothing about intelligence itself. Instead, they tell how someone scored on an IQ test relative to others who took the same test. Early in the 20th century, it was decided that the average IQ score should be 100 and the standard deviation of IQ should be 15 or 16, regardless of how many items are on the test. Given the Z score, it is easy to obtain IQ with a simple formula: $IQ = 100 + 15Z$. The result is a score that is familiar to many people: Average IQ is about 100, very few people score above 130, and few are below 70. On the surface of things, it really does not matter just what items are on the IQ test; the average IQ will come close to 100 with the aid of the little formula. It really is convenient and compact. But the *scale transformation* from raw score to IQ score effectively conceals just what it is that the test measures in terms of real psychological processes.

Example—Hypothetical

Specific examples should help to understand how the process of data transformation works. Fig. 6.4 presents hypothetical intelligence test scores for 20 people in the form of a dot plot where the score of each person is shown by a black dot. If two people received the same test score, one dot is plotted directly over another. The maximum possible score on the test is 80, and the minimum is 0, but nobody scored below 30 in the example. For those 20 scores, the arithmetic average or *mean* is 50.6. The *mode*, the most common score, is 52, and the *median*, the score with half of scores above and below it, is 51. The mean

a. X = raw score of one person

b. N = # of people taking the test

c. sum(X) or ΣX = total of all scores

d. mean score $M = \dfrac{\text{sum}(X)}{N}$

e. deviation score of one person $d = X - M$

f. squared deviation score $d^2 = (X - M)^2$

g. sum of squared deviations $SS = \Sigma (X - M)^2$

h. variance = average squared deviation = SS/N

i. standard deviation $S = \sqrt{\text{variance}}$

j. standard score $Z = (X - M)/S$

k. $IQ = 100 + 15(Z)$

FIG. 6.3 Steps in calculating an IQ score from the raw score X on a psychological test that evaluates intelligence.

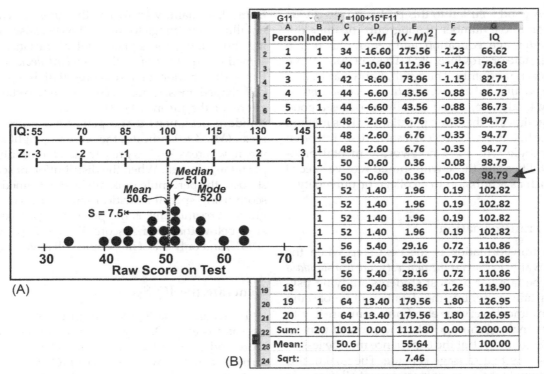

G11		f_x = 100+15*F11				
A	B	C	D	E	F	G
Person	Index	X	X-M	$(X-M)^2$	z	IQ
1	1	34	-16.60	275.56	-2.23	66.62
2	1	40	-10.60	112.36	-1.42	78.68
3	1	42	-8.60	73.96	-1.15	82.71
4	1	44	-6.60	43.56	-0.88	86.73
5	1	44	-6.60	43.56	-0.88	86.73
6	1	48	-2.60	6.76	-0.35	94.77
	1	48	-2.60	6.76	-0.35	94.77
	1	48	-2.60	6.76	-0.35	94.77
	1	50	-0.60	0.36	-0.08	98.79
	1	50	-0.60	0.36	-0.08	98.79
	1	52	1.40	1.96	0.19	102.82
	1	52	1.40	1.96	0.19	102.82
	1	52	1.40	1.96	0.19	102.82
	1	52	1.40	1.96	0.19	102.82
	1	56	5.40	29.16	0.72	110.86
	1	56	5.40	29.16	0.72	110.86
	1	56	5.40	29.16	0.72	110.86
18	1	60	9.40	88.36	1.26	118.90
19	1	64	13.40	179.56	1.80	126.95
20	1	64	13.40	179.56	1.80	126.95
Sum:	20	1012	0.00	1112.80	0.00	2000.00
Mean:		50.6		55.64		100.00
Sqrt:				7.46		

(B)

FIG. 6.4 (A) Hypothetical example of 20 test raw scores that are then transformed into Z and IQ scores. (B) Portion of a spreadsheet for doing the computations. The highlighted value in cell G11 is calculated using the formula in the box at the top indicated as $f_x = 100 + 15 \times F11$, where F11 is the Z score for the observation in cell C11.

is used in computing IQ. The spreadsheet shows the data and calculations. Columns are represented by letters A–G and rows by numbers 1–24. The highlighted IQ score of 98.08 is in cell G11. A formula for cell G11 is entered into the box termed f_x and is the basis for the computation. Sums of scores in a column are in row 22. When the index is set at 1 for each person, the sum of the 1s is simply the number of people, a quantity needed in other calculations, such as finding the mean by dividing total for X by N and finding variance by dividing SS = 1112.8 by N. The standard deviation S = 7.46, the square root of variance, is used to calculate Z, which is used to find IQ. The transformed scales for Z and IQ are shown above the dots. Converting the raw score into IQ does not change the dot plot, just the measurement scale. Who scores best and worst is not affected by conversion to IQ.

CHANGING STANDARDS

Certain tests of mental ability have used the same test items for many years and thus can readily detect a general improvement in performance on the same test in the whole population. This has happened with the Raven's Progressive Matrices test. The same stimulus cards were in use since the test was originally published in 1936. When the test was administered to a large sample

of boys and girls in Alberta, Canada, the average number of items answered correctly had risen substantially by 1977 compared with the same test given in 1957, and far more students achieved perfect or nearly perfect scores than had been the case 20 years earlier (Clarke, Nyberg, & Worth, 1978). Substantial changes in test scores were noted in the 1938 vs 1979 standardization results in the United Kingdom, and improvements in performance of more recent samples have been observed in several other countries (Raven, 2000). A recent review of several hundred published studies from population samples of 45 countries indicated that, on average, the gradual increase in performance on the Raven's test was the equivalent of about 2 IQ points per decade or 10 IQ points in 50 years (Brouwers, Van de Vijver, & Van Hemert, 2009).

For mental tests that report the score as IQ, calibrating a test against a standardization sample yields a standard that is not fixed or constant; it estimates the abilities of a population of people in a specific year. Norms for the test are published in a manual that tells how to convert raw scores to IQ scores based on the original standardization, and the same manual is used for many years until a new version of the test is produced. If the psychological entity that the test measures increases in the general population, average raw scores and IQ test scores will rise too, and if this continues for several years, the mean IQ score may rise to a value well above the original standardization sample. A large amount of data has shown that this has indeed happened (Flynn, 1984, 1987). It therefore is important to restandardize a test every few years so that the mean IQ score is restored to 100. In order to insure that the new version of the test given to a new sample is measuring the same thing as the older test, makers of the major IQ tests also give the older version of the test to the new sample. This has been done several times from 1947 to 1978. For example, the Stanford-Binet test was standardized in 1932 when mean IQ was set at 100, and then, it was revised in 1971. People who took the 1932 version of the test in 1971 scored almost 10 IQ points higher than people of the same age who took it in 1932. For the Wechsler Intelligence Scale for Children (WISC), the standardization sample of people who took the new test in 1947 also took the old Stanford-Binet and scored 6.6 IQ points higher than the 1932 standardization sample, and in 1972, when a new version of the WISC was standardized, the children scored 6.8 IQ points higher than did the 1947 sample of children. Flynn (1987) estimated that the abilities measured by several major IQ tests had been increasing at about 3 IQ points per decade for 50 years.

The SAT

The **SAT test** is a rite of passage for many high-school graduates in the United States and Canada who aspire to higher education; it is sometimes used in admission decisions. The test was originally named the Scholastic Aptitude Test, then the Scholastic Assessment Test, and now simply the SAT. A scale transformation converts the raw number of items correct into the SAT score so that the mean is about 500, the standard deviation is about 100, and the range of scores on the reading/writing and math tests is from 200 to 800. A formula to do this is simply $SAT = 500 + 100Z$, wherein the standard score (Z) is computed from the raw test scores. Prior to 1941–42, the scaled score was computed afresh using data for those who took the test that year, which made it impossible to detect changes occurring over several years. Beginning in 1941–42, scores in future years were referenced to the 10,600 students who took the test in 1941–42 so that a student with a particular raw score in 1957, for example, would get the same scaled score as a student with the same raw score in 1942. One result was a gradual decline in average SAT scores over five decades. This did not necessarily mean that students in recent years were not as smart as those taking it decades earlier or that schools were getting worse. During the period after WWII, there was a substantial growth in the proportion of the population graduating from high school and continuing on to college or university after high school, such that more students with somewhat lower high-school

performances continued with their educations. Shortly before the test was revised in 1995, average SAT verbal scale scores had declined to 428 and math scores to 482 (National Center for Education Statistics, 1997, Table 129; Winerip, 1994), so in 1994, the norms were adjusted to bring the mean scores back to 500.

Some organizations exclusively for people with very high IQs have in the past allowed the use of the SAT score as a proxy for IQ. But the SAT is not an IQ test of the kind exemplified by the WAIS or Stanford-Binet (see Chapter 15). Only a fraction of young people in many states in the United States seek admission to a college or university and take the SAT. Merely because the scale transformation $SAT = 500 + 100Z$ looks so much like $IQ = 100 + 15Z$ does not make the SAT an IQ test. The test items are designed to challenge students who are just completing high school and seeking higher education.

Educational psychologists became increasingly critical of the misuse of the SAT as an aptitude test, and there were many disputes about whether the test was "really" an *aptitude test* that measured the ability to learn or instead an *achievement test* that measured the amount that has been learned. Anastasi (1984) pointed out that achievement tests clearly are sensitive to what has been learned, whereas in the early days of psychological testing, aptitude was held to be something innate and unlearned. Thus, the aptitude/achievement distinction was rooted in a simplistic nature-nurture dichotomy (discussed in Chapter 14). Large volumes of research on tests revealed that all so-called aptitude tests reflect learned knowledge. In 1988, the College Entrance Examination Board, author of the SAT, appointed an expert committee to consider further changes in the test. In its report, the committee recommended changing the name of the test to Scholastic Assessment Test (Bok et al., 1990). Several other recommended changes were incorporated into the 1994–95 version of the new SAT. American Mensa, an organization for people with high IQ scores, lists more than 200 psychological tests (see "qualifying test scores" web page) that can support an application for membership and notes that after January 31, 1994, it no longer accepted results of the SAT.

It had been recognized for some time that coaching could raise SAT scores, especially on the math test (Dear, 1958), and that IQ tests themselves yield higher scores when the person has taken the same test twice or been coached on how to do the test (Vernon, 1954). A more recent survey of mental and academic ability test results showed substantial effects of repeated testing and coaching (Hausknecht, Halpert, Di Paolo, & Moriarty Gerrard, 2007). A thriving coterie of advice and tutoring services to enhance SAT scores has found many customers via the Internet.

The College Board made unequivocal statements about the purpose and design of its latest version of the SAT

(College Board, 2015). After extensive discussions with students, parents, college admission officers, and teachers of kindergarten through grade 12 school classes, the Board stated that the SAT measures "college and career readiness" and "student achievement." The revised test is designed to be "better and more clearly aligned to best practices in classroom instruction." Some examples provided by the Board emphasize the close connection of test items with material encountered in the classroom (College Board, 2015). Most high-school dropouts would be totally baffled by those highbrow items. Despite the extensive discussion provided by the Board, it is difficult to know just what the test measures because of the way it is scored. The new version includes three component tests (reading, writing, and math) that have 52, 44, and 58 questions, respectively. Each component-test raw score is then scaled to place the transformed scores into a range from 200 to 800, and the scaled scores are provided to the students. Knowing the scaled score, there is no convenient way to know how many test items were answered correctly. The scaled score rates students relative to each other, not relative to a scale of real knowledge. This test is useful for making decisions about admission to a college or university because the performance of students relative to each other is of paramount importance. Most universities receive applications from far more students than they can admit, and they usually accept those who rank most highly on several indicators, including admission test scores such as the SAT that provide a clear basis for rank ordering the students.

TESTS WITH NO STANDARDS

In the Internet age, new psychological tests appear almost daily, sometimes sponsored by therapists or companies as a convenient way to attract new clients. On the surface, they are much like familiar IQ tests or the SAT, in that there are several test items or questions that can be answered by choosing the best answer from a limited set of options. On its face, a test often seems to be relevant to the advertised topic of the test. In the parlance of psychological testing, it has *face validity*.

A Google search of the term "test of anxiety" returns several self-rating scales by different institutes, associations, or even magazines. There is a 17-item test with Yes-No answers, a 20-question test on a 5-point scale (Never to Always), a 22-item test on a 5-point scale (Never to Usually), and a 30-item test on a 4-point scale (None to High). The magazine *Psychology Today* offers a test "to find out if you're too anxious" and suffer from a common "mental illness." After taking the test, a person can receive a free synopsis or a full report for a fee, along with a list of therapists who offer their services to reduce anxiety. Seeking and finding data that show population-based norms for anxiety levels detected by

these tests or correlations between scores on different tests is an exercise in frustration. Not one of them published data on actual test scores. There is no way to determine which of them yields the highest frequency of people judged to be too anxious and needing treatment.

There are published self-rating tests of anxiety that have been in use in medical settings for several years. The states of anxiety-depression (sAD) portion of the Delusions-Symptoms-States Inventory (DSSI) provides seven items to rate anxiety, each evaluated on a 4-point scale from 0 to 3, such that the highest possible score is 21 (Bedford & Deary, 1997; Bedford, Foulds, & Sheffield, 1976; Christensen et al., 1999). The Hospital Anxiety and Depression Scale (HADS) also has seven items pertinent to anxiety that requires only 2–5 min to complete. Each item is scored on a 4-point scale from 0 to 3 with a maximum possible score of 21 (Snaith, 2003; Stern, 2014). In the initial DSSI, 96 psychiatric patients were interviewed by psychiatrists and classified according to the strength of evidence of anxiety and depression using the categories 0 = "none," 1 = "a little," 2 = "a lot," and 3 = "outstanding." DSSI mean scores were indeed higher in those rated in categories 2 and 3 by a psychiatrist. The DSSI scale administered to nonpatients revealed that 90% of them scored 0, 1, or 2 and none was above 8. Authors of the scale considered score ranges from 8 to 10 to be indicative of mild anxiety, from 11 to 14 moderate, and from 15 to 21 severe anxiety. In a similar study comparing HADS scores with psychiatric judgments (Zigmond & Snaith, 1983), cases seen as definite cases of anxiety by a psychiatrist almost all had high HADS scores, and those judged to be noncases mostly had low HADS scores, but there were some remarkable discrepancies.

The frequency of clinically significant anxiety in a country can be assessed with a representative sample of that population. Large samples have been tested in several countries using structured, brief clinical interviews. This approach assesses *current prevalence* of anxiety and other mental problems, the percentage of people in the population who currently express symptoms of a mental disorder in an interview. When the interview also explores each person's history of symptoms and possible treatments for a mental disorder, it is possible to assess *lifetime prevalence*, the frequency of showing a disorder at some time in a person's adult life. If the duration of symptoms is relatively short, current prevalence will be much lower than lifetime, whereas for a disorder that tends to persist for years after symptoms first appear, current and lifetime frequencies will be similar. An extensive review found that rates differed substantially in different countries (Martin, 2003). Current prevalence ranged from 1.2% in the United States to 2.8% in Italy, whereas lifetime prevalence ranged from about 5% in the United States and Italy, to 9% in France and Hong Kong, to 22% in Iceland.

CLINICAL INTERVIEW AND DIAGNOSIS

Formal diagnosis of a mental disorder is done by a professional clinical psychologist or psychiatrist who interviews the person and considers the case history. Diagnosis assigns the person to a category in an official scheme approved by a professional organization or a government. There are many possibilities for the decision-maker, and ambiguities abound. Attempts have been made to formulate clear and concise diagnostic criteria, so that different professionals will arrive at the same conclusion about a particular person (concordance), but universal agreement remains an elusive quality. Two major classifications are in common use today—the Diagnostic and Statistical Manual (DSM) of the American Psychiatric Association (APA) and the International Classification of Diseases (ICD) sponsored by the World Health Organization (WHO). The first version of DSM was published in 1952, and it has seen four major revisions, leading to the current DSM-5. Version 6 of the ICD was adopted in 1948 when the WHO became part of the United Nations in the post-WWII period. It has been extensively revised four times to create ICD-10 and will soon give rise to ICD-11 in 2018.

In his review of the origins of DSM-1, Grob (1991) observed that classification systems "emerge from the crucible of human experience" and "reflect specific historical circumstances." When the US Bureau of the Census in 1908 asked the Medico-Psychological Association, forerunner of the APA, to advise on categories for tabulating patients in mental hospitals, a committee was struck and 9 years later reported that the situation was "chaotic" and there were no uniform standards across different states or countries. Two world wars and the need to classify the many mental disorders of soldiers and veterans eventually gave rise to the first version of the DSM. At about the same time, the ICD listed many categories but offered little guidance on how to assign a case to a specific disorder, and it altogether lacked several categories that were in common use in the United States (Gaines, 1992). The APA worked closely with those crafting a new version of the ICD (ICD-8) but opted for its own criteria in DSM-2 in 1968 because clinicians in many other countries had not observed some of the American classes and other things listed in ICD-8 were unknown in the United States. DSM-3 marked a major turning point in which the APA adopted a disease model of mental illness that embraced biological explanations and new kinds of drug therapy while downplaying cultural factors. For the first time, the DSM proposed definite durations and frequencies of symptoms that warranted a specific diagnosis (Shorter, 2015). Gaines (1992) noted that a classification system "is constantly coming into being, a perpetually unfinished system." This assertion is amply confirmed by the latest work to revise both DSM and ICD yet again.

Comparing Systems

Andrews, Slade, and Peters (1999) compared diagnoses of more than 1300 people from a representative sample in Australia according to DSM-4 and ICD-10. They observed different criteria for the same diagnosis in the two systems "in almost every category" but considered many of them trivial. Table 6.2 summarizes concordance rates for several categories. The reasons for the discrepancies were apparent in several instances. Under ICD, a diagnosis of obsessive-compulsive disorder requires that the person attempt to resist *both* the obsessions and the compulsions, whereas DSM requires resistance only to the obsessions. For social phobia, DSM requires fear of social situations *and* feelings of humiliation *and* avoidance, whereas ICD requires fear of social situations *or* humiliation *or* avoidance. The authors protested the "unnecessary dissonance" between systems and called for the adoption of identical wording for many definitions.

Another group of investigators evaluated 109 stroke patients for the presence of vascular dementia resulting from the stroke according to five different diagnostic systems (Pohjasvaara, Mantyla, Ylikoski, Kaste, & Erkinjuntti, 2000). In Table 6.3, the obligatory symptoms required for the diagnosis are shown as "Yes," whereas those not needed are "No." Not all criteria are set forth in the table. The frequency of the diagnosis of vascular dementia varied widely. Relaxed criteria under DSM-4 resulted in a major increase in diagnoses compared with DSM-3. These

TABLE 6.2 Percent Concordance According the ICD-10 and DSM-4 Criteria[a]

Disorder name	Frequency in sample (%)		Diagnostic concordance (%)
	ICD-10	DSM-4	
Dysthymia	3.4	3.1	87
Generalized anxiety disorder	9.9	10.1	77
Social phobia	12.9	10.5	66
Obsessive-compulsive disorder	1.9	2.7	64
Panic disorder	6.1	4.1	56
Post-traumatic stress disorder	6.9	3.0	35
Substance harmful use or abuse	8.4	8.6	33

[a] *Sample of more than 1300 people was not only representative but also enriched with people known to have some kind of mental problem. Concordance represents cases where both were given the same diagnosis among those diagnosed with that disorder in either system.*
Adapted from Andrews, G., Slade, T., & Peters, L. (1999). Classification in psychiatry: ICD-10 versus DSM-IV. British Journal of Psychiatry, 174, 3–5, reproduced with permission from Cambridge University Press.

TABLE 6.3 Criteria for Vascular Dementia According to Five Diagnostic Systems[a]

Criterion	DSM-3	ADDTC	ICD-10	NINDS	DSM-4
Stepwise deterioration	Yes	No	No	Yes	No
Focal neurological symptoms	Yes	No	No	No	Yes
Two or more ischemic strokes	No	Yes	No	Yes	No
Temporal relationship between stroke and dementia	No	Yes[b]	No	Yes	No
Frequency among 109 patients	36%	86%	36%	33%	92%

[a] ADDTC is Alzheimer's Disease Diagnostic and Treatment Centers, and NINDS is National Institute of Neurological Disorders and Stroke. Yes means criterion is mandatory, and No means it is not. Excerpts from Pohjasvaara et al. (2000), Tables 1 and 2.
[b] Must be two or more instances for the same person.

and related observations have prompted calls for further revision of methods for judging dementia (Perneczky et al., 2016).

Agreement Between Psychiatrists

Table 6.3 shows how disagreement on diagnosis can arise from the use of different criteria or systems of classification. Another source is evident when the same patient is interviewed on separate occasions by different psychiatrists who are guided by the same set of criteria. *Interobserver agreement* is indicated by the kappa (Greek κ) coefficient, which is close to 1.0 when there is perfect agreement. When two psychiatrists evaluated patients on separate occasions using DSM-5 criteria, kappa was a respectable 0.69 for post-traumatic stress disorder but less than 0.4 for major depressive disorder, borderline personality disorder, and several other categories (Regier et al., 2013). The disagreement could arise from differing emphasis on subtle aspects of the symptomology or genuine changes in the manifestation of symptoms on different interview occasions. A further complication was apparent from *comorbidity*, in which one person can express more than one psychiatric disorder at the same time. Among the 133 patients who met the criteria for post-traumatic stress disorder (PTSD), 32 showed only PTSD, whereas 56 showed PTSD plus one other major psychiatric disorder (major depression, alcohol use disorder, or generalized anxiety disorder), another 34 showed PTSD and 2 other disorders, and 11 unfortunates met the criteria for PTSD and 3 other disorders. In this respect, psychiatric disorders are like many other medical maladies. A person with Type II diabetes might also suffer from liver disease and prostate cancer at the same time. Additional facts would be needed to determine if one disorder was in part a consequence of another. For example, PTSD might be a

major contributing factor to alcohol use disorder. As will be discussed at greater length in Chapters 16 and 17, there are likely to be common underlying factors both genetic and environmental shared by many mental disorders.

The Interview

Another approach to achieving better agreement is to standardize the psychiatric interview itself (Spitzer, Williams, Gibbon, & First, 1992). The MINI (mini-international neuropsychiatric interview) interview is structured in accord with DSM-4 (Sheehan et al., 2006), and CIDI is based on ICD-10. The MINI is set up like a decision tree or branching flowchart that leads to the specific diagnosis only if several criteria are met. For the category termed dysthymia, the person (a) must have felt depressed most of the time for at least 2 years, (b) never felt OK for 2 or more months during that time period, (c) experienced at least two from a list of six symptoms, and (d) experienced significant distress because of them. Lacking any one of the four criteria would exit the dysthymia portion of the interview. The person would be left with his or her problems that are judged not severe enough to warrant a DSM diagnosis. The process can be quite brief when the clinician exits from that portion of the interview as soon as any major criterion is not met.

The dysthymia example illustrates how specific time periods and co-occurrence of symptoms are stipulated, such as 2 years or 2 months and two rather than three symptoms from a list of six. This is done in an effort to enhance agreement among psychiatrists about the occurrence of clinically significant dysthymia that warrants treatment. No evidence is presented to indicate a firm scientific basis for two rather than three symptoms. Implicitly, it acknowledges that mood exists on a continuum, as recognized by the Beck Depression Inventory, and there is no "natural" cutoff on the scale that denotes pathology. There is also no stipulation that a patient who does not meet the minimal criterion for a mental disorder according to DSM or ICD should not be treated by the psychiatrist. The sharp boundaries between categories can be important for research projects that compare data across times and places, whereas the decision of whether and how to treat the patient clearly involves several other factors that are not included in the strictly applied decision tree.

Changing Criteria and Categories

Periodic changes in the criteria reveal the somewhat arbitrary features of those that are to be applied in a particular year. The transition from one version of DSM to the next can challenge a clinician to suddenly abandon the old and implement the new criteria after years of practice on the old. DSM-3 and DSM-4 differ on several criteria, some of which are of major importance, whereas

DSM-4 and DSM-5 appear to be very similar on most criteria, with the large exception occasioned by merging DSM-4 dysthymia with DSM-4 chronic major depression to create DSM-5 persistent depressive disorder. Five separate disorders in DSM-4 were merged into the new autism spectrum disorder in DSM-5 (autistic disorder, Asperger's syndrome, childhood disintegrative disorder, Rett syndrome, and pervasive developmental disorder not otherwise specified). Undoubtedly, there were good reasons for many changes, sufficient to win majority support on the review committees, but not one of the categories represents a finished product. Names and criteria evolve as society and the profession of psychiatry change. Human mental qualities do not express universal constants the way the kilogram does.

DSM Critiques—Internal

Despite many revisions and refinements of the DSM system, the field of mental health shows little evidence of progress over several decades and clearly lags behind most other fields of medical science (Cuthbert & Insel, 2013). It is generally acknowledged that mortality from mental disorders has not declined, nor has the general prevalence of DSM diagnoses. There are no effective clinical tests for mental disorders and no preventive interventions. Serious doubts exist among experts about the clinical validity and utility of many DSM diagnostic categories, doubts that have not been relieved by periodic changes in definitions. In his history of DSM, Shorter (2015) observed that "questions are starting to be asked about whether this massive venture is on the right track."

The DSM is conceptualized as a collection of distinct, nonoverlapping categories of disease with sharp boundaries between illness and normal variation. This approach is challenged by the common observation of comorbidity. Furthermore, genetic studies sometimes find much in common among the more frequent diagnoses (discussed in Chapter 16). There is also abundant evidence that cutoffs for diagnoses lack support from clinical studies (Hyman, 2011). Although many changes have made wording in DSM more precise and should have enhanced agreement among psychiatrists on a diagnosis (reliability), serious problems with validity persist. Consequently, the NIMH has put forward an alternative approach, The Research Diagnostic Criteria (RDoC), that views the mind as having several overlapping dimensions with many gradations and no sharp boundaries (Cuthbert & Insel, 2013). The RDoC proposes that there are five primary domains of mental function, each consisting of three to five key constructs, and it now fosters work to devise new methods for measuring each construct. The announced goal of this project is to devise a biologically based diagnostic system wherein there are demonstrable connections between genetic abnormalities, brain functions, and diagnoses.

This is likely to be a long-term project because at the present time, "we don't have rigorously tested, reproducible, clinically actionable biomarkers for any psychiatric disorder" (Insel, 2014). The promises of individualized and precision medicine in the field of mental health as a result of the explosion of genomic knowledge have not yet been realized. The frontiers of knowledge appear to be as rugged and undisciplined as in 2000 before the human genome was sequenced.

DSM Critiques—External

Many mental health professionals outside the ranks of American psychiatry have expressed misgivings about several changes implemented in DSM-5. The British Psychological Association and the Society for Humanistic Psychology communicated their concerns to the DSM-5 task force (Elkins, 2012; Elkins, Robbins, & Kamens, 2011), and their efforts gave rise to a petition signed by more than 15,000 mental health professionals (Robbins, 2011). Their concerns went beyond issues that were of concern to the NIMH-DSM insiders. Four principal issues were emphasized:

1. The changes tend to lower the diagnostic threshold, which will inflate the prevalence of many disorders. This will lead to "excessive medicalization and stigmatization" of transient distress.
2. The changes emphasize biological causes and encourage overuse of medications.
3. New diagnostic categories are introduced without substantiating scientific evidence.
4. Language in DSM-5 de-emphasizes social and psychological explanations while regarding sociopolitical deviance as a form of neuropathology.

The petition won support from the former chair of the DSM-4 committee and the leader of the effort to formulate DSM-3 (Frances, 2011). The net result of the many criticisms was that a few of the planned additions to DSM-5 that provoked the most outrage were withdrawn, whereas several contentious categories were retained and now have official blessing from the APA. The massive tome, with 947 pages describing more than 300 different diagnoses, is the result of several schools of thought and practice contending for primacy over a period of several decades. Somewhat surprisingly, there are no references to the scientific literature in DSM-5. The diagnostic system is presented to the profession and the lay public on a "take it or leave it" basis. Many will accept the authority of the APA, but doubts persist and seem to be growing.

Steps Toward New Diagnostic Systems

Following the publication of the Open Letter and extensive commentary on issues raised therein, a Global

Summit on Diagnostic Alternatives was convened online from 2013 to 2015 and then at a meeting on August 5 and 6, 2014. The discussions are available as blog posts. Several groups are now working on diagnostic systems to replace the DSM. The *Psychodynamic Diagnostic Manual* (*PDM-2*) appeals to practitioners of psychoanalysis and offers an elaborate classification with case studies to illustrate its applications (Lingiardi & McWilliams, 2017). A group of clinical investigators has devised a novel Hierarchical Taxonomy of Psychopathology (HiTOP) that derives quantitative dimensions using elaborate statistical methods based on multiple factor analysis (Kotov et al., 2017).

Whether any of the newer systems can displace or supplement the DSM in a major way remains to be seen. The director of the National Institute of Mental Health (NIMH) commented that "the inertia of diagnostic orthodoxy exerts a powerful hegemony over any alternative approaches" (Cuthbert & Insel, 2010).

CONCLUSIONS

This synopsis of differing ways of identifying mental and behavioral problems highlights disagreements among professionals and suggests that the study of behavior and the human mind is still in its infancy. This impression is not entirely accurate. Psychology has devised many kinds of tests and explored ways to analyze the data with sophisticated statistical models that regard human variation as something that exists on a continuum. In the field of behavioral psychology, distinct categories with arbitrary boundaries are viewed with skepticism. The practice of medical science that classifies and categorizes patients and disorders in order to prescribe treatments has achieved remarkable advances in the diagnosis and treatment of many distinct disorders of bodily function. Disorders of the mind and brain are proving to be a major challenge for medical models of disease. As anticipated in Chapter 1 on levels, explaining thought and behavior that occur at the level of a whole individual living in society on the basis of processes and defects at the molecular level is not easily done. The nervous system is of course made up of molecules, and many aspects of neural function are well understood at the molecular level, as elucidated in Chapters 4 and 5. Nevertheless, the leap from molecules to diagnostic categories as embodied in DSM/ICD criteria demands almost superhuman capabilities. Despite immense effort and expense, the quest for biological defects underlying most kinds of troublesome behaviors has not revealed the secrets of maladaptive behavior. Instead, the study of behavior is currently an active research frontier.

A defect in just one gene certainly can alter a behavior and mental function. The next section explores several examples in which the genetic basis of an abnormality of behavior is fairly well understood. Lessons learned from those examples can then be applied to our attempts to understand the role of genetics in the much more common human characteristics in Chapters 14–17 about complex traits.

HIGHLIGHTS

- Behavior is a complex phenotype and cannot arise solely from genes or be determined by a gene. A test of behavior yields a measure of a phenotype.
- Different people can achieve the same score on a test of behavior for different reasons.
- Psychology is keenly interested in differences among people in the normal range of behavior, whereas psychiatry is primarily concerned with identifying behaviors that represent pathology outside the normal range.
- Many tests of behavior report the results as a transformed score such as percentile rank, rather than a raw score such as the number of test items answered correctly.
- Many tests seek to measure some unseen psychological construct such as anxiety. The extent to which a test does this is termed the test's validity.
- A test is standardized by giving it to a representative sample of a population. Then, a person's score on a future occasion can be expressed relative to the standardization sample.
- Intelligence, as indicated by common IQ tests, has been gradually increasing over a period of five decades or more. This trend has often been obscured by the practice of updating and restandardizing the test every few years to maintain the average IQ score near 100.
- Tests of school achievement or readiness to begin university or college (e.g., the SAT) are not intelligence tests. They are designed to assess learned knowledge and can benefit considerably from coaching and practice.
- Diagnosis of a mental or behavioral disorder is often done according to written criteria (e.g., DSM or ICD) established by a professional or international organization of experts.
- Criteria for many mental disorders have changed substantially since 1950, and many of those changes have resulted in more people being diagnosed with a disorder.
- A good diagnostic system tends to yield the same opinion about a particular patient when he or she is

interviewed and appraised on different occasions by different clinicians.

- At the present time, no mental disorder can be diagnosed by biochemical or genetic testing.
- Several alternatives to the DSM have been proposed and are now being refined and evaluated in clinical trials.

References

American Psychiatric Association. (2013). *Diagnostic and statistical manual of mental disorders* (5th ed.). Arlington, VA: American Psychiatric Association.

Anastasi, A. (1984). Aptitude and achievement tests: the curious case of the indestructible strawperson. In B. S. Plake (Ed.), *Social and technical issues in testing: implications for test construction and usage*. Hillsdale, NJ: Lawrence Erlbaum Associates.

Andrews, G., Slade, T., & Peters, L. (1999). Classification in psychiatry: ICD-10 versus DSM-IV. *British Journal of Psychiatry, 174*, 3–5.

Bailey, I. L., & Lovie-Kitchin, J. E. (2013). Visual acuity testing. From the laboratory to the clinic. *Vision Research, 90*, 2–9.

Beck, A. T., Steer, R. A., & Brown, G. K. (1996). *Beck depression inventory* (2nd ed.). San Antonio, TX: The Psychological Corporation.

Beck, A. T., Steer, R. A., & Carbin, M. G. (1988). Psychometric properties of the Beck depression inventory: twenty-five years of evaluation. *Clinical Psychology Review, 8*, 77–100.

Beck, A. T., Ward, C. H., Mendelson, M., Mock, J., & Erbaugh, J. (1961). An inventory for measuring depression. *Archives of General Psychiatry, 4*, 561–571.

Bedford, A., & Deary, I. J. (1997). The personal disturbance scale (DSSI/sAD): development, use and structure. *Personality and Individual Differences, 22*, 493–510.

Bedford, A., Foulds, G., & Sheffield, B. (1976). A new personal disturbance scale (DSSI/sAD). *British Journal of Clinical Psychology, 15*, 387–394.

Bok, D. C., Gardner, D. P., Binder, S., Bond, L., Cole, T., Harris, J. J., … Goldberg, M. (1990). *Beyond prediction*. New York: College Entrance Examination Board.

Bregman, E. O. (1925). *Revision of army alpha examination, form A*. New York: The Psychological Corporation.

Brouwers, S. A., Van de Vijver, F. J., & Van Hemert, D. A. (2009). Variation in Raven's progressive matrices scores across time and place. *Learning and Individual Differences, 19*, 330–338.

Christensen, H., Jorm, A., Mackinnon, A., Korten, A., Jacomb, P., Henderson, A., & Rodgers, B. (1999). Age differences in depression and anxiety symptoms: a structural equation modelling analysis of data from a general population sample. *Psychological Medicine, 29*, 325–339.

Clark, C. M., & Ewbank, D. C. (1996). Performance of the dementia severity rating scale: a caregiver questionnaire for rating severity in Alzheimer disease. *Alzheimer Disease and Associated Disorders, 10*, 31–39.

Clarke, S. C. T., Nyberg, V., & Worth, W. H. (1978). *Technical report on Edmonton grade III achievement 1956-1977 comparisons*. Edmonton: Alberta Advisory Committee on Educational Studies.

College Board. (2015). *Test specifications for the redesigned SAT*. New York: The College Board.

Cruel Camera. (1982). *The fifth estate*. CBC. May 5.

Cuthbert, B. N., & Insel, T. R. (2010). Toward new approaches to psychotic disorders: the NIMH research domain criteria project. *Schizophrenia Bulletin, 36*, 1061–1062.

Cuthbert, B. N., & Insel, T. R. (2013). Toward the future of psychiatric diagnosis: the seven pillars of RDoC. *BMC Medicine, 11*, 126.

Dear, R. E. (1958). *The effect of a program of intensive coaching on SAT scores*. (Vols. 58–5) (pp. 319–330). Princeton, NJ: Educational Testing Service, RB.

Elkins, D. N. (2012). A brief overview of the DSM-5 reform effort. *Society for Humanistic Psychology Newsletter*. October Retrieved from http://www.apadivisions.org/division-32/publications/newsletters/humanistic/2012/10/dsm-5-reform.aspx.

Elkins, D. N., Robbins, B., & Kamens, S. (2011). Response to letter from DSM-5 task force and the American Psychiatric Association. *Society for Humanistic Psychology*,(November 7). Retrieved from http://societyforhumanisticpsychology.blogspot.com/2011/11/response-to-letter-from-dsm-5-task.html.

Flynn, J. R. (1984). The mean IQ of Americans: massive gains 1932 to 1978. *Psychological Bulletin, 95*, 29–51.

Flynn, J. R. (1987). Massive IQ gains in 14 nations: what IQ tests really measure. *Psychological Bulletin, 101*, 171–191.

Folger, T. (2017). The race to replace the kilogram. *Scientific American*. Retrieved from https://www.scientificamerican.com/article/the-race-to-replace-the-kilogram/.

Frances, A. J. (2011). Why psychiatrists should sign the petition to reform DSM-5. *Psychology Today*,(November 4). Retrieved from https://www.psychologytoday.com/blog/dsm5-in-distress/201111/why-psychiatrists-should-sign-the-petition-reform-dsm-5.

Gaines, A. D. (1992). From DSM-I to III-R; voices of self, mastery and the other: a cultural constructivist reading of US psychiatric classification. *Social Science and Medicine, 35*, 3–24.

Grob, G. N. (1991). Origins of DSM-I: a study in appearance and reality. *The American Journal of Psychiatry, 148*, 421–431.

Hausknecht, J. P., Halpert, J. A., Di Paolo, N. T., & Moriarty Gerrard, M. O. (2007). Retesting in selection: a meta-analysis of coaching and practice effects for tests of cognitive ability. *Journal of Applied Psychology, 92*, 373–385.

Hyman, S. E. (2011). Diagnosing the DSM: diagnostic classification needs fundamental reform. *Cerebrum, 2011*, 6–15.

Insel, T. R. (2014). The NIMH research domain criteria (RDoC) project: precision medicine for psychiatry. *American Journal of Psychiatry, 171*, 395–397.

Kaplan, R. M., & Saccuzzo, D. P. (2017). *Psychological testing: Principles, applications, and issues*. Boston, MA: Cengage Learning.

Kotov, R., Krueger, R. F., Watson, D., Achenbach, T. M., Althoff, R. R., Bagby, R. M., & Zimmerman, M. (2017). The hierarchical taxonomy of psychopathology (HiTOP): a dimensional alternative to traditional nosologies. *Journal of Abnormal Psychology, 126*, 454–477.

Lasa, L., Ayuso-Mateos, J. L., Vazquez-Barquero, J. L., Diez-Manrique, F. J., & Dowrick, C. F. (2000). The use of the Beck depression inventory to screen for depression in the general population: a preliminary analysis. *Journal of Affective Disorders, 57*, 261–265.

Lingiardi, V., & McWilliams, N. (2017). *Psychodynamic diagnostic manual* (2nd ed.). New York: Guilford Publications.

Martin, P. (2003). The epidemiology of anxiety disorders: a review. *Dialogues in Clinical Neuroscience, 5*, 281–298.

Moelter, S., Glenn, M., Xie, S. X., Chittams, J., Arnold, S., & Clark, C. (2012). The dementia severity rating scale for dementia: relationships with the CDR sum of boxes, clinical diagnosis and CSF biomarkers. *Alzheimer's & Dementia, 8*, 132.

Perneczky, R., Tene, O., Attems, J., Giannakopoulos, P., Ikram, M. A., Federico, A., … Middleton, L. T. (2016). Is the time ripe for new diagnostic criteria of cognitive impairment due to cerebrovascular disease? Consensus report of the International Congress on Vascular Dementia working group. *BMC Medicine, 14*, 162.

Pohjasvaara, T., Mantyla, R., Ylikoski, R., Kaste, M., & Erkinjuntti, T. (2000). Comparison of different clinical criteria (DSM-III, ADDTC, ICD-10, NINDS-AIREN, DSM-IV) for the diagnosis of vascular dementia. *Stroke, 31*, 2952–2957.

Polgar, M. (2003). Beck Depression Inventory. In vol.1. *Gale encyclopedia of mental disorders* (pp. 111–112).

Raven, J. (2000). The Raven's progressive matrices: change and stability over culture and time. *Cognitive Psychology, 41*, 1–48.

Regier, D. A., Narrow, W. E., Clarke, D. E., Kraemer, H. C., Kuramoto, S. J., Kuhl, E. A., & Kupfer, D. J. (2013). DSM-5 field trials in the United States and Canada, Part II: test-retest reliability of selected categorical diagnoses. *American Journal of Psychiatry, 170*, 59–70.

Robbins, B. (2011). *Open letter to the DSM-5. Retrieved from https://www.ipetitions.com/petition/dsm5/.*

Sheehan, D., Janavs, J., Baker, R., Harnett-Sheehan, K., Knapp, E., Sheehan, M., … Lepine, J. P. (2006). *The Mini International Neuropsychiatric Interview (MINI) English Version 5.0.0.* Tampa, FL: For the DSM-IV.

Shorter, E. (2015). *What psychiatry left out of the DSM-5: Historical mental disorders today.* New York: Routledge.

Snaith, R. P. (2003). The hospital anxiety and depression scale. *Health and Quality of Life Outcomes, 1*, 29.

Spitzer, R. L., Williams, J. B., Gibbon, M., & First, M. B. (1992). The structured clinical interview for DSM-III-R (SCID). I: history, rationale, and description. *Archives of General Psychiatry, 49*, 624–629.

Sprinkle, S. D., Lurie, D., Insko, S. L., Atkinson, G., Jones, G. L., Logan, A. R., & Bissada, N. N. (2002). Criterion validity, severity cut scores, and test-retest reliability of the Beck Depression Inventory-II in a university counseling center sample. *Journal of Counseling Psychology, 49*, 381–385.

Stern, A. F. (2014). The hospital anxiety and depression scale. *Occupational Medicine, 64*, 393–394.

Veerman, J. L., Dowrick, C., Ayuso-Mateos, J. L., Dunn, G., & Barendregt, J. J. (2009). Population prevalence of depression and mean Beck Depression Inventory score. *British Journal of Psychiatry, 195*, 516–519.

Vernon, P. (1954). Practice and coaching effects in intelligence tests. *The Educational Forum, 18*, 269–280.

Wang, Y. P., & Gorenstein, C. (2013). Psychometric properties of the Beck Depression Inventory-II: a comprehensive review. *Revista Brasileira de Psiquiatria, 35*, 416–431.

Winerip, M. (1994). S.A.T. increases the average score, by Fiat. *The New York Times*, 001001. June 11.

Woodford, R. (2003). Lemming suicide myth: disney film faked bogus behavior. *Alaska Fish & Wildlife News.* Retrieved from http://www.adfg.alaska.gov/index.cfm?adfg=wildlifenews.view_article&articles_id=56.

Zigmond, A. S., & Snaith, R. P. (1983). The hospital anxiety and depression scale. *Acta Psychiatrica Scandinavica, 67*, 361–370.

Single-Gene and Chromosomal Disorders

The next six chapters review examples wherein a specific abnormality in one identified gene or chromosome changes nervous system function and behavior. Lessons drawn from these relatively simple examples then inform the discussions in Chapters 14–20 of more complex traits involving variants of several genes and environments that interact in myriad ways. We will see that simple genetic effects are not so simple when explored in depth. It is apparent that the way a gene works is highly specific and often surprising. Where the gene is expressed, its relation with the nervous system, behaviors that are altered when it is disabled by a mutation, and things that help to compensate for the genetic defect—all these features must be discovered through study in depth of that one gene and its mutant forms. Despite the many challenges of this work, much progress has been made, and several single-gene disorders are now well understood. Examples chosen for Chapters 8 through 13 involve different modes of genetic transmission and expression.

ALLELES

As discussed in Chapter 2, most of us carry several new genetic mutations somewhere in the genome that can be passed to children through egg or sperm. Nevertheless, for any particular gene, the vast majority of us have a form that works quite well and can be considered "normal," but there are sometimes variants in the DNA sequence of a gene that can alter a phenotype or even cause a troublesome disease. The variants in DNA sequence of a gene are termed *alleles*. Now that it is possible to decode the DNA sequence of an individual, large numbers of alleles are being discovered and cataloged for many genes.

The albinism phenotype, for example, has been observed in families around the world. A child lacking pigmentation in hair, skin, and eyes appears in an extended family in which everyone else has normal pigmentation. People notice this, and sometimes, medical geneticists assess the person's DNA. Fig. 2.6 shows how a change in just one nucleotide base of the tyrosine hydroxylase or tyrosinase (TYR) gene can result in albinism. Fig. 7.1 shows a list of 239 alleles for which a change in the DNA leads to a change in one amino acid at a specific site in the tyrosinase enzyme, a large protein having 529 amino acids encoded in five exons (Kalahroudi et al., 2014). Most alleles are very rare and often have been seen in only one or a few individuals. Some alleles involve deletion or insertion of a base that can also change the amino acid sequence. To qualify as an official allele, a mutation does not need to cause a change in phenotype or pigmentation; it just needs to be a deviation from the official reference DNA sequence for that gene. In some instances, a new mutation in a known gene, especially one in an intron, does not change the phenotype at all and is known as a "silent" or neutral mutation. However silent it may be, it nevertheless qualifies as a new allele and can be cataloged.

Alleles of the TYR gene can be found by entering the gene name into the search box in the OMIM database and scrolling down to the allele section, then requesting the table view. This returns a list of 38 alleles. Requesting the ClinVar option takes one to the clinical variation database at NCBI, where 167 alleles are currently listed, most of which involve some clinically significant variant in the TYR gene, primarily albinism. The official designation of an allele presents the gene symbol, followed by the base change at a specific position in the DNA and then the amino acid substitution caused by that change. For example, (TYR):c.61C > T(p.Pro21Ser) indicates replacement (>) of cytosine by thymine at position 61 in the exon DNA that results in substitution of a serine for proline at position 21 in the tyrosinase enzyme. This conventional way of symbolizing an allele is cumbersome but precise. Experts in molecular genetics are the most likely users of that kind of information. The numbering for location in DNA is based on just the exons that are represented in the finished mRNA (Fig. 3.1). The ClinVar database states that about 1/20,000 people in most populations carry that rare allele.

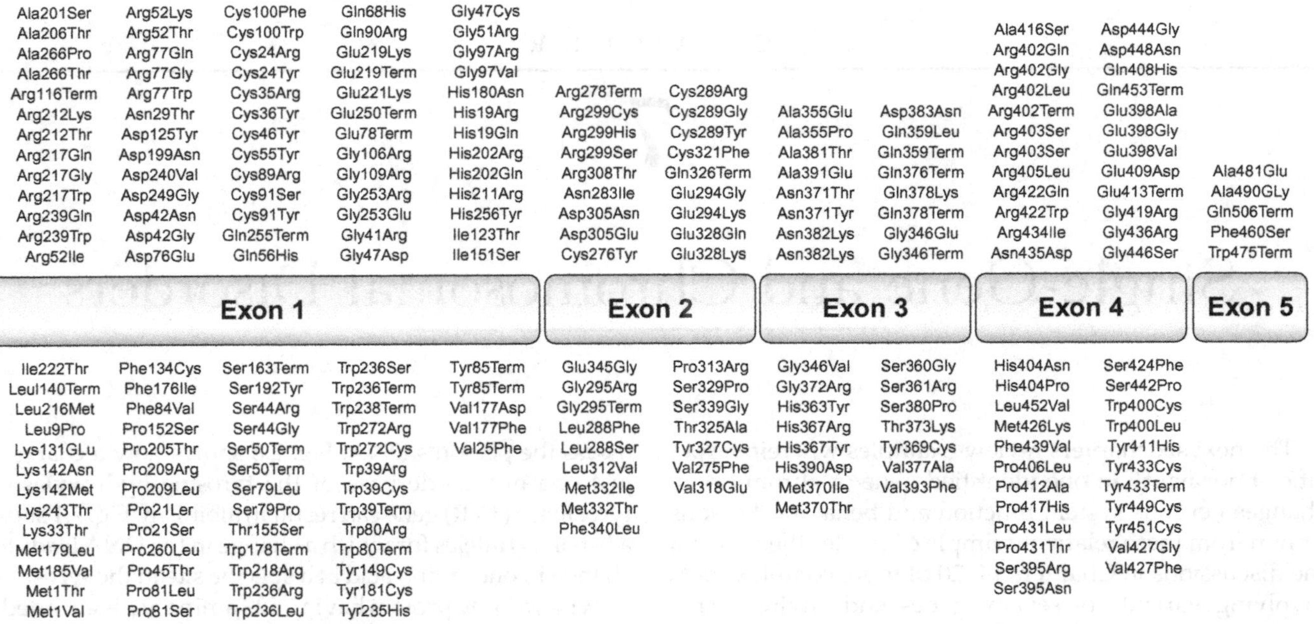
Ala201Ser
Ala206Thr
Ala266Pro
Ala266Thr
Arg116Term
Arg212Lys
Arg212Thr
Arg217Gln
Arg217Gly
Arg217Trp
Arg239Gln
Arg239Trp
Arg52Ile
Arg52Lys
Arg52Thr
Arg77Gln
Arg77Gly
Arg77Trp
Asn29Thr
Asp125Tyr
Asp199Asn
Asp240Val
Asp249Gly
Asp42Asn
Asp42Gly
Asp76Glu
Cys100Phe
Cys100Trp
Cys24Arg
Cys24Tyr
Cys35Arg
Cys36Tyr
Cys46Tyr
Cys55Tyr
Cys89Arg
Cys91Ser
Cys91Tyr
Gln255Term
Gln56His
Gln68His
Gln90Arg
Glu219Lys
Glu219Term
Glu221Lys
Glu250Term
Glu78Term
Gly106Arg
Gly109Arg
Gly253Arg
Gly253Glu
Gly41Arg
Gly47Asp
Gly47Cys
Gly51Arg
Gly97Arg
Gly97Val
His180Asn
His19Arg
His19Gln
His202Arg
His202Gln
His211Arg
His256Tyr
Ile123Thr
Ile151Ser
Arg278Term
Arg299Cys
Arg299His
Arg299Ser
Arg308Thr
Asn283Ile
Asp305Asn
Asp305Glu
Cys276Tyr
Cys289Arg
Cys289Gly
Cys289Tyr
Cys321Phe
Gln326Term
Glu294Gly
Glu294Lys
Glu328Gln
Glu328Lys
Ala355Glu
Ala355Pro
Ala381Thr
Ala391Glu
Asn371Thr
Asn371Tyr
Asn382Lys
Asn382Lys
Asp383Asn
Gln359Leu
Gln359Term
Gln376Term
Gln378Lys
Gln378Term
Gly346Glu
Gly346Term
Ala416Ser
Arg402Gln
Arg402Gly
Arg402Leu
Arg402Term
Arg403Ser
Arg403Ser
Arg405Leu
Arg422Gln
Arg422Trp
Arg434Ile
Asn435Asp
Asp444Gly
Asp448Asn
Gln408His
Gln453Term
Glu398Ala
Glu398Gly
Glu398Val
Glu409Asp
Glu413Term
Gly419Arg
Gly436Arg
Gly446Ser
Ala481Glu
Ala490GLy
Gln506Term
Phe460Ser
Trp475Term
Exon 1
Exon 2
Exon 3
Exon 4
Exon 5
Ile222Thr
Leul140Term
Leu216Met
Leu9Pro
Lys131GLu
Lys142Asn
Lys142Met
Lys243Thr
Lys33Thr
Met179Leu
Met185Val
Met1Thr
Met1Val
Phe134Cys
Phe176Ile
Phe84Val
Pro152Ser
Pro205Thr
Pro209Arg
Pro209Leu
Pro21Ler
Pro21Ser
Pro260Leu
Pro45Thr
Pro81Leu
Pro81Ser
Ser163Term
Ser192Tyr
Ser44Arg
Ser44Gly
Ser50Term
Ser50Term
Ser79Leu
Ser79Pro
Thr155Ser
Trp178Term
Trp218Arg
Trp236Arg
Trp236Leu
Trp236Ser
Trp236Term
Trp238Term
Trp272Arg
Trp272Cys
Trp39Arg
Trp39Cys
Trp39Term
Trp80Arg
Trp80Term
Tyr149Cys
Tyr181Cys
Tyr235His
Tyr85Term
Tyr85Term
Val177Asp
Val177Phe
Val25Phe
Glu345Gly
Gly295Arg
Gly295Term
Leu288Phe
Leu288Ser
Leu312Val
Met332Ile
Met332Thr
Phe340Leu
Pro313Arg
Ser329Pro
Ser339Gly
Thr325Ala
Tyr327Cys
Val275Phe
Val318Glu
Gly346Val
Gly372Arg
His363Tyr
His367Arg
His367Tyr
His390Asp
Met370Ile
Met370Thr
Ser360Gly
Ser361Arg
Ser380Pro
Thr373Lys
Tyr369Cys
Val377Ala
Val393Phe
His404Asn
His404Pro
Leu452Val
Met426Lys
Phe439Val
Pro406Leu
Pro412Ala
Pro417His
Pro431Leu
Pro431Thr
Ser395Arg
Ser395Asn
Ser424Phe
Ser442Pro
Trp400Cys
Trp400Leu
Tyr411His
Tyr433Cys
Tyr433Term
Tyr449Cys
Tyr451Cys
Val427Gly
Val427Phe

FIG. 7.1 Alleles of the TYR gene that alter amino acid sequence of the tyrosinase enzyme, positioned above and below the exon where the allele occurs, in alphabetic order by amino acid that is replaced. The original amino acid and its position in the protein are followed by the new amino acid. Most are very rare and localized in certain regions of the world. *From Kalahroudi, V. G., Kamalidehghan, B., Kani, A. A., Aryani, O., Tondar, M., Ahmadipour, F., … Houshmand, M. (2014). Two novel tyrosinase (TYR) gene mutations with pathogenic impact on oculocutaneous albinism type 1 (OCA1).* PloS One, *9, e106656, Fig. 6.*

Generally speaking, what we designate as phenotypic albinism of the OCA1 type (oculocutaneous albinism type 1; see Table 2.3) occurs when a person inherits two defective TYR alleles. In most cases, a person with albinism has two different TYR mutations, each of which generates a nonfunctional tyrosinase enzyme. There is no such thing as an albino gene or a gene for albinism. Instead, there are numerous alleles of the TYR gene that, in certain combinations, can result in phenotypic albinism. Some alleles of the TYR gene reduce the intensity of melanin pigmentation without abolishing it, and they play important roles in normal variation in human skin colors (Chapter 19).

We want to understand not only what gene is altered in a particular case but also how it malfunctions. The study of malfunction provides valuable insights into the normal functions of a gene. At the present time, there is often no way to examine the DNA sequence of a little-known gene and deduce exactly how it works in healthy cells and nervous systems. It is essential to compare phenotypes of individuals who are known to have different genotypes involving different alleles. Mutation can serve as a kind of miniature surgery in which a minute portion of the inherited gene is excised or altered in a way that changes a characteristic of the nervous system or a behavior. Allelic variants provide important clues about how a network of molecules functions. In some instances, the gene products may appear to form a single pathway from a beginning at a precursor to an end with a finished product (e.g., Fig. 2.5 about melanin), but in many situations, the "genetics may be more aptly represented by a network than a pathway" (Barsh, 1995). Diagrams are often simplified by omitting alternative paths and branches.

Functions of a gene are best discerned in real people living ordinary lives, not just from a chemical path diagram. The TYR mutations that generate phenotypic albinism have a number of unexpected effects. For example, in people with normal pigmentation, fibers from the optic nerve of one eye divide into two bundles at the optic chiasm and travel to the left and right hemispheres, so that an image on the left half of the retina of the right eye projects to the visual cortex of the left hemisphere, while fibers from the right half of the retina of the right eye project to the right hemisphere. Roughly 50% of the optic nerve fibers from one eye cross to the opposite hemisphere. In albinism, however, the fraction of axons that project to the same side is greatly reduced and disorganized as well (Creel, 2015; Creel, Summers, & King, 1990), which can change phenotypes such as stereoscopic depth perception and visual acuity. Thus, albinism reveals the important role of pigmentation in the retina for the organization of the visual system of the brain. Furthermore, myopia or nearsightedness is more common in albinism. Risk of skin cancers (melanoma) is also elevated. Those phenotypic changes effected by a TYR mutation reveal *pleiotropic* gene action; defects in a single kind of gene have multiple phenotypic effects. The phenotype in albinism even extends into the realm of social experience of the affected person, because in many countries, someone with albinism is subjected to not only impolite stares but also cruel discrimination and even murder ("Tanzania's albino community: 'Killed like animals", 2014; Bever, 2015).

GENETIC TRANSMISSION

Textbooks of genetics and many scholarly articles describe three kinds of inheritance: dominant transmission, recessive transmission, and sex-linked transmission. These concepts combine two distinctly different phenomena—the transmission of genes or alleles from parent to offspring and the expression of a person's genotype in his or her phenotype. In fact, for what are commonly termed dominant and recessive traits, the modes of genetic transmission are identical, and X-linked transmission is similar.

Autosomal transmission involves genes that are located on chromosome pairs 1 through 22 (the autosomes), whereas genes located on the X or Y chromosome show *sex-linked transmission*. A parent has two copies of each autosomal gene, one derived from each of his or her parents, and the copies are located at the same place (*locus*) on the two chromosomes of the same kind (Fig. 7.2A). Those two copies of the same gene can consist of either

two different alleles or two copies of the same allele. Consider the case in which almost all people have two copies of a normal form of a gene, often symbolized as the + allele, and a few in the population also carry a mutation in that gene, an allele we can symbolize as *m*. (Some texts show a recessive gene as lower case *m* and a dominant gene as upper case *M*). Thus, when there are only two kinds of alleles, + and *m*, there can be only three kinds of genotypes or pairs of alleles: +/+, +/*m*, and *m*/*m*. For most autosomal genes, genotypes +/*m* and *m*/+ work the same way, except that one genotype gets the *m* from the mother and the other has the paternal *m*, so it is convenient to show both as just +/*m*. Some authorities always place the allele obtained from the mother before the "/," but this is not a universally applied rule.

Autosomal transmission gives rise to the three kinds of genotypes in the offspring via three simple rules.

- First, when a germ cell (ovum or sperm) forms, it receives just one allele from the cell in the ovary or testis that generates the germ cells (Fig. 7.2A). If the mother's genotype is +/*m*, her ovum can have allele + or allele *m* but not both. The process whereby a cell with two copies of each chromosome (diploid) gives rise to a germ cell with just one copy of each chromosome (haploid) is termed *meiosis*.

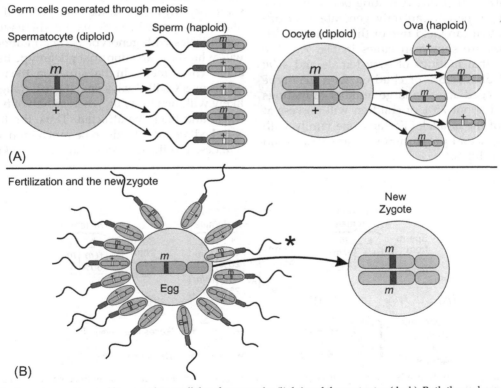

FIG. 7.2 Example shows one autosomal locus with two alleles, the normal + (light) and the mutant *m* (dark). Both the male and female parents are heterozygous at that locus. (A) A single allele is conveyed to a haploid germ cell from the diploid spermatocyte or oocyte in the gonads. Which allele it receives is a random event. (B) A single egg leaves the ovary and enters the oviduct, where it encounters vast numbers of sperm cells, each carrying just one allele of a gene. Which sperm actually fertilizes the egg is also a random event.

- Second, the probability that a particular allele is acquired by a single germ cell is the same for the two alleles, namely, 1/2 or 50%. Which allele inhabits a specific germ cell is a chance event, much like flipping a coin to see whether you get a head or tail.
- Third, which alleles are carried by the egg and sperm that unite to form the new individual (Fig. 7.2B) is also a matter of chance. The germ cells cannot reveal to each other which alleles they carry. In the sperm, the DNA in the nucleus is usually tightly coiled and cannot be expressed in protein until after fertilization. In effect, there are two coin flips, one to decide which allele the embryo gets from the mother and one to decide which it gets from the father. Those are *independent events*, and the probability of any pair of outcomes can be found simply by multiplying the probabilities of each outcome of a coin toss. If the genotypes of mother and father are known, we can determine the probability of obtaining each of the three possible genotypes in an offspring. Genetic transmission is inherently probabilistic, unpredictable, and uncontrollable. The new individual's actual genotype can only be observed after the fact.

For certain pairs of parent genotypes in a mating, the result is obvious. If both parents are +/+, all the offspring must also be +/+, whereas parents who are both m/m can generate only m/m offspring. A mating between a +/+ parent and a m/m parent can only generate +/m offspring. Matings that can yield two or three different offspring genotypes are shown in tables in Fig. 7.3. The example reveals that matings of +/m female and +/m male will yield about 25% of children with genotype m/m, when averaged across a large number of matings of that kind, and about 75% of children will possess at least one copy of the m allele. For any one family with few children, the actual proportion who are m/m could range from 0% to 100%.

Technical terms are often used in scientific articles to describe some of these situations. When a person inherits two identical alleles (+/+ or m/m), the genotype is said to be *homozygous*, whereas a genotype with two different alleles (+/m) is termed *heterozygous*. When there are several kinds of mutant alleles in a population (e.g., m_1, m_2, and m_3), a genotype consisting of two different mutations (e.g., m_1/m_2) is sometimes termed a *compound heterozygote*.

Sex-linked transmission involves a female with two X chromosomes, one from each parent, and a male with just one X and also a Y chromosome. The male always gets his X from his mother and the Y from his father. If a sperm carries the Y, the result will usually be a boy, while a sperm with the X will usually conceive a girl. Any gene that resides on the Y chromosome will be transmitted from father to son every generation. If the father has a Y-related genetic disease, the son will also tend to have it. Such abnormalities of the Y are rare, and very few behavioral characteristics are believed to be related to a Y chromosome gene (Chapter 13). Males are particularly vulnerable to genetic defects that reside on the X. If the female has X chromosome genotype +/m, she may not show an abnormal phenotype, whereas a male who gets just the aberrant m allele is likely to have problems. The male will always pass his allele on his X to his daughter, whereas the alleles from the mother will follow the same pattern as for autosomal genes. If the mutant m allele is very rare in a population, there will be extremely few females with genotype m/m and almost all individuals who show an abnormal phenotype because of the m allele will be males. In that situation, the male with the m on his one X will always pass it to his daughter who most likely will then be +/m, whereas half of her sons will get the m and likely be afflicted. Thus, the frequency of an X-linked disorder tends to be much higher in males than females, and the disorder also tends to skip generations.

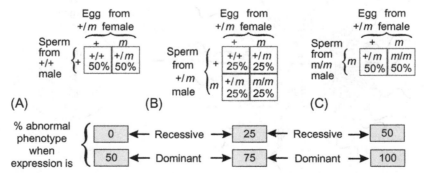

FIG. 7.3   Three examples of the transmission of genes from parents to offspring. (A) Father is normal, and mother is a carrier of a mutation. (B) Both parents are carriers. (C) Father has two copies of the mutation, and mother is a carrier. Probabilities or percentages of each offspring genotype follow from principles of transmission genetics. Both +/m and m/+ genotypes are shown as +/m. Phenotypes, on the other hand, depend on the mode of expression. If the gene is recessive, the probabilities of a child who has genotype m/m and shows the phenotypic anomaly are 0%, 25%, and 50% in the three cases. If the mutation is a dominant allele, probabilities of showing the phenotypic anomaly are 50%, 75%, and 100% in the three cases.

Mitochondrial transmission occurs when a gene is part of the small genome of the mitochondria in the cytoplasm (designated as mtDNA) that is entirely separate from the DNA in the nucleus (Chapter 11). The mitochondrial genes are critically important parts of chemical networks that provide the energy to all cells. The sperm conveys virtually no cytoplasm to the new zygote. Instead, the zygote obtains all of its cytoplasmic organelles from the mother via the egg. A phenotypic defect related to a mutation of a mitochondrial gene will often appear in both sons and daughters in the next generation.

These patterns of genetic transmission and probabilities of several possible genotypes in offspring are especially important for clinical application in genetic counseling. When a harmful mutation has made its presence known in a family, other prospective parents among the relatives might want to know how likely the same problem is to affect their future offspring.

THE RELATION OF GENOTYPE TO PHENOTYPE: GENE EXPRESSION

Recessive Expression

Phenotype of an individual is a question of gene expression. There are many instances in which a rare autosomal mutation proves to be *recessive*. By definition, this means that the phenotypes of genotypes $+/+$ and $+/m$ are the same; both appear to be normal, and having just one good copy of the gene in each cell provides enough normal protein to function well. Those with the $+/m$ genotype are said to be unaffected *carriers* of the mutation. For a recessive mutation, only those with two copies of the defective gene have genotype m/m and show an abnormal phenotype, as occurs for the TYR mutations that cause albinism. This can happen only if both parents are carriers or one is a carrier and the other is m/m. If the defect is very rare, those with the m/m genotype will seldom appear, and almost all defective alleles will exist in unaffected carriers. Suppose that about 1% of the people in a population are unaffected carriers $+/m$ and 99% are $+/+$. To obtain the m/m genotype, two carriers must mate. This will be a chance occurrence because neither person will know that the other is a carrier. The probability of that happening is approximately $(.01)(.01) = (.0001) = 1/10,000$. On average, about 25% of children from such a mating combination will have the m/m genotype, so the frequency of affected individuals in the population should be $(.0001)(.25) = .000025$ or $1/40,000$. There would be 400 times more unaffected carriers in the population ($1/100$) than affected persons ($1/40,000$).

Thus, one hallmark of autosomal recessive gene expression for an uncommon allele is the *lack of family history* of the abnormal phenotype, because ancestors with the $+/m$ genotype did not happen to mate with someone who was also $+/m$. Then, in the current generation, two people with the $+/m$ genotype get together. Neither knows he or she carries the m mutation, until the defect pops up in the midst of two seemingly normal family histories. The affected family will tend to have about 25% of children with the abnormal phenotype when there are many children, and equal numbers of males and females will be affected on average.

Dominant Expression

If the mutant allele is *dominant*, having just one copy is enough to give the carrier $+/m$ an abnormal phenotype. The abnormal allele generates an abnormal protein that interferes with cellular function. In that case, there is usually a family history and roughly half of the children from someone with genotype $+/m$ will show the defect if the other parent is $+/+$. As shown in Fig. 7.3B, if both parents are $+/m$, then 75% of their children will be afflicted to some extent.

Two variations on the theme of dominant expression sometimes occur. If the mutant allele m is quite harmful, genotype m/m may not be viable at all and may perish unnoticed in the early embryo. In such a case, $2/3$ of live-born children will have the $+/m$ genotype and abnormal phenotype. In some instances, the $+/m$ genotype clearly shows an abnormal phenotype that is much less affected than the m/m, but the m/m genotype is still viable. This is sometimes termed *intermediate inheritance*.

MULTIPLE ALLELES, GENOTYPES, AND PHENOTYPES

Multiple alleles and genotypes occur for many genes, and the basic rules of genetic transmission are the same. There are just more possibilities. But phenotypic expression depends on how the particular gene works in the biological system. The ABO gene provides an example that played an important role in the history of human genetics. It determines our familiar blood types. Most of us have blood type A, B, AB, or O. Those are phenotypes. They are detected by obtaining a few drops of blood and adding them to small amounts of test solution containing antibodies.

The ABO system has existed in animals for millions of years and has no observable phenotype in most circumstances. Only with the advent of blood transfusions in the twentieth century did medical staff learn that some people cannot tolerate certain types of blood from someone else. After tests for blood types A, B, AB, and O were devised, studies of parents and children revealed that the blood types are determined by a single gene with

TABLE 7.1 Four Common Alleles of the ABO Gene

Allele	Frequency[a]	DNA variant	Function
A1	17%	Reference sequence	Transfer acetylgalactosamine to H antigen
A2	7%	1-BP DEL, C1056[b]	Same as A1, has leucine at position 156
B	9%	7 bases differ, 4 change amino acids	Transfer galactose to H antigen
O	67%	1-BP DEL, G258[b]	Not modify H antigen

[a] *Average of allele frequencies in European and African populations (Patnaik, Helmberg, & Blumenfeld, 2014).*
[b] *These single-base-pair deletions involve nucleotide base C at position 1056 and G and 258 in the ABO (A-transferase) gene.*
Source: OMIM.

multiple alleles. The four most common alleles are shown in Table 7.1. They are very common in many populations and have not been culled by natural selection because they all function well enough in the body. Over 180 minor variants of the major types of alleles have been compiled in the Blood Group Antigen Mutation Database (Patnaik et al., 2014).

Because of its historical origins, the gene is now named ABO, but it does not code for blood types directly. Instead, it codes for the structure of the enzyme that attaches a small carbohydrate molecule to a protein termed H antigen that is expressed on the surface of red blood cells and many other cells in the body. The basic H antigen is encoded in a separate gene (fucosyltransferase 1 (FUT1)). The A allele of the ABO gene codes for an enzyme named alpha 1-3-*N*-acetylgalactosaminyltransferase, mercifully abbreviated as A-transferase (OMIM). It transfers one molecule of acetylgalactosamine to the H antigen as the last step in the synthesis of the active molecule. B-transferase transfers galactose to H. For allele O, a mutation disables the transferase enzyme, and nothing is added to H.

The alleles possessed by an individual become apparent only through blood tests for antibodies. Early in life, the immune system of a child "learns" that proteins present in the body at that time are part of the self. Later, invasion of the body by a foreign or nonself protein will be met with an attack by antibodies generated by the person's immune system. The phenotype of an individual is the kind of antigen that will be attacked and the kinds of blood types that will be accepted in a transfusion, summarized in Table 7.2. Gene transmission for the ABO alleles follows all the usual rules of transmission, whereas expression in a phenotype is actually generated via an intricate system of molecules that comprise the immune system. Allele A1 does not code for antigen type A; it codes for an enzyme that helps to generate antigen A from the H antigen. This may seem like a fine distinction, but it can matter greatly when one wants to make sense of how a gene works and then make use of that knowledge.

The distinction between transmission and expression is shown for two kinds of matings in Fig. 7.4. Because there are four alleles, there can be $4 \times 4 = 16$ possible allele pairs, but for this particular kind of blood antigen, it does not matter whether a specific allele comes from the mother or father. Allele combination A1/B would yield

TABLE 7.2 Phenotypes of Genotypes Involving Alleles of the ABO Gene

Genotype	Antibodies	Accept blood	Genotype	Antibodies	Accept blood
A1/A1	to B	A, O	A/O	to B	A, O
A1/A2	to B	A, O	B/B	to A	B, O
A2/A2	to B	A, O	B/O	to A	B, O
A/B	Neither A nor B	A, B, O	O/O	to A and B	Only O

FIG. 7.4 Genotypes of offspring at the ABO locus and their phenotypes from two kinds of mating combinations of parents. (A) The two parents have four different alleles. (B) The parents have three different alleles. The first 2×2 table shows probabilities of genetic transmission, while the second shows the phenotypes expressed by those genotypes. Many other combinations of parental genotypes are possible.

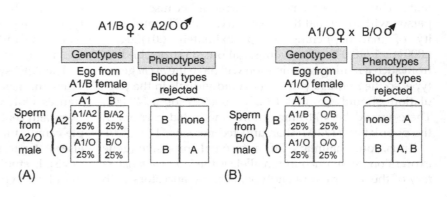

the same phenotypes as B/A1. There can therefore be 10 distinct kinds of matings, each with a 2 × 2 table similar to Fig. 7.4. If the alleles possessed by the parents are known, one can then know the probabilities of each kind of genotype among the offspring; the rules of genetic transmission are the same for all mating combinations.

For allele expression, on the other hand, the kinds of antibodies synthesized by the immune system are the direct opposite of the kinds of ABO alleles the person possesses. Being A1/B prevents the synthesis of antibodies to antigens A and B because the immune system recognizes them as "self," and this allows the person to accept any of the three major blood types in a transfusion. The O allele acts like a recessive in reverse; being O/O causes the immune system to reject both A and B types of antigens and accept only type O. Thus, type O can accept only type O blood, whereas type O blood can be accepted by all ABO allele combinations and therefore serves as the "universal donor." To make sense of the blood type and transfusion phenotypes, we therefore must know how the larger immune system functions. Only by knowing this can we make sense of the facts that the case in Fig. 7.4A involves two parents with four different alleles that result in just three phenotypes, whereas the case in Fig. 7.4B entails parents with three different alleles that generate four phenotypes.

INSTRUCTIVE EXAMPLES

Studies of specific genetic defects presented in the next few chapters followed a similar course of investigation. First, some kind of medical malady or behavioral disorder comes to the attention of scientists, who then seek to diagnose it in a way that renders it distinct from similar abnormalities. So, the first step is to define the phenotype. Then, the occurrence of the abnormality is assessed in different kinds of relatives to see what kind of pattern it exhibits. In situations wherein the facts suggest that just one major gene is the source of the abnormality, the investigation then follows one of several paths: (a) In some cases, the abnormality can be assessed biochemically, and the enzyme that is abnormal can be identified without knowing the exact location of the gene on a chromosome or its DNA sequence. Effective treatments for a disorder can in some cases be devised without knowing anything about the specific defect in the DNA. This occurred for phenylketonuria (Chapter 8). (b) When virtually nothing is known about the kind of protein product of the likely gene, efforts are made to localize the gene on a specific chromosome and then narrow the search to a relatively short sequence of DNA in that region. When a clear difference in the DNA sequence is associated with the normal and abnormal phenotypes, this strongly suggests that the gene responsible for the defect has been

located. Knowing the DNA sequence of the newly discovered gene, researchers can then deduce the nature of its protein and study where it is expressed and how it functions. This course was followed for Huntington disease (Chapter 9). (c) Nonautosomal patterns of transmission give clues that lead in some cases to a gene on the X chromosome, as happened with certain forms of androgen insensitivity and color blindness (Chapter 10), or a gene in the mitochondria, as for Leber's optic neuropathy (Chapter 11).

It turns out that several distinctive anomalies arise not from a defect in the specific DNA sequence but by the duplication or even absence of an entire chromosome. Down syndrome (Chapter 12) and the XYY male (Chapter 13) provide interesting examples. Diagnosis of an extra or absent chromosome requires only a good picture of stained chromosomes using a microscope. A chromosomal disorder does not behave like a single gene, and in many situations, the extra chromosome arises anew and is not passed on to the next generation.

There are now thousands of genetic disorders known to be of clinical significance wherein the cause resides in a mutation of a single gene. The genes and alleles chosen for examples here have especially large and consistent effects that make their workings easier to discern. There are now good reasons to expect that most genes exist in many different allelic forms, the effects of which can range from a severe, even fatal defect, to a mild deficit, to a silent mutation with no apparent effect at all. What is termed the "normal" form of a gene is more realistically viewed as a collection of alleles that seem to work well enough and result in no measurable phenotypic differences in most circumstances. When doing probability computations, it is convenient to lump them all into a "+" category with a frequency equal to the sum of frequencies of all those alleles. Nevertheless, we should be aware that some of those alleles very likely do have a detrimental effect on some phenotype but the effect is often too small to be noticed in most studies. The world of genes is not divided into two classes—the perfectly normal and the disastrously mutated. Genes with more subtle effects appear to have roles especially in the domain of complex traits (Chapters 14–19).

HIGHLIGHTS

- A gene is a segment of DNA that codes for a protein, and the specific *locus* for that gene is the place where it occurs in the long DNA molecule of a chromosome.
- *Alleles* are variants of a gene that have slightly different DNA sequences. New alleles can arise through mutation. The *genotype* of an individual at a locus is the pair of alleles inherited from the parents. If the alleles are the same, the person is said to be *homozygous*,

whereas someone with two different alleles is *heterozygous*.

- Genome sequencing of an individual is revealing large numbers of previously unknown alleles of the same gene. For many genes, more than 100 alleles are known that can significantly impair gene function. Most of them are quite rare in a population.
- Many genes show *pleiotropy*, whereby the same gene alters many different phenotypes. Pleiotropic gene action is observed by studying mutations that alter the phenotypes.
- Germ cells get only one allele as a result of meiosis. Which allele is transmitted to offspring via the egg or sperm is a random or chance event. Which two alleles are paired in the new individual is also a chance event. Which allele is carried by the sperm and which is carried by the egg are independent events.
- Laws of genetic transmission allow us to estimate the probabilities of obtaining various genotypes among the offspring of two parents having specific genotypes. These laws are important for genetic counseling in which risks of having an affected child are estimated.
- In recessive gene expression, a person must have two aberrant alleles of a gene in order to show a phenotypic deficit, whereas in dominant expression, only one aberrant allele is sufficient to alter the phenotype.

- The relation between the genotype and the kind of phenotypes it influences depends on how that specific gene works and the network of other molecules of which it is a part.

References

Barsh, G. S. (1995). Pigmentation, pleiotropy, and genetic pathways in humans and mice. *American Journal of Human Genetics, 57,* 743–747.

Bever, L. (2015). Where albino body parts fetch big money, albinos still get butchered. *The Washington Post.* (2015). https://www. washingtonpost.com/news/morning-mix/wp/2015/03/13/how-tanzanias-upcoming-election-could-put-albinos-at-risk-for-attack/?noredirect=on&utm_term=.aa1843067683 March 13.

Creel, D. J. (2015). Visual and auditory anomalies associated with albinism. In H. Kolb, R. Nelson, E. Fernandez, & B. Jones (Eds.), *Webvision: The organization of the retina and visual system.* Salt Lake City, UT: University of Utah Health Sciences Center.

Creel, D. J., Summers, C. G., & King, R. A. (1990). Visual anomalies associated with albinism. *Ophthalmic Paediatrics and Genetics, 11,* 193–200.

Kalahroudi, V. G., Kamalidehghan, B., Kani, A. A., Aryani, O., Tondar, M., Ahmadipour, F., … Houshmand, M. (2014). Two novel tyrosinase (TYR) gene mutations with pathogenic impact on oculocutaneous albinism type 1 (OCA1). *PloS One, 9,* e106656.

Patnaik, S. K., Helmberg, W., & Blumenfeld, O. O. (2014). BGMUT database of allelic variants of genes encoding human blood group antigens. *Transfusion Medicine and Hemotherapy, 41,* 346–351.

Tanzania's albino community. (2014). *Killed like animals.* London: BBC. December 9.

8

Phenylketonuria (PKU)

On April 22, 1930, a boy was born in Norway to parents whose first child, a girl, suffered mental deficiency and slow development. The mother breastfed this child for several months, as she had done for her daughter, and she became distressed when his development was very poor. She also noticed that the boy had a peculiar musty odor. This odor was an important clue to the cause of her son's troubles, but the woman's physician did not believe her and instead concluded there was something mentally wrong with the mother. She was diagnosed as suffering from an obsession and sent to New York for treatment by a psychiatrist, who was unable to cure her. Upon her return to Norway, the woman convinced her cousin, the veterinarian Asbjörn Fölling, that she was telling the truth, and Fölling began the pioneering research that led to the discovery of a distinct biochemical defect in certain mentally deficient children, a relatively simple genetic cause, and a remarkably effective treatment to prevent that kind of mental deficiency (Jervis, 1953). Apart from the initial misdiagnosis of the mother, research on the disorder proceeded in a logical sequence, each successful step leading to the next.

Fölling (1934) approached the problem as a biochemist. He took samples of urine from the boy and the girl and tried many chemical tests in search of something that would distinguish their urine from that of normal children. For some reason, a solution of iron chloride ($FeCl_3$) did the job. When it was added to a sample of urine, the urine from the two children turned a deep green color, whereas urine from normal people did not. Fölling applied this test to urine samples from 430 mentally deficient children in Norway, many living in institutions, and detected eight more instances of the green color reaction. Of the 10 positive cases, nine were clearly mentally deficient to various degrees (one boy, the brother of an afflicted girl, was too young to be sure about the future course of his development). Fölling did the painstaking analytic work that was necessary to find the substance causing the green reaction. He purified it through many steps that concentrated it to such a degree it could be crystallized, an indication of the purity of his sample.

Its chemical formula was $C_9H_8O_3$, and its structure was the same as phenylpyruvic acid (ppa), an organic compound never before detected in humans (Fig. 8.1). In addition, Fölling noticed the close chemical relation of ppa to the amino acid molecule phenylalanine, an important component of proteins. He observed that if one of his patients was given a diet high in phenylalanine or protein, the amount of ppa in urine increased. He correctly concluded that the conversion of phenylalanine to ppa, although it did occur in his patients, is not the normal way phenylalanine is metabolized, and he noted that phenylalanine is probably converted to the amino acid tyrosine in normal children. Some of the phenylpyruvic acid was itself converted to phenylacetic acid and excreted in the urine, and the phenylacetic acid gave the child a peculiar musty odor. Thus, in 1934, Fölling came within a hair's breadth of discovering the actual metabolic defect in this form of mental deficiency.

The iron chloride test for phenylpyruvic acid provided an objective way to diagnose a specific disorder among hundreds of cases of mental deficiency that had diverse and unknown causes. Fölling observed that some of his patients were more severely disabled than others. In a terse case report of an 11-year-old girl, he wrote, "Speaks coherently, but the words are sometimes confused. Slow apprehension and expression. Confused and without interest. Difficulties in reading and arithmetic. Writes neatly." This was one of the least impaired cases. A 27-year-old woman was in a bad state: "She has to be led by her hand to all places always. She is a sweet person and clean. She utters a few inarticulate sounds and is not able to reply yes or no." The mental impairment was part of an array of symptoms (pleiotropy) and by itself was not useful as a diagnostic criterion. It was a necessary part of the story but not sufficient to identify a child as having "Fölling's disease." Furthermore, there was no striking array of physical features to aid a diagnosis. The odor of phenylacetic acid in the urine was not a reliable indicator because many people could not smell it. Only a chemical indicator added to a urine sample clearly distinguished Fölling's subset of patients.

Genes, Brain Function, and Behavior
https://doi.org/10.1016/B978-0-12-812832-9.00008-7

Two names were at one time in common use for the biochemical defect seen in "Fölling's disease"—phenylpyruvic acidemia (or oligophrenia) and phenylketonuria. Phenylpyruvic acid is one of a class of chemicals termed phenyl ketones by organic chemists. Only one kind of phenyl ketone, namely, phenylpyruvic acid, was detected in the children studied by Fölling, so the term phenylketonuria is less precise than phenylpyruvic acidemia. Nevertheless, phenylketonuria was widely adopted and abbreviated PKU, as the disorder is known today.

TRANSMISSION AND FREQUENCY

Equipped with the simple iron chloride test, investigators in other countries searched for mentally deficient children in institutions and detected more cases of "Fölling's disease." One objective was to determine the mode of its transmission in families. When a new case was found, the child was designated an index case or "proband," and when possible, all the living brothers and sisters as well as the parents and other relatives of the proband were also tested for the green reaction. There were no children of the probands because people with the disorder were unable to reproduce.

A definite pattern of results emerged. Quite frequently, there was another sibling with the same disorder, especially in large families, but neither parent nor other close relative ever showed the chemical defect. In the vast majority of cases, the mentally disabled children were conceived by parents with mental abilities in the normal range. In one survey, a family with five children was found, three of whom showed the green reaction and mental deficiency. This suggested that the disorder might have a simple genetic cause. If so, the gene would be recessive. A large survey published in 1939 reported on 266 families with at least one case of PKU (Jervis, 1939). All parents were normal, but 433 of their 1094 children showed the green reaction with iron chloride. Among those 433, about 51% were female, and 49% were male. The investigators concluded that the chemical defect and the mental disability were caused by a single recessive gene located on an autosome. This was the first demonstration of a metabolic defect and mental disability produced by single-gene Mendelian transmission.

For a recessive gene, when there is a mating of two unaffected carriers, close to 25% of the children should be afflicted (Fig. 7.3B). If the observed proportion of afflicted individuals agrees closely with the expectation from genetic theory, then the hypothesis of a single recessive gene is embraced, but if observation and prediction are too far apart, the hypothesis should be kept at arm's length and viewed with skepticism. The data from the 1939 study were perplexing. There were 433/1094 (39.6%) afflicted children in the 266 families, which is

TABLE 8.1 Frequency of PKU in Families of Different Size[a]

# of children in family	Total children	# with PKU	% with PKU
1	6	6	100
2	14	8	57
3	18	10	56
4	20	8	40
5	35	13	37

[a] Excerpt from Jervis (1939).
From Jervis, G. A. (1939). *The genetics of phenylpyruvic oligophrenia*. The British Journal of Psychiatry, 85, 719–762.

quite a few more than expected. Why the discrepancy? A hint is apparent when data are viewed separately according to the number of siblings (Table 8.1). As family size increased, the proportion of mentally deficient children decreased. This is peculiar and makes no sense at all from the standpoint of genetic transmission. The crux of the problem can be seen in the six families with only one child: every child was afflicted. We expect that among all matings of two carriers that produce only one child, 3/4 should have a normal child and 1/4 should have an affected child. So, the problem was in the method of choosing people to be in the study. Because the researchers started by finding mentally deficient probands whose urine showed the green reaction, all the families wherein two carriers mated but produced only one normal child were missed.

The problem of *biased sampling* is inherent in the *proband method* and does not raise doubts about the integrity or skills of the researchers. Textbook examples of single-locus Mendelian inheritance that yield neat 3:1 ratios of normal to afflicted persons assume that all of the families of two carriers are known and included in the study, but this is often not true in research with human beings. The scientists studying the genetics of PKU were aware of this problem and adjusted their figures to correct for the sampling bias. There are several ways to do this. The simplest but least accurate is to delete all the probands from the calculations and look only at brothers and sisters tested after the proband was detected. About 1/4 of these people should show the defect. For the 266 families mentioned above, there must have been 266 probands, and the adjusted proportion is $(433-266)/(1094-266) = 167/828 = 20.2\%$, which is a little low but is reasonably close to the expected 25%. A more sophisticated mathematical method of adjusting for the bias (Weinberg-Lenz) yielded an estimate of 27.4%, which is very close to 25%.

To summarize, four facts led researchers to proclaim discovery of a simple genetic cause of PKU. (1) There was a clear distinction between afflicted and normal

FIG. 8.1 Chemical pathways involved in PKU. Defect in PAH gene prevents conversion of phenylalanine to tyrosine in the liver. Tetrahydrobiopterin (BH4) serves as cofactor. When phenylalanine rises to high concentrations, some of it is converted to phenylpyruvic acid and phenylacetic acid that are then excreted in urine.

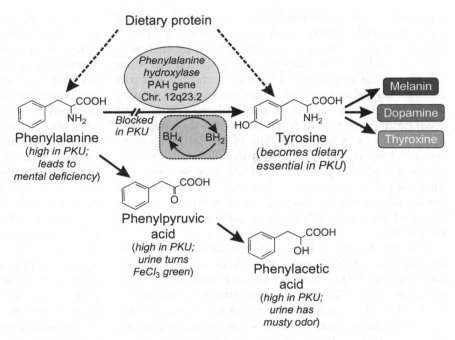

children as defined by the response of their urine to a solution of iron chloride. Children having other sources of mental disability did not show the green reaction product. The dichotomous distribution of phenotypes is one indicator of transmission of a single gene having a large effect. (2) No parents of affected children were themselves affected, which indicates recessive gene expression. (3) The proportion of affected children was close to 25%, typical of a recessive gene. (4) Equal numbers of males and females were affected, which indicates the gene is on one of the 22 autosomes, not a sex chromosome. Evidence pointed to a biochemical abnormality of the metabolism of the amino acid phenylalanine, something that might very well arise from a defect in an enzyme encoded by a gene.

CHEMICAL PATHWAY AND PREVENTION

One year later, a group of researchers found that the level of phenylalanine in the blood of children who had phenylpyruvic acid in their urine was about 30 times the level in normal people (Jervis, Block, Bolling, & Kanze, 1940). The measurement of phenylalanine levels, being a quantity expressed in terms of milligrams of phenylalanine per 100 mL of blood serum (mg%), was much better for making a diagnosis than a simple presence or absence of a green reaction product in the iron chloride solution. Among normal people, none had a blood concentration of more than 3 mg%, whereas all known cases

of PKU registered levels greater than 10 mg%. This discovery prepared the way for new methods of treating the disorder. In 1947, it was established that in normal people, the compound phenylalanine is converted into tyrosine and transported widely throughout the body via the bloodstream (Jervis, 1947). Tyrosine is then the starting point for the synthesis of the skin pigment melanin, the thyroid hormone thyroxine, and the neurotransmitter molecules norepinephrine and dopamine (Fig. 8.1). In children with PKU, however, the phenylalanine could not be converted into tyrosine; there was a complete block of this metabolic step, and phenylalanine concentration rose to high levels that damaged the brain. In 1953, the biochemical cause of PKU was finally identified: there is an enzyme in the cells of the *liver* that normally converts phenylalanine from the blood into tyrosine, which in turn is secreted back into the bloodstream and circulated to other cells in the body. They named the enzyme phenylalanine hydroxylase because it added one hydroxyl group (OH) to the phenylalanine molecule, making it tyrosine (Fig. 8.1). The activity of the enzyme in the livers of people with PKU was so low that it could not be detected (Jervis, 1953). Researchers were confident that the unique enzyme was encoded in a gene and the mutant form of the gene was recessive. The gene was then named the phenylalanine hydroxylase gene (PAH).

From the time when Fölling developed the iron chloride test until the 1940s when the more sensitive phenylalanine assay was adopted, thousands of children with PKU were detected in Canada, the United States, and countries in Europe. Some of those cases were detected soon after birth because a previous sibling was known to have PKU, and hence, a test was ordered for each subsequent child of the two carriers. It was noticed that some infants whose development was not retarded at birth began to lag behind normal children after a few months. Furthermore, their phenylalanine levels soon after birth were normal but then rose quickly when they began to drink mother's milk. The decline of mental functioning was precipitous, mental disability in childhood was usually profound, and persons with PKU often required lifetime custodial care in an institution. At the time, the doctor who made the diagnosis was powerless to save the child from his or her fate or relieve the sorrow of the parents.

Once this much was known, investigators asked the following: where does the excess phenylalanine come from in the first place? This was already well understood from nutrition science. The doctrine of intermediate metabolism teaches that all animals including humans ingest complex foods from the environment that consist of large molecules of carbohydrates, fats, and proteins, and then, the digestive system breaks them down into small molecules termed the central intermediates. Those molecules enter the blood stream and are transported to organs and tissues throughout the body, where they are absorbed and then utilized. At their destinations, the small molecules are sometimes joined together in new ways or converted into some other kind of molecule that is not present in the diet. For example, amino acids (Table 2.1) are central intermediates obtained from dietary protein. When protein is eaten, it is first broken down into its constituent amino acids by enzymes in the stomach (pepsin) and small intestine (trypsin), and then, the amino acid molecules are carried into the bloodstream and circulated to every cell in the body, where they are assembled into new kinds of proteins (Chapter 3) that are the characteristic of the animal that ate the food. Some amino acids are also used to manufacture neurotransmitters (Chapter 4). An amino acid that cannot be manufactured in the body must be obtained via the diet, and it is termed a dietary essential.

Humans cannot synthesize phenylalanine in the body and must get it via the diet, whereas tyrosine can be made from phenylalanine and is not normally a dietary essential. The primary source of both amino acids is protein. A rich source of protein containing phenylalanine is milk, which of course is the main food given to an infant during the first few months of life. Another rich source is red meat. It was known that a lab animal could live quite well with no protein in the diet at all, provided it is given enough of the right kinds of amino acids. Pondering these various facts, scientists wondered: if a child with PKU were fed an artificial diet with no phenylalanine, would this prevent the buildup of high phenylalanine levels in the blood and prevent the mental decline?

A few infants known to have PKU were placed on a synthetic diet very low in Phe, and results were encouraging. Unfortunately, several infants soon developed symptoms of protein malnutrition. The cause for this was quickly found and corrected. Because phenylalanine is a necessary part of thousands of different proteins in the body, every person, even one with PKU, must have *some* phenylalanine in the diet in order to be able to synthesize proteins in the muscles, heart, brain, and other organs. A diet was needed that had enough phenylalanine to permit the child's cells to make important proteins but not enough to allow high levels of phenylalanine to build up in the bloodstream.

This was achieved in 1955 by closely monitoring the phenylalanine level in the blood of each infant and adjusting the amount in the diet accordingly (Armstrong & Tyler, 1955; Bickel, Gerrard, & Hickmans, 1953). Results were nothing short of spectacular. Infants born with PKU but nurtured on the special diet from an early age instead of mother's or cow's milk never developed any of the symptoms of PKU. Blood phenylalanine levels were near normal; there was no phenylpyruvic acid in the urine; there was no peculiar odor; and most important of all, brain growth and behavioral development were quite normal. Children with PKU who were placed on the low-phenylalanine diet within 3 months of birth were later found to have an average IQ close to 100 (Horner, Streamer, Alejandrino, Reed, & Ibbott, 1962; Levy & Waisbren, 1987). If the diet was not started until 6 or more months after birth, there was a lasting deficit in mental test scores, depending on when the diet began, but decline into the range of severe mental disability was prevented. The consequence of a rare but devastating form of mental disability, which at one time caused suffering for many children and their families, had been overcome. The children still had the abnormal form of the PAH gene and phenylalanine hydroxylase enzyme, but they no longer had PKU. PKU is a phenotype, a set of symptoms. The special diet prevented those symptoms.

NEWBORN SCREENING, PREVENTION AND A CRITICAL PERIOD

When it was realized that a heavy financial burden on the state and suffering of children and their families could be averted by preventing PKU, many governments sponsored programs to detect and treat children with the disorder. In 1962, several states in the United States initiated newborn screening programs in which each baby born in a public hospital was given a test for a high level of phenylalanine in the blood (Guthrie & Susi, 1963). Before long, the tests were adopted in several provinces of Canada and countries in Europe. Several low-phenylalanine synthetic formulas for PKU infants were devised, such as

Lofenalac that is still in use, and natural foods were assayed for their phenylalanine levels so that after weaning from the synthetic milk, the infants could safely eat a nutritious meal of solid food. A survey done in 1971 of 117 institutions for the mentally deficient in Canada and the United States revealed that after the newborn screening programs began, there was *not even one* new case of PKU that led to hospitalization for mental disability (MacCready & Levy, 1972). For those who treat mental disability today, PKU has literally become a famous textbook example with which few have had contact in their professional careers. It was not simply that the PKU children no longer needed support from an institution. The treatment was so effective that they no longer had PKU at all.

PKU is a cluster of phenotypes. There is a very high level of phenylalanine in the blood, a presence of phenylpyruvic acid in the urine that turns a solution of iron chloride green, phenylacetic acid in the urine that gives the infant a musty odor, a requirement to get tyrosine from the diet, and general retardation of development and mental disability that worsens as the time without treatment increases. The genotype, on the other hand, is two abnormal alleles of the PAH gene that code for ineffective forms of the enzyme phenylalanine hydroxylase.

The new dietary treatment was not a panacea for all those afflicted with PKU. When children who had suffered the disorder for several years were then fed a low-phenylalanine diet, their blood levels of phenylalanine declined as expected, but their mental functions remained poor. Apparently, the brain damage had already been done and could no longer be corrected through a simple change in diet. There seemed to be a *sensitive period* of brain development beginning soon after birth, when levels of phenylalanine in the blood are critically important. The earlier the child goes on the diet, the better the outcome will be. Starting the diet after the age of 16 months seemed to help some children to a moderate degree but could not raise them above the criterion for mental disability. A study that placed children with PKU on the diet after the age of 6 years found that the patients averaged a dismal 38 on an IQ test after several months on the low-phenylalanine regimen (Levy & Waisbren, 1987). For them and the millions of other children who must have suffered from PKU in the course of human history, the discoveries of science had come too late.

With the advent of the newborn screening programs, which examined literally millions of infants, many new facts and further complexity were uncovered. For the first time, a really accurate estimate of the frequency of PKU was possible because all children in a district were assessed by newborn screening. There are different frequencies of PKU in various ethnic groups. PKU tends to be relatively common in Ireland, Scotland, and Wales, whereas it is relatively rare among Africans.

An approximation of the rate among people of European ancestry is about one case in every 10,000–14,000 births. Using that figure, geneticists estimated that about 1 out of every 60 people in a population having mainly European ancestry must be carriers of one defective allele of the PAH gene but show no symptoms. If the probability of randomly choosing a carrier as a mate is $1/60$, then the probability of pairing two carriers is $(1/60)(1/60) = 1/3600$. For a child of two carriers, the probability of getting two mutant alleles is $1/4$, so the probability of any child having PKU is $(1/3600)(1/4) = 1/14,400$.

CARRIER DETECTION

Because the measurement of phenylalanine in the blood is so sensitive, it became possible to find out who carries just one defective gene by using the *phenylalanine tolerance test*. A person is given a large dose of phenylalanine by mouth (about 100 mg per kilogram of body weight). Small samples of blood are then taken at 1, 2, 3, and 4h or more after drinking the phenylalanine (Hsia, Driscoll, Troll, & Knox, 1956; Jervis, 1960). In carriers, the concentration of phenylalanine in the blood rises to a higher level and takes longer to return to the predrink level than in noncarriers. The one normal gene possessed by carriers is sufficient to keep blood phenylalanine levels low when the person consumes a normal diet, but it is not enough to metabolize quickly a sudden, high dose of phenylalanine. Because the test is somewhat inconvenient and expensive, it has never been used for screening people in an entire population but instead was sometimes used when there had been a previous case of PKU in a family that would make it likely other carriers would be detected. Once a carrier is known, the only information of any possible value is whether that person's partner is also a carrier. A couple planning to have children might want to know if there is a high probability of having a child with PKU and be prepared to commence the diet at an early age. With the advent of newborn screening programs, such a child would be detected with routine screening. Given the rarity of mutant PAH alleles in the population, it is unlikely that a known carrier would seek to screen all prospective mates for PAH genotype. Genetic testing has not yet become part of the cultural rituals of dating and mating. The rationale for carrier detection will be discussed at greater length in Chapter 20.

OTHER CAUSES OF ELEVATED PHENYLALANINE

The newborn screening programs uncovered a few cases of high blood phenylalanine or hyperphenylalaninemia (HPA) that do not behave like typical PKU

(Kaufman, 1983). In rare instances, the high level of phenylalanine after birth is transient and for unknown reasons returns to an acceptable level without recourse to a special diet. This category can be detected by placing a child, who has been on the special diet, on a high-protein diet for a few days at 6 months or 1 year of age and measuring the blood phenylalanine level. If it remains low, the child does not have PKU and can stay on a normal diet. This condition, known as transient HPA, is not a simple genetic disorder. In a few cases of HPA, the level of phenylalanine in blood is high but not as high as in classical PKU. This can happen when a mutant allele of the PAH gene reduces but does not abolish enzyme activity. Dietary treatment is still required, but foods with high levels of phenylalanine need not be avoided altogether, provided blood levels are closely monitored. A number of patients with high phenylalanine levels nevertheless have normal activity of phenylalanine hydroxylase in the liver. Something else is impeding the conversion of phenylalanine to tyrosine. It is now known that the metabolism of phenylalanine requires the participation of a cofactor called tetrahydro-biopterin (BH4). If there is not enough active BH4 present in the liver, the phenylalanine level in the blood will rise, just as in PKU (Blau, 2010). The BH4 molecule is produced from a precursor called dihydrobiopterin (BH2), and this conversion is done with the aid of the enzyme dihydropteridine reductase. The level of BH4 can be deficient for two reasons—defective dihydropteridine reductase (a genetic defect itself in the BH4-deficient hyperphenylalaninemia (HPABH4C) gene) or faulty synthesis of BH2. About 2% of cases of HPA detected by infant screening have BH4 deficiency. They can be treated by low-phenylalanine diet plus BH4 supplements.

Thus, a high level of phenylalanine in the blood of an infant fed milk that is high in protein is a nonspecific phenotype with the name "hyperphenylalaninemia." HPA is a heterogeneous category with several different causes, and each distinct disorder with a unique cause needs to be treated in a different way. The category comprises (a) transient HPA, (b) classical PKU with no PAH activity, (c) milder forms of HPA wherein there is some PAH enzyme activity, (d) dihydropteridine reductase deficiency that impairs BH4, and (e) deficiency of BH2. To establish which specific disorder a child with HPA has, the medical geneticist must apply differential diagnosis to exclude the other possibilities.

DIET TERMINATION

The low-phenylalanine diet has been adopted very widely in countries with advanced medical systems, but it is inconvenient for the child and the family, and there are costs to the family and governments. A single year of Lofenalac can cost a family up to $8500 (Khamsi, 2013), and some medical insurance plans do not reimburse those costs for what is classed as a medical food rather than a drug. Therefore, people want to know if and when it is safe for a child to quit the bland special diet and return to eating hamburgers and peanut butter washed down with cow's milk. It is certain that returning to a conventional diet will lead to a return of high blood phenylalanine levels for those born with inactive PAH enzyme. The question is whether this will impair mental functioning later in life. As reviewed by several authorities (Kaufman, 1983; Scriver & Clow, 1980), available data in the 1980s did not point to an obvious time for stopping the diet. Results from a clinic in Boston indicated that there was no significant decline in IQ for 30 children who had gone on the special diet within 6 weeks of birth and returned to a regular diet at between 4 and 6 1/2 years of age. In a study of 140 children, all were on the special diet until 6 years of age, and then, each child was randomly assigned to one of two groups; one group remained on the diet, and the other returned to a normal diet. After 2 years off the diet, the normal diet group showed a significantly lower IQ than those who stayed on the special diet. Another study indicated better outcomes when the diet was continued until 8 years; average IQ scores of children who stopped the diet before 72 months, from 72 to 95 months, and 96 months or later were 94, 100, and 104, respectively. Many authorities recommended that a person who showed PKU in early childhood should stay on a low-phenylalanine diet throughout adulthood, whereas Koch and Wenz (1987) consented to diet discontinuation after the age of 12 years. Considering the conflicting opinions of experts, there was a need for more and better data to decide the issue.

Given that the child stays on the low-Phe diet long enough, prospects for normal intellectual and social functions are very good (Koch, Yusin, & Fishler, 1985). In many countries, there now are thousands of people who lack an effective PAH enzyme but are able to lead productive lives and even have families of their own.

MATERNAL PKU

The challenges do not end when an adult with two mutant alleles of PAH goes off the low-phenylalanine diet and experiences elevated phenylalanine levels once again. Several adult women with PKU returned to high blood phenylalanine levels and had children who did not have PKU. But their children in almost every case were seriously mentally impaired and had small heads and brains or "microcephaly" (Fisch, Doeden, Lansky, & Anderson, 1969). In one family, the IQ scores were obtained for the parents and five of their children, and all family members were given a phenylalanine test

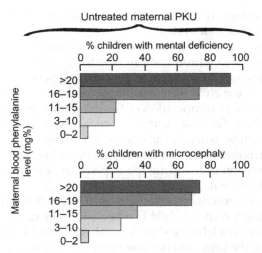

FIG. 8.2 Rates of mental deficiency and microcephaly in children in relation to mother's blood level of phenylalanine during pregnancy for cases of untreated maternal PKU. The children themselves did not have PKU and possessed just one mutant allele of the PAH gene. *Reprinted from Lenke, R. R., & Levy, H. L. (1982). Maternal phenylketonuria—results of dietary therapy.* American Journal of Obstetrics and Gynecology, 142, 548–553. Copyright 1982, with permission of Elsevier.

(Blomquist, Gustavson, & Holmgren, 1980). The father had a normal phenylalanine level and mental ability. The mother showed a high phenylalanine level of 30.4 mg% in her blood, but her scores on several IQ tests were 80–90, in the normal range. The couple had six children, five of whom survived infancy. All five were microcephalic and mentally deficient with IQ scores from 30 to 40, but they all had normal phenylalanine levels. A large survey of over 300 adult women with HPA in several different countries (Lenke & Levy, 1982) found that percentages of their children with mental abnormalities varied with the level of phenylalanine in the mother's bloodstream during pregnancy (Fig. 8.2). It was the high phenylalanine level in the maternal environment that caused the mental disability in the children. The low IQ scores of the children were not caused by their own genotypes. The mother's genotype generated a maternal environment that harmed the children's brain development.

Since this problem was recognized, physicians have routinely treated a number of PKU women by putting them back on the low-phenylalanine diet during pregnancy in order to bring the blood level down to normal and avoid brain damage to the fetus. Data from several studies indicate that the fetus and child will be spared noteworthy deficits if the woman with PKU starts the special diet before she becomes pregnant and faithfully adheres to it until the child is born. After birth, the child does not need to avoid milk and other sources of phenylalanine because it has one normal PAH allele and does not have PKU. Several women who went back on the diet before pregnancy gave birth to healthy children who scored above average on standardized mental tests,

and those who resumed the diet earlier in pregnancy had healthier children.

SOCIAL FACTORS AND COMPLIANCE

At a distance, it is clear why feeding a child or a pregnant mother with PKU using a low-phenylalanine diet is good for the child and its family. In practice, however, compliance with the diet is not easily achieved. When the infant is subsisting on milk, it is not too difficult to use bottle-feeding and replace natural milk with a special solution, such as Lofenalac, that is manufactured to be low in phenylalanine. When the child begins to consume solid food and milk, taste and palatability become major issues, and they are multiplied by a variety of social factors. Several authors have noted that many families are unable to adjust to the newly required maintenance of dietary therapy. It is clear that in many families, the reasons for remaining on the diet and potential consequences of going off it were not well understood or appreciated. Rather than being helped by peers in complying with the diet, it was noted that "The majority of the patients would not tell their friends that they had PKU" (Koch et al., 1985). In maternal PKU, when an adult woman returned to the diet, there were several challenges. Levy (1987) observed, "Increasingly, it is becoming obvious that optimal therapy for maternal PKU may be impossible unless the psychosocial environment of these young women is understood and their needs relative to this environment are met."

In the early stages of research on PKU treatment regimens, the wide range of ages when diet control was lost and when diet was resumed in maternal PKU was not too surprising because even scientific experts on PKU disagreed on ages when a normal diet would do little harm. As evidence accumulated, it became clear that the best medical advice was to stay on a special diet low in phenylalanine until and throughout adulthood. But "playing it safe" was easier said than done. Better education and social supports were needed to achieve medical goals.

LESSONS LEARNED 30 YEARS AGO

By 1984, the genetic cause of PKU, a noteworthy metabolic defect, was well understood. It was and still is presented as a textbook example, a prototypical recessive Mendelian genetic mutation in the PAH gene. Briefly stated, among the millions of people in a country, a few of them carry a mutant allele that codes for an ineffective form of the phenylalanine hydroxylase enzyme, but they do not realize this and do not suffer from its effects because their other allele works well enough. On rare occasions, two such carriers happen to mate with each other and produce one or more children. On average, about 3/4 of their children are normal, although many are also carriers of the mutant allele, but about 1/4 receive two copies of the defective gene, one from each parent. Because there is little or no active phenylalanine hydroxylase in the liver of a child who has two defective alleles, phenylalanine in the bloodstream cannot be converted into tyrosine. When the infant begins to drink milk, the concentration of phenylalanine in the bloodstream rises to extremely high levels that are toxic for the developing brain and lead to mental disability. Some of the phenylalanine is converted into phenylacetic acid and excreted via the urine, which gives the urine a musty smell. Rearing a child on a low-phenylalanine diet can prevent the high blood levels of phenylalanine and avoid the harmful consequences of the mutation. It can also prevent harm to a genetically normal fetus carried by a pregnant woman who has two defective PAH alleles. Thus, knowledge of the gene and the enzyme for which it codes provides a good explanation for the abnormal phenotypes associated with PKU and its sensitivity to diet and the pattern of occurrence in families. This illuminating account of PKU was formulated before the era of molecular genetics, when there was no information about the nature of the defect in the DNA or the location of the gene on a chromosome.

Several important lessons had been learned by 1990 from the story of PKU.

HIGHLIGHTS PART 1

- PKU is an array of symptoms or phenotypes, unusual features of the blood, urine, and brain. When the child is reared on a special diet low in phenylalanine, he or she does not show those symptoms and no longer has PKU. There is no gene "for" PKU or a "PKU gene."
- PKU itself is not hereditary. The parents almost never show symptoms of PKU. A gene that codes for the enzyme phenylalanine hydroxylase (PAH) is transmitted from parents to offspring.
- The PAH gene does not code for any value of IQ or even the level of phenylalanine in the blood. The gene contains a code for one thing only—the sequence of amino acids in a large enzyme.
- It is wrong to assert in a general sense that nothing can be done to treat a disorder if its cause is genetic. Instead, it is our ignorance of the causes that often leads to poor outcomes.
- Although the mutant gene can have consequences that impair brain development and intelligence, it acts mainly in the liver, not the brain. This does not mean that intellectual activity takes place in the liver. The effect on the brain is indirect. The brain is part of a physiological system. PKU makes sense only in the context of that system.
- Consequences of the mutant alleles depend strongly on the child's environment. In fact, having two mutant alleles that disable phenylalanine hydroxylase render the child *more sensitive* to variations in the amino acid composition of the diet than normal children. This is a clear instance of *gene–environment interaction*, wherein the effects of a changed environment depend on the genotype, and the expression of the genotype depends on the rearing environment.
- The environmental effects involve a *sensitive period* during which the child is especially vulnerable to a high-phenylalanine diet. Several years after birth, when the brain has become relatively mature, a return to a normal phenylalanine diet no longer results in seriously impaired mental functioning.
- Compliance with the special diet for a person with PKU depends strongly on social relations with other people. The social context of a person with the genetic defect matters greatly. Social skills and motivation of the person with PKU are important, while guidance and support by a family and professionals for the person with PKU influence adherence to an unpleasant dietary regime year after year.

30 YEARS LATER

More recent reviews of PKU and PAH function (Blau, 2016; Blau, van Spronsen, & Levy, 2010; Scriver, 2007) support the earlier conclusions and provide deeper insights into the disorder. Now that we have entered the era of DNA, the genetic basis of PKU is better understood. A partial DNA sequence of the PAH gene was determined in 1983 (Woo, Lidsky, Güttler, Chandra, & Robson, 1983), and it was mapped to a narrow region of bands 22–24 on the q arm of chromosome 12 (Lidsky et al., 1985). The current listings of the PAH gene in OMIM, GeneCards, and UniProt state the PAH enzyme has 452 amino acids encoded in 13 exons. The mRNA from the PAH gene and the enzyme itself are strongly expressed mainly in the liver and kidney, which boosts the credibility of the argument that harmful effects on brain development are consequences of high blood phenylalanine, rather than expression of the PAH gene in the brain itself.

Update: Genotype

Knowing more about the DNA sequence of the PAH gene has led to the discovery of numerous alleles that alter the protein structure of phenylalanine hydroxylase

(Rouse et al., 2000). It turns out that a child with PKU always has two abnormal alleles but rarely are they copies of the same allele. Provided each allele codes for a defective form of the enzyme, the child's liver will not be able to manufacture functional phenylalanine hydroxylase. Each mutant allele shows recessive expression because, when paired with a normal PAH allele, it has no negative effect on the child.

The PAHdb Phenylalanine Hydroxylase Locus Knowledgebase at McGill University (Scriver et al., 2003) in 2009 presented data on 567 PAH mutations classified by the type of mutation and country of origin, while the Human Gene Mutation Database maintained at the Institute of Medical Genetics in Cardiff in 2015 listed 703 known mutations in the PAH gene. OMIM entry 612,349 for PAH in 2016 lists 612 mutations classified by phenotype, but there are many for which information on the clinical phenotype is currently lacking. The PAHvdb database listed 957 allelic variants of PAH in 2016 and boosted the total to 1044 in April of 2018. The BIOPKU database provides information about phenotypes arising from many genotypes and based on more than 10,000 PKU patients (Blau, 2016). New alleles that continue to be discovered are typically very rare and localized in one or a few populations.

There is evidence that a clinical phenotype of HPA with moderately increased levels of phenylalanine in infant blood can result from specific alleles of PAH that do not completely eliminate PAH activity (Blau, 2016; Blau et al., 2010). To qualify clinically as PKU, PAH activity must be very low or totally absent. For example, the OMIM ClinVar database indicates that the PAH mutation c.1243C > T(p.Asp415Asn) results in the milder HPA, whereas allele c.1240 T > C(p.Tyr414His) that is just one amino acid away, at position 414, results in classical PKU. UniProt lists PAH amino acid substitution Ser16Pro as a cause of classical PKU and Phe39Leu as a form of the milder HPA. At present, DNA data are not routinely used to diagnose PKU or tetrahydrobiopterin (BH4) deficiency, and the full DNA sequence of the PAH gene is still not known (Blau, 2016).

Detailed investigation of phenylalanine and tyrosine metabolism has revealed considerable complexity in what was once regarded as a simple, single-gene disorder (Rocha & MacDonald, 2016; Scriver & Waters, 1999; Weiss & Buchanan, 2003). One consistent finding is that the level of phenylalanine in blood cannot be explained solely by knowing the specific allele of the PAH gene. Scriver (2007) pointed out that "mutant genotype…will not always predict the finer features of the corresponding mutant phenotype." He concluded, "An important practical lesson to take from this aspect of the PKU story and its counterparts is obvious: *treat the actual phenotype not the one predicted from genotype at the major locus.*" (Emphasis in the original.)

Update: Diagnosis

The most dependable sign of PKU remains the level of phenylalanine and other amino acids in the blood. The specific times when blood should be measured and the specific levels considered harmful and warranting treatment vary with the child's age and even the country where the tests are mandated (Blau et al., 2010). In Europe, after an infant with a very high phenylalanine level has been identified, further tests to assess possible effects of BH4 supplementation are given on days two and three after birth to assess the role of BH4, whereas in the United States, levels of biopterins in urine are sometimes measured on days 1, 7, and 14, and then, response to extra BH4 is assessed at weeks 3 and 4.

Update: Treatment

As noted by Blau (2010), "The restriction of dietary phenylalanine remains the mainstay of phenylketonuria management." Thus, 60 years after the metabolic defect in PKU was identified and a low-phenylalanine diet was devised, diet "remains the primary treatment" (Singh et al., 2014). Accumulated evidence supports the recommendation that a child with PKU should begin the diet shortly after birth and remain on it through adulthood and that a woman who returns to a normal diet and therefore expresses PKU before pregnancy should return to the low-phenylalanine diet prior to conception and throughout gestation (Blau, 2010; Koch et al., 2002; Rouse et al., 2000). Cleary (2010) refers to the best practice as "diet for life." Several alternative treatments or supplements to the usual dietary treatment are being considered (Rocha & MacDonald, 2016).

Update: Diet Formulation

Guidelines established by the American College of Medical Genetics and Genomics provide extensive recommendations about frequency of assessing blood amino acids and amounts of nutrients, especially protein, needed to achieve safe phenylalanine levels throughout the life span (Singh et al., 2014; Vockley et al., 2014). They emphasize that the nutrients should be equivalent to a normal diet with the sole exception of phenylalanine and tyrosine levels, and levels of those amino acids need to be monitored often and adjustments in dietary intake made in order to keep levels in a narrow range. A cookbook is available to aid people with PKU in the family to provide an appetizing diet that can keep phenylalanine levels under control without or with minimal recourse to expensive prescription medications known as "medical foods" (Schuett, 1997). The amount of protein and phenylalanine in each unit of a serving is provided. Rocha and MacDonald (2016) stress that there is still an "urgent need" to learn more about food consumption patterns of people struggling to remain true to the low-phenylalanine diet.

Update: Compliance

All authorities in recent writings emphasize the social difficulties in achieving compliance with the medically recommended diets. Cleary (2010) commented on the way "the strict diet creates awkward social occasions." Blau et al. (2010) observed that "Long-term compliance with treatment remains a key challenge," and that there is a "social burden" that impairs the quality of life. Thus, it appears that more thorough investigation of family dynamics and social supports may help to achieve better phenotypic outcomes. It is unlikely that young adults seeking independence from family will carry their copy of Shuett and adhere to its recipes at all times, especially in a peer environment dominated by fast food. Better education of young people and their parents about how much departure from the ideal diet is actually harmful should be helpful. Families would also like to know what can be done to compensate for the occasional binge on a high-phenylalanine diet. The "diet for life" mantra may not be a realistic option for many people. It may also help to have better education of the wider public about the nature of genetic defects like the mutant PAH gene. This could lessen the stigma of having a genetic defect in an enzyme where consequences are easily ameliorated.

Update: Newborn Screening

Many jurisdictions now provide blood testing for infants in order to detect not only PKU but also a wide range of disorders related to genetic defects. The test requires only a simple heel prick to draw a drop of blood. Table 8.2 shows the kinds of tests and frequency of each condition detected in the Province of British Columbia in Canada. All can be detected by levels of compounds in the blood, and all can be treated effectively. For disorders that can be caused by more than one kind of genetic defect, diagnosis is done on phenotype, not genotype. Reasons for this and prospects for genetic screening in the future are discussed in Chapter 20.

THE FUTURE

Frequency of Mutant Forms of PAH

Today, newborn screening can effectively identify infants with PKU and then quickly switch them onto a low-phenylalanine diet. They then become persons capable of having families of their own, whereas previously, their copies of the mutant PAH alleles would have disappeared from the population. Is there cause to worry that

TABLE 8.2 Newborn Screening for Genetic Disorders in British Columbia[a]

Disorder	Rate	Disorder	Rate
Cystic fibrosis	1/3600	Methylmalonic acidemia	1/40,000
Congenital hypothyroidism[b]	1/4000	Sickle cell disease	1/45,000
Phenylketonuria[b]	1/12,000	Propionic acidemia	1/50,000
Medium-chain acyl-CoA dehydrogenase deficiency[b]	1/12,000	Long-chain hydroxyacyl-CoA dehydrogenase deficiency[b]	1/80,000
Congenital adrenal hyperplasia	1/16,000	Very-long-chain Acyl-CoA dehydrogenase deficiency	1/90,000
Argininosuccinic acidemia	1/30,000	Glutaric aciduria type 1[b]	1/120,000
Citrullinemia	1/30,000	Isovaleric acidemia	1/150,000
Galactosemia[b]	140,000	Maple syrup urine disease	1/185,000
Cobalamine	1/40,000	Homocystinuria	1/200,000

[a] *Arranged by rate. Source: Perinatal Services BC, URL: www.perinatalservicesbc.ca.*
[b] *On the original list of six disorders. Others added in 2010.*

frequency of the mutant PAH allele will increase in the population? To some extent, this might happen, but how large will the change be? A mathematical theory of genetics was devised long ago to predict changes in allele frequencies when people with certain genotypes produce more or fewer children than others. A numerical example can demonstrate the principle in the case of PKU.

In this exercise, let the normal allele of the PAH gene be shown as + and a mutant form that eliminates phenylalanine hydroxylase activity in the liver be shown as m. Here, we will combine all those very rare mutant alleles into one group. The proportion of normal alleles in a population can be symbolized as P(+) and for the inactive alleles combined it is then Q(m), such that $P+Q=1$. For a gene on an autosome, it is expected that the value of P will be the same for males and females. It is reasonable to assume that parents mate randomly with respect to their PAH genotype. Of course, mating itself is not random; many personal preferences, family pressures, and cultural rituals are involved. Nevertheless, all those many social factors are unrelated to PAH genotype. The net result is that the probability that a child gets the + allele from the mother is P, and it is also P for the allele from the father. Because the outcome for the allele from the father is unrelated to what the child gets from the mother, we say they are independent events. Probability theory teaches that the probability of getting one + from dad and another + from mom is the product obtained by multiplying $(P)(P) = P^2$. The same idea leads us to expect the probability of two m alleles to be $(Q)(Q) = Q^2$. To get a carrier, the child can get + from dad and m from mom with probability $(P)(Q)$, and m from dad and + from mom with $(Q)(P)$. Because it does not matter whether the mutant allele comes from the father or mother, the total probability of being a carrier is the sum $PQ + QP = 2PQ$.

Next, we need to estimate P and Q. If we can find Q, then $P = 1 - Q$. To know the frequency of carriers, we would need to run a phenylalanine tolerance test on every member of a population, a daunting and expensive task. Newborn screening of all infants in a population, however, uncovers virtually every instance of the m/m genotype, and the proportion of afflicted infants should be Q^2. If we know Q^2, just take its square root (SQRT) and then we have a good guess about the value of Q. If there are about 1 per 10,000 infants that show high levels of phenylalanine in the blood, then $Q = SQRT(0.0001) = 0.01$ or $1/100$, and $P(+) = 0.99$. The relative frequencies of the three genotypes $+/+$, $+/m$, and m/m in the population are then $P^2 = (0.99)(0.99) = 0.9801$, $2PQ = 2(0.99)(0.01) = 0.0198$, and $Q^2 = (0.01)(0.01) = 0.0001$. To check our calculations, note that the probabilities of these three possibilities should add to 1, and indeed, they do: $0.9801 + 0.0198 + 0.0001 = 1.0000$.

With good estimates of the frequencies of the alleles and genotypes, let us now consider a large population with 1 million people who interbreed to make the next generation. To keep things simple, suppose each couple has just two children, so that population size is stable across generations. Knowing the probabilities of each genotype, we can estimate the number of people with each genotype and the total number of alleles in the population, based on the idea that each person contributes two alleles to the next generation, one for each child.

Scenario 1. Genotypes result from random pairing of genes with no difference in fertility:

Genotype $+/+$	980,100 people	$2(980,100) = 1,960,200$ + alleles, no m alleles
Genotype $+/m$	19,800 people	$19,800 +$ alleles, $19,800\,m$ alleles
Genotype m/m	100 people	$200\,m$ alleles

Check calculations: $1,960,200 + 19,800 + 19,800 + 200 = 2,000,000$ alleles (1,000,000 people).

Scenario 2. None with PKU reproduce:

If the children with PKU are untreated, none will be able to have children, and all 200 mutant alleles possessed by afflicted people with genotype m/m will be lost forever. This will lead to a decrease in the value of Q in the next generation:

| Genotype $+/+$ | 980,100 people | $2(980,100) = 1,960,200$ + alleles, no mutant alleles |
| Genotype $+/m$ | 19,800 people | $19,800 +$ alleles, $19,800\,m$ alleles |

Totals: $1,960,200 + 19,800 + 19,800 = 1,999,800$ alleles.

Among the 1,999,800 alleles that are transmitted to the next generation, the number of the normal + allele will be $1,960,200 + 19,800 = 1,980,000$, while there will be $19,800\,m$ alleles. Thus, the gene frequencies in the next generation will be $P_1 = (198,000)/(1,999,800) = 0.9901$ and $Q_1 = (19,800)/(1,999,800) = 0.0099$. The net result is that the frequency of the mutant allele will decline from 0.01 to 0.0099. That appears to be a rather small difference. Let us look at it in terms of proportions of people in the next generation with the three genotypes, using $P_1 = 0.9901$ and $Q_1 = 0.0099$. Let us assume they generate 1,000,000 children, in order to keep the population size stable:

Genotype $+/+$	$(0.9901)(0.9901) = 0.980298$ or 980,298 people with no m alleles
Genotype $+/m$	$2(0.9901)(0.0099) = 0.019604$ or 19,604 people with $19,604\,m$ alleles
Genotype m/m	$(0.0099)(0.0099) = 0.000098$ or 98 people with $196\,m$ alleles

The net result in a population of 1 million people wherein all with PKU are unable to reproduce would be a decline in the number of PKU cases from 100 to 98, a paltry two people fewer with a treatable genetic disorder. Two important conclusions about a recessive gene with very harmful effects are warranted from this example. First, when the recessive gene is rare, the vast majority of the mutant alleles are possessed by carriers who show no symptoms. Second, if those with two copies of the mutant allele do not reproduce, the change in gene frequency and the reduction in number of people afflicted with the disorder in the next generation will be extremely small.

If infants with PKU can later have children and they have normal numbers of offspring, about the same numbers as those without the defect in phenylalanine metabolism, the gene and genotype frequencies will tend to remain the same from one generation to the next. There would be no increase in the frequency of the mutant form of PAH. An increase in defective PAH alleles would occur only if new treatments resulted in affected people having more children and healthier children than those lacking even one mutant PAH allele. This seems most unlikely.

Although the change in allele frequency from just one generation wherein everyone with PKU fails to reproduce is small, when this process continues generation after generation, it will eventually lead to a very low frequency of the mutant alleles and instances of PKU. It is likely that this indeed happened over the course of human history and among our simian ancestors.

There are also likely to be new mutations in the PAH gene that are transmitted to the next generation. In almost every instance, the mutation will occur in someone who has two normal copies of the PAH gene in almost all body cells, and the child who inherits the new mutation via a germ cell will then be a carrier, not someone affected with PKU. Very gradually over many generations, there will be an increase in the overall frequency of defective PAH alleles in the population if the new mutations are not purged by infertility or inviability. For the first few generations when people born with PKU can later have families of their own, we expect that the frequency of PKU would rise by about two cases per million people each generation. It would require centuries to even notice the difference at the phenotypic level.

Consequences of the New Molecular Genetics

To date, the new genetics based on knowledge of DNA and the human genome sequence has had no significant impact on the diagnosis or treatment of PKU. Almost all major discoveries that find application today in modern medical practice with PKU occurred before 1990. Knowing which specific alleles of the PAH gene a child with PKU possesses does not yet help to formulate a better diet in comparison with what can be done by knowing just the levels of phenylalanine, tyrosine, and BH4. Advanced genetic knowledge does not help to achieve better compliance with the low-phenylalanine diet or keep a person on that diet for life.

There are new advances in genetic engineering every year, and it is conceivable that some clever and technically skilled lab worker will find a way to snip out a defective PAH allele and replace it with a good one. Something like this can already be done in lab mice, albeit at great expense with many failures along the way. At the present time, a world in which all the mutant forms of the PAH gene have been cut out and flushed away is a realm of science fiction. New discoveries could change this situation. There is no reason to believe that today's state of the art in diagnosis and treatment of PKU will be the best we can do over the coming centuries. Nevertheless, what exists today is really quite effective. Not perfect, but darn good.

Compliance and the Social Context

All authorities reviewing the situation with PKU remark on the difficulties of adhering to a bland, unappetizing diet in the face of social pressures to conform to a modern youth lifestyle. Part of the challenge arises from the fact that consequences of dietary relapse are not immediate; effects on mental function require months

and years to become manifest. Without doubt, some of the families of children born with PKU are well educated, affluent, ambitious, and highly motivated to achieve the best possible outcome for their children. Nevertheless, there must be many situations in which family life is unstable and lacking in some of the essentials for a child to thrive, even for a child without the added liability of a genetic mutation like the PAH defect. Studies of family life and its social and economic context deserve priority. Several of the studies cited in this chapter report that poor compliance with the optimal diet in childhood can reduce IQ test scores by 10 points or more, a large effect by contemporary standards in psychology research. The cost to their families and to society is not an easy thing to assess, but it could be measured using established tools of epidemiology and developmental psychology.

For young scholars in the early stages of their careers, the lure of advanced biotechnology is great. When investigating a genetic defect like the kind involved in PKU, the genetic approach seems so obvious and unquestionable. Nevertheless, a geneticist should be first and foremost a scientist. Not every question can be answered fully at the molecular level (Chapter 1). Loyalty to one's discipline can blind a person to alternative approaches that may be far more productive. The importance of social factors in compliance was noted by medical researchers and nutritionists who were qualified to identify the problem but not capable of solving it. All genetic defects exist in a social context. Scientific studies of that context may lead to better results for the excellent diet already designed by expert nutritionists. In the early days of research on PKU, biochemists and physiologists led the way. Now that they have accomplished so much and have found methods to treat and prevent PKU, the forefront of research on this disorder perhaps will shift into the social sciences.

HIGHLIGHTS PART 2

- More than 1000 alleles of the PAH gene have been identified worldwide. Many of these impair the function of the PAH enzyme and show recessive expression. About 76% of people showing PKU early in life are heterozygous for different mutant alleles. It is now apparent that the degree of impairment of amino acid metabolism depends on the specific *combination* of alleles.
- Now that people with two mutant PAH alleles can avoid most of the symptoms of PKU and raise families of their own, the frequency of PAH mutations in the population is likely to increase over many generations, but the amount of the change will be trivially small.
- Compliance with the low-phenylalanine diet remains a major challenge. There is still a need for palatable and

socially acceptable diets to prevent symptoms of PKU in children and adults. Social factors are a major part of the story of PKU.

• PKU was the first specific biochemical cause of mental disability to be discovered and the first genetic condition to be treated successfully. It was once thought to be a simple single-gene disorder, but it turned out to be remarkably complex both genetically and phenotypically. This fact has important implications for attempts to understand what are now termed complex traits (Chapter 14).

References

Armstrong, M. D., & Tyler, F. H. (1955). Studies on phenylketonuria. I. Restricted phenylalanine intake in phenylketonuria. *The Journal of Clinical Investigation, 34,* 565–580.

Bickel, H., Gerrard, J., & Hickmans, E. M. (1953). Influence of phenylalanine intake on phenylketonuria. *Lancet, 265,* 812–813.

Blau, N. (2010). Disorders of tetrahyrdobiopterin metabolism presenting with and without hyperphenylalaninemia. In *Vol. 99. Molecular Genetics and Metabolism* (pp. 194–195). Elsevier Science.

Blau, N. (2016). Genetics of phenylketonuria: then and now. *Human Mutation, 37,* 508–515.

Blau, N., van Spronsen, F. J., & Levy, H. L. (2010). Phenylketonuria. *The Lancet, 376,* 1417–1427.

Blomquist, H., Gustavson, K.-H., & Holmgren, G. (1980). Severe mental retardation in five siblings due to maternal phenylketonuria. *Neuropediatrics, 11,* 256–261.

Cleary, M. (2010). Phenylketonuria. *Pediatrics and Child Health, 21,* 61–64.

Fisch, R. O., Doeden, D., Lansky, L. L., & Anderson, J. A. (1969). Maternal phenylketonuria: detrimental effects on embryogenesis and fetal development. *American Journal of Diseases of Children, 118,* 847–858.

Fölling, A. (1934). Über ausscheidung von phenylbrenztraubensäure in den harn als stoffwechselanomalie in verbindung mit imbezillität. *Hoppe-Seyler's Zeitschrift für Physiologische Chemie, 227,* 169–181.

Guthrie, R., & Susi, A. (1963). A simple phenylalanine method for detecting phenylketonuria in large populations of newborn infants. *Pediatrics, 32,* 338–343.

Horner, F. A., Streamer, C. W., Alejandrino, L. L., Reed, L. H., & Ibbott, F. (1962). Termination of dietary treatment of phenylketonuria. *New England Journal of Medicine, 266,* 79–81.

Hsia, D. Y., Driscoll, K. W., Troll, W., & Knox, W. E. (1956). Detection by phenylalanine tolerance tests of heterozygous carriers of phenylketonuria. *Nature, 178,* 1239–1240.

Jervis, G. A. (1939). The genetics of phenylpyruvic oligophrenia. *The British Journal of Psychiatry, 85,* 719–762.

Jervis, G. A. (1947). Studies on phenylpyruvic oligophrenia: the position of the metabolic error. *Journal of Biological Chemistry, 169,* 651–656.

Jervis, G. A. (1953). Phenylpyruvic oligophrenia: deficiency of phenylalanine oxidizing system. *Proceedings of the Society for Experimental Biology and Medicine, 82,* 514–515.

Jervis, G. A. (1960). Detection of heterozygotes for phenylketonuria. *Clinica Chimica Acta, 5,* 471–476.

Jervis, G. A., Block, R. J., Bolling, D., & Kanze, E. (1940). Chemical and metabolic studies on phenylalanine II. The phenylalanine content of the blood and spinal fluid in phenylpyruvic oligophrenia. *Journal of Biological Chemistry, 134,* 105–113.

Kaufman, S. (1983). Phenylketonuria and its variants. *Advances in Human Genetics, 13,* 217–297.

Khamsi, R. (2013). Rethinking the formula. *Nature Medicine, 19,* 525–529.

Koch, R., Burton, B., Hoganson, G., Peterson, R., Rhead, W., Rouse, B., ... Azen, C. (2002). Phenylketonuria in adulthood: a collaborative study. *Journal of Inherited Metabolic Disease, 25,* 333–346.

Koch, R., & Wenz, E. (1987). Phenylketonuria. *Annual Review of Nutrition, 7,* 117–135.

Koch, R., Yusin, M., & Fishler, K. (1985). Successful adjustment to society by adults with phenylketonuria. *Journal of Inherited Metabolic Disease, 8,* 209–211.

Lenke, R. R., & Levy, H. L. (1982). Maternal phenylketonuria—results of dietary therapy. *American Journal of Obstetrics and Gynecology, 142,* 548–553.

Levy, H. (1987). Maternal phenylketonuria. Review with emphasis on pathogenesis. *Enzyme, 38,* 312–320.

Levy, H. L., & Waisbren, S. (1987). The PKU paradigm: the mixed results from early dietary treatment. In S. Kaufman (Ed.), *Amino acids in health and disease: New perspectives* (pp. 539–551). New York: Alan R. Liss.

Lidsky, A. S., Ledley, F. D., DiLella, A. G., Kwok, S. C., Daiger, S. P., Robson, K. J., & Woo, S. L. (1985). Extensive restriction site polymorphism at the human phenylalanine hydroxylase locus and application in prenatal diagnosis of phenylketonuria. *American Journal of Human Genetics, 37,* 619–634.

MacCready, R. A., & Levy, H. L. (1972). The problem of maternal phenylketonuria. *American Journal of Obstetrics and Gynecology, 113,* 121–128.

Rocha, J. C., & MacDonald, A. (2016). Dietary intervention in the management of phenylketonuria: current perspectives. *Pediatric Health, Medicine and Therapeutics, 7,* 155–163.

Rouse, B., Matalon, R., Koch, R., Azen, C., Levy, H., Hanley, W., & de la Cruz, F. (2000). Maternal phenylketonuria syndrome: congenital heart defects, microcephaly, and developmental outcomes. *Journal of Pediatrics, 136,* 57–61.

Schuett, V. E. (1997). *Low protein cookery for phenylketonuria.* Madison, WI: University of Wisconsin Press.

Scriver, C. R. (2007). The PAH gene, phenylketonuria, and a paradigm shift. *Human Mutation, 28,* 831–845.

Scriver, C. R., & Clow, C. L. (1980). Phenylketonuria: epitome of human biochemical genetics (first of two parts). *New England Journal of Medicine, 303,* 1336–1342.

Scriver, C. R., Hurtubise, M., Konecki, D., Phommarinh, M., Prevost, L., Erlandsen, H., ... McDonald, D. (2003). PAHdb 2003: what a locus-specific knowledgebase can do. *Human Mutation, 21,* 333–344.

Scriver, C. R., & Waters, P. J. (1999). Monogenic traits are not simple—lessons from phenylketonuria. *Trends in Genetics, 15,* 267–272.

Singh, R. H., Rohr, F., Frazier, D., Cunningham, A., Mofidi, S., Ogata, B., & Van Calcar, S. C. (2014). Recommendations for the nutrition management of phenylalanine hydroxylase deficiency. *Genetics in Medicine, 16,* 121–131.

Vockley, J., Andersson, H. C., Antshel, K. M., Braverman, N. E., Burton, B. K., Frazier, D. M., & Genomics Therapeutics Committee (2014). Phenylalanine hydroxylase deficiency: diagnosis and management guideline. *Genetics in Medicine, 16,* 188–200.

Weiss, K. M., & Buchanan, A. V. (2003). Evolution by phenotype: a biomedical perspective. *Perspectives in Biology and Medicine, 46,* 159–182.

Woo, S. L., Lidsky, A. S., Güttler, F., Chandra, T., & Robson, K. J. (1983). Cloned human phenylalanine hydroxylase gene allows prenatal diagnosis and carrier detection of classical phenylketonuria. *Nature, 306,* 151–155.

9

Huntington Disease

For many years, Huntington disease (HD) has served as the prototypical, textbook example of a simple dominant genetic disorder, one that leads to degeneration of nerve cells in the brain and progressive loss of motor coordination ability and cognitive functioning. It was first described by Huntington in 1872, before the principles of genetic transmission or even genes themselves were part of science. Numerous studies of HD were done before the gene involved in the neurological disorder was localized to band 16 of the p arm of chromosome 4 in 1983 (Gusella et al., 1983). The HTT gene itself was identified and sequenced in 1993 by a large group of investigators forming the Huntington Disease Collaborative Research Group (MacDonald et al., 1993). Knowledge of the gene helped to clarify some puzzling facts about the occurrence of HD in families and instances for which there was no family history of HD. Research on this gene continues to uncover surprises, such as the finding that disease-related alleles are much more common in the general population than was once believed (Kay et al., 2016). A large trove of information about HD is now available (Bates, Tabrizi, & Jones, 2014), yet an effective treatment, preventive intervention, or cure remains elusive.

HISTORY

In the 19th century, neurologists recognized several kinds of "chorea," peculiar patterns of motor coordination that to some observers resembled dance movements. Huntington (1872) reported a rare "medical curiosity," a subset of patients in whom there was always a family history and for whom the disease became evident before 40 years of age. Whereas many cases of chorea that he reviewed were thought to respond to treatments such as the purgative calomel plus rhubarb and castor oil, spirits of turpentine, or blistering of the spinal column by tartar emetic, nothing seemed to help those suffering from the familial form. Huntington stressed that if a parent shows the problem, one or more children always show it, and it never skips a generation. He also observed that symptoms include insanity, exemplified by the cases of two men aged 50 who could hardly walk but who "never let an opportunity to flirt with a girl go past unimproved."

After 1900, when the principles of Mendelian genetics were understood, HD became widely known as a dominant genetic disorder, as portrayed in a family tree in which every generation shows the defect (Fig. 9.1A). The disease gained prominence when folk singer Woody Guthrie was found to have it. After his death in 1967, his widow Marjorie helped to found the Committee to Combat Huntington Disease and wrote a vivid account of her husband's struggles with HD.

Accurate diagnosis of HD was difficult until the gene was identified and a genetic test became widely available. Initial symptoms were often very similar to those of Parkinson's disease or multiple sclerosis, disorders that sometimes appeared in a parent and child. As more cases of HD were investigated, it became clear that a substantial fraction did not involve an affected parent. It was of course possible that a parent who carried the HD allele died young, before symptoms emerged, but cases were found in which both parents of an HD patient remained together and symptom-free until old age. This was a major challenge to the simplified story of a single dominant gene. Researchers considered several causes that might have this effect, such as the legal father not being the biological father. However, once the gene was known, most of the facts fell neatly into place, and many faithful partners were exonerated.

Genes, Brain Function, and Behavior
https://doi.org/10.1016/B978-0-12-812832-9.00009-9

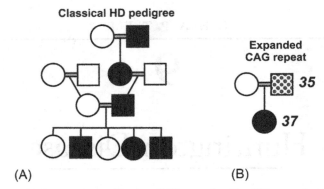

FIG. 9.1 Transmission of gene causing HD from parent to offspring. (A) In most families, HD occurs in every generation, and on average, half of children show symptoms. (B) In a few cases, neither parent is affected by HD, but the size of the gene passed to a child expands into a range that generates symptoms.

AN EXPANDABLE MUTATION

The genetic defect in HD is radically different from a single-base substitution or deletion that is typical in recessive disorders (Fig. 2.6). When the gene was discovered (MacDonald et al., 1993), the large protein for which it codes was unknown to science. The protein was then named huntingtin after George Huntington, and the gene was designated HTT. Affected persons with HD have multiple copies of a short sequence of the three nucleotides CAG that code for the amino acid glutamine. Normal individuals also have several copies of CAG beginning at position 54 in the DNA of exon 1 of the HTT gene, a large gene that has 67 exons and codes for a long chain of 3142 amino acids. The database UniProtKB - P42858 (HD_HUMAN) provides the amino acid sequence for a normal person having 21 repeats (bold), in which Q represents the amino acid glutamine and P is proline that shows shorter repeats.

The first 300 amino acids are

Most HD patients show 40 or more CAG repeats in succession and a very long chain of glutamines in the huntingtin protein. There are many alleles of the gene that differ in the number of CAG repeats, and the probability of exhibiting symptoms of HD depends on the number of repeats (Table 9.1). For many individuals with 36–39 repeats, HD symptoms never appear (Kay et al., 2016). With 40 or more repeats, the age of onset depends strongly on the number of CAG repeats (Fig. 9.2). There is no sharp dichotomy between adult and childhood or adolescent onset. When there are 27–35 repeats and the person is phenotypically normal, there is an elevated risk of producing an offspring who has >35 repeats and shows symptoms of HD. This is a kind of mutation that generates a new allele, but it is not entirely random because expansion to >35 repeats usually does not occur in people having fewer than 27 CAG repeats. The increased number of repeats can thus lead to the abnormal HD phenotype in a child when neither parent shows symptoms of HD.

MATLEKLMKA FESLKSF**QQQ QQQQQQQQQQ QQQQQQQQ**PP PPPPPPPPPQ

LPQPPPQAQP LLPQPQPPPP PPPPPPGPAV AEEPLHRPKK ELSATKKDRV

NHCLTICENI VAQSVRNSPE FQKLLGIAME LFLLCSDDAE SDVRMVADEC

LNKVIKALMD SNLPRLQLEL YKEIKKNGAP RSLRAALWRF AELAHLVRPQ

KCRPYLVNLL PCLTRTSKRP EESVQETLAA AVPKIMASFG NFANDNEIKV...

TABLE 9.1 Number of CAG Repeats in the HTT Gene Related to HD Phenotype

CAG repeats	Phenotype	Expansion of repeats in next generation
9–26	Normal	No
27–35	Normal	Occasionally; 10% for 35 repeats
36	<0.1% show HD	Often
37	About 0.3% show HD	Often
38	About 1.2% show HD	Often
39	About 2.8% show HD	Often
40 or more	100% show HD; the age of onset depends on the number of repeats	Yes, >90% of cases; usually fatal with very long repeat length

Sources: OMIM; Brinkman, R. R., Mezei, M. M., Theilmann, J., Almqvist, E., & Hayden, M. R. (1997). The likelihood of being affected with Huntington disease by a particular age, for a specific CAG size. American Journal of Human Genetics, 60, 1202–1210; *Kay, C., Collins, J. A., Miedzybrodzka, Z., Madore, S. J., Gordon, E. S., Gerry, N., Hayden, M. R. (2016). Huntington disease reduced penetrance alleles occur at high frequency in the general population.* Neurology, 87, 282–288; *Leeflang, E. P., Zhang, L., Tavare, S., Hubert, R., Srinidhi, J., MacDonald, M. E., et al. (1995). Single sperm analysis of the trinucleotide repeats in the Huntington's disease gene: quantification of the mutation frequency spectrum.* Human Molecular Genetics, 4, 1519–1526; *Walker, F. O. (2007). Huntington's disease.* The Lancet, 369, 218–228.

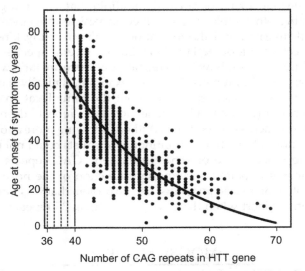

FIG. 9.2 The age of onset of noteworthy symptoms of HD in diagnosed cases with different numbers of CAG repeats in the HTT gene. Thick curve is average age at each allele size. Although the age of onset depends strongly on CAG repeat size, there is a wide range of values at a single CAG repeat length, which tells us that other factors are involved. *Adapted from Gusella, J. F., & MacDonald, M. E. (2009). Huntington's disease: the case for genetic modifiers.* Genome Medicine, 1, 80.81–80.86.

The approximate rates of HD in people with 36–39 repeats were already known from clinical cases, but the rate of people having those sizes of CAG repeats but *no* HD symptoms was not known. The recent study by Kay et al. (2016) examined a large sample of 7315 people from three populations of mainly European ancestry, and they found 18 having 36 or more CAG repeats. The data indicated a population frequency of 0.25% for people with 36–39 repeats, about one in 400, a rate far higher than expected. About 6.2% of people had from 27 to 35 repeats but never showed HD themselves.

MUTANT GENOTYPE, NORMAL PHENOTYPE

Thus, it is not fully accurate to say that HD is directly and simply caused by a single dominant gene. Although any person with one allele having 40 or more CAG repeats will almost certainly develop symptoms of HD eventually, many of them show no symptoms until they are >40 years old. A genetic defect does not automatically mean a phenotypic defect, a fact that provides hope for eventually finding ways to preclude the onset of symptoms altogether. Most people having an HTT gene with 36–39 repeats never show symptoms at any age, but a few show full symptoms of classical HD. In technical parlance, the HTT gene is said to show *incomplete penetrance*. Penetrance is the frequency of those with the genotype who show an abnormal phenotype. Clearly, there is more to the story of HD than just the one defective copy of the HTT gene. That defect, transmitted from parent to child, is necessary for having HD, but it is not a sufficient explanation. For these reasons and more, it is not good practice to define a defective allele of HTT as the HD gene. It does not necessarily code *for* the expression of Huntington disease. Instead, it codes for a huntingtin protein having an explicit number of glutamine repeats in a row. Knowing the number of CAG repeats, we can then know the *probability* of expressing the HD phenotype.

A few cases have been reported wherein both parents developed HD after having had children, and a child received two abnormal HTT alleles, one from each parent (Squitieri et al., 2003). Although the age of onset is about the same whether there are one or two abnormal alleles, progression of the disease appears to be more rapid when there are two.

Another peculiar feature of some cases is the sex of the parent who transmits the defective allele. For an autosomal gene, we expect this to happen equally often via the mother or father. There is a clear exception, however, for HD patients who develop symptoms at an early age because of a very large CAG repeat. >90% of cases of HD in childhood were transmitted via an HTT allele from the father, and a large majority also came from the father for HD patients first diagnosed as adolescents (Quarrell, 2014). Evidently, something happens more often in males that can generate large CAG repeats.

DIAGNOSIS

Prior to 1993, when the HTT gene was identified, diagnosis of HD in its early stages was prone to error and ambiguity. The travails of folk singer Woody Guthrie provide a vivid example. According to his biography (https://www.woodyguthrie.org), "Becoming more and more unpredictable during a final series of road trips, Woody eventually returned to New York where he was hospitalized several times. Mistakenly diagnosed and treated for everything from alcoholism to schizophrenia, his symptoms kept worsening. Picked up for 'vagrancy' in New Jersey in 1954, he was admitted into the nearby Greystone Psychiatric Hospital, where he was finally diagnosed with Huntington disease."

Early diagnosis based solely on a behavioral phenotype relies on mental and motor symptoms that are shared with several other disorders. Knowing there is a family history increases confidence that a patient does indeed have HD, but there are other rare neurological defects with dominant expression, and HD itself does not always show a family history.

Medical geneticists distinguish between three stages of the disease: (a) when symptoms are clearly manifest and HD can be diagnosed with confidence, (b) a *prodromal stage* when there may be peculiarities of behavior or thought, none of which would warrant a diagnosis of HD, and (c) a *premanifest stage* when the mutant HTT gene is present but symptoms are not. The picture is blurred by the common occurrence of "symptom hunting" among those who believe they are at risk for HD. Ever vigilant for the first wayward twitch or bizarre thought, the person and her relatives often "notice" telltale signs they believe are portents of things to come. In many cases in which definite HD eventually appears, it is not clear exactly when the first bona fide symptoms related specifically to the mutant HTT gene arose. When in doubt, the future course of the disease can today be foretold in many cases by the definitive DNA test for CAG repeat number. If someone is at risk for HD, being the child or sibling of another person with definite HD, the test can usually end the guesswork, if those at risk want to know. They can be confirmed as carriers of the HTT mutation during the premanifest stage when they do not yet actually have HD.

The expanded huntingtin protein with >40 glutamines in a row results in death of many neurons in the cerebral cortex of adults, especially in the striate area that is so important for motor function (Chapter 4). This leads to changes that can be seen with magnetic resonance imaging (MRI) scans of the brain. The cortex becomes thinner, and the size of the fluid-filled ventricles increases. Research on large samples of persons at risk for HD has found that changes in brain structure during the premanifest stage often precede changes in behavior by as much as 10 years (Wild & Tabrizi, 2014). It is remarkable how much deterioration can occur in the brain without substantially eroding behavioral function. Evidently, neural and behavioral plasticity can compensate for the early loss of so many neurons (Kloppel et al., 2015).

THE FUNCTION OF HUNTINGTIN PROTEIN

The structure of the novel huntingtin protein discerned from the base sequence of the HTT gene DNA casts little light on the array of symptoms seen in HD. The massive huntingtin is unlike any other proteins that have been studied, and its resemblance to better understood genes and proteins provides few telltale clues about function. It is not an enzyme responsible for just one discrete step in a chemical reaction in a limited array of cells having well-documented functions in the body. It is not a distinct component of an elegantly crafted structure such as the sarcomere of muscle fibers (Fig. 4.8). HTT gene expression is ubiquitous—present in almost all cells of the body, not just the brain or muscle. Furthermore, huntingtin genes very similar to human HTT have been found in virtually all kinds of animals but are lacking in plants and fungi, which implies that the huntingtin protein has been present for a very long time on Earth (Zuccato & Cattaneo, 2016). The normal huntingtin protein seems to encode a kind of molecular scaffold that aids in organizing other molecules involved in vesicle movements, cell division, and the formation of cilia in many kinds of cells (Saudou & Humbert, 2016). Several species of mammals also show CAG repeats, but there are far fewer than seen in humans. Mice, for example, typically have only seven CAG repeats at about the same location as seen in the human HTT gene, and their repeats do not expand across generations. Without knowing more about the normal function of huntingtin, scientists could not be sure whether the presence of the long CAG repeats in humans impaired an important function of the normal protein or instead generated a toxic protein that was functionally inert when there were fewer than 35 repeats.

MOUSE MODELS

Research on gene function took a large step forward when ways to alter genes of mice were discovered in the 1980s. Unlike the effects of radiation or mutagenic chemicals, the new methods gave precise control over the change in an animal's DNA. What seemed like science fiction at the time has now become a routine procedure in many labs that do genetic engineering. Animal models have played important roles in psychology and neurobiology since the pioneering studies of rat behavior by John B. Watson and the physiology of learning by Ivan Pavlov with dogs >100 years ago. The new molecular genetic methods were developed using special strains of the humble lab mouse (*Mus domesticus*, Table 1.2), and before long, neuroscience labs were virtually overrun by mice.

There are good reasons to use lab animals for these kinds of exploratory studies of gene function. The methods utilize embryos grown in a glass dish, and they involve tampering with the DNA code. Things could go wrong, and in fact, many experiments are deliberately designed to make things go wrong in order to learn what a gene normally does. For several reasons, human embryos are not suitable for this kind of invasive interference with the normal course of development. The hope is that research with mice will light the path to a better understanding of the human condition and eventually to new therapies that will cure genetic diseases.

Three general kinds of alterations in the mouse genome are commonly employed:

Gene *knockout* technology targets a specific gene and inserts a segment of foreign DNA into it that disables it.

Transgenic methods insert a segment of foreign DNA, even an entire gene from a human, into the mouse DNA so that the human form of the protein is synthesized in a mouse. Transgenic animals often end up with more than one copy of the foreign gene in different places in their DNA.

Knock-in technology places the foreign gene into a place near the comparable mouse gene so that its time and place of expression are the same as for the mouse gene.

The work of three teams of investigators led by Capecchi, Evans, and Smithies sparked a revolutionary change in the way genetics is done, and it earned them the Nobel Prize in Physiology and Medicine in 2007. Capecchi's (2007) Nobel lecture provides an illuminating account of the rapid progress that culminated in a scientific article published in 1988 (Mansour, Thomas, & Capecchi, 1988) that set forth a general method of knocking out or disabling a known gene so that it could no longer be expressed in protein. Their gene-targeting technology was soon adopted in many other labs to delete specific genes from the mouse genome, including the huntingtin (*Htt*) gene in 1995 (Duyao et al., 1995; Nasir et al., 1995; Zeitlin, Liu, Chapman, Papaioannou, & Efstratiadis, 1995), only 2 years after it was discovered in humans.

The new methods reversed the customary sequence of events in genetics. Prior to the 1980s, genetic analysis of the brain and behavior began with the observation of an interesting phenotypic difference, such as slow development and mental deficiency in PKU or motor and mental abnormalities in Huntington disease. Researchers then set off on a long quest to identify the gene or genes responsible for the phenotypic difference. With "reverse genetics," the process began when the DNA sequence of the normal allele of a known gene was determined, a sequence often ascertained without knowing much about the function of the gene. The process described by Capecchi and colleagues involved several steps outlined below. At the time, that elaborate protocol was regarded by many scholars as miraculous. As the technology spread throughout mouse genetics and neuroscience, more scientists learned it and incorporated its lessons well.

The recipe combines several technologies. The reader is not expected to grasp fully how the entire process works. It is typically applied in large labs equipped with advanced machines operated by scientists with PhD degrees. The recipe illustrates the extraordinary effort that is being devoted to the analysis of genetic defects and the search for ways to prevent or ameliorate them.

Production of targeted mutations in mice:

Step 1. Remove cells from an early mouse embryo and grow them in a glass dish.

Step 2. Synthesize a piece of DNA (the vector) that contains a bacterial gene for resistance to the antibiotic neomycin (neo^r) inserted between two known sequences of a mouse gene.

Step 3. Add the vector to the dish with embryonic cells.

Step 4. Allow time for the vector to recombine with the DNA of those cells. This inserts the vector into the mouse gene in a small fraction of those cells. See Fig. 9.3.

Step 5. Expose the population of embryonic cells to the antibiotic neomycin, which kills cells that lack the neomycin resistance gene (neo^r), leaving a dish with many cells in which the new DNA sequence has been incorporated into the targeted gene.

Step 6. Suck up a few of those transformed cells into a fine glass needle and inject them into another early mouse embryo. Allow time for some of them to become incorporated into the small mass of cells in that embryo. The embryos become *chimeras*, a mix of cells from two different genetic ancestors (Fig. 5.1)—one the source of most of the normal embryo cells and the other from few genetically engineered cells with the neo^r insert.

Step 7. Introduce the chimeric embryos into the uterus of a surrogate female mouse when her maternal environment is conducive to further development of the embryos.

Step 8. Allow her to give birth to viable mice and rear them to weaning. The chimeric mice will usually be phenotypically normal because the altered cells are in a small minority. Let the mice grow to maturity, when they are ready for breeding.

Step 9. Mate the chimeric mice to a genetically normal mouse. Most of the offspring will be normal, but a few of them will have one normal allele (+) and one targeted mutation (*tm*), making them carriers of the mutation (+/*tm*).

Step 10. At this point, a test can be done to determine which of the mice indeed harbor the altered allele. The mice with the (+/*tm*) genotype at the gene locus of interest then become the parents of a new strain of mice that

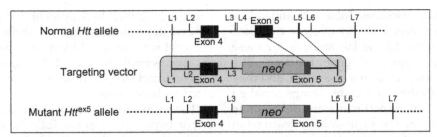

FIG. 9.3   A portion of the normal mouse *Htt* gene containing exons 4 and 5 is shown, along with seven locations (L1–7) that are defined by unique sequences of nucleotide bases. A targeting vector is synthesized with a neomycin resistance gene inserted between L3 and the end of exon 5. When this is incubated with the normal gene, it sometimes recombines in a way that inserts the vector into the *Htt* gene. The new mutation lacking most of exon 5 is then introduced into mouse embryo cells that grow into adult mice having a mixture of normal cells and others with the mutant *Htt*ex5. *Adapted from Nasir, J., Floresco, S. B., O'Kusky, J. R., Diewert, V. M., Richman, J. M., Zeisler, J., ... Hayden, M. R. (1995). Targeted disruption of the Huntington's disease gene results in embryonic lethality and behavioral and morphological changes in heterozygotes. Cell, 81, 811–823, Fig. 1A, Copyright 1995, with permission from Elsevier.*

differs from the original strain at only one gene. At all other genes throughout the genome, they are the same as the ancestral strain.

Step 11. Mate two carriers (+/*tm*) × (+/*tm*). Among the offspring, there should be 25% normal (+/+) mice, 50% carriers of the mutation (+/*tm*), and 25% "knockouts" (*tm*/*tm*) with two copies of the inserted vector. Because the inserted sequence contains a long segment of DNA specifying the bacterial *neo*r gene, the DNA can be transcribed but cannot be translated into a normal protein because the mRNA contains a premature STOP codon (Table 2.1). The (*tm*/*tm*) mouse will lack the protein altogether. Then, phenotypes can be observed to determine what kinds of things must depend on that gene in normal mice.

The studies of knockouts of the mouse *Htt* gene established that it is essential for normal development from a very early age. Mouse embryos with two mutant *Htt*ex5 alleles perish before day 9 of embryonic life (Nasir et al., 1995). Those with just one mutant allele appear to be normal in most respects, although they exhibit some deficits in tests of memory.

The method of targeted mutations has grown into a massive enterprise that seeks to disable every one of the >20,000 known mouse genes in a quest to determine which are essential for normal development and which are of little importance or can be compensated by other genes and physiological processes when they are lost. A large study of 1751 knockouts of mouse genes found that 410 of them are essential for development, so that mice lacking them died (Dickinson et al., 2016). More than 50 genes were found to be relevant to specific human genetic diseases. That study was part of a larger project (the International Mouse Phenotyping Consortium) aiming to knock out 5000 different mouse genes. The second phase of this work has recently been awarded a grant of $28.3 million by the National Institutes of Health in the

United States (Trotter, 2016). This is biological science on a massive scale.

Methods are now available to disable a gene at specific stages of development rather than from the very beginning of the embryo. A new kind of mouse was generated with a special molecular sequence in the *Htt* gene that made it possible to block production of the huntingtin protein by injecting the mouse with the drug tamoxifen. When this was done in mice older than 4 months, the loss of *Htt* function did not cause death of brain neurons or lead to early death of the mouse (Wang, Liu, Gaertig, Li, & Li, 2016). It appears that the huntingtin protein is therefore not essential for normal brain function in adult mice, although it is important for early development in embryos.

There is a serious shortcoming in knockout studies of mouse *Htt* as they pertain to humans, because the defect in human HD is not a lack of HTT but instead the presence of the expanded CAG repeat. A knockout of *Htt* is typically a recessive allele, whereas the troublemaker in HD is a dominant mutation. This problem was addressed by designing a targeting vector that incorporated an entire mutant human HTT gene with a large CAG repeat. When this was incorporated into a mouse embryo and then bred to create a new strain of transgenic or knock-in mice, the animals possessed the human disease gene and human huntingtin protein with a long glutamine repeat. The new kinds of mice had a form of human HD. When the CAG repeat was sufficiently long, the mice exhibited several of the symptoms typical of human HD (Menalled, 2005). Researchers tried many ways to generate mice that expressed symptoms very similar to human HD, reasoning that the closer is the symptomology, the better the model will be for assessing ways to halt disease progression. A newly created knock-in mouse having the human HTT gene with 175 CAG repeats utilizes a mouse strain that is prone to neuron loss, and the resulting strain

Q175FDN mice shows great promise as targets for improved therapies (Southwell et al., 2016).

Extensive studies have revealed large numbers of things that go wrong in an animal when its *Htt* gene has so many CAG repeats. Having created a mouse model that shows many of the features of human HD, the next logical step is to find ways to prevent those symptoms from appearing. If it can be done successfully in a mouse that shows many symptoms of human HD because it carries the mutant human HTT allele, this should help to find ways to arrest the decline in humans with HD. That is the promise of mouse models.

TREATMENT OF HUMAN HD

At the present time, medical doctors treat the symptoms of HD such as difficulties with motor coordination, depression, and sometimes dementia. The symptoms present themselves in a wide variety of ways in different HD patients, and accordingly, treatment is tailored to the individual. No medication has been proved effective in reducing cognitive impairment in HD (Nance & Guttman, 2014). Physical therapy can be helpful, as can a program of home-based exercise (Khalil et al., 2013), and alterations of the home environment can make it safer and easier to live with HD. A variety of antipsychotic drugs are in common use with HD patients, but not one of them is specifically formulated for use with HD patients.

In most cases of HD, the patient lives for 10–20 years with the diagnosis. In principle, the average duration of overt disease can be estimated from its prevalence and incidence. *Prevalence* represents the number of people in a population at one point in time, perhaps a specific day, who show sufficient symptoms to warrant a diagnosis of HD. *Incidence*, on the other hand, represents the number of new cases that occur within a specific period of time such as 1 year. Incidence equals prevalence rate divided by average duration in years, and average duration equals prevalence divided by incidence. Accurate estimates of each of these three quantities are difficult to obtain. There is evidence in several regions of the world that prevalence rates of HD are increasing substantially (Kay, Fisher, & Hayden, 2014). Several factors likely combine to generate this trend, including longer life expectancy in the entire population, longer survival times for those with HD, and more complete reporting of cases of HD. The cause of death is often a malady distinct from HD, such as heart failure or stroke. The life of a person with HD may be extended by improved medical treatment of other conditions that may afflict the patient.

PREVENTION OF HD IN CURRENT GENERATION

There is not yet any effective means to stop or retard the progression of human HD from the premanifest to fully manifest stages (Aguiar, van der Gaag, & Cortese, 2017; Saudou & Humbert, 2016). Many experiments are being done to achieve this in animal models of human HD. Several kinds of therapies have been evaluated, including >128 different drugs, with mixed success to date (Bates & Landles, 2014). Some appear promising, but the road from lab bench to bedside is a long one, and many of those drugs are experimental and have not yet been tested for safety with humans.

There is much evidence suggesting that loss of function at the phenotypic level in human HD arises from toxic effects of the long glutamine repeat in the huntingtin protein. Great benefits are expected if the mutant form of the gene could somehow be suppressed so that the aberrant protein is not expressed at all. Knocking out the normal mouse *Htt* gene in adults does little harm, which suggests a complete suppression of the HTT allele would not likely do any harm to the transgenic or knock-in mice.

One approach to silencing the mutant HTT allele is *RNA interference* (RNAi) with HTT expression. The problem in HD arises from the presence of the huntingtin protein with long glutamine repeats. If the translation of the protein from the HTT mRNA can be blocked to a substantial extent, perhaps, the reduced quantities of aberrant huntingtin will delay the progression of the disease. The RNAi approach has been used with considerable success in mice (Harper et al., 2005). Alternatively, it is possible to disrupt transcription of the mutant HTT allele itself, and this too can markedly improve neurological symptoms in mouse models (Carroll et al., 2011). Recent research with humans has identified a number of DNA sequence variations in mutant HTT alleles that can serve as targets for short nucleotide probes that suppress transcription specifically of the mutant HTT (Kay et al., 2015). There have been some promising results from the RNAi approach applied to cells in tissue culture that have not yet been translated into treatments for human patients.

A formidable challenge to drug therapies that utilize short sequences of DNA or RNA to target the HTT gene or its mRNA is that they do not readily cross the blood-brain barrier, and the new designer drugs must be infused directly into the brain or the cerebral spinal fluid. When introduced into the striatum of a mouse with a human HTT gene having 128 CAG repeats, RNAi probes greatly reduced expression of the mutant huntingtin protein (Stanek et al., 2014). Transferring this technology to treat humans with HD is no easy task. A recent review of promising therapies cautions that "complex metabolic derangements in HD remain under study, but no clear therapeutic strategy has yet emerged" (Wild & Tabrizi, 2014).

PREVENTION OF HD IN FUTURE GENERATIONS

The eugenics movement of the 19th century advocated that anyone with HD should be forcibly sterilized to cleanse the population of the malady. That cruel and ham-fisted approach was based on bad biology and had virtually no effect because most of those who develop HD conceive children before a diagnosis has been made. Today, it is possible to determine who carries a long CAG repeat before the person shows overt signs of HD. Children and siblings of a person with HD are persons at risk for carrying the faulty gene. They *could* have a test done to learn if they are carriers. In fact, most people at risk decline to have the test done. Many choose to live years of good life and have a family of their own before knowing their genetic fate in later life (Scrivener, 2013). If they take the test as a young adult and learn they carry a long CAG repeat, their troubles commence long before there are actual symptoms that impair the quality of life.

There are ways to address this dilemma (Stern, 2014), and these have already become a common practice at a large number of fertility clinics (Baruch, Kaufman, & Hudson, 2008). People who are confirmed carriers of a long CAG repeat can have normal children of their own by utilizing in vitro embryo fertilization (Fig. 5.1) and genetic testing of the embryos (Fig. 9.4). When a

woman at the right stage of her reproductive cycle is given an injection of the drug clomiphene citrate or luteinizing hormone and follicle-stimulating hormone, her ovaries will shed several ripe eggs instead of the usual one egg. These can be removed from the oviducts and added to a glass dish containing her partner's sperm. Regardless of whether the mother or the father carries the mutant HTT allele, about half of the embryos conceived in this way will themselves possess the mutant allele. When the embryo has grown to about eight cells, one of the cells can be removed and a test done to assess whether it has the mutation. Those embryos with the mutation are then discarded, and the healthy ones are returned to the uterus of the mother, where, with a little luck, one of them will implant and grow into a normal child. Because more than one embryo is usually transferred back to the woman who provided the eggs, the birth of twins is a common consequence.

The methods used for *embryo selection* are widely available in private clinics, but the procedures can be quite expensive, as described in clinic websites. One full cycle of collecting eggs and sperm, fertilizing, checking embryo genotype, returning embryos to the womb, and then monitoring progress until birth typically costs from $15,000 to $20,000. Success rate is not high, especially when the mother is older than 40 years, and two cycles may be needed to accomplish birth of a normal child.

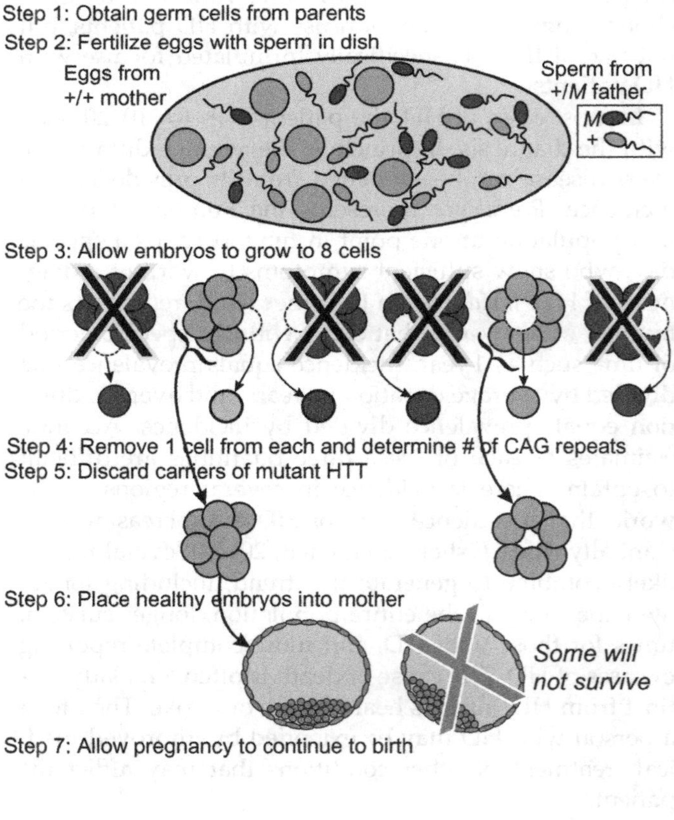

FIG. 9.4 Production and selection of genetically normal embryos from a normal mother and a father who carries a mutant allele of the HTT gene. Dark cells represent those with a large number of CAG repeats. An embryo with one of eight cells removed can survive and grow when implanted back into the woman from whom the egg was obtained. Because not all embryos will survive the procedures, it is common practice to generate extra embryos.

Step 1: Obtain germ cells from parents
Step 2: Fertilize eggs with sperm in dish

Eggs from +/+ mother

Sperm from +/M father

Step 3: Allow embryos to grow to 8 cells

Step 4: Remove 1 cell from each and determine # of CAG repeats
Step 5: Discard carriers of mutant HTT

Step 6: Place healthy embryos into mother

Some will not survive

Step 7: Allow pregnancy to continue to birth

For a young couple hoping to have two children, the cost just of this medical procedure could amount to $60,000–$80,000. That would be a major financial burden for people who must also pay interest on a mortgage and student loans. The financial reward would not be realized for another 40 or 50 years when their child might otherwise begin to show symptoms of HD and eventually need nursing home care.

Prospects for embryo selection and drug treatment of HD pose a real conundrum. If genotyping and embryo selection are practiced extensively, there would be fewer patients afflicted by HD who would need those expensive drugs, once they become available. On the other hand, if 40 or 50 years into the future there is indeed a convenient way to treat carriers of the HTT mutation so that symptoms never appear, the travails of embryo selection might seem less appealing. Add the reality of private enterprise vying for customers, and the best course of action is not so clear. Even now, there is visible hype on many websites and exaggeration of benefits of one or the other approach. Furthermore, there is political opposition in some jurisdictions to any kind of tampering with human embryos. These issues are discussed at length in Chapter 20.

Social factors will likely play a major role in HD in the future. Embryo selection would probably have the greatest impact in a country where universal health care is provided by government and funded by taxes. Families of limited means would otherwise not be able to afford the procedures. The savings to a universal health-care system over the long term could be quite large because costs of years of care for those who eventually develop HD would be vastly curtailed. Religious and political considerations are of relevance for public policies regarding reproductive choices. HD arises at the molecular level, whereas impacts on family life and health-care options are things that happen at the level of society (Table 1.1). Contemporary embryo selection differs from the abuses of the eugenics movement in two important ways: It is based on good science, and it would be voluntary. Informed consent would be needed from both prospective parents before the procedure is performed. Whether there is voluntary selection against embryos with the HTT mutation or compulsory extermination of carriers of the mutation by an all-powerful state reflects the kind of civilization that prevails in a country.

Although embryo selection might greatly reduce the frequency of HD in future generations, it would not eliminate all cases because of continuing expansion of CAG repeats from one generation to the next. There is a cycle of birth and extinction of an HTT mutation within a family line. First, there is gradual repeat expansion across generations until someone exceeds the threshold of 36–40 repeats and a new case of HD appears de novo. The repeat expands further in the progeny of that person until eventually, the allele becomes so large that the onset of symptoms happens very early and curtails reproduction, thereby extinguishing the mutation in that family line. The process tends to be self-limiting, which is why HD today is a rare occurrence.

HIGHLIGHTS

- When there is incomplete penetrance, someone who has the genotype that commonly leads to the neurological disease never develops the disease phenotype. The person can nevertheless transmit the troublesome gene to the next generation, and the disease can skip a generation.
- Late age of onset can also complicate the pattern. Someone with the disease-related genotype can lead a full life and not show neurological symptoms until old age. They often have children before learning about their own genotype.
- In Huntington disease (HD) that arises from the HTT gene, neurological deterioration occurs when there is a long string of >40 consecutive repeats of the CAG nucleotide triplet in the DNA and the amino acid glutamine in the protein.
- When people have a form of the HTT gene with 27–40 CAG repeats, the number of repeats tends to increase spontaneously from one generation to the next, until a case of HD occurs in a family.
- For HD, the unusual pattern of transmission across generations and the nature of the protein were discovered only because the gene itself (HTT) was identified.
- Because several neurological disorders show a similar phenotype with the loss of motor coordination, diagnosis of HD is best done from the DNA, not the behavior.
- The HTT gene is ubiquitous; it is expressed in almost all kinds of cells in most organs from the early embryo to the adult, and eventually, the aberrant huntingtin protein causes cell degeneration in all organs.
- At the present time, some symptoms of HD can be lessened with psychiatric drugs, but there is no way to arrest the disease process itself.
- People who carry the HTT allele with many CAG repeats can have normal children using embryo selection to propagate only those embryos with normal HTT alleles.

References

Aguiar, S., van der Gaag, B., & Cortese, F. A. B. (2017). RNAi mechanisms in Huntington's disease therapy: siRNA versus shRNA. *Translational Neurodegeneration, 6*, 30.

Baruch, S., Kaufman, D. J., & Hudson, K. L. (2008). Preimplantation genetic screening: a survey of in vitro fertilization clinics. *Genetics in Medicine, 10*, 685–690.

Bates, G., & Landles, C. (2014). Preclinical experimental therapeutics. G. Bates, S. J. Tabrizi, & L. Jones (Eds.), Vol. 4 *Huntington's disease.* (4th ed.), New York: Oxford University Press.

Bates, G., Tabrizi, S., & Jones, L. (2014). *Huntington's disease.* New York: Oxford University Press.

Capecchi, M. R. (2007). *Prize presentation—The nobel prize in physiology or medicine 2007. Nobelprize.org. Retrieved from:* (2007). *http://www. nobelprize.org/nobel_prizes/medicine/laureates/2007/capecchi-prize-present.html.*

Carroll, J. B., Warby, S. C., Southwell, A. L., Doty, C. N., Greenlee, S., Skotte, N., ... Hayden, M. R. (2011). Potent and selective antisense oligonucleotides targeting single-nucleotide polymorphisms in the Huntington disease gene/allele-specific silencing of mutant huntingtin. *Molecular Therapy, 19,* 2178–2185.

Dickinson, M. E., Flenniken, A. M., Ji, X., Teboul, L., Wong, M. D., White, J. K., ... Adissu, H. (2016). High-throughput discovery of novel developmental phenotypes. *Nature, 537,* 508–514.

Duyao, M. P., Auerbach, A. B., Ryan, A., Persichetti, F., Barnes, G. T., McNeil, S. M., ... Joyner, A. L. (1995). Inactivation of the mouse Huntington's disease gene homolog Hdh. *Science, 269,* 407–410.

Gusella, J. F., Wexler, N. S., Conneally, P. M., Naylor, S. L., Anderson, M. A., Tanzi, R. E., ... Sakaguchi, A. Y. (1983). A polymorphic DNA marker genetically linked to Huntington's disease. *Nature, 306,* 234–238.

Harper, S. Q., Staber, P. D., He, X., Eliason, S. L., Martins, I. H., Mao, Q., ... Davidson, B. L. (2005). RNA interference improves motor and neuropathological abnormalities in a Huntington's disease mouse model. *Proceedings of the National Academy of Sciences of the United States of America, 102,* 5820–5825.

Huntington, G. (1872). On chorea. *The Medical and Surgical Reporter, 26,* 317–321.

Kay, C., Collins, J. A., Miedzybrodzka, Z., Madore, S. J., Gordon, E. S., Gerry, N., ... Hayden, M. R. (2016). Huntington disease reduced penetrance alleles occur at high frequency in the general population. *Neurology, 87,* 282–288.

Kay, C., Collins, J. A., Skotte, N. H., Southwell, A. L., Warby, S. C., Caron, N. S., & Hayden, M. R. (2015). Huntingtin haplotypes provide prioritized target panels for allele-specific silencing in Huntington disease patients of European ancestry. *Molecular Therapy, 23,* 1759–1771.

Kay, C., Fisher, E. R., & Hayden, M. R. (2014). Epidemiology. In G. P. Bates, S. J. Tabrizi, & L. Jones (Eds.), *Huntington's disease.* (4th ed.). New York: Oxford University Press.

Khalil, H., Quinn, L., van Deursen, R., Dawes, H., Playle, R., Rosser, A., & Busse, M. (2013). What effect does a structured home-based exercise programme have on people with Huntington's disease? A randomized, controlled pilot study. *Clinical Rehabilitation, 27,* 646–658.

Kloppel, S., Gregory, S., Scheller, E., Minkova, L., Razi, A., Durr, A., ... Track-On, i. (2015). Compensation in preclinical Huntington's disease: Evidence from the track-on HD study. *eBioMedicine, 2,* 1420–1429.

MacDonald, M. E., Ambrose, C. M., Duyao, M. P., Myers, R. H., Lin, C., Srinidhi, L., ... Groot, N. (1993). A novel gene containing a trinucleotide repeat that is expanded and unstable on Huntington's disease chromosomes. *Cell, 72,* 971–983.

Mansour, S. L., Thomas, K. R., & Capecchi, M. R. (1988). Disruption of the proto-oncogene int-2 in mouse embryo-derived stem cells: a general strategy for targeting mutations to non-selectable genes. *Nature, 336,* 348–352.

Menalled, L. B. (2005). Knock-in mouse models of Huntington's disease. *NeuroRx, 2,* 465–470.

Nance, M., & Guttman, M. (2014). A09 the enroll-HD care improvement committee: how enroll-HD can improve quality of care in HD. *Journal of Neurology, Neurosurgery and Psychiatry, 85,* A3–A4.

Nasir, J., Floresco, S. B., O'Kusky, J. R., Diewert, V. M., Richman, J. M., Zeisler, J., ... Hayden, M. R. (1995). Targeted disruption of the Huntington's disease gene results in embryonic lethality and behavioral and morphological changes in heterozygotes. *Cell, 81,* 811–823.

Quarrell, O. W. J. (2014). Juvenile Huntington's disease. In G. P. Bates, S. J. Tabrizi, & L. Jones (Eds.), *Huntington's disease* (4th ed.). (pp. 66–85). New York: Oxford University Press.

Saudou, F., & Humbert, S. (2016). The biology of huntingtin. *Neuron, 89,* 910–926.

Scrivener, L. (2013). Huntington's disease: why those at risk choose not to be tested. *The Toronto Star,* Retrieved from (2013). https://www. thestar.com/news/insight/2013/02/10/huntingtons_disease_ does_keep_mount_sinai_scientist_from_brilliant_work.html.

Southwell, A. L., Smith-Dijak, A., Kay, C., Sepers, M., Villanueva, E. B., Parsons, M. P., ... Hayden, M. R. (2016). An enhanced Q175 knock-in mouse model of Huntington disease with higher mutant huntingtin levels and accelerated disease phenotypes. *Human Molecular Genetics, 25,* 3654–3675.

Squitieri, F., Gellera, C., Cannella, M., Mariotti, C., Cislaghi, G., Rubinsztein, D. C., ... Simpson, S. A. (2003). Homozygosity for CAG mutation in Huntington disease is associated with a more severe clinical course. *Brain, 126,* 946–955.

Stanek, L. M., Sardi, S. P., Mastis, B., Richards, A. R., Treleaven, C. M., Taksir, T., ... Shihabuddin, L. S. (2014). Silencing mutant huntingtin by adeno-associated virus-mediated RNA interference ameliorates disease manifestations in the YAC128 mouse model of Huntington's disease. *Human Gene Therapy, 25,* 461–474.

Stern, H. J. (2014). Preimplantation genetic diagnosis: prenatal testing for embryos finally achieving its potential. *Journal of Clinical Medicine, 3,* 280–309.

Trotter, B. (2016). Jackson lab gets $28.3 million grant for mouse genome project. *Bangor Daily News,* August 8.

Wang, G., Liu, X., Gaertig, M. A., Li, S., & Li, X.-J. (2016). Ablation of huntingtin in adult neurons is nondeleterious but its depletion in young mice causes acute pancreatitis. *Proceedings of the National Academy of Sciences, 113,* 3359–3364.

Wild, E. J., & Tabrizi, S. J. (2014). Targets for future clinical trials in Huntington's disease: what's in the pipeline? *Movement Disorders, 29,* 1434–1445.

Zeitlin, S., Liu, J.-P., Chapman, D. L., Papaioannou, V. E., & Efstratiadis, A. (1995). Increased apoptosis and early embryonic lethality in mice nullizygous for the Huntington's disease gene homologue. *Nature Genetics, 11,* 155–163.

Zuccato, C., & Cattaneo, E. (2016). The Huntington's paradox. *Scientific American, 315,* 56–61.

Further Reading

Brinkman, R. R., Mezei, M. M., Theilmann, J., Almqvist, E., & Hayden, M. R. (1997). The likelihood of being affected with Huntington disease by a particular age, for a specific CAG size. *American Journal of Human Genetics, 60,* 1202–1210.

Gusella, J. F., & MacDonald, M. E. (2009). Huntington's disease: the case for genetic modifiers. *Genome Medicine, 1,* 80.81–80.86.

Leeflang, E. P., Zhang, L., Tavare, S., Hubert, R., Srinidhi, J., MacDonald, M. E., et al. (1995). Single sperm analysis of the trinucleotide repeats in the Huntington's disease gene: quantification of the mutation frequency spectrum. *Human Molecular Genetics, 4,* 1519–1526.

Walker, F. O. (2007). Huntington's disease. *The Lancet, 369,* 218–228.

10

Androgen Insensitivity Syndrome (AIS)

This chapter is about sex, its biological and genetic determinants, and societal definitions of what male or female is. Like so many simple dichotomies, the binary view of sex and gender is challenged by many facts. One such fact is Androgen Insensitivity Syndrome (AIS, OMIM 300068), something that occurs in about one per 20,000 births. The person with AIS usually not only has the XY karyotype but also has a mutation in a gene on the X chromosome that renders cells throughout the body insensitive to androgen hormones secreted into the blood. That insensitivity extends from the period of organ formation in the embryo into adulthood. One common consequence is that the typical male genitalia do not form in the fetus, and the newborn infant is often assigned the female sex by doctors and family alike. Eventually the X-linked mutation may be discovered when she does not commence menstrual periods during the teen years and lacks pubic hair. As personal histories describe so well, the discovery can come as a shock.

The jazz singer Eden Atwood described her experience: "The truth unfolded, at least in my memory, like this. The doctor came in and nervously took a seat. He informed me and my mother that we were both scheduled to see a psychologist that day, separately and together. And then, he proceeded to use terms like Müllerian duct regression factor and chromosomal abnormality. The toughest thing to hear by far, though, was the name of my condition, my disease. Testicular Feminization Syndrome. If I could have chosen a super power it would have been evaporation." (https://oii.org.au/author/admin/18, February 2010). Many other case histories have been described in the medical literature (Pizzo, Lagana, Borrielli, & Dugo, 2013; Yellappa & Venkatesh, 2015), including one family in which two children had AIS, one being raised as a boy and the other as a girl (Spătaru et al., 2013).

HISTORY

It was known for many years that sex of an infant at birth can sometimes be unclear or that a person raised as one sex proves at autopsy or during abdominal surgery to have internal organs that do not match his or her external genitalia. A review of 82 cases by Morris (1953) suggested a distinct syndrome in "patients who are essentially normal-appearing women but who have undescended testes in place of ovaries." At the time the syndrome was termed "testicular feminization," a label later replaced by the more clinically accurate AIS (Lee, Houk, Ahmed, & Hughes, 2006). It was not apparent whether the syndrome involved a deficiency in androgens or an inability to respond to their presence. Evidence accumulated that the syndrome is transmitted as a sex-linked gene on the X chromosome. That belief was strengthened by discovery in rats and mice of similar characteristics related specifically to a gene on the X chromosome of the female that abolished sensitivity of male rodents to androgen hormones (reviewed by Bardin & Bullock, 1974).

Further studies demonstrated that the syndrome was specifically caused by a defect in the androgen receptor itself (Wilson, 1985). It was soon thereafter mapped to the q11–12 band of human chromosome X (Brown et al., 1989), and the DNA sequence of the gene was determined (Chang, Kokontis, & Liao, 1988; Lubahn et al., 1988).

MODES OF TRANSMISSION AND EXPRESSION

Being on the X, there are always two copies of the AR gene in females, whereas XY persons have only one—the AR gene transmitted by the mother via her X. The mutant form of AR is rare, about 1/20,000, and the probability of a female receiving two mutant AR genes is remote. Furthermore, almost all AIS persons are infertile because of malformations of the gonads, so they cannot pass the mutation to the next generation. Expression of the AR gene is dominant in XY persons who need only one mutant form to show AIS and recessive in XX. Females who carry one AR mutation have a 50% probability of passing it to their chromosomally XY offspring, so that on average, about half of those with the XY karyotype will show AIS if the XX mother is a carrier.

The reader will note a difficulty in describing persons with AIS as male. They do indeed have a Y chromosome that is usually associated with being male, but AIS reveals that the association is not an unbroken rule. Nature breaks it more often than we realize. Most XY persons with AIS are designated female at birth, grow up being girls, and see themselves as feminine as adults. They are not male or men, despite carrying a Y chromosome. It is important to express clearly whether the term "sex" is being used in reference to chromosomal sex, anatomical sex, or culturally assigned and self-identified gender. These do not always travel together.

As will be addressed further in Chapter 13 on the XYY male, there are many genes on the Y chromosome that have little or nothing to do with sexual functioning and there are several genes on an autosome that can cause a mismatch between anatomical and chromosomal sex. It is helpful to understand how anatomical sex develops and how sex hormones function. Then, the different pathways of development that occur in AIS can make sense.

HORMONE SYNTHESIS AND RECEPTORS

Sex hormones are steroids (Fig. 10.1) synthesized from the mother of all steroids, cholesterol that is obtained from the diet. Unlike neurotransmitters that carry a signal from one neuron to an adjacent neuron, hormones are often secreted into the bloodstream and circulate widely throughout the body, where they influence the functioning of many organs. Two broad classes of hormones are vitally important for sex and reproduction. Androgens include androstenedione, testosterone, and dihydrotestosterone (DHT). Estrogens include estradiol, estrone, and estriol. Androgens are sometimes called the male hormones and estrogens the female type, but this dichotomy is not accurate. Males produce both androgens and

estrogens, and many of their cells have both androgen and estrogen receptors. The same is true for females. Furthermore, when androgens are produced in copious amounts, they can be converted to estrogens. Males and females differ mainly in the amounts produced and the distribution of receptors in certain kinds of tissues. In women, the levels of estrogens change dramatically with the menstrual cycle and especially during pregnancy and then at menopause.

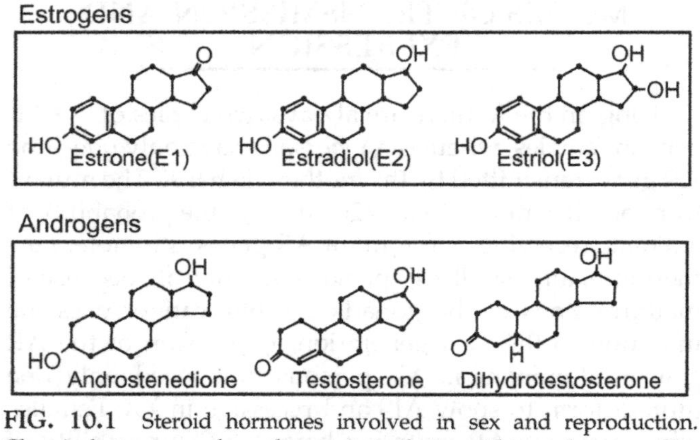

The steroid hormones are synthesized from cholesterol in discrete steps via a series of enzymes encoded by genes located on different autosomes (Fig. 10.2). The hormones are usually released into the bloodstream and then have a wide range of effects, including the building of muscle and bone. Just as neurotransmitters have receptors, so the steroid hormones have receptors that are sensitive to only two or three kinds of molecules among the thousands of different chemicals that bathe each cell. The androgens bind to the androgen receptor (AR) that is itself encoded by the AR gene located on the X chromosome. Likewise, the estrogens bind to the estrogen receptors that are encoded by the ESR1 or ESR2 gene. Fig. 10.1 shows how very similar are the members of one class of hormones, differing in some cases by the addition or removal of a single hydrogen (H) or hydroxyl (OH) group. All of the androgens can bind to and activate the AR receptor because of their very similar shapes, although the dihydrotestosterone form has a much greater affinity for the AR than does testosterone. When the AR receptor is not functional, the person will nevertheless be responsive to estrogens because they activate a different receptor.

Despite their critical importance for sex and reproduction, the androgens and estrogens depend almost entirely on genes that are not located on the X or Y. The AR is the sole exception, being present on the X. It happens that males can get along in life quite well with just one normal copy of the AR gene. The same is true of females, because an XX person who carries just one mutant allele of the AR gene shows no phenotypic effects.

AR GENE STRUCTURE AND FUNCTION

Much is known about androgens, estrogens, and the AR receptor because they play important roles in common health problems such as breast and prostate cancers and menopause in women and sexual dysfunction in men. Decades of medical research have accumulated an extensive body of knowledge that is now helping to understand AIS, even though it is a rare occurrence. Much of the knowledge involves things that transpire at the molecular level.

The AR gene has eight exons, and the protein encoded therein has 920 amino acids (Fig. 10.3) with three

Estrogens

Androgens

FIG. 10.1 Steroid hormones involved in sex and reproduction. Chemical structures show the great similarity within each group. The *black dot* at each intersection is a carbon atom. Carbon typically has four chemical bonds, but most bonds with a single hydrogen are not shown.

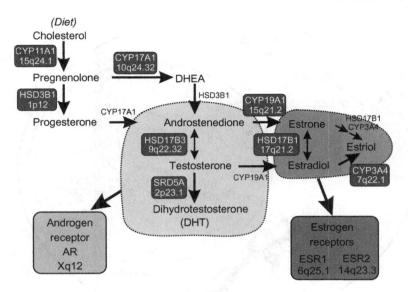

FIG. 10.2 Synthesis of steroid hormones from cholesterol. Each step is mediated by a specific gene shown in upper case. All but the AR gene are on an autosome. The androgens bind to the AR protein, and estrogens bind to ESR1 and ESR2 proteins. *GeneCards and OMIM*

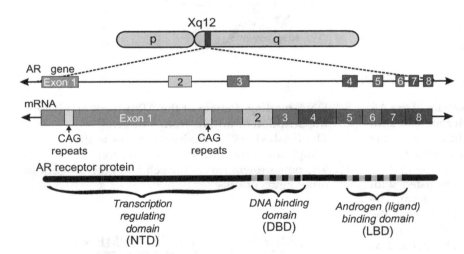

FIG. 10.3 Location and structure of the AR gene, mRNA, and protein. Exon 1 contains two regions of multiple CAG base repeats. When the first one has more than 40 repeats, Kennedy's disease results in motor neuron loss. Mutations in any of the three major domains can cause AIS. The DNA-binding domain recognizes and attaches to genes whose expression is controlled by androgens. Androgens are ligands that normally attach to the LBD, unless a mutation disrupts this function. *GeneCards, OMIM, NCBI*

important domains. The first, encoded by exon 1, is crucial for regulating the action of the gene. Its technical title is the N-terminal domain (NTD). The second domain, encoded by exons 2 and 3, has a special DNA-binding sequence that enables it to bind to the regulatory part of the DNA of certain kinds of other genes in the cell nucleus. It is termed the DNA-binding domain (DBD). Exons 3–8 encode two structures that work closely in unison. First, there is a hinge-like structure that enables the protein to bend sharply when an androgen molecule binds to the adjacent androgen-binding domain. The molecule that binds to a specific receptor is termed a "ligand," and the portion of the receptor protein that binds it is termed the ligand-binding domain (LBD).

An intricate series of molecular steps is initiated when a testosterone molecule from the blood enters a cell via small pores in the cell membrane (Fig. 10.4). Whereas

neurotransmitter receptors are usually embedded in the cell membrane of a neuron (Chapter 4), ARs are large but mobile proteins located in the cytoplasm of many kinds of cells. They work together with several kinds of smaller accessory proteins that are themselves encoded by different genes. The names and specific roles of the accessory proteins are not mentioned here. The picture provided in Fig. 10.4 is sufficiently complicated to convey a good idea of how the AR gene works. Only the cell itself needs to "know" exactly how this is done. The net result of this elaborate dance of molecules is that the arrival of testosterone at the pores of a cell that contains ARs triggers the transcription of androgen-sensitive genes that then synthesize their protein products that are used to control diverse functions of many other kinds of cells. Stated briefly, the AR regulates the transcription of other genes. The presence of testosterone initiates the synthesis

FIG. 10.4 Steps in expression of a gene controlled by testosterone (T) and its AR receptor. The AR in the cytoplasm forms a complex with accessory proteins (acc) that are released when DHT made from T binds to the AR protein. AR with DHT then enters the nucleus and forms a dimer that combines with other accessory proteins to form a transcription complex that makes mRNA from a gene whose expression is controlled by the AR. The mRNA is then translated into a protein that usually leaves the cell and influences other cell functions in the body. *OMIM, UniProt; Galani, A., Kitsiou-Tzeli, S., Sofokleous, C., Kanavakis, E., & Kalpini-Mavrou, A. (2008). Androgen insensitivity syndrome: clinical features and molecular defects. Hormones, 7, 217–229; Matsumoto, T., Sakari, M., Okada, M., Yokoyama, A., Takahashi, S., Kouzmenko, A., & Kato, S. (2013). The androgen receptor in health and disease. Annual Review of Physiology, 75, 201–224.*

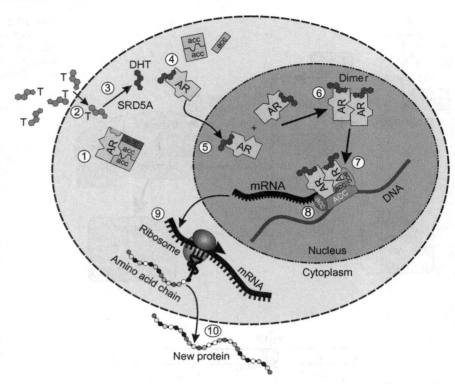

of other proteins by binding to the LBD domain of the AR protein. Likewise, the absence of testosterone can shut down the production of those proteins. In this way, the rate of synthesis of many other proteins can be turned up or down like a thermostat.

Steps in the dance involving the AR are listed with numbers that correspond to parts on the diagram. The diagram thus conveys a good impression of what the AR gene does via its product, the AR. In this example, the gene itself and the protein are both abbreviated as AR. The AR molecule (Fig. 10.4) in the cytoplasm is the protein, whereas the AR gene itself is located somewhere in the labyrinth of the nucleus. Fig. 10.4 does not show steps in the synthesis of the AR protein.

1. AR in cell binds to accessory proteins (acc) that prevent it from entering the nucleus.
2. Testosterone molecules (T) released into the blood then enter the target cell.
3. They are converted to DHT via an enzyme encoded by the SRD5A2 (steroid 5 alpha-reductase 2) gene.
4. DHT binds to the LBD region on the AR receptor molecule, which releases accessory proteins.
5. The AR with DHT attached then enters the cell nucleus.
6. Two AR/DHT molecules join to form a dimer.
7. The dimer combines with other accessory proteins to form a transcription complex.
8. The complex attaches to a segment of DNA of an androgen-responsive gene that corresponds to the

DNA-binding domain of the AR protein and begins transcription of mRNA from that gene.

9. The finished mRNA molecule moves to the cytoplasm and is translated into a protein.
10. The new protein leaves the cell and goes to work in other kinds of cells

EMBRYO SEX DEVELOPMENT

One of the earliest functions of the AR receptor occurs at 7–8 weeks after the conception of the embryo before there are uniquely male or female organs (Rey & Grinspon, 2011). There is a primordial, undifferentiated gonadal structure that can become part of either a male or female person, depending on the chemical signals it receives. That gonad has two kinds of tubes connected to it, the Müllerian and Wolffian ducts, named after the scientists who first observed them while dissecting a chicken embryo under a microscope. Fig. 10.5 shows the pathways the embryo can take as it develops (Tanagho & McAninch, 2000).

In a male with a normal Y chromosome, the sex-determining region Y (SRY) gene on the Y chromosome produces a protein that influences the gonad to become a testis capable of making testosterone. The testis also contains Sertoli cells that secrete anti-Müllerian hormone (AMH) that causes the Müllerian ducts to degenerate, so that no uterus, cervix, or a complete vagina forms in the

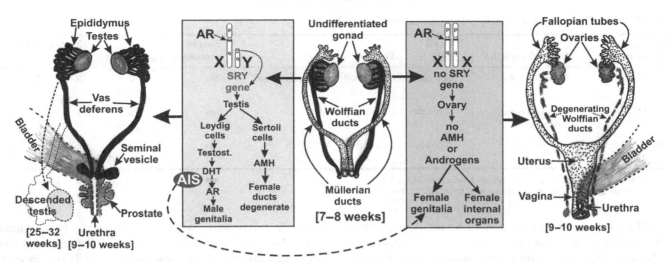

FIG. 10.5 Development of gonads and reproductive structures in the early human embryo. In the absence of the SRY gene on the Y chromosome, the ovaries and a uterus form, and the Wolffian ducts degenerate. The female XX embryo does not produce high levels of androgens, and consequently, the female type of genitalia will form. In an XY embryo, the SRY gene stimulates formation of a testis that makes AMH that causes the Müllerian ducts to degenerate. The Leydig cells make testosterone and DHT that stimulate the formation of male genitalia and descent of the testes into the scrotum. In AIS, DHT cannot activate the AR, and male genitalia cannot form, so that there is commonly a female type of genitalia in the absence of ovaries and uterus. *Galani, A., Kitsiou-Tzeli, S., Sofokleous, C., Kanavakis, E., & Kalpini-Mavrou, A. (2008). Androgen insensitivity syndrome: Clinical features and molecular defects.* Hormones, 7, 217–229; Rey, R. A., & Grinspon, R. P. (2011). Normal male sexual differentiation and aetiology of disorders of sex development. Best Practice & Research Clinical Endocrinology & Metabolism, 25, 221–238; Tanagho, E.A., and McAninch, J.W., (2000). Smith's General Urology (15th ed.). New York: McGraw-Hill.*

fetus. Instead, the Wolffian ducts mature into the epididymis that is connected by the vas deferens to the prostate gland and eventually the penis. The testosterone secreted by the male testis then stimulates the ARs in nearby cells, which promotes the growth of a penis and descent of the testes to the level of the scrotum.

In XX embryos lacking a Y chromosome with the SRY gene, the gonad matures into an ovary, and the Müllerian ducts form a fallopian tube connected to a uterus with a cervix connected to a vagina. The female ovary does not make androgens, so there is no formation of an external penis of typical male size. Instead, there is a clitoris that has sensory endings similar to those of the male penis.

In many cases of AIS, the genital tissues cannot detect the presence of testosterone, and this prevents the formation of the typical male external genitalia. Instead of a penis, there is the beginning of a vagina with attached labial folds, but the vagina is usually short and has a blind ending that cannot join to a cervix. Lacking ovaries and a uterus, the person is not able to menstruate in the teen years or conceive a child.

PHENOTYPES AND GENOTYPES IN AIS

Being caused by a mutation in the AR gene, AIS can present itself in varying degrees, which are related to the specific kind of AR mutation. In almost all cases of genuine AIS, regardless of external genitalia, there are no ovaries or uterus. Instead, there are testes, and the condition of the external genitals in a child is thus a phenotype that has some relation with a genotype as part of a larger syndrome. Clinicians recognize three types of AIS according to the extent of difference from the typical male situation with a fully formed penis and scrotum containing testes making sperm, as well as phenotypes involving breasts (Table 10.1).

The Androgen Receptor Gene Mutations Database is maintained at McGill University in Montreal (Gottlieb et al., 2012). A total of 1029 mutations were described in 2012. More than 500 AR mutations involving more than 800 individual cases were clearly associated with AIS, whereas many others were de novo mutations from prostate cancers and were not transmitted in families. As has happened with several other syndromes, knowing about a wide range of mutations has added some complexity to our understanding of AIS. One surprise was that some individuals with definite variants of an AR gene exon and protein do not show symptoms of AIS at all. This raises a question about exactly which AR gene sequence should be regarded as the prototypical "normal" situation. Even more surprising is that about 40% of persons showing a clinical AIS phenotype do not carry a mutation in the AR gene. Two other genes that can cause AIS when mutated are steroid 5 alpha reductase 2 (SRD5A2), which converts testosterone to DHT, and hydroxysteroid 17-beta dehydrogenase 3 (HSD17B3), which catalyzes the conversion of androstenedione to testosterone (Fig. 10.2).

TABLE 10.1   Three Types of Androgen Insensitivity Syndrome (AIS)[a]

Type and abbreviation	External genitalia	Pubic hair	Testes	Sperm formation	Breasts	Mutations by region[b]
Complete CAIS	Typical female	Sparse	Often inside abdomen or fused with labia	No	At puberty	89/49/157
Partial PAIS	Variable, often ambiguous	Variable	Variable	No	At puberty	13/21/94
Mild MAIS	Typical male	Normal	Descended	No	At puberty	22/4/17

[a] Source: *Excerpts from Gottlieb, Beitel, Nadarajah, Paliouras, & Trifiro, 2012*
[b] *Number of AR mutations in database with change in NTD/DBD/LBD (Fig. 10.3).*

AIS is one of several disorders of sexual development that can alter genital anatomy and reproductive function. A 2006 review of the literature identified mutations in 28 different genes that could alter sex organs, only two of which were located on the X chromosome and one on the Y (Lee et al., 2006). A subsequent review expanded the list to 32 different genes, including 5 on the X and 1 on the Y (Rey & Grinspon, 2011). Thus, in accord with the steroid synthesis pathways outlined in Fig. 10.2, the vast majority of genes involved in sexual development and function are not located on the sex chromosomes. The genes are organized in intricate networks, and the phenotypic consequences of a mutation depend strongly on its place in a network.

Any of the three categories of AIS can be caused by a mutation in the DNA of the AR gene NTD, which controls gene expression, the DBD, or the LBD where the AR dimer attaches to begin transcription of another gene (Table 10.1, Fig. 10.3). The diversity of AIS was emphasized by the discovery of several instances in which a single specific mutation could be associated with complete AIS (CAIS), partial AIS (PAIS), or even mild AIS (MAIS) in different individuals (Gottlieb et al., 2012). For example, a single-base substitution of adenine for guanine at base 2343 in exon 6 of the AR gene gave rise to the replacement of methionine by isoleucine at amino acid 781 in the AR protein, and this was associated with either CAIS, PAIS, or MAIS in different people. Thus, the manifestation of a gene mutation in a phenotype is not at all a simple one-to-one correspondence; many other factors must be involved to yield such a wide range of phenotypes from the same mutation.

KENNEDY DISEASE

In fewer than one per 100,000 persons, an individual can show both the mild form of AIS with breast development and the absence of sperm plus a degenerative disorder of motor neurons known as Kennedy's disease (MIM 313200) or spinal and bulbar muscular atrophy of Kennedy. It happens that there are normally two regions of the AR gene exon 1 where repeated sequences of the three nucleotides CAG are located (Fig. 10.3), one having 8–35 CAG repeats that give rise to a chain of glutamines in the AR protein (Casella, Maduro, Lipshultz, & Lamb, 2001), similar to what occurs in the huntingtin protein. There is evidence that the CAG repeat length is related to infertility and motor symptoms (Pan et al., 2016). When there are more than 35–40 repeats, the person can show both the milder form of MAIS and Kennedy's disease, a disorder much like Huntington's disease (Galani, Kitsiou-Tzeli, Sofokleous, Kanavakis, & Kalpini-Mavrou, 2008). Initially, there is tremor of the hands and muscle twitches, leading eventually to widespread muscle weakness and slurred speech (MalaCards). Onset of motor symptoms occurs from 20 to 60 years of age. This rare defect is not transmitted to the next generation by the XY person, who is infertile.

GENE EXPRESSION AND FUNCTION

Studies of gene expression provide two kinds of important information about androgen receptor function: the kinds of tissues where the AR receptor itself is expressed and the kinds of genes whose expression is mediated or facilitated by the AR. The first kind of information is relatively easy to obtain with current technology, whereas the second is a vast realm broad and deep. It is hoped that eventually this knowledge will somehow help to treat symptoms of AIS. Presently, much of what is known about the AR protein has been learned through work on prostate cancer, a major cause of mortality in men that is often treated by reducing androgen levels and blocking ARs with drugs. AR gene expression can be detected either by the mRNA that corresponds to exons in the AR gene or by direct assays of the AR protein with antibodies (Chapter 3).

The Human Protein Atlas provides rich details about expression of many genes using both methods on a wide range of adult human tissues. It presents summary data

TABLE 10.2 Numbers of Genes Highly Expressed in Four Kinds of Human Tissues[a]

Tissue	Percentage of all 19,628 genes expressed	Number of genes with enhanced expression	Number of genes group enriched	Number of genes highly enriched in one tissue
Brain	74	783	239	415
Ovary	65	101	18	6
Testis	82	802	363	1035
Prostate	73	100	48	20

[a] Source: *Human Protein Atlas, 2016. Definitions of enhanced and enriched are provided in the text and the source.*

on overall patterns of gene expression in 37 different kinds of tissue. Table 10.2 shows the number of genes normally expressed in four kinds of organs of special interest here—the brain, testis, ovary, and prostate. The atlas documents results for 19,628 known human genes and their associated proteins. A large majority of these is expressed in each kind of tissue. There are 7367 genes expressed in all 37 kinds of tissues that are essential for a wide variety of cell functions, sometimes termed "housekeeping" genes because all cells need them all the time. There are also 1098 genes that evidently are not expressed in any of the tissues assayed in adults. Many of them may instead be involved in the early embryo. Those with *enhanced* expression are at least five times more abundant in one tissue than the average of all tissues, whereas those termed *enriched* are five times more abundant in just one kind of tissue than in any other tissue. *Group enriched* are highly abundant in two to seven kinds of tissue. However one looks at these molecular data, every kind of organ and tissue is immensely varied in its gene activity and population of proteins.

The AR gene is expressed locally in embryonic gonadal tissue, but it is also widely expressed in major organs throughout the adult body, with the exception of the digestive system. Fig. 10.6 based on the Human Protein Atlas shows the degree of gene expression in several organs for the genes AR, ESR1, and ESR2. All three kinds of receptors are found in most kinds of tissues, although abundance varies greatly. A closer look at six kinds of steroid receptor expression revealed major differences among 12 different regions within the mouse brain (Mahfouz et al., 2016). New technologies now allow gene expression to be assessed for thousands of different genes within a single cell (Darmanis et al., 2015; Moffitt et al., 2016; Wills et al., 2013), which gives a more detailed portrait of features that are masked when whole tissue samples are studied. These methods should reveal whether androgen and estrogen receptors are sometimes coexpressed in the same cell, not just the same tissue.

Given that the AR plays a major role in the expression of other genes (Fig. 10.4), it is of interest to know how many and which specific genes are responsive to androgens in the bloodstream. This question has been addressed in several different ways, and the published literature presents diverse opinions. Computerized scanning of published articles has been used to establish databases of genes that are responsive to androgens and estrogens. In 2009, the Androgen-Responsive Gene Database reported 1785 human genes whose expression levels are sensitive to androgen levels (Jiang et al., 2009). In 2004, the Estrogen-Responsive Gene Database (Tang, Han, & Bajic, 2004) reported 797 human genes that were responsive to estrogen levels. New whole-genome sequencing technologies confirm that thousands of genes are regulated by estrogen levels, many of which have relevance for breast cancers (Ikeda et al., 2015), and more than 1000 genes are regulated by the AR, many of which are involved in prostate cancer (Chen et al., 2010; Takayama et al., 2011).

In this current context of complexity, it is unrealistic to expect anyone to formulate a concise narrative about what the AR does. It is involved in very many functions throughout the body. Even in a single kind of organ such as the breast or the prostate, the expression levels of hundreds or even thousands of genes are regulated by steroid hormones. It is therefore most remarkable that a woman with CAIS can lack sensitivity to testosterone and DHT entirely, yet be so normal in so many ways. In extensive reviews (Galani et al., 2008; Gottlieb, Beitel, & Trifiro, 2014; Hughes et al., 2012), the only medical issue that was mentioned, apart from genital anatomy and reproductive function, was a low risk of cancerous tumors of the remnants of the testicular cells. These facts do not imply that the AR has no important functions outside of the reproductive organs. They do suggest that what is ordinarily accomplished via the androgen pathways and the AR receptor can probably be done in other ways in AIS. There must be alternative pathways in steroid hormone functions that are utilized in CAIS.

SEX DIFFERENCES IN GENE EXPRESSION

Androgen levels are generally higher in typical males, whereas estrogens are higher in typical females, especially before menopause, although both sexes produce and utilize both androgens and estrogens. A comprehensive assessment of male-female differences in gene expression in the postmortem human adult brain (Trabzuni et al., 2013) compared expression of 17,501 genes in 12 different brain regions (Chapter 4) in a large sample of 36 female and 101 male brains. The authors emphasized in the title of their article that they found "widespread sex differences" in most brain regions.

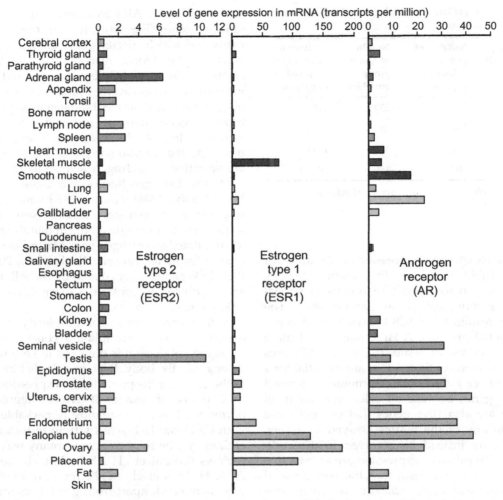

FIG. 10.6 Level of expression of androgen and two estrogen receptors in several kinds of organs and tissues in postmortem samples. A gene expression array detects mRNA from thousands of different genes, and a data processor determines the number of mRNA transcripts for each gene. Adding those numbers across all genes yields millions of transcripts. Then, the level for each gene is divided by the total for all genes to yield transcripts per million, as provided in the HPA dataset. *https://www.proteinatlas.org/ENSG00000091831-ESR1/tissue; https://www.proteinatlas.org/ENSG00000169083-AR/tissue; https://www.proteinatlas.org/ENSG00000140009-ESR2/tissue.*

Indeed, they did find differences in most brain regions for dozens of genes. On the other hand, their data can be seen as a striking demonstration of the great similarity of the male and female brains in levels of gene expression. In the occipital cortex, only 1.7% of genes showed a significant male-female difference. About 1% differed in the thalamus, and all other brain regions showed a sex difference in expression levels in only 0.3% or fewer of all genes. Thus, for almost all regions of the human brain, levels of gene expression were very similar in males and females for more than 99.5% of the 17,501 identified genes. Furthermore, 85% of the sex differences they detected arose from different modes of splicing the mRNA sequences from different exons of the same gene (Chapter 3). Two things thus appear to be true at the same time: Many genes show significant sex differences in expression somewhere in the brain, while at the same

time, the overwhelming majority of genes do not differ in levels of expression between the sexes. When there are more than 17,000 genes under consideration, even a small percentage of genes showing a sex difference can amount to 100 or more genes. How those few genes might be involved in sex-typical behaviors is not yet known. It will be interesting to learn how many of them have expression levels mediated by androgen and estrogen receptors.

OTHER PHENOTYPES

AIS is rare in the broad population, and there are few extensive registers of cases similar to what exists for chromosome anomalies in many countries. There is an International Disorders of Sex Development Registry (I-DSD) that serves to recruit participants for research projects and provides a repository of information (see https://www.i-dsd.org/; also Ahmed, Rodie, Jiang, & Sinnott, 2010). It contains only a fraction of total instances of AIS, mainly those whose families sought specialized medical help with matters related to reproduction. Among the 649 cases contained in the database in 2013, several involved chromosome anomalies, while about 170 were evidently AIS, but only nine of them had a confirmed AR gene mutation (Cox et al., 2014). Even with the new registries, it will be very difficult to obtain a truly representative sample of AIS cases for comparison with typical XY males and XX females on a wide variety of phenotypes. There is still a substantial degree of shame that makes people reluctant to share their diagnoses with researchers or friends. At the present time, we simply do not know if there are small differences arising from AIS in average scores on psychological tests or indicators of behaviors such as motor coordination, speed, or strength. Many women with CAIS are leading normal lives, apart from their inability to conceive children. The sample of 42 European women with CAIS reported by Doehnert, Bertelloni, Werner, Dati, and Hiort (2015) consisted of patients at several clinics who sought medical help with regard to a lack of menstruation as they entered adulthood. As part of that clinical evaluation, blood samples were analyzed for a wide range of hormones in order to better understand their current situations. Testosterone and estradiol levels and free androgen indices were in the normal adult male range. Testosterone levels were mostly in the adult reference range for typical males, far above the range for XX females, and estradiol levels not only were also mostly in the range for typical males but also overlapped the range for typical females.

WOMEN IN SPORTS

Today, there is an abundance of unfettered speculation on the Internet about the involvement of AIS in instances of exceptional prowess in women's sports and among high fashion models. Several websites and blogs suggest that women with CAIS are employed by Victoria's Secret to model underwear because of some degree of masculinization of prototypical female characteristics. There seems to be no verifiable evidence, however, of AIS in the ranks of the supermodels.

There is a long and varied history of official attempts to exclude women with masculinized characteristics or hints of male biology from competing in Olympic sports (Ritchie, Reynard, & Lewis, 2008; Schweinbenz & Cronk, 2010). The original rationale presented by Olympic officials was to prevent men from gaining an unfair advantage in the single-minded pursuit of medals by covertly competing as women, but there seems to be no well-

documented case in which this ever happened. Instead, scrutiny was reserved for women who did not appear to be sufficiently feminine. In 1960, women hoping to compete in the Olympics were required to appear naked for inspection by a team of gynecologists. Then, in 1968, a modified chromosome test was introduced. Then, they switched to DNA evidence of the presence of the SRY gene. At the 1992 Barcelona games, all results of sex tests were kept secret. Eight women athletes showed signs of SRY before the Atlanta Olympics in 1996 but were allowed to compete anyway. In 1999, the International Olympic Committee ceased the practice of official gender and sex verification for all potential women contestants. Then, in 2011, the IOC proposed to use testosterone levels to decide who can compete as a woman (Marchant, 2011). Shortly before the Rio Olympics in 2016, the Court of Arbitration for Sport suspended the testosterone rules (Gleeson & Brady, 2016).

As far as women with CAIS are concerned, it is difficult to imagine how their high testosterone levels could confer advantage when their AR mutations make them insensitive to any amount of testosterone or DHT. Nevertheless, the mass media continue to have a field day playing with the reputations of foremost women athletes. Most of the targets of their speculations have never had a confirmed diagnosis of any genetic anomaly made public.

TREATMENT

CAIS

When life does not proceed in the typical way, medical professionals will seek means to shift it back onto the usual track. Those with CAIS are almost always raised as females and self-identify as female. When they fail to menstruate in the teen years and a physician is consulted, it will be discovered that the vagina ends abruptly, and there is no cervix or uterus. Those are organs that cannot be replaced. There is no effective treatment for the principal symptom of AIS. The woman will never be able to conceive and give birth to a child. In many cases, the woman nevertheless forms an intimate relationship and engages in pleasurable sexual activity because she has a clitoris that can be highly sensitive. When the vagina is short, the typical form of intercourse can be uncomfortable. There are several ways to enlarge the vagina through exercise or surgery. The short vagina is sometimes rectified by a difficult and risky surgery that converts a portion of colon into an artificial vagina. The surgery would usually be done on an adult woman and under current rules would never take place without her informed consent in most countries. Some experts recommend that the remnant gonadal tissue be removed surgically early in postnatal life in cases of CAIS, but the risk of the surgery itself must be weighed against the low risk of a gonadoblastoma tumor in an adult, which is about 2% for CAIS (Lee et al., 2006). If the testes are removed, hormone replacement therapy will be needed to induce changes involved in puberty.

An alternative, low-risk approach was pioneered by Frank (1938) who devised a series of progressively larger inserts that dilated the small vagina over a period of several months until it reached normal dimensions. The technique has been routinely applied with good results for many years to treat agenesis of the vagina arising from Mayer-Rokitansky-Küster-Hauser syndrome, which afflicts some chromosome XX females (Gargollo et al., 2009), and it can also be used with AIS (Fliegner et al., 2014; Ismail-Pratt, Bikoo, Liao, Conway, & Creighton, 2007). Inexpensive vaginal dilator apparatus can be obtained from medical supply houses or even Amazon along with detailed online instructions on use.

PAIS

The situation with PAIS can be more complicated. The wide range of genital anatomies poses a special challenge. In many cases, the attending obstetrician will notice at the time of birth that the genitalia are not typical or that there are partially descended testes. Infant sex is one phenotype that always gets recorded on a birth certificate, so the issue cannot just be swept aside. A consultation with parents will take place, and in some instances there is discussion about the possibility of surgical alteration of the infant. The testes may be removed for cosmetic or gender identity reasons when the decision is to make the child appear more like a girl, and estrogen supplements may be given at the time of puberty. The ethical basis of such a choice by parents and doctors has been contested, and some activists regard involuntary surgical alteration of sex organs as a serious violation of the individual. Instead, they recommend that the decision be a choice for the individual with PAIS to make when he or she becomes an adult.

Infants with PAIS who have a penis and descended testes are usually raised as boys. Little is known about events at puberty for such cases (Hughes et al., 2012). They are infertile, and failure to form sperm cells is a difficult condition to remedy. Some boys also have an open groove along the penis (hypospadias) instead of a closed tube of the urethra that may warrant surgical repair. This can happen for conditions other than AIS, but evidence suggests that this relatively minor surgery is often not successful in PAIS and needs several operations to complete the repair (Lucas-Herald et al., 2016), possibly because the defect in the AR interferes with healing of the surgical wounds. The surgery can be much more difficult when genitalia are ambiguous and intermediate between male and female. Parents may find it very difficult to decide on gender assignment.

In the form of PAIS in which the child has female genitals, the situation is much like CAIS, except that the person grows pubic and underarm hair after puberty, something that is quite typical of most people.

MAIS

Mild cases have a penis and descended testes, but they often experience growth of breasts. This situation can be treated by surgical breast removal and high doses of androgens.

For all forms of AIS, psychological support and counseling should help to cope with the challenges of informing friends and colleagues about the AIS condition. For parents, there is the added burden of deciding when to inform the child. A young person with AIS may experience a high level of shame about unusual genital anatomy. This can be such a serious stressor in some instances that experts recommend photography of the genitalia be done when the person is under anesthesia for a surgical procedure so that the person does not realized she was closely observed (Lee et al., 2006). It sometimes appears that the greatest challenges arise not from the medical dimensions of a genetic anomaly but the social implications of being seen as different (Diamond & Watson, 2004), especially when the difference involves sexuality.

At the present time, virtually all forms of treatment for AIS are symptomatic. Knowing the nature of the specific AR gene mutation helps to understand the origin of the symptoms, but it provides no guidance on how to treat them. One of the most effective treatments in the case of CAIS is a simple plastic dilator that is inexpensive and easy to use at home.

TERMINOLOGY, RESPECT AND HUMAN RIGHTS

From a narrow medical perspective, AIS is something unto itself, a mutation in a single gene that affects the development of certain organs. How the phenomenon is presented to the person with AIS, the family and the larger society entail value judgments about words used to describe it and which little boxes on official forms should be checked. In most countries, the laws and government records presume there are only two sexes and socially defined genders (girl, woman, boy, and man) that equate neatly with biological sex (female and male). AIS defies conventional dichotomies. There are people with the XY karyotype who are clearly women, but are they biologically female or male? Biological anatomy (Fig. 10.5, Table 10.1) does not offer an unequivocal answer. Someone having the XY chromosome set and a mutant AR gene might show male internal organs such as a testis and vas deferens but female external genitalia. Furthermore, genes intimately involved in sex hormone synthesis and estrogen signaling are not clustered on the X and Y chromosomes (Fig. 10.2). As will be discussed further in Chapters 13 and 18, the X chromosome is really not a sex chromosome at all. It functions much like an ordinary autosome in XX females. The Y chromosome definitely has much to do with male sexual development (Chapter 13), but the AR gene involved in AIS is not on the Y; it is on the X. The representation of human sex and gender in language long predated a valid biological understanding of sex development. Our ancestors oversimplified.

There are three broad categories wherein an individual does not fit neatly into the dichotomy of typical male or typical female. A *transsexual* person is someone whose gender identity does not correspond with his or her sex assigned at birth. As an adult, that person may choose to change genders and request that public protections and facilities be provided that recognize their choices. They may even opt for surgical alteration of external genitals and breasts. Most cases of CAIS described in the medical literature are raised as girls and embrace that identity, so they are not transsexual. *Intersex* is a condition that exists between the typical male and female, and there are many kinds of anatomies that might qualify a person for that label. Until recently, most cases of AIS were considered intersex, and there was even an Intersex Society of North America. A *hermaphrodite* is an organism having both male and female reproductive organs in the same body. This can happen sequentially when an anatomical male fish turns into an anatomical female or vice versa, things that are the normal course of events in many species of fish. Or it can happen simultaneously, as seen in many worms and mollusks, in which one individual has both male and female gonads. In past years, AIS was often termed hermaphrodism, but it became clear that this is not accurate from a biological standpoint, and usage of the term has been phased out. There are instances of PAIS when portions of both an anatomical ovary and an anatomical testis may be present, forming the so-called ovotestis. In some animal species, the term ovotestis may be warranted because the animal can form both eggs and sperm, but in human cases of AIS, the person can form neither eggs nor sperm.

As more was learned about AIS, many became dissatisfied with the medical terms "testicular feminization" and "pseudohermaphrodism," and requests were made to change the official names. The old medical terms implied that a person with AIS was some kind of a freak, an implication that drained away a portion of her

humanity in the eyes of others. A consensus group of 50 pediatric endocrinologists from several countries drafted new guidelines to be used in diagnosis and best practices for managing different kinds of cases. The published statement (Hughes, Houk, Ahmed, Lee, & Society, 2006) did away with the old terminology and adopted the broad name "disorders of sexual development" (DSD) that is now widely employed by medical practitioners. The DSD diagnosis is usually accompanied by a more specific description of a particular case, such as 46,XY DSD when the person has 46 chromosomes including a Y chromosome but also DSD, which is characteristic of AIS. Three years after the new guidelines were published, the term DSD, which was almost unknown prior to 2006, had become the most common term in the literature, but intersex, hermaphrodite, and pseudohermaphrodite persisted (Pasterski, Prentice, & Hughes, 2010). A search of the published literature with Google scholar indicated that from 2011 to 2016, DSD was the most common term used with AIS, but instances of "testicular feminization" and "pseudohermaphrodism" continue to appear in the scientific literature, especially for articles originating outside of Europe and North America.

Those having AIS can now get in touch with others sharing similar diagnoses and access a large trove of information via the Androgen Insensitivity Syndrome Support Group. After its formal inception in the United Kingdom in 1993, the AISSG has helped to establish support groups in several other countries, including Australia, the United States, Canada, Germany, the Netherlands, and Spain. It has also worked to educate and influence the medical profession and governments about AIS.

There is some discomfort even with the phrase *disorder* of sexual development. Most people with CAIS are leading normal lives and have found ways to accommodate their infertility, such as adoption. They usually have normal testes that produce estradiol that is essential for female puberty and normal or slightly elevated amounts of testosterone and DHT. After puberty, women with CAIS lack pubic and underarm hair, which hardly warrants designation as a medical disorder in a society that often expunges those same hairs in the pursuit of beauty. CAIS does not require continuing medical treatment. The Swiss National Commission on Biomedical Ethics (Höffe et al., 2012) recently recognized that the term "disorder" can be stigmatizing, and they proposed use of the phrase "*difference* in sexual development" (DSD).

It has been noted that the psychological problems arising from having AIS often have greater impact on the person than purely medical factors (Diamond & Watson, 2004). When parents and doctors conceal a diagnosis of AIS from a child and young adult, the secrecy, when it is inevitably discovered, accentuates a sense of shame and reinforces a stigma that is difficult to erase. Many people with AIS have remarked that the diagnosis is bad, but having it hidden from them is worse. It can shatter the trust that is so important in a parent-child or patient-physician relationship. When a young woman who seeks medical help because she is not yet menstruating learns the truth about her AIS, one of the first questions she asks is whether her parents knew. Although almost all women with CAIS accept being reared as female, many of those diagnosed and treated as PAIS eventually opt to change genders as adults, and they sometimes experience anger over childhood surgery done to alter their genitals without their consent. The psychological consequences of secrecy, shame, and stigma can be devastating, and effects can be amplified when there was surgery.

Guidelines adopted by the Swiss Commission emphasize the need to inform the child about her condition early, as she gradually becomes able to understand the situation. Current standards of medical ethics require full disclosure to the patient of information about her in the medical file. In Switzerland, this is typically done from 10 to 14 years, considerably before the legal age of consent at 16 years. It is highly recommended that any nontrivial and potentially irreversible surgical procedures to alter the genitals be delayed as long as possible, ideally until the person is able to consent or refuse. A policy of "watchful waiting" is sometimes the wisest.

AIS and other differences in sexual development call into question the validity of the customary male/female dichotomy. Sound arguments can be made for a third category of sex that might be termed "other." Categories of socially constructed gender beyond man/woman definitely exist now, although there is no universal agreement about what names should be applied. The Swiss Commission deliberated on the legal and insurance implications, and they decided the best course of action is for a country to make it relatively easy for an adult to enter a change of gender from one to the other into official records, a kind of flexible dichotomy.

HIGHLIGHTS

- An X-linked gene usually behaves as a recessive gene in chromosomally XX persons and as a dominant gene in XY persons. Consequently, those with the XY karyotype usually show a much higher frequency of any unusual trait that is related to a mutation in the X-linked gene.
- The AR gene is located on the X chromosome.
  A mutation in that gene can result in an XY person who is raised as a girl and identifies as a woman when an adult, although she will not have ovaries or a uterus. Her organs can make testosterone but cannot detect it.

- The AR is expressed in at least 37 kinds of tissue in many organs, and it influences transcription levels of more than 1000 other genes.
- Androgen insensitivity exists in different degrees, ranging from complete (CAIS) to partial (PAIS) to mild (MAIS), and the degree of insensitivity is probably related to the specific allele of the AR gene. More than 1000 mutations in the AR gene are known.
- Given the important role of testosterone in the function of so many organs, it is remarkable that many women with AIS are so phenotypically normal, apart from an absence of the uterus and ovaries.
- Surgical alterations of the genitals in AIS are a highly controversial practice, especially when performed on a child without her consent. Evolving ethical standards are now urging that no irreversible steps be taken until the child is mature enough to understand the procedures and give full and informed consent.

References

Ahmed, S., Rodie, M., Jiang, J., & Sinnott, R. (2010). The European disorder of sex development registry: a virtual research environment. *Sexual Development, 4*, 192–198.

Bardin, C. W., & Bullock, L. P. (1974). Proceedings: testicular feminization: studies of the molecular basis of a genetic defect. *Journal of Investigative Dermatology, 63*, 75–84.

Brown, C. J., Goss, S. J., Lubahn, D. B., Joseph, D. R., Wilson, E. M., French, F. S., & Willard, H. F. (1989). Androgen receptor locus on the human X chromosome: regional localization to Xq11-12 and description of a DNA polymorphism. *American Journal of Human Genetics, 44*, 264–269.

Casella, R., Maduro, M. R., Lipshultz, L. I., & Lamb, D. J. (2001). Significance of the polyglutamine tract polymorphism in the androgen receptor. *Urology, 58*, 651–656.

Chang, C., Kokontis, J., & Liao, S. (1988). Molecular cloning of human and rat complementary DNA encoding androgen receptors. *Science, 240*, 324–326.

Chen, M., Feuerstein, M. A., Levina, E., Baghel, P. S., Carkner, R. D., Tanner, M. J., ... Buttyan, R. (2010). Hedgehog/Gli supports androgen signaling in androgen deprived and androgen independent prostate cancer cells. *Molecular Cancer, 9*, 89.

Cox, K., Bryce, J., Jiang, J., Rodie, M., Sinnott, R., Alkhawari, M., ... Ahmed, S. F. (2014). Novel associations in disorders of sex development: findings from the I-DSD registry. *Journal of Clinical Endocrinology and Metabolism, 99*, E348–E355.

Darmanis, S., Sloan, S. A., Zhang, Y., Enge, M., Caneda, C., Shuer, L. M., ... Quake, S. R. (2015). A survey of human brain transcriptome diversity at the single cell level. *Proceedings of the National Academy of Sciences of the United States of America, 112*, 7285–7290.

Diamond, M., & Watson, L. A. (2004). Androgen insensitivity syndrome and Klinefelter's syndrome: sex and gender considerations. *Child and Adolescent Psychiatric Clinics, 13*, 623–640.

Doehnert, U., Bertelloni, S., Werner, R., Dati, E., & Hiort, O. (2015). Characteristic features of reproductive hormone profiles in late adolescent and adult females with complete androgen insensitivity syndrome. *Sexual Development, 9*, 69–74.

Fliegner, M., Krupp, K., Brunner, F., Rall, K., Brucker, S. Y., Briken, P., & Richter-Appelt, H. (2014). Sexual life and sexual wellness in individuals with complete androgen insensitivity syndrome (CAIS) and

Mayer-Rokitansky-Kuster-Hauser syndrome (MRKHS). *The Journal of Sexual Medicine, 11,* 729–742.

Frank, R. T. (1938). The formation of an artificial vagina without operation. *American Journal of Obstetrics and Gynecology, 35,* 1053–1055.

Galani, A., Kitsiou-Tzeli, S., Sofokleous, C., Kanavakis, E., & Kalpini-Mavrou, A. (2008). Androgen insensitivity syndrome: clinical features and molecular defects. *Hormones, 7,* 217–229.

Gargollo, P. C., Cannon, G. M., Jr., Diamond, D. A., Thomas, P., Burke, V., & Laufer, M. R. (2009). Should progressive perineal dilation be considered first line therapy for vaginal agenesis? *Journal of Urology, 182,* 1882–1889.

Gleeson, S., & Brady, E. (2016). With rules suspended intersex athlete to take center stage at Rio Olympics. *USA Today,* August 1.

Gottlieb, B., Beitel, L. K., Nadarajah, A., Paliouras, M., & Trifiro, M. (2012). The androgen receptor gene mutations database: 2012 update. *Human Mutation, 33,* 887–894.

Gottlieb, B., Beitel, L. K., & Trifiro, M. A. (2014). Androgen insensitivity syndrome. In R. A. Pagon, M. P. Adam, & H. H. Ardingeret al.*GeneReviews®*. Seattle, WA: University of Washington.

Höffe, O., Baumann-Hölzle, R., Boehler, A., Bondolfi, A., Ebneter-Fässler, K., Foppa, C., ... Tag, B. (2012). On the management of differences of sex development: ethical issues relating to "intersexuality" In: *Swiss National Advisory Commission on Biomedical Ethics NEK-CNE.*

Hughes, I. A., Davies, J. D., Bunch, T. I., Pasterski, V., Mastroyannopoulou, K., & MacDougall, J. (2012). Androgen insensitivity syndrome. *Lancet, 380,* 1419–1428.

Hughes, I. A., Houk, C., Ahmed, S. F., Lee, P. A., & Society, L. W. P. E. (2006). Consensus statement on management of intersex disorders. *Journal of Pediatric Urology, 2,* 148–162.

Ikeda, K., Horie-Inoue, K., Ueno, T., Suzuki, T., Sato, W., Shigekawa, T., ... Inoue, S. (2015). miR-378a-3p modulates tamoxifen sensitivity in breast cancer MCF-7 cells through targeting GOLT1A. *Scientific Reports, 5,* 13170.

Ismail-Pratt, I. S., Bikoo, M., Liao, L. M., Conway, G. S., & Creighton, S. M. (2007). Normalization of the vagina by dilator treatment alone in complete androgen insensitivity syndrome and Mayer-Rokitansky-Kuster-Hauser syndrome. *Human Reproduction, 22,* 2020–2024.

Jiang, M., Ma, Y., Chen, C., Fu, X., Yang, S., Li, X., ... Li, Y. (2009). Androgen-responsive gene database: integrated knowledge on androgen-responsive genes. *Molecular Endocrinology, 23,* 1927–1933.

Lee, P. A., Houk, C. P., Ahmed, S. F., & Hughes, I. A. (2006). Consensus statement on management of intersex disorders. *Pediatrics, 118,* e488–e500.

Lubahn, D. B., Joseph, D. R., Sullivan, P. M., Willard, H. F., French, F. S., & Wilson, E. M. (1988). Cloning of human androgen receptor complementary DNA and localization to the X chromosome. *Science, 240,* 327–330.

Lucas-Herald, A., Bertelloni, S., Juul, A., Bryce, J., Jiang, J., Rodie, M., ... Ahmed, S. F. (2016). The long-term outcome of boys with partial androgen insensitivity syndrome and a mutation in the androgen receptor gene. *Journal of Clinical Endocrinology and Metabolism, 101,* 3959–3967.

Mahfouz, A., Lelieveldt, B. P., Grefhorst, A., Van Weert, L. T., Mol, I. M., Sips, H. C., ... Reinders, M. J. (2016). Genome-wide coexpression of steroid receptors in the mouse brain: identifying signaling pathways and functionally coordinated regions. *Proceedings of the National Academy of Sciences of the United States of America, 113,* 2738–2743.

Marchant, J. (2011). Women with high male hormone levels face sport ban. *Nature News,* April 14. Retrieved from (2011). http://www.nature.com/news/2011/110414/full/news.2011.237.html.

Moffitt, J. R., Hao, J., Wang, G., Chen, K. H., Babcock, H. P., & Zhuang, X. (2016). High-throughput single-cell gene-expression profiling with multiplexed error-robust fluorescence in situ hybridization. *Proceedings of the National Academy of Sciences of the United States of America, 113,* 11046–11051.

Morris, J. M. (1953). The syndrome of testicular feminization in male pseudohermaphrodites. *American Journal of Obstetrics and Gynecology, 65,* 1192–1211.

Pan, B., Li, R., Chen, Y., Tang, Q., Wu, W., Chen, L., ... Xia, Y. (2016). Genetic association between androgen receptor gene CAG repeat length polymorphism and male infertility: a meta-analysis. *Medicine, 95,* e2878.

Pasterski, V., Prentice, P., & Hughes, I. (2010). Impact of the consensus statement and the new DSD classification system. *Best Practice & Research Clinical Endocrinology & Metabolism, 24,* 187–195.

Pizzo, A., Lagana, A. S., Borrielli, I., & Dugo, N. (2013). Complete androgen insensitivity syndrome: a rare case of disorder of sex development. *Case Reports in Obstetrics and Gynecology, 2013,* 232696.

Rey, R. A., & Grinspon, R. P. (2011). Normal male sexual differentiation and aetiology of disorders of sex development. *Best Practice & Research Clinical Endocrinology & Metabolism, 25,* 221–238.

Ritchie, R., Reynard, J., & Lewis, T. (2008). Intersex and the Olympic games. *Journal of the Royal Society of Medicine, 101,* 395–399.

Schweinbenz, A. N., & Cronk, A. (2010). Femininity control at the Olympic games. *Thirdspace: A Journal of Feminist Theory & Culture, 9.*

Spătaru, R., Costea, G., Spiridon, L., Procopiuc, C., Dumitriu, N., Sîrbu, D.-G., ... Dan, I. (2013). Partial androgen insensitivity syndrome. Multidisciplinary approach-genetic, endocrinological, surgical, psychological, psychiatric, social, ethical and forensic. *Romanian Journal of Legal Medicine, 21,* 201–206.

Takayama, K., Tsutsumi, S., Katayama, S., Okayama, T., Horie-Inoue, K., Ikeda, K., ... Ikeo, K. (2011). Integration of cap analysis of gene expression and chromatin immunoprecipitation analysis on array reveals genome-wide androgen receptor signaling in prostate cancer cells. *Oncogene, 30,* 619–630.

Tanagho, E. A., & McAninch, J. W. (2000). *Smith's General Urology* (15th ed.). New York: McGraw-Hill.

Tang, S., Han, H., & Bajic, V. B. (2004). ERGDB: estrogen responsive genes database. *Nucleic Acids Research, 32,* D533–D536.

Trabzuni, D., Ramasamy, A., Imran, S., Walker, R., Smith, C., Weale, M. E., ... North American Brain Expression, C (2013). Widespread sex differences in gene expression and splicing in the adult human brain. *Nature Communications, 4,* 2771.

Wills, Q. F., Livak, K. J., Tipping, A. J., Enver, T., Goldson, A. J., Sexton, D. W., & Holmes, C. (2013). Single-cell gene expression analysis reveals genetic associations masked in whole-tissue experiments. *Nature Biotechnology, 31,* 748–752.

Wilson, E. M. (1985). Interconversion of androgen receptor forms by divalent cations and 8 S androgen receptor-promoting factor. Effects of Zn2+, Cd2+, Ca2+, and Mg2+. *Journal of Biological Chemistry, 260,* 8683–8689.

Yellappa, S. C., & Venkatesh, N. (2015). Psychological aspects of androgen insensitivity syndrome—a case report. *Journal of Psychology & Psychotherapy, 5,* 1.

11

Leber's Optic Neuropathy

The German ophthalmologist Theodore Leber published a description in 1871 of a peculiar form of blindness that occurred suddenly in young adults (Leber, 1871). Within just a few weeks, there was major loss of central vision in one eye, followed within a few more weeks by loss in the other. He noted that the disorder clustered in certain families and males were more often affected than females. He also discerned that the disorder could be transmitted to a son by a mother who did not show symptoms herself. Other investigators compiled pedigrees of families hosting Leber's form of blindness, and it became apparent that the disease was always transmitted via the mother (Imai & Moriwaki, 1936) and probably involved some feature of the cytoplasm of the egg. A man with Leber's disease never had sons or daughters with the same disorder. The pattern resembled that of an X-linked gene like the androgen receptor, wherein XY persons always obtain it from the mother, but some facts did not fit with X-linkage. The frequency of affected daughters was quite high in some families, and in large pedigrees, almost all daughters passed the defect to some of their offspring. The proportions of affected sons and daughters differed greatly among families but did not seem to fit the expected proportions for a single gene on the X. One author observed that the disease was a "clinical and genetic puzzle" (Waardenburg, 1969).

The disorder became known as Leber's hereditary optic neuropathy (LHON). The H term was added to distinguish it from other forms of blindness, but that really was not necessary. Using Leber's name clearly meant a form of blindness transmitted via the mother. Huntington disease, AIS (see Chapter 10 on Androgen Insensitivity Syndrome), and Down syndrome never include the H term. Nevertheless, traditional usage prevails today. Some authors refer to LHON as Leber's neuritis rather than neuropathy. Neuritis indicates inflammation of the optic nerve, whereas neuropathy is a more general term that can involve a significant deterioration and even death of tissue, something that occurs in many but not all cases. Discovery of a genetic cause clarified these quandaries of diagnosis and naming.

While scientists were struggling to solve the puzzle, a new phenomenon was discovered among sheep in England where a disease called scrapie was shown to be passed from mother to offspring by an infective agent, a strange kind of virus resident in the cell cytoplasm that worked very slowly to degrade the nervous system (Dickinson, Young, Stamp, & Renwick, 1965). A similar disease sometimes affected cattle—bovine spongiform encephalopathy (BSE) or mad cow disease. It appeared that Creutzfeld-Jakob disease (CJD), a rare but fatal neurological disorder in humans, also arose from a virus that was similar to BSE. In a few cases, the infective agent was passed to unrelated persons via surgical instruments used in brain surgery on CJD patients. It was then proposed that Leber's optic neuropathy is transmitted by an infective agent that resides in the cytoplasm (Erickson, 1972; Wallace, 1970), which might explain why it is transmitted via the mother but does not seem to occur in definite ratios of affected and unaffected offspring the way a classical Mendelian gene would behave. A lively controversy ensued.

One author proposed that the infective agent is a protein called a "prion" (Prusiner, 1982), while others argued forcefully that transmission across generations must involve the nucleic acids RNA and DNA (Taubes, 1986). The reality of a novel form of protein that can transmit CJD and BSE has been confirmed (Collinge, 2001), but doubts linger. The precise structure of the infectious agent has not yet been elucidated (Requena & Wille, 2014). The official fact sheet about CJD by the National Institute for Neurological Diseases and Stroke in the United States in 2017 designates the prion as "the leading scientific theory" about CJD rather than established fact.

Further study of LHON, on the other hand, showed definitively that it is not related to CJD or BSE and is not caused by an infectious protein. Instead, it arises from genetic mutations of a conventional kind, except that the DNA is located in an unusual place.

THE PUZZLE SOLVED: MITOCHONDRIA

New discoveries pointed to genes in a small organelle in the cytoplasm, the mitochondrion. It occurs in all cells and has its own DNA that is transmitted to both male and female children via the maternal cytoplasm (Borst, 1977). It even has its own special variant of the genetic code (Table 2.1) in which the triplet UGA codes for tryptophan instead of STOP and AUA codes for methionine instead of isoleucine. The mitochondrial genome is much smaller than that of the chromosomes in the cell nucleus, and it lacks introns. A major effort was made to determine its DNA sequence. This was achieved (Anderson et al., 1981) more than 25 years before the immense chromosomal genome sequence was finished in 2006. The mtDNA has precisely 16,569 base pairs arranged in a circle (Fig. 11.1), and it contains 37 genes, 13 of which code

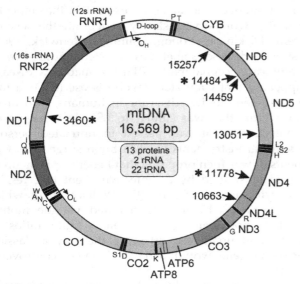

FIG. 11.1 Thirty-seven genes comprising the circular DNA of mitochondria. Thirteen code for proteins that act as enzymes (ATP6 and ATP8; CO1, CO2, and CO3; CYB; and ND1 to ND6), two code for ribosomal RNAs (RNR1 and RNR2), and 22 code for tRNA (A to Y). Full gene symbols and names are given in Table 11.2. The DNA is arranged into two circular chains termed light and heavy that are superimposed in this diagram. Sites where transcription is started for the light and heavy chains are shown as O_L and O_H. Locations of mutations known to give rise to LHON are shown inside the circle at the base number where a base substitution occurs. The three most common mutations listed in Table 11.1 are shown by *. *Sources: http://www.mitomap.org 2017 "Morbid map of the human mtDNA genome"; Anderson, S., Bankier, A. T., Barrell, B. G., de Bruijn, M. H., Coulson, A. R., Drouin, J., et al. (1981). Sequence and organization of the human mitochondrial genome. Nature, 290, 457–465; Chinnery, P. F., & Hudson, G. (2013). Mitochondrial genetics. British Medical Bulletin, 106, 135–159; Taanman, J. W. (1999). The mitochondrial genome: structure, transcription, translation and replication. Biochimica et Biophysica Acta (BBA) – Bioenergetics, 1410, 103–123; Tuppen, H. A., Blakely, E. L., Turnbull, D. M., & Taylor, R. W. (2010). Mitochondrial DNA mutations and human disease. Biochimica et Biophysica Acta (BBA) – Bioenergetics, 1797, 113–128.*

for enzymes with critical roles in metabolism and energy production in all cells. Two genes code for an essential part of the mechanism for synthesis of all proteins in every kind of cell in the body—the RNA involved in making a ribosome (rRNA)—and 22 others specify the structures of the small transfer RNAs (tRNA) that help to assemble a new protein (Fig. 3.1). One segment of mtDNA known as the D loop contains no genes and is a hotspot for mutations because they have no phenotypic effects and therefore are not pruned from a family tree.

Knowing the normal DNA sequence of human mitochondria, organelles that are always transmitted via the mother, researchers assessed several pedigrees of families in which there were several cases of LHON. Wallace et al. (1988) examined 33 people, 23 of them blind, from nine families and found that all of those with LHON had the A nucleotide substituted for G at position 11778 that resulted in the amino acid arginine instead of histidine at position 340 in the protein encoded by the MT-ND4 gene. This was the first point mutation in any gene that was shown to cause a neurological disorder. But in two pedigrees, that mutation was not the cause of the blindness; instead, mutations elsewhere in the mtDNA gave rise to almost identical clinical phenotypes of blindness. Furthermore, there was a wide range of severity of symptoms in those with the same mutation. Thus, discovery of the 11778G > A mutation was a big step forward but did not explain everything about LHON.

Medical geneticists around the world then tested many other cases of LHON and found even more mutations in mitochondrial genes associated with the maternally transmitted factor causing blindness. Table 11.1 summarizes data for 12 mutations involving six different mitochondrial genes that cause LHON in different families. The MitoMaps database now lists 18 mutations known to cause LHON in several members of different families and an additional 17 mutations identified in only a single family, sometimes involving just one individual case. The MalaCards database lists 43 different mitochondrial mutations, most of which have been observed in just one family. The overwhelming majority of all cases worldwide involve just one of three mutations, those at positions 3460, 11778, and 14484 in genes MT-ND1, MT-ND4, and MT-ND6, respectively. The 11778 mutation is most common, especially in Asia where it accounts for more than 90% of LHON cases. All of the mitochondrial genes coding for enzymes or parts of them are intimately involved in a biochemical process that enables all cells to derive energy from food. It is clear from the example of LHON that the same abnormal phenotype, a clinical syndrome, can be caused by different mutations in different genes.

INCOMPLETE PENETRANCE, SEX, SMOKING, RECOVERY

Having an unambiguous biochemical means to identify mitochondrial mutations, researchers soon learned that many people in affected families carry a mutation but never show vision loss at any age. This phenomenon denotes *incomplete penetrance* in the parlance of genetics. The term was coined at a time when many believed that a gene codes for some interesting trait or character such as vision or blindness. When every person having the mutation shows the abnormal phenotype, it is said that the information in the gene penetrates to the observable level of a phenotype; it is fully penetrant. If only some with the abnormal genotype show the neurological defect, penetrance is said to be incomplete. Penetrance is a matter of gene expression, not gene transmission. That is why a mother can pass a mutation causing LHON to her offspring while she herself has normal vision.

The data in Table 11.1 show the percentages of all male family members with a specific mitochondrial mutation who also show the symptoms of LHON. None of the better known mutations has a penetrance close to 100% among people in middle age. This fact generates hope for the future. If the reasons for some people being symptom-free can be discovered, they may point to things that can be done to prevent others from developing LHON.

An even more potent source of incomplete penetrance is genetic sex. Females are far less likely to show LHON when they possess a mutation listed in Table 11.1. In situations in which about half of males with a mutation show LHON, 10% or fewer females usually show vision loss when they carry the same mutation. The reason for this dramatic sex difference is uncertain.

Another possible source of low versus high penetrance might be the features of the environment. Many studies have found hints and suggestions that environmental toxins contribute to LHON. It is conceivable that a particular toxin has little harmful effect on vision in most people, whereas, among those carrying a mitochondrial gene mutation, the same toxin could harm cells in the retina. In a large family in Brazil with many cases of LHON extending over seven generations, all were descended from just one immigrant woman from Verona, Italy, and every one of 265 cases showing LHON carried the 11778 mutation (Sadun et al., 2003). The decline in penetrance over generations was striking (Fig. 11.2). In the second and third generations living in Brazil, 50%–70% of LHON cases were male, whereas in generations four to six, all were male.

In the Brazilian family, the rate of smoking tobacco was 70% among those with LHON but only 16% among people carrying the 11778 mutation who did not show vision loss. LHON was also more common among those who consumed alcohol. The effect of smoking was similar to an earlier study (Tsao, Aitken, & Johns, 1999) that found more than 90% penetrance among male smokers versus <50% penetrance among nonsmokers in families showing LHON. A large study of LHON in the United Kingdom corroborated the large sex difference and also noted that penetrance in male smokers exceeded 90% if they lived long enough. The verdict on smoking and LHON is not yet final, because several research groups have not detected the same pattern in their study populations (e.g., Kerrison et al., 2000).

TABLE 11.1 Mutations Causing LHON

Mutation	Gene	Percentage of all cases	Penetrance in males	Recovery rate	Vision loss
3460G>A	MT-ND1	13%	40%–80%	22%	Can see some light
11778G>A	MT-ND4	69%	82%	4%	Can see no light
14484T>C	MT-ND6	14%	68%	37%–65%	Can count fingers
15257G>A	MT-CYB	Rare	72%	28%	Can see hand motion
3635G>A	MT-ND1	Rare	54%	Low	?
3733G>A	MT-ND1	Rare	36%–44%	Yes	?
4171C>A	MT-ND1	Rare	47%	Yes	?
10663T>C	MT-ND4L	Rare	60%	?	?
13051G>A	MT-ND5	Rare	63%	?	?
14482C>A	MT-ND6	Rare	89%	Yes	?
14459G>A	MT-ND6	Rare	10%	No	Severe loss
14502T>C	MT-ND6	Rare	11%	?	?

? = unknown or not specified.
Sources: MitoMaps, OMIM.

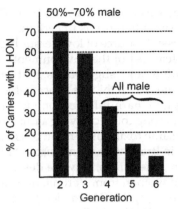

FIG. 11.2   The percentage of LHON cases among confirmed carriers of a mutation in a mitochondrial gene that is associated with LHON. Data from a large pedigree of 265 members of a multigeneration family in Brazil who were all descended from a single female immigrant born in Italy in 1861. All possessed the same 11778 mutation in the same gene (ND4). Overall penetrance declined substantially over generations, and the proportion of male LHON cases rose to 100%. Among carriers with LHON, 70% were smokers, while smoking occurred in only 16% of carriers who did not show LHON. *Adapted from Sadun, A. A., Carelli, V., Salomao, S. R., Berezovsky, A., Quiros, P. A., Sadun, F., et al. (2003). Extensive investigation of a large Brazilian pedigree of 11778/haplogroup in Leber's hereditary optic neuropathy.* American Journal of Ophthalmology, 136, *231–238, Copyright 2003, with permission from Elsevier.*

The typical course of LHON begins with mild blurring of vision in the central part of the visual field of one eye and then progresses over several weeks to severe loss of vision in both eyes. Most notably affected are the retinal ganglion cells that gather information from across the retina and send it along the optic nerve. In many cases of LHON, large numbers of those cells die and degenerate, followed by the degeneration of the myelin sheaths and axons in the optic nerve, similar to what is seen in multiple sclerosis (Chalmers & Schapira, 1999; Tran, Bhargava, & MacDonald, 2001). Loss of vision is severe when many cells and axons perish. Nevertheless, for reasons that are not well understood, many cases experience gradual recovery of visual capacity after reaching a nadir, especially for the 14484T>C mutation. Recovery further reduces penetrance.

PREVALENCE AND FOUNDER EFFECTS

Because of the rather low penetrance for several mutations, the large sex difference in expression, and the wide range of age of onset of LHON, it is difficult to give a precise estimate of how many people in a population carry a specific mutation in the mtDNA. No population surveys of mtDNA mutations have yet been published. Almost all data have been collected in families having one or more index cases who experienced some kind of medical problem and sought help. The frequency of LHON itself has

been variously estimated at 1 per 8500 in England (Man et al., 2003) and 1 per 25,000 to 1 in 50,000 in Northern Europe (Genetics Home Reference; Man, Turnbull, & Chinnery, 2002). The population frequency of the mutations listed in Table 11.1 is likely to be considerably higher in entire populations but is currently unknown. In Finland, about 1 per 50,000 people shows LHON, whereas roughly 1 per 9000 carries one of the three major mutations (Puomila et al., 2007). Much of the discrepancy arises from incomplete penetrance and the sex difference.

Mutation frequencies in local populations in the Americas can deviate greatly from those in their countries of origin in Europe because of the specific individuals who first immigrated. When most people in a modern population descend from one or a few immigrant women, the mutation frequency will tend to reflect the genotypes of the founders. For example, researchers in Montreal, Canada, examined mtDNA of 42 patients diagnosed as LHON and their extended families, testing for the three most common mutations listed in Table 11.1. Complete data were obtained from 31 index cases and 91 relatives for a total sample of 122 individuals (Macmillan et al., 1998). The large majority (89%) of index cases and relatives carried the 14484T>C mutation, whereas in Europe, the 11778G>A mutation is by far the most common (about 70%) and in Asia it accounts for more than 90% of LHON cases. Analysis of eight other mtDNA mutations in the D-loop zone that lacks genes revealed that all individuals in the Quebec sample with the 14484T>C mutation had nearly identical sets of mutations at those eight sites, none of which is related to LHON. This made it highly likely that all cases of LHON in the study sample descended from just one immigrant woman who moved to Quebec more than 200 years ago. She was the founder of a large group of people with a common ancestry on the female side of the pedigree. The founder effect thus generated a local population with a high frequency of the 14484T>C mutation embedded in a surrounding population with mainly 11778G>A mutations. The large discrepancy in allele prevalence rate was attributed to a *founder effect*.

WHAT THE GENES DO

The small size of the mitochondrial genome with relatively few genes involved in common processes should make it easier to understand what those individual genes do and how the knowledge might aid the search for a means of curing or preventing LHON. Mitochondria are present in every cell in the body and perform two kinds of functions that are absolutely essential in all kinds of cells. Because of their central place in the biochemistry of life, mitochondrial genes have been studied extensively and are relatively well understood (Chinnery &

Hudson, 2013; Tuppen, Blakely, Turnbull, & Taylor, 2010).

The first kind of function is protein synthesis. As described in Chapter 3, mRNA is translated into protein at the ribosome in the cytoplasm. The ribosome is made of several subunits, including two kinds of RNA that are encoded in the MT-RNR1 and MT-RNR2 genes, and the translation of each set of three nucleotides in mRNA into an amino acid is accomplished via small tRNA molecules whose structures are encoded in mitochondrial genes (Table 11.2). Any major mutation in any one of these genes will usually be fatal. No known variants have any involvement in LHON.

The second function is the production of energy for use in all kinds of metabolic processes throughout the body (Lodish, Berk, & Zipursky, 2000). The key molecule in this process is adenosine triphosphate or ATP. It is manufactured from adenosine diphosphate or ADP in the mitochondria and then transported to the cytoplasm of the cell. One might imagine a world in which just one enzyme could convert ADP to ATP, similar to the way phenylalanine is converted to tyrosine in the liver by the enzyme phenylalanine hydroxylase (Fig. 8.1). But the world we inhabit has devised a more intricate mechanism that depends on numerous proteins arranged in a series of complexes—the respiratory chain—and is encoded by multiple genes, some of which are part of the mitochondrial genome. Through this process, the digestion of dietary sugar and other nutrients enables the cell to create an abundance of ATP. It has been estimated that one molecule of glucose, when fully metabolized to CO_2 and water (H_2O), enables the synthesis of 30 molecules of ATP (Berg, Tymoczko, & Stryer, 2002). Mutations of the subunits of the large protein complexes can impair the efficiency of this process and cause local shortages of energy to power cellular processes. Certain of these mutations can cause LHON.

Five kinds of protein complexes are arranged in a series along the inner membrane of the mitochondrion. Certain of the complexes are tightly linked to other processes that metabolize carbohydrates, fats, and proteins. The term "complex" is well chosen, because the structures are both marvelous and intimidating. Why so many components are needed to make a single complex work properly is an unanswered question. Mutations in the protein subunits prove beyond doubt that things must be arranged in this way if the whole is to work properly. The net effect is that hydrogen ions or protons (H^+) are generated at each step and cause a cascade of changes that deliver three such ions to complex V where ADP is converted to ATP.

Two special features of this process are relevant to the discussion of LHON. First, it is well established that the molecular complexes I–V (or 1–5) utilize proteins that are mostly encoded in genes residing in the cell nucleus (Table 11.2) and then transported into the mitochondria where they work hand in hand with proteins derived from mtDNA. This division of tasks has evolved over many hundreds of millions of years. It is apparent that mitochondria originated as bacteria that were then incorporated into larger cells having additional genes (Martin, Garg, & Zimorski, 2015), although specifics of just how this happened long ago are lacking. Gradually, the two life forms merged into one integral whole.

Second, the genes and proteins involved in mitochondrial function are ubiquitous (present and necessary in all cells) and have nothing to do in any specific way with nervous system function or vision. Nevertheless, mutations at certain sites in the mtDNA molecule can impair the functioning of the cascade of energy along the complex I–V chain in a way that impairs the functioning of retinal ganglion cells more than most other kinds of cells. Certain specific mitochondrial mutations result in what is now called LHON.

TABLE 11.2 Genes Specifying Parts of Mitochondrial Complexes That Are Encoded in Mitochondrial and Nuclear DNA

Complex	MT genes	Genes in nucleus	MT gene symbols	MT gene names
1	7	38	MT-ND1, MT-ND2, MT-ND3, MT-ND4, MT-ND4L, MT-ND5, MT-ND6	NADH-ubiquinone oxidoreductase subunits 1, 2, 3, 4, 4L, 5, 6
2	0	4	None	None
3	1	9	MT-CYB	Cytochrome b subunit III
4	3	16	MT-CO1, MT-CO2, MT-CO3	Cytochrome c oxidase subunit 1, 2, 3
5	2	17	MT-ATP6, MT-ATP8	Adenosine triphosphate ATP synthase 6, 8
Ribosome	2	>150	MT-RNR1, MT-RNR2	Ribosomal RNA 12S, 16S
tRNA	22	0	MT-TA to TY	22 Transfer RNAs, coding for alanine (A) to tyrosine (Y)

Sources: Human Gene Nomenclature, Mitochondrial respiratory chain complex, HGNC; Gene Cards; OMIM; Anderson, S., Bankier, A. T., Barrell, B. G., de Bruijn, M. H., Coulson, A. R., Drouin, J., et al. (1981). Sequence and organization of the human mitochondrial genome. Nature, 290, 457–465; Wallace, D. C. (1994). Mitochondrial DNA mutations in diseases of energy metabolism. Journal of Bioenergetics and Biomembranes, 26, 241–250.

Are the multiple proteins arranged into intricate complexes unique to mitochondrial metabolism? Probably not. The five complexes appear to have many more subunits than, for example, the sarcomere in striped muscles shown in Fig. 4.8, but that diagram is a simplification of an organelle that involves many more kinds of proteins and genes that are not shown. The apparent complexity of a structure reflects to a large degree the knowledge we have about it based on extensive research. Recall the evidence cited in Chapter 3 on gene expression that indicates hundreds and even thousands of genes are expressed in many kinds of tissues and cells. Those diverse proteins in a cell are not just floating randomly in a cytoplasmic soup. They must be arranged into structures wherein many kinds of proteins adhere to each other in order to make things work properly. Discerning the nature of those structures is not an easy task. Each mitochondrial complex is itself the focus of many talented scientists working in teams and exchanging information daily (e.g., www.complexi.org and www.atpsynthase.info). The mtDNA sequence was decoded in 1981, and LHON was ascribed to a specific mutation in the MT-ND4 gene in 1988. Thirty years later, it is easy enough to point to the place in a mitochondrial protein complex where a mutation can cause LHON, but it is still not understood just how the altered protein impairs functioning of retinal ganglion cells in particular or how some people can carry the identical mutation but not show any symptoms at all.

TREATMENT AND CURE

There is no generally accepted treatment for LHON, according to a review of the medical literature (Yu-Wai-Man, Griffiths, Hudson, & Chinnery, 2009) and the 2016 edition of the EyeWiki website of the American Academy of Ophthalmology. The Genetic and Rare Disease Information Center (GARD) website in 2017 notes that there is no cure, but ongoing studies aim to find a treatment. One option is oral idebenone, a synthetic analog of ubiquinone that is involved in the cascade that generates ATP in mitochondria. Idebenone, originally intended to treat dementia and now marketed as Raxone, is said to be the only clinically proved treatment option for LHON (Gueven, 2016). There are facts to support the claim about efficacy of idebenone, yet the picture is not so clear. The European Medicines Agency in 2013 refused to license it on the grounds that the clinical benefit appeared to be too small to warrant approval (Refusal of the marketing authorisation for Raxone (idebenone), 2013).

In view of this continuing uncertainty about idebenone, it is illuminating to examine the principal evidence of its benefits. A study reported by Klopstock et al. (2011) involved 85 patients diagnosed with LHON and carrying one of the three most common mtDNA mutations (Table 11.1). They were randomly assigned to receive either a placebo or idebenone for 24 weeks. Group assignment was constrained to give equal proportions of men and women and the three genotypes in the two treatment conditions. Tests of visual acuity were given before and after 4, 12, and 24 weeks of treatment by people who did not know which patients were receiving the drug. Because neither the patients nor those giving the tests knew who was getting idebenone, it was a well-controlled double-blind study. The researchers also declared in advance that a particular outcome measure would serve as the primary end point of the study: best recovery score. The recovery of vision score was the difference between baseline acuity measures at the outset and after the 24-week period when using the Early Treatment Diabetic Retinopathy Study (ETDRS) chart (Fig. 6.1). Patients in both conditions scored better on average on the 24-week test than on the initial baseline test. Average recovery scores were higher for the idebenone-treated patients than those receiving placebo, but the difference was not very large. As the authors remarked, "the difference between groups did not reach statistical significance" (p. 2680). The authors of the study were very forthright about the implications of their findings: "considerable advances in our understanding of the molecular and biochemical basis of mitochondrial DNA-associated diseases have not yet translated into treatments of proven efficacy."

Being a drug with "antioxidant" effects on mitochondrial metabolism, idebenone has been advocated or at least mentioned on several websites (e.g., WebMD, HBC, and SDFT) as a treatment for a large number of conditions: dementia, Alzheimer's disease, Friedreich's ataxia, Parkinson's disease, multiple sclerosis, liver and heart disease, and LHON. The drug is moving through the approval process in the United States as an agent to combat Duchenne muscular dystrophy. Even though it has not been formally approved by governments to treat these conditions, at the present time, idebenone is readily available via the Internet as a "dietary supplement" and as an "antiaging" skin wrinkle therapy. The widespread publicity and easy availability of idebenone create further challenges for researchers seeking conclusive evidence that it really works. Being a relatively rare disease, LHON requires a major effort to find enough patients willing to take part in a clinical trial. Not just willing but drug-free patients are needed. Authors of the double-blind study lamented that "patients often find the prospect of taking placebo unacceptable and self-medicate, using Internet-based suppliers of ... unapproved medication" (Klopstock et al., 2011). Those who have already been taking idebenone on their own will probably not be suitable subjects for a well-controlled clinical trial.

Prevention of LHON is not yet possible. A person can carry an undetected mutation in mtDNA for years before symptoms appear in young adults. If there is a family history of LHON, genetic testing may uncover that same mutation in other relatives, some of whom may still be asymptomatic. If only there were an effective treatment for those who already show symptoms of LHON, that treatment might be worth a try as a preventive medicine, provided side effects are minimal. Indeed, it might be more effective at prevention than as a treatment for a retina that has already suffered major damage. Several medical authorities recommend that those at risk for LHON because of a family history refrain from smoking tobacco and drinking large amounts of alcohol, which is good advice to all of us to prevent disease.

HIGHLIGHTS

- The mitochondria, minute organelles in the cytoplasm with their own DNA (mtDNA with 16,569 base pairs encoding 37 genes), are transmitted from mother to both sons and daughters.
- Mutations in several mitochondrial genes can lead to a form of blindness known as LHON that is much more prevalent in males than females.
- Most mutations causing LHON show incomplete penetrance, whereby many people who carry the mutation show no symptoms at all. A woman with no symptoms can nevertheless transmit an mtDNA mutation to her children.
- Specific mutations sometimes show a striking founder effect in which dozens of people descended from the same immigrant woman carry the identical mutation.
- Unknown environmental factors clearly influence the degree of penetrance. Smoking appears to increase the risk of LHON in males.
- The mitochondrial genes involved in LHON are part of large molecular complexes consisting of proteins derived from dozens of genes in both the nucleus and the mitochondria that combine into complexes to derive energy from glucose to power many cell processes.

References

Anderson, S., Bankier, A. T., Barrell, B. G., de Bruijn, M. H., Coulson, A. R., Drouin, J., et al. (1981). Sequence and organization of the human mitochondrial genome. *Nature, 290*, 457–465.

Berg, J. M., Tymoczko, J. L., & Stryer, L. (2002). Section 18.3, The respiratory chain consists of four complexes: three proton pumps and a physical link to the citric acid cycle. In *Biochemistry* (5th ed.). New York: W.H. Freeman.

Borst, P. (1977). Structure and function of mitochondrial DNA. *Trends in Biochemical Sciences, 2*, 31–34.

Chalmers, R. M., & Schapira, A. H. (1999). Clinical, biochemical and molecular genetic features of Leber's hereditary optic neuropathy. *Biochimica et Biophysica Acta (BBA)—Bioenergetics, 1410*, 147–158.

Chinnery, P. F., & Hudson, G. (2013). Mitochondrial genetics. *British Medical Bulletin, 106*, 135–159.

Collinge, J. (2001). Prion diseases of humans and animals: their causes and molecular basis. *Annual Review of Neuroscience, 24*, 519–550.

Dickinson, A. G., Young, G. B., Stamp, J. T., & Renwick, C. C. (1965). An analysis of natural scrapie in Suffolk sheep. *Heredity, 20*, 485–503.

Erickson, R. P. (1972). Leber's optic atrophy, a possible example of maternal inheritance. *American Journal of Human Genetics, 24*, 348–349.

Gueven, N. (2016). Idebenone for Leber's hereditary optic neuropathy. *Drugs Today, 52*, 173–181.

Imai, Y., & Moriwaki, D. (1936). A probable case of cytoplasmic inheritance in man: a critique of Leber's disease. *Journal of Genetics, 33*, 163–167.

Kerrison, J. B., Miller, N. R., Hsu, F., Beaty, T. H., Maumenee, I. H., Smith, K. H., et al. (2000). A case-control study of tobacco and alcohol consumption in Leber hereditary optic neuropathy. *American Journal of Ophthalmology, 130*, 803–812.

Klopstock, T., Yu-Wai-Man, P., Dimitriadis, K., Rouleau, J., Heck, S., Bailie, M., et al. (2011). A randomized placebo-controlled trial of idebenone in Leber's hereditary optic neuropathy. *Brain, 134*, 2677–2686.

Leber, T. (1871). Über hereditäre und congenital-angelegte Sehnervenleiden. *Graefe's Archive for Clinical and Experimental Ophthalmology, 17*, 249–291.

Lodish, H., Berk, A., & Zipursky, S. L. (2000). *Molecular cell biology* (4 ed.). New York: W. H. Freeman & Company.

Macmillan, C., Kirkham, T., Fu, K., Allison, V., Andermann, E., Chitayat, D., et al. (1998). Pedigree analysis of French Canadian families with T14484C Leber's hereditary optic neuropathy. *Neurology, 50*, 417–422.

Man, P., Griffiths, P., Brown, D., Howell, N., Turnbull, D., & Chinnery, P. (2003). The epidemiology of Leber hereditary optic neuropathy in the North East of England. *The American Journal of Human Genetics, 72*, 333–339.

Man, P. Y. W., Turnbull, D., & Chinnery, P. (2002). Leber hereditary optic neuropathy. *Journal of Medical Genetics, 39*, 162–169.

Martin, W. F., Garg, S., & Zimorski, V. (2015). Endosymbiotic theories for eukaryote origin. *Philosophical Transactions of the Royal Society of London Series B: Biological Sciences, 370*, 20140330.

Prusiner, S. B. (1982). Novel proteinaceous infectious particles cause scrapie. *Science, 216*, 136–144.

Puomila, A., Hämäläinen, P., Kivioja, S., Savontaus, M.-L., Koivumäki, S., Huoponen, K., & Nikoskelainen, E. (2007). Epidemiology and penetrance of Leber hereditary optic neuropathy in Finland. *European Journal of Human Genetics, 15*, 1079–1089.

Refusal of the marketing authorisation for Raxone (idebenone) (2013). *European Medicines Agency* (pp. 1–2). .

Requena, J. R., & Wille, H. (2014). The structure of the infectious prion protein: experimental data and molecular models. *Prion, 8*, 60–66.

Sadun, A. A., Carelli, V., Salomao, S. R., Berezovsky, A., Quiros, P. A., Sadun, F., et al. (2003). Extensive investigation of a large Brazilian pedigree of 11778/haplogroup J Leber hereditary optic neuropathy. *American Journal of Ophthalmology, 136*, 231–238.

Taubes, G. (1986). The game of the name is fame. But is it science? *Discover, 7*, 28–52.

Tran, M., Bhargava, R., & MacDonald, I. M. (2001). Leber hereditary optic neuropathy, progressive visual loss, and multiple-sclerosis-like symptoms. *American Journal of Ophthalmology, 132*, 591–593.

Tsao, K., Aitken, P. A., & Johns, D. R. (1999). Smoking as an aetiological factor in a pedigree with Leber's hereditary optic neuropathy. *British Journal of Ophthalmology, 83*, 577–581.

Tuppen, H. A., Blakely, E. L., Turnbull, D. M., & Taylor, R. W. (2010). Mitochondrial DNA mutations and human disease. *Biochimica et Biophysica Acta (BBA) – Bioenergetics, 1797*, 113–128.

Waardenburg, P. J. (1969). Some remarks on the clinical and genetic puzzle of Leber's optic neuritis. *Journal de Génétique Humaine, 17,* 479–495.

Wallace, D. (1970). A new manifestation of Leber's disease and a new explanation for the agency responsible for its unusual pattern of inheritance. *Brain, 93,* 121–132.

Wallace, D. C., Singh, G., Lott, M. T., Hodge, J. A., Schurr, T. G., Lezza, A. M. S., et al. (1988).). Mitochondrial DNA mutation associated with Leber's hereditary optic neuropathy. *Science, 242,* 1427–1430.

Yu-Wai-Man, P., Griffiths, P. G., Hudson, G., & Chinnery, P. F. (2009). Inherited mitochondrial optic neuropathies. *Journal of Medical Genetics, 46,* 145–158.

Further Reading

Taanman, J. W. (1999). The mitochondrial genome: structure, transcription, translation and replication. *Biochimica et Biophysica Acta (BBA) – Bioenergetics, 1410,* 103–123.

Wallace, D. C. (1994). Mitochondrial DNA mutations in diseases of energy metabolism. *Journal of Bioenergetics and Biomembranes, 26,* 241–250.

12

Down Syndrome

Many readers of this book will have had contact with a person expressing Down syndrome (DS), maybe even a family member. It is one of the most common anomalies of the chromosomes, one that often leads to a wide variety of physical and mental challenges. Unlike mutations in a single gene such as seen in phenylketonuria and Huntington disease, the anomaly in the chromosomes of a person with DS is not transmitted across generations. Instead, it arises de novo; neither parent has the chromosome defect that afflicts the child, and the child almost never passes it to offspring later in life. It reflects a change in the chromosomes but is not itself hereditary.

HISTORY

It was the total configuration of physical features that led the physician J. L. H. Down to describe the syndrome in 1866, long before the cause of the disorder was known. He was not the first to describe the syndrome, nor was his description the most accurate, but his report (Down, 1866) became widely known, and his name is now generally used to identify the disorder.

Down worked with mentally disabled children in England, and he tried to devise a scheme to classify them into a "natural system" on the basis of their different faces. About one group of children, he wrote, "The face is flat and broad … The cheeks are roundish … The eyes are obliquely placed … The palpebral fissure is very narrow … The lips are large and thick with transverse fissures. The tongue is long, thick, and much roughened. The nose is small …" People working with disabled children in other institutions were able to perceive a similar combination of features in some of their own patients, and eventually, it became accepted practice to diagnose "Down syndrome" in an infant even before clear signs of mental disability appeared.

In 1866, the causes of mental disability were almost completely unknown, whereas today, we know that there are a great many different causes, including genetic defects, birth trauma, and malnutrition. The work of Down was fruitful in one respect: He identified a distinct group of children with something in common, and that grouping eventually led to the discovery almost 100 years later of the cause of a relatively frequent form of mental disability.

Down claimed to have found "natural" groupings, but these proved to be figments of his imagination. A hint of a gross error lingers in his name for the syndrome—"mongolism"—that was in common use until the 1970s. He claimed that "the great Caucasian family" was biologically the most highly evolved of all human groups, that other groups were mentally inferior to Caucasians, and that an abnormal child born to white parents must be a reversion (an atavism) to an earlier stage of human evolution (see Gould, 1981). He wrote that "A very large number of congenital idiots are typical mongols," and he claimed to have found children of Caucasian parents that qualified as "white Negroes" and others of the "Malay variety." The theory of the "ethnic classification of idiots" was rejected by science long ago. Only the name for the syndrome described by Dr. Down survived him.

An extensive and authoritative history of the syndrome, its signs and names, was provided by Penrose and Smith (1966), who opted for the term "Down's anomaly" and expressed skepticism about the pejorative term "mongolism." Neri and Opitz (2009) revisited this history and lamented that the direction of recent research on Down syndrome has taken with its emphasis on prenatal screening.

TERMINOLOGY—THE DEMISE OF "MONGOLISM"

In genetics, it is common practice to name a disease or even a gene after the person who first described or discovered it. That is why the term "Down syndrome" is in common use today. In fact, the use of Down syndrome is recent. A search of published articles with Google Scholar to identify those using various terms in the title is summarized in Table 12.1. Prior to 1965, the term "mongolism" was dominant, and "Down" in the title

Genes, Brain Function, and Behavior
https://doi.org/10.1016/B978-0-12-812832-9.00012-9

TABLE 12.1 Title Terms Used in Articles Listed in Google Scholar From 1960 to 1999

Term	60–64	65–69	70–74	75–79	80–84	85–89	90–94	95–99
Mongolism	110	116	65	23	4	3	2	3
Down's syndrome	39	228	369	409	450	623	688	763
Down syndrome	3	3	5	65	226	436	811	1,250

TABLE 12.2 Frequency of Certain Features in DS and Non-DS Newborn Children

Physical feature	DS children	Control children
Oblique palpebral fissure	46/57	2/86
Excess skin at the back of the neck	46/57	1/86
Flattened facial features	50/56	1/85
Four-finger palmar crease	30/56	8/86
Weak muscle tone	43/56	3/86

Denominator is number evaluated.
Source: *Penrose, L. S., & Smith, G. F. (1966). Down's Anomaly. London: J & A Churchill (selections from Table 47).*

without the apostrophe was a rarity. The 1960s was a period when opposition to racist thinking in science and society at large was growing. There was a complete absence of scientific evidence for the biological similarity of children with DS and ethnic groups from a large geographic region in Asia classified by Western anthropologists as "Mongoloid." Once the chromosomal basis of DS was recognized, it was found that the disorder could and did occur in all peoples around the globe regardless of their ancestries. The persistence of the term "mongolism" in Western science was a figment of historical precedent and racial prejudice. It faded rapidly and almost disappeared from scientific discourse by 1980.

In recent years, the term "Down syndrome" has gradually grown to dominate the field, but "Down's syndrome" is still in common use. The inclusion of an apostrophe after the physician's name does not provoke disputes over noble principles. The term "Down" is simply shorter. It is now common practice to denote most medical disorders named for a discoverer without an apostrophe, for example, Turner syndrome or Huntington disease.

THE SYNDROME

Just as for people with the usual set of 46 chromosomes, those with DS have a wide range of physical features. Although certain of these features are more common in DS than normal infants, it has always been difficult to make a firm diagnosis based solely on external appearance. Some authors described the situation in stereotypical terms, claiming DS involves slanted eyes, a rough and protruding tongue, a long crease across the palm, and mental disability. Terminology also differed among authorities in the past. Gibson (1978), for example, called the syndrome "mongolism" and the physical signs "stigma," including the "simian crease" that equated a common feature of the palm to that of a monkey. Penrose and Smith (1966) in their authoritative treatise instead tabulated "physical signs" and referred to a "four-finger palmar crease." Even the recent summary on the Mayo Clinic website refers to the "distinct facial appearance" and gives a long list of common features while cautioning that all DS children do not show those features.

Penrose and Smith (1966) cited actual frequencies and compared them with non-DS children. As shown in Table 12.2, some DS children did not show a typical feature, while some ostensibly normal children did. They reported that about half of DS children showed the stereotypical protruding and furrowed tongue, short neck, and excessive space between the toes. They provided a diagram of the epicanthal fold of the eye but noted in Table 47 that it occurs in less than half of DS cases. They cited experienced observers who claimed they could easily make a diagnosis at birth from physical appearance alone but then cautioned that this claim "should be treated with certain reservations." Gibson (1978) in Table 3 also showed that many of the typical "stigmata" were present in half or fewer of DS children. Perusing Google Images today, it is evident that DS takes many forms with no external feature present in all instances. Most of the DS children studied prior to 1980 did indeed show some degree of mental disability, but that indicator was not of diagnostic value because so many other causes of the same phenotype were known.

As indicated in the entry for DS in OMIM (#190685), several common phenotypes seen in DS have great significance for the health of the children because they involve important internal organs, especially the heart, not just surface anomalies. A survey based on the California Birth Defects Monitoring Program tabulated frequencies of a wide range of malformations observed at birth or within the first year after birth for all children, including the vast majority that were not diagnosed with DS (Torfs & Christianson, 1998). The frequencies of anomalies seen in DS children were much higher than the frequencies in non-DS children wherein most anomalies were quite rare, but the overall frequencies for most features seen in DS cases were not at all high, except for the heart defects (56%) and digestive system problems (11%). Many of the heart defects could imperil life and would need to be corrected by surgery.

Thus, it appears that *some* children with DS do indeed show an unusual array of physical signs that are indicative of DS, but others do not. When there is a syndrome, most affected children will show some of the signs, but the array of signs present in specific children can range very widely. Some may be severely afflicted, whereas others show only a few anomalies of little consequence. The situation is similar to what was seen with Huntington disease before the CAG repeat was discovered. Phenotype is not an infallible indicator of genotype. Instead, some genotypes that we now realize are typical of the disorders do not show the typical phenotypes. With advances in genetics came clarity about the diagnosis of both DS and HD.

TRISOMY 21

For many years, children with Down syndrome (DS) were common among residents of homes for the disabled, but the cause of the disorder remained a mystery until 1959 when three French scientists made a careful examination of the chromosomes from nine DS children and discovered that each child had an extra one in cells taken from the skin (Lejeune, Gauthier, & Turpin, 1959); they had 47 instead of the usual 46 chromosomes in a single cell. The researchers noticed that the extra chromosome was always a small one that appeared much like other small ones seen in normal people. The discovery was possible because of a technological advance in human chromosome staining and counting. After 1959, some uncertainty remained about the nature of the defect in DS cases. Precisely which chromosome was present in triplicate could not be certain because the one given number 22 appeared very similar under the microscope to number 21. Chromosomes 21 and 22 were often lumped into group G. Some authors considered DS to be a case of "trisomy 21–22" (Hall, 1963). Lejeune (1970) himself maintained the distinction between 21 and 22 was "clear-cut," and new methods proved he was correct. There were also a few cases of DS in which there were 46 chromosomes but one seemed to have an extra piece attached to one end (a *translocation*).

A breakthrough occurred when new stains made it possible to see distinctive patterns of banding on each chromosome (Drets & Shaw, 1971). The bands (Fig. 2.2) revealed a clear difference between 21 and 22 (O'Riordan, Robinson, Buckton, & Evans, 1971). DS was definitely trisomy 21. Most DS cases (about 93%) had an extra copy of chromosome number 21 and were trisomic, but about 5% had a translocation in which all or an extra piece of 21 attached to another copy of 21 (de Grouchy & Turleau, 1977; Hassold & Hunt, 2001).

MATERNAL AGE AND PARENTAL ORIGIN OF THE EXTRA CHROMOSOME

An extra chromosome causes DS, but what causes the extra chromosome to appear? A definitive answer cannot

yet be given, but there are some good clues. One fact was known long before the chromosome defect was discovered (Penrose, 1933): older mothers tend to have a much higher frequency of DS babies than younger mothers. Fig. 12.1 shows the rates for mothers ranging in age from 15 to 45 years from several different samples. The shape of the curve is remarkably similar in many countries (Hecht & Hook, 1996), rising dramatically with age in women older than 35. The overall frequency of DS in a society therefore depends very much on the age at which most women have their children. For example, in one study where the proportion of all mothers younger than 26 years increased substantially from 1961 to 1975, the prevalence of DS births declined from 1.7 to 0.8 per 1000 births (Fryers & Mackay, 1979). In countries in which the proportion of older women giving birth rose, so too did the rate of DS. The large effect of maternal age suggested that the problem arises in the mother, but detailed analysis of the duplicated chromosome showed that in about 5% of cases, it came from the father (Gaulden, 1992; Hassold et al., 1996; Hassold & Hunt, 2001). Later studies indicated that there was little or no age effect for cases derived from the father (Allen et al., 2009).

When more sensitive methods of karyotyping were applied to spontaneous abortions and stillbirths as well as live births (Hassold et al., 1996), trisomies were found for every chromosome except the large chromosome 1. More than 35% of spontaneous abortions showed a defect in chromosome number, usually an extra copy, but in almost all instances, fetuses with trisomies perished before birth. The same fate befell virtually all monosomies except Turner syndrome. Only a few with extra copies of chromosomes 13, 18, 21, X, or Y survived, trisomy 21 being by far the most common pattern among live-born infants. Thus, DS in live-born infants is

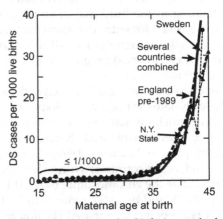

FIG. 12.1 Rate of DS cases among live births in samples from different countries plotted for each maternal age. Data for Sweden and New York State taken from Hook and Lindsjö (1978). Data for England and Wales based on Morris, Mutton, and Alberman (2002). Data for combined samples from several countries are based on Hecht and Hook (1996).

relatively uncommon, about one per 1000 births in many jurisdictions, whereas the total burden of variation in chromosomal number is quite large; almost half of conceptions have some kind of serious defect in the chromosomes. Because an extra copy or loss of an entire chromosome creates many difficulties for embryonic development, most of the mistakes are spontaneously eliminated long before birth, often before the woman realizes she is pregnant.

SCREENING FOR DOWN SYNDROME

The dramatic maternal age effect plus new methods of karyotyping gave birth to programs for screening conceptions for chromosome defects prior to birth. Screening has now become a common practice in many jurisdictions. There are often two stages in the process. The first is a relatively broad screen of pregnancies at risk to determine which are most likely to harbor a chromosome anomaly. This might be done for mothers older than 35 years, in the United States, for example. In the United Kingdom, it is common practice to assess the fetus with an ultrasound device to assess the thickness of the skin at the back of the neck ("nuchal translucency") and other features of the face and limbs that tend to be unusual in DS (Nicolaides, Brizot, & Snijders, 1994; Tapon, 2010). The procedure is relatively low cost and painless and poses little risk to the fetus. It detects only about 85% of DS cases and also identifies about 5% of normal fetuses as having signs of DS (false positives).

The second step is taken if measures obtained from the ultrasound scan are outside the range of values seen in normal fetuses. A specialist may then be asked to perform a diagnostic test that directly examines the fetal chromosomes. A common test uses amniocentesis, whereby a sample of amniotic fluid is withdrawn under anesthesia and cells floating in the fluid that have the genotype of the fetus are checked for extra chromosomes. Amniocentesis is more costly and entails a small risk of damage to the fetus or even an aborted pregnancy, but the results of the test are usually definitive.

When an extra chromosome is detected, there may be a decision to terminate the pregnancy. Some surveys found that most pregnancies where DS was detected in a prenatal test were then terminated (Cocchi et al., 2010; Mansfield, Hopfer, & Marteau, 1999). Screening began in Sweden and other countries with advanced medical systems soon after the chromosomal basis of DS became known (Fig. 12.2), and by 1980, more than half of all DS conceptions were being terminated (Iselius & Lindsten, 1986). A survey of terminations in the United States from 1995 to 2011 reported that 67% of all pregnancies involving a DS fetus were terminated, whereas another study of the period 2006–10 in the United States reported about

30% of DS conceptions were terminated after a prenatal test and more than 20% were aborted spontaneously, resulting in live births for less than half of DS conceptions (de Graaf, Buckley, & Skotko, 2015). When mothers older than 35 years are screened, diagnostic tests of chromosomes are more likely to be done than for younger mothers. One result of this practice is that a very large fraction of DS conceptions in older women is detected and then terminated (Morris et al., 2002). Termination may not be the only option when DS is detected in a fetus. Knowing a DS birth is likely, some parents may be better able to prepare themselves emotionally and logistically, and some may also use the lead time to arrange an adoption (Tapon, 2010).

At first glance, it might appear that many more births of DS infants could easily be prevented by extending the screening to younger mothers. What the numbers in Fig. 12.2 do not show, however, is the number of pregnancies that were in fact subjected to the specialized screening and diagnostic testing procedures. Terminated pregnancies almost certainly involved diagnostic testing that detected trisomy 21. Live births could have resulted either from the lack of testing or a decision not to terminate after receiving a positive test. For women 41–45 years of age in that study, 878 terminated the pregnancy because of an indication of DS, whereas 380 gave birth to a DS child. The total number of pregnancies in that age range was 58,481 over the same period (Morris et al., 2002). More than 60 older women likely received the ultrasound screening for every one case of DS detected. Most of the women giving birth to a DS infant were younger than 35, an age range when it might require 1,000 specialized ultrasound tests to detect one case of DS among a population of more than 6 million expectant mothers.

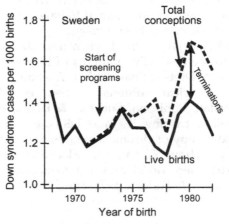

FIG. 12.2 Frequency of DS live births and total DS per 1000 births or conceptions. The difference between conceptions and live births indicates terminated pregnancies. *Reprinted by permission from Springer/ Nature, Iselius, L., & Lindsten, J. (1986). Changes in the incidence of Down syndrome in Sweden during 1968–1982. Human Genetics, 72, 133–139, Copyright 1986.*

Ultrasound scans and chromosome assays can not only have benefits but also entail costs. The technology exists to scan and test every new pregnancy and detect every new case of DS, but the strain on already overextended ultrasound clinics and health-care budgets would be substantial. Ultrasound scans are used for many kinds of medical problems, and waiting lists always seem to be long and growing. Different countries set policies on prenatal testing in their own ways (Tapon, 2010), and private insurance companies offer policies that may or may not cover screening and genetic testing. An analysis of costs of prenatal screening in the United States rated several screening methods for the number of DS cases detected, damage to normal fetuses, and cost of the procedures. The optimal procedure costs about $45,000 to avert one case of DS (Odibo, Stamilio, Nelson, Sehdev, & Macones, 2005). Furthermore, about one normal fetus was lost through screening for every four DS cases detected. The risk of damage to a normal fetus was low, but many more normal pregnancies were subjected to the tests. Several sociopolitical factors shape the decision about whom to screen and whether or not to terminate a pregnancy. These difficult questions arise for a wide range of genetic abnormalities, and they will be discussed at greater length in Chapter 20.

A recent discovery may have a major impact on screening. It happens that a number of cells from the fetal side of the placenta often enter the mother's bloodstream, and the fetal DNA can then be assayed by taking a sample of the mother's blood, an inexpensive and safe procedure. The detection rate of DS with this method is 98% or better, and the false-positive rate is less than 0.2%, markedly better than previous screening methods (Norton et al., 2015; Palomaki, Eklund, Neveux, & Lambert Messerlian, 2015). The method can be used to screen for a large number of different anomalies of chromosome number. It is already being incorporated into screening protocols in some jurisdictions, for example, the Province of British Columbia in Canada where all pregnant women regardless of age are now offered screening (Perinatal Services B.C., 2016). Those that result in an indication of DS are then followed by amniocentesis to confirm the presence of trisomy 21.

PREVALENCE AND MORTALITY

Data on pregnancy termination suggest that DS infants are not highly valued in modern society. Researchers have noted how DS children at one time were often neglected and failed to receive treatment for a wide range of medical conditions such as heart disease (Sondheimer, Byrum, & Blackman, 1985) and hearing impairments (Cunningham & McArthur, 1981). One good indicator of the quality of care is life expectancy. Of course, having a serious chromosomal defect and a heart defect is likely to shorten life, but treating symptoms that can readily be repaired may help to prolong life and enhance its quality. Earlier in the twentieth century, the life span of DS children averaged only 10 years, and mortality was more than 50% in the first year after birth (Fryers & Mackay, 1979). Many cases of DS were sent to institutions for disabled children where they received only rudimentary care and did not thrive. It was like a self-fulfilling prophecy. Some authorities believed that harboring a chromosome defect sealed the child's fate and nothing could be done to ameliorate a genetic mistake. As a consequence of that belief, little was done, and that inaction indeed sealed the fate of many DS children. Gradually, societal and medical attitudes toward DS began to soften, and treatment improved. Increasingly, the DS infant was reared at home, attended school, and had regular visits to a physician. The result was a gradual and very dramatic increase in life expectancy from 10 to 60 years (Bittles & Glasson, 2004).

GENES IN THE DS CRITICAL REGION

Chromosome 21 is the smallest among human autosomes, consisting of about 46 million base pairs (Mb) in its DNA. The finished sequence of all portions of DNA believed to contain genes was reported by Hattori et al. (2000) and amounted to 33,546,361 base pairs in the q arm. The DNA coded for 127 verified genes, another 98 that were predicted to be real genes, and 59 pseudogenes. Sixteen years later, the finished sequence involved 38,038,835 base pairs coding for 231 genes, 225 of which are verified, and 184 pseudogenes (Vega genome browser). Thus, children with DS typically have three copies of about 225 different genes in each cell. For some of these, having an extra copy probably has little influence on cell function. Precisely how many of those 225 are pertinent to the DS phenotype is an important question. If the number is large, there is probably little to be gained from an analysis of so complex an array of genes, if such a task is even feasible. On the other hand, if there are just a few genes for which an extra copy creates troubles for cellular functions and if they can be identified, then it may be possible to use this knowledge to design treatments to lessen the impact of an extra copy. For example, gene silencing that is being developed to treat HD (Chapter 9) might someday be extended to suppress the extra copies in DS children.

The sequencing of chromosome 21 inspired a burst of optimism about the future of DS and a large amount of research on molecular genetics of DS. Hattori et al. (2000) saw that the full sequence would help to diagnose a variety of abnormalities of chromosome 21, which certainly did happen, and they anticipated that their work "will aid in identifying genes involved in mechanisms of disease development" and "eventually lead to individualized treatments." Antonarakis, Lyle, Deutsch, and Reymond (2002) stated, "Patients with trisomy 21 and their families will almost certainly benefit greatly from the completion of the sequence of HC21." (HC21 is human chromosome 21.) It is illuminating to assess progress toward these goals.

From the pioneering work of Lejeune et al. (1959) and afterward, it was clear that DS usually involved an extra copy of an entire chromosome, which made it a trisomy, but there were intriguing reports of "incomplete trisomy" (Ilbery, Lee, & Winn, 1961) and "partial trisomy 21" (Aula, Leisti, & Koskull, 1973) in which only a portion of a chromosome was duplicated or translocated and attached to the end of an entire 21. Those preliminary observations suggested that three copies of just some of the genes on chromosome 21 were sufficient to generate phenotypic DS. Relying on the new chromosome banding methods, Niebuhr (1974) proposed that an extra copy of band 21q22 was sufficient for DS, and Rahmani et al. (1989) narrowed the suspect region to a segment less than 3 Mb long, which they termed the DS "critical region." Other groups joined the search and mapped a large number of DNA markers across chromosome 21 to help pinpoint the genes most strongly related to the DS phenotypes. Sinet et al. (1994) analyzed the DNA from eight people with partial trisomy 21, and Ohira et al. (1996) generated a detailed map of the region around band

21q22 and determined the limit of the extra material in a family with partial trisomy. Combining their results with data from other research groups, they proposed a small region with 1.6 million bases (Mb) on chromosome 21.

The early efforts suffered from small sample sizes and a lack of knowledge of what genes were located where on the chromosome. The situation changed for the better after Hattori et al. (2000) determined the full sequence of chromosome 21. It was then possible to determine where real genes were located, although the functions of many of them were unknown. A cluster of formerly unknown genes was designated DSCR1–DSCR10 because they were located in the DS "critical region," not because they had any clear relevance to DS phenotypes. Several later proved to be pseudogenes, and others were renamed when their functions were better known.

Researchers applied the new genetic data to larger samples of people with partial trisomies. Only a few dozen cases had been identified among more than 5 million people who had developed DS over several decades, but partial trisomies were the only means available in humans to dissect those 225 genes into smaller groups. A detailed map of DNA segments related to defects of internal organs was produced by Korbel et al. (2009), who based the results on 30 confirmed cases of partial trisomy with DS. As shown in Fig. 12.3, they identified several regions around 21q22 that harbored genes that

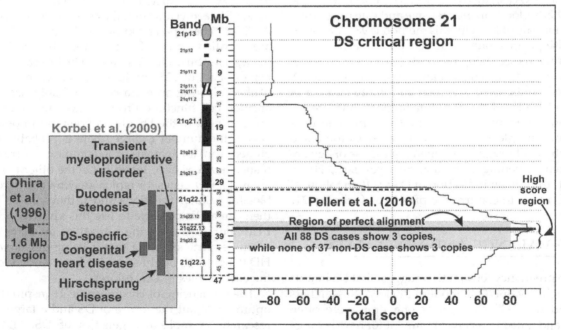

FIG. 12.3   Estimates of locations of the Down syndrome critical region that leads to DS when it is present in three copies. All are based on rare cases of partial trisomy 21. Ohira et al. (1996) examined several members of just one family. Korbel et al. (2009) assessed 30 cases from several families that showed DS with different kinds of defects of internal organs. Pelleri et al. (2016) found that almost all of 88 cases of DS with partial trisomy had a very narrow region involving band 21q22.13 present in triplicate. Graph of total score adapted from Pelleri et al. (2016). Total score represents the number of cases sharing a specific region in triplicate, minus the number of cases not sharing the region. Low negative scores suggest that regions in question are not part of the DS critical region.

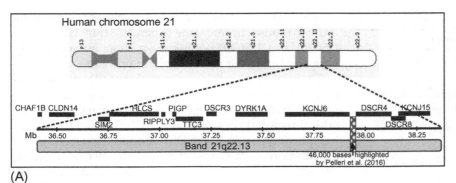

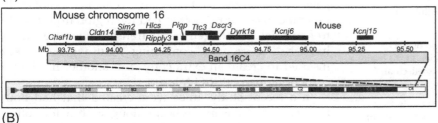

(A)

(B)

FIG. 12.4 Genes located in the DS critical region in human and mouse. (A) Human chromosome band 21q22.13 includes 13 genes shown in capital letters. Gene abbreviations and locations are provided in the NCBI website. The very narrow region implicated in the study by Pelleri et al. (2016) occurs between the genes KCNJ6 and DSCR4. (B) Locations of mouse gene counterparts on mouse chromosome 16 band C4 occur in a very similar order and occupy about the same number of bases in the DNA. *Mouse data taken from the Mouse Genome Database at MGI and the Vega Genome Browser.*

altered several phenotypes when present in triplicate. Their regions were broader than the one proposed by Ohira et al. on the basis of just one family, and they implied that different genes present in triplicate were likely involved in different phenotypes, although there might also be genes in common.

A recent article reported an analysis of all 88 known cases of partial trisomy of chromosome 21 published in the scientific literature from 1973 to 2015 in which DS had been diagnosed and another 37 partial trisomies without DS (Pelleri et al., 2016). The specific segment of chromosome 21 that was present in three copies differed for most cases, so the researchers aligned them along known base sequences and searched for common segments of DNA that corresponded with the most common phenotypes of DS—facial appearance and mental disability. There happened to be a relatively short segment of chromosome 21 in band 21q22.13 extending from base pair #37,929,229 to #37,975,580 that was present in all 88 cases of DS and none of the 37 non-DS cases (Fig. 12.3). This was much like finding the proverbial needle in a haystack, and there was one very sobering complication. The interval with 46,351 bases contained *no gene* or at least nothing that resembled any kind of gene known to biomedical genetics. Furthermore, it seemed to exclude the possible roles of several nearby genes that had been seen as good candidates related to important DS phenotypes, for example, DSCAM that had been named the "Down syndrome cell adhesion molecule" (Yamakawa et al., 1998) and dual specificity tyrosine phosphorylation regulated kinase 1A (DYRK1A).

Thus did a study with unprecedented power to peer deeply into the human genome zoom in so closely that

it stranded the investigators in a zone of doubt. That zone can be seen in a detailed map of band 21q22.13 that encompasses most of the DS critical region (Fig. 12.4A). The study by Pelleri et al. placed their best guess as to the location of a critical gene right in the middle of a gene-free zone. If one broadens the view of their data to include regions with very high scores above 80 units but fewer than a perfect 88, the whole of band 21q22.13 is included.

It is informative to consider what those genes in 21q22.13 do in the body. Knowledge about gene functions is summarized briefly in Table 12.3. It is apparent that those genes do substantially different things, but the names can be mystifying. SIM2 is "single-minded family bHLH transcription factor 2." This does not refer to a bizarre personality trait that travels in families! The gene was originally discovered in fruit flies during a period when playful fly researchers liked to give new genes cute names. It plays a role in brain development and may be relevant to DS. Some are probably involved in many kinds of cells throughout the body (e.g., chromatin assembly factor 1 subunit B (CHAF1B) and holocarboxylase synthetase (HLCS)) and seem unlikely to cause specific DS phenotypes. Others are poorly understood and still carry the DSCR moniker. Then, there are six that have clear relevance to heart and nervous system development or function. It cannot be said for certain than any one or a few of them in concert explain DS phenotypes. Separating the contributions of genes that in almost all cases of DS are joined on a single chromosome is a daunting task. From time to time, there will be new cases of partial trisomy 21 that may cast stronger light on a region of genes now known only dimly.

TABLE 12.3   Genes in the DS Critical Region 21q22.13

Gene	Name	Function(s)
CHAF1B	Chromatin assembly factor 1 subunit B	DNA replication
CLDN14*	Claudin 14	Watertight seals around cell margins, cause of deafness
DSCR3	Down syndrome critical region 3	In region linked to DS
DSCR4	Down syndrome critical region 4	In region linked to DS
DSCR8	Down syndrome critical region 8	In region linked to DS
DYRK1A*	Dual specificity tyrosine phosphorylation regulated kinase 1A	Regulation of cell proliferation in the brain, a cause of intellectual disability
HLCS	Holocarboxylase synthetase	Binding of biotin to histones, metabolism of fatty acids and amino acids
KCNJ6*	Potassium voltage-gated channel subfamily J member 6	Promotes inward flow of potassium ions to cells of the heart and brain
KCNJ15	Potassium voltage-gated channel subfamily J member 15	Regulates flow of potassium ions into cells
PIGP*	Phosphatidylinositol glycan anchor biosynthesis class P, formerly DSCR5	Synthesis of glycolipid anchor to blood cells, tongue development
RIPPLY3	Ripply transcriptional repressor 3, formerly DSCR6	No data
SIM2*	Single-minded family bHLH transcription factor 2	Regulates the generation of nerve cells, discovered in fruit flies
TTC3*	Tetratricopeptide repeat domain 3	Expressed in embryonic neuron development

BOLD* indicates effects similar to phenotypes seen in DS.
Source: *NCBI Gene database.*

MOUSE MODELS TO THE RESCUE?

Lab mice are very similar to humans at the genetic level. Gene symbols in humans consist of uppercase letters, whereas in mice, italicized lowercase is conventional. Otherwise, the same abbreviation and gene name is employed if the DNA sequence and exon structures are highly similar in the two species. The full DNA sequence of human chromosome 21 revealed 178 definite genes, most of which had counterparts in mice (Hattori et al., 2000). Toyoda et al. (2002) found that almost all of the human genes in the DS

critical region in band 21q22.13 are present in the same order in mouse chromosome 16 band C4 (Fig. 12.4B). Only DSCR4 and DSCR8 of unknown function seem to be uniquely human or present only in primates.

Some interesting things can be done with mice that are quite impossible with humans. Genetic engineering was used to make mice having three copies of genes in the DS critical region (Dierssen et al., 2001). Some of those mice had phenotypes similar to DS. One new mouse strain had a 5 Mb portion of mouse chromosome 16 duplicated on one strand of the chromosome up to and including the gene potassium voltage-gated channel subfamily J member 6 (*Kcnj6*). Another had the region from *Dyrk1a* to *Kcnj6* deleted altogether. Six transgenic mouse strains have had a segment containing mouse gene *Dyrk1a* inserted to create an extra copy, and another has had a copy of human DYRK1A inserted.

In a recent study, mice were created that had both *Dyrk1a* and *Kcnj6* in triplicate along with a few other genes on an adjacent segment (Jiang et al., 2015). The result was animals with deficits in learning and memory. When a closely related strain was created that had just two copies of those genetic elements, the deficits were greatly reduced, as would be expected. In mice with just two of the three genes in triplicate, results depended on the specific pair of genes. The authors commented that they observed "a highly complex picture of gene actions and interactions" (Jiang et al., 2015). The specific *combination* of genes present in triplicate at different DNA positions was important.

It appears that certain genes can help to compensate for an excess of gene product generated by a nearby gene or a gene that is part of the same biochemical pathway. This makes good sense. For a recessive gene having a mutant allele that codes for an ineffective enzyme, as in PKU, the one normal allele possessed by a carrier may be able to make enough enzyme to get the job done. Then doubling the number of normal alleles does no damage at all; the system adjusts itself. Similar mechanisms may act to adjust for the presence of three copies of a normal allele, at least in most instances, but there may be limits to this process when too many genes are present in three copies.

INTO THE MAINSTREAM

While molecular biologists have been working hard to solve the DS puzzle, sweeping changes in the lives of DS children have occurred. Not so long ago, it was commonplace for a DS child to be kept at home or sent to an institution for custodial care and minimal education. Beginning in the 1980s and accelerating through the 1990s, many more children with DS were sent to the same schools as their siblings and neighbors, and the old institutions far from their homes were closed. Part of the motive by governments was of course to save money, but education in better schools had benefits for the children too. There was a new perspective on DS in which the children were viewed by educators "as having special needs to be met rather than a problem to be overcome" (Cuckle, 1999). In many jurisdictions, 75% or more of DS children are now attending regular classes near their homes.

Several research projects compared DS children in separate "special education" classes with those sent into the "mainstream" classrooms with their nondisabled peers. The special classes themselves saw considerable progress in the capabilities of DS children, but inclusion in normal classrooms had a substantial impact on communication skills (Buckley, Bird, Sacks, & Archer, 2006). A recent study in the Netherlands found much larger improvements for those in the mainstream classes, and they noted that the "special" classes actually involved half as much teaching time for subjects such as reading, writing, and mathematics (de Graaf, van Hove, & Haveman, 2013). Well-controlled studies of improved teaching methods for DS children noted modest advantages over more traditional instruction (Burgoyne et al., 2012).

It has been observed in virtually every study in recent years that DS children of school age continue to improve substantially in a wide range of skills both academic and practical. Just how far they will be able to advance under optimal conditions is not yet known. Despite improved teaching, most DS children still perform substantially below the levels typical for their peers in the classroom (de Sola et al., 2015), which requires many adjustments by their classmates and instructors. Considering the impact of the extra chromosome on so many phenotypes beginning prior to birth and continuing into old age, it may not be realistic to expect that people with DS will grow to rank equally with their peers. Nevertheless, recent experience has demonstrated that they can improve and will very likely continue to improve. Inclusion in normal classrooms appears to benefit many DS children, and they are coming to expect and even demand it (Smith, 2010).

THE CONTINUING QUEST FOR A CURE

Researchers continue to explore ways to treat cognitive deficits seen in DS and even to prevent the trisomy altogether. One group affiliated with a biotechnology company reported that the insertion of the ZSCAN4 (zinc finger and SCAN domain containing 4) gene into triploid cells growing in a glass dish results in more cells with just two copies of each chromosome (Amano et al., 2015). Another group has administered a compound from green tea called epigallocatechin-3-gallate, an inhibitor of the DYRK1A enzyme, to a special mouse strain with a trisomy and observed improvements in memory (De la Torre et al., 2014). A recent clinical trial of a numbered compound with a small group of DS adults (ClinicalTrials.gov no. NCT01699711) found small improvements that may warrant a larger study needed to obtain government approval (de la Torre et al., 2016), although others viewed the effects as too small to be clinically important (Fernandez & Edgin, 2016). Meanwhile, the green tea extract is now available at pharmacies in Spain.

Others are evaluating possible benefits to DS children from the antidepressant drug fluoxetine (Prozac) or choline given as a dietary supplement. Inspired by work with pregnant trisomic mice (Guidi et al., 2014), a clinical trial is now underway in the United States where fluoxetine is being given to pregnant women carrying a DS fetus during a rapid period of brain growth (Rochman, 2016). The drug is not licensed for treating DS but can be prescribed "off list" by physicians at their discretion. New genes and remedies are sometimes announced and raise hopes that are not yet bolstered by scientific evidence. For example, the gene C21ORF91 (chromosome 21 open reading frame 91), which is not in the 21q22.13 region, was recently proclaimed to be important for the formation of neural circuits in mice (Li et al., 2016) and heralded as a cause of intellectual disabilities like DS that can be treated with novel therapies (Dovey, 2016).

Current wisdom on drug therapies for DS in the United States is provided in the official NIH website: "Some people with Down syndrome take amino acid supplements or drugs that affect their brain activity. However, many of the recent clinical trials of these treatments were poorly controlled and revealed adverse effects from these treatments. Since then, newer psychoactive drugs that are much more specific have been developed. No controlled clinical studies of these medications for Down syndrome have demonstrated their safety and efficacy, however. Many studies of drugs to treat symptoms of dementia in Down syndrome have included only a few participants. The results of these studies have not shown clear benefits of these drugs, either" (NICHHD, 2016).

The road from a bright idea about a chemical fix for DS to an effective and safe remedy that can be prescribed by a family physician is not shown on any map. The task of the scientist has become more difficult because some families of people with DS now argue against treating what they see as an anomaly, not a defect to be cured. Some urge that we should simply accept DS children as they are and grant them the same rights as every other child. It is not altogether clear how acceptance of those with DS as fully human with the same rights as everyone else can be an argument against finding treatments to enhance the quality of their lives.

HIGHLIGHTS

- Having an extra copy of a chromosome (trisomy) is a rare condition that occurs de novo—neither parent is trisomic, and the affected person does not transmit it to offspring.

- Down syndrome (DS) is a trisomy involving the small chromosome number 21. Trisomies involving larger chromosomes are almost always fatal, but someone with DS now has a life expectancy of about 60 years, compared with about 10 years in 1930.

- Phenotypes associated with DS are highly variable and involve many organs. Some of the more common features include mental disability, heart defects, thick skin at the back of the neck, a flattened face, and weak muscle tone. DS is now diagnosed by genotype, not phenotype.

- The risk of DS increases greatly when the mother is more than 35 years old. Evidence of a paternal age effect is not consistent.

- Prenatal screening is often done to detect DS in a fetus, and there is sometimes a decision to terminate the pregnancy.

- In some cases of DS, only a small portion of chromosome 21 is present in triplicate owing to a chromosome translocation. DS usually occurs when there is an extra copy of the Down syndrome critical region that spans band 21q22.13. That region contains about 13 genes, but it is not yet certain which of those leads to DS or which is related to specific aspects of the phenotype.

- Many DS children are now joining their non-DS peers in mainstream school classrooms, and this educational and social enrichment clearly improves the social communication skills of DS teenagers.

References

Allen, E. G., Freeman, S. B., Druschel, C., Hobbs, C. A., O'Leary, L. A., Romitti, P. A., ... Sherman, S. L. (2009). Maternal age and risk for trisomy 21 assessed by the origin of chromosome nondisjunction: a report from the Atlanta and National Down Syndrome Projects. *Human Genetics, 125,* 41–52.

Amano, T., Jeffries, E., Amano, M., Ko, A. C., Yu, H., & Ko, M. S. (2015). Correction of Down syndrome and Edwards syndrome aneuploidies in human cell cultures. *DNA Research, 22,* 331–342.

Antonarakis, S. E., Lyle, R., Deutsch, S., & Reymond, A. (2002). Chromosome 21: a small land of fascinating disorders with unknown pathophysiology. *International Journal of Developmental Biology, 46,* 89–96.

Aula, P., Leisti, J., & Koskull, H. v. (1973). Partial trisomy 21. *Clinical Genetics, 4,* 241–251.

Bittles, A. H., & Glasson, E. J. (2004). Clinical, social, and ethical implications of changing life expectancy in Down syndrome. *Developmental Medicine and Child Neurology, 46,* 282–286.

Buckley, S., Bird, G., Sacks, B., & Archer, T. (2006). A comparison of mainstream and special education for teenagers with Down syndrome: implications for parents and teachers. *Down's Syndrome, Research and Practice, 9,* 54–67.

Burgoyne, K., Duff, F. J., Clarke, P. J., Buckley, S., Snowling, M. J., & Hulme, C. (2012). Efficacy of a reading and language intervention for children with Down syndrome: a randomized controlled trial. *Journal of Child Psychology and Psychiatry, and Allied Disciplines, 53,* 1044–1053.

Cocchi, G., Gualdi, S., Bower, C., Halliday, J., Jonsson, B., Myrelid, A., ... Annerén, G. (2010). International trends of Down syndrome 1993-2004: births in relation to maternal age and terminations of pregnancies. *Birth Defects Research. Part A: Clinical and Molecular Teratology, 88,* 474–479.

Cuckle, P. (1999). Getting in and staying there: children with Down syndrome in mainstream schools. *Down's Syndrome, Research and Practice, 6,* 95–99.

Cunningham, C., & McArthur, K. (1981). Hearing loss and treatment in young Down's syndrome children. *Child: Care, Health and Development, 7,* 357–374.

de Graaf, G., Buckley, F., & Skotko, B. G. (2015). Estimates of the live births, natural losses, and elective terminations with Down syndrome in the United States. *American Journal of Medical Genetics Part A, 167,* 756–767.

de Graaf, G., van Hove, G., & Haveman, M. (2013). More academics in regular schools? The effect of regular versus special school placement on academic skills in Dutch primary school students with Down syndrome. *Journal of Intellectual Disability Research, 57,* 21–38.

de Grouchy, J., & Turleau, C. (1977). *Atlas des maladies chromosomiques* (2nd ed.). Paris, France: Expansion Scientifique Française.

de la Torre, R., de Sola, S., Hernandez, G., Farre, M., Pujol, J., Rodriguez, J., ... TESDAD study group. (2016). Safety and efficacy of cognitive training plus epigallocatechin-3-gallate in young adults with Down's syndrome (TESDAD): a double-blind, randomised, placebo-controlled, phase 2 trial. *Lancet Neurology, 15,* 801–810.

de la Torre, R., De Sola, S., Pons, M., Duchon, A., de Lagran, M. M., Farre, M., ... Dierssen, M. (2014). Epigallocatechin-3-gallate, a DYRK1A inhibitor, rescues cognitive deficits in Down syndrome mouse models and in humans. *Molecular Nutrition & Food Research, 58,* 278–288.

de Sola, S., de la Torre, R., Sánchez-Benavides, G., Benejam, B., Cuenca-Royo, A., Del Hoyo, L., ... Dierssen, M. (2015). A new cognitive evaluation battery for Down syndrome and its relevance for clinical trials. *Frontiers in Psychology, 6,* 708.

Dierssen, M., Fillat, C., Crnic, L., Arbonés, M., Flórez, J., & Estivill, X. (2001). Murine models for Down syndrome. *Physiology & Behavior, 73,* 859–871.

Dovey, D. (2016). *Down syndrome update: gene associated with intellectual disabilities is identified; could lead to genetic treatment. In Under the hood.* Retrieved from: https://www.medicaldaily.com/down-syndrome-update-gene-associated-intellectual-disabilities-identified-402038.

Down, J. L. H. (1866). Observations on an ethnic classification of idiots. *London Hospital Reports, 3,* 259–262.

Drets, M. E., & Shaw, M. W. (1971). Specific banding patterns of human chromosomes. *Proceedings of the National Academy of Sciences of the United States of America, 68,* 2073–2077.

Fernandez, F., & Edgin, J. O. (2016). Pharmacotherapy in Down's syndrome: which way forward? *Lancet Neurology, 15,* 776–777.

Fryers, T., & Mackay, R. (1979). Down syndrome: prevalence at birth, mortality and survival. A 17-year study. *Early Human Development, 3,* 29–41.

Gaulden, M. E. (1992). Maternal age effect: the enigma of Down syndrome and other trisomic conditions. *Mutation Research, 296,* 69–88.

Gibson, D. (1978). *Down's syndrome: The psychology of mongolism.* Cambridge: Cambridge University Press.

Gould, S. J. (1981). *The mismeasure of man.* New York: W. W. Norton & Company.

Guidi, S., Stagni, F., Bianchi, P., Ciani, E., Giacomini, A., De Franceschi, M., ... Bartesaghi, R. (2014). Prenatal pharmacotherapy rescues brain development in a Down's syndrome mouse model. *Brain, 137*, 380–401.

Hall, B. (1963). Mongolism and other abnormalities in a family with trisomy 21-22 tendency. *Acta Paediatrica. Supplement, 52*, 77–91.

Hassold, T., Abruzzo, M., Adkins, K., Griffin, D., Merrill, M., Millie, E., ... Zaragoza, M. (1996). Human aneuploidy: incidence, origin, and etiology. *Environmental and Molecular Mutagenesis, 28*, 167–175.

Hassold, T., & Hunt, P. (2001). To err (meiotically) is human: the genesis of human aneuploidy. *Nature Reviews Genetics, 2*, 280–291.

Hattori, M., Fujiyama, A., Taylor, T. D., Watanabe, H., Yada, T., Park, H. S., ... Yaspo, M. L. (2000). The DNA sequence of human chromosome 21. *Nature, 405*, 311–319.

Hecht, C. A., & Hook, E. B. (1996). Rates of Down syndrome at livebirth by one-year maternal age intervals in studies with apparent close to complete ascertainment in populations of European origin: a proposed revised rate schedule for use in genetic and prenatal screening. *American Journal of Medical Genetics, 62*, 376–385.

Hook, E. B., & Lindsjö, A. (1978). Down syndrome in live births by single year maternal age interval in a Swedish study: comparison with results from a New York State study. *American Journal of Human Genetics, 30*, 19–27.

Ilbery, P. L., Lee, C. W., & Winn, S. M. (1961). Incomplete trisomy in a mongoloid child exhibiting minimal stigmata. *Medical Journal of Australia, 48*, 182–184.

Iselius, L., & Lindsten, J. (1986). Changes in the incidence of Down syndrome in Sweden during 1968–1982. *Human Genetics, 72*, 133–139.

Jiang, X., Liu, C., Yu, T., Zhang, L., Meng, K., Xing, Z., ... Yu, Y. E. (2015). Genetic dissection of the Down syndrome critical region. *Human Molecular Genetics, 24*, 6540–6551.

Korbel, J. O., Tirosh-Wagner, T., Urban, A. E., Chen, X. N., Kasowski, M., Dai, L., ... Korenberg, J. R. (2009). The genetic architecture of Down syndrome phenotypes revealed by high-resolution analysis of human segmental trisomies. *Proceedings of the National Academy of Sciences of the United States of America, 106*, 12031–12036.

Lejeune, J. (1970). Chromosomes in trisomy 21. *Annals of the New York Academy of Sciences, 171*, 381–390.

Lejeune, J., Gauthier, M., & Turpin, R. (1959). Human chromosomes in tissue cultures. *Comptes Rendus Hebdomadaires des Séances de l'Académie des Sciences, 248*, 602–603.

Li, S. S., Qu, Z., Haas, M., Ngo, L., Heo, Y. J., Kang, H. J., ... Heng, J. I. (2016). The HSA21 gene EURL/C21ORF91 controls neurogenesis within the cerebral cortex and is implicated in the pathogenesis of Down syndrome. *Scientific Reports, 6*, 29514.

Mansfield, C., Hopfer, S., & Marteau, T. M. (1999). Termination rates after prenatal diagnosis of Down syndrome, spina bifida, anencephaly, and Turner and Klinefelter syndromes: a systematic literature review. *Prenatal Diagnosis, 19*, 808–812.

Morris, J. K., Mutton, D. E., & Alberman, E. (2002). Revised estimates of the maternal age specific live birth prevalence of Down's syndrome. *Journal of Medical Screening, 9*, 2–6.

Neri, G., & Opitz, J. M. (2009). Down syndrome: comments and reflections on the 50th anniversary of Lejeune's discovery. *American Journal of Medical Genetics Part A, 149A*, 2647–2654.

Nicolaides, K. H., Brizot, M. L., & Snijders, R. J. (1994). Fetal nuchal translucency: ultrasound screening for fetal trisomy in the first trimester of pregnancy. *British Journal of Obstetrics and Gynaecology, 101*, 782–786.

NICHHD (2016). *National Institute of Child Health and Human Development.* https://www.nichd.nih.gov/health/topics/down/conditioninfo/pages/treatments.aspx.

Niebuhr, E. (1974). Down's syndrome. *Human Genetics, 21*, 99–101.

Norton, M., Jacobsson, B., Swamy, G., Laurent, L., Ranzini, A., Brar, H., ... Wapner, R. (2015). Cell-free DNA analysis for noninvasive examination of trisomy. *New England Journal of Medicine, 372*, 1589–1597.

Odibo, A. O., Stamilio, D. M., Nelson, D. B., Sehdev, H. M., & Macones, G. A. (2005). A cost-effectiveness analysis of prenatal screening strategies for Down syndrome. *Obstetrics and Gynecology, 106*, 562–568.

Ohira, M., Ichikawa, H., Suzuki, E., Iwaki, M., Suzuki, K., Saito-Ohara, F., ... Sakaki, Y. (1996). A 1.6-Mb P1-based physical map of the Down syndrome region on chromosome 21. *Genomics, 33*, 65–74.

O'Riordan, M. L., Robinson, J. A., Buckton, K. E., & Evans, H. J. (1971). Distinguishing between the chromosomes involved in Down's syndrome (trisomy 21) and chronic myeloid leukaemia (Ph1) by fluorescence. *Nature, 230*, 167–168.

Palomaki, G. E., Eklund, E. E., Neveux, L. M., & Lambert Messerlian, G. M. (2015). Evaluating first trimester maternal serum screening combinations for Down syndrome suitable for use with reflexive secondary screening via sequencing of cell free DNA: high detection with low rates of invasive procedures. *Prenatal Diagnosis, 35*, 789–796.

Pelleri, M. C., Cicchini, E., Locatelli, C., Vitale, L., Caracausi, M., Piovesan, A., ... Strippoli, P. (2016). Systematic reanalysis of partial trisomy 21 cases with or without Down syndrome suggests a small region on 21q22.13 as critical to the phenotype. *Human Molecular Genetics, 25*, 2525–2538.

Penrose, L. S. (1933). The relative effects of paternal and maternal age in mongolism. *Journal of Genetics, 27*, 219–224.

Penrose, L. S., & Smith, G. F. (1966). *Down's anomaly.* London: J & A Churchill.

Perinatal Services B.C. (2016). *Obstetric guideline: Prenatal screening for down syndrome, trisomy 18 and open neural tube defects.* Perinatal Services BC.

Rahmani, Z., Blouin, J. L., Creau-Goldberg, N., Watkins, P. C., Mattei, J. F., Poissonnier, M., et al. (1989). Critical role of the D21S55 region on chromosome 21 in the pathogenesis of Down syndrome. *Proceedings of the National Academy of Sciences of the United States of America, 86*, 5958–5962.

Rochman, B. (2016). Parents turn to prozac to treat Down syndrome. *MIT Technology Review,*(January 12).

Sinet, P. M., Théophile, D., Rahmani, Z., Chettouh, Z., Blouin, J. L., Prieur, M., ... Delabar, J. M. (1994). Mapping of the Down syndrome phenotype on chromosome 21 at the molecular level. *Biomedicine and Pharmacotherapy, 48*, 247–252.

Smith, P. (2010). Whatever happened to inclusion? In *Vol.7. In The place of students with intellectual disabilities in education.* New York: Peter Lang.

Sondheimer, H. M., Byrum, C. J., & Blackman, M. S. (1985). Unequal cardiac care for children with Down's syndrome. *American Journal of Diseases of Children, 139*, 68–70.

Tapon, D. (2010). Prenatal testing for Down syndrome: comparison of screening practices in the UK and USA. *Journal of Genetic Counseling, 19*, 112–130.

Torfs, C. P., & Christianson, R. E. (1998). Anomalies in Down syndrome individuals in a large population-based registry. *American Journal of Medical Genetics, 77*, 431–438.

Toyoda, A., Noguchi, H., Taylor, T. D., Ito, T., Pletcher, M. T., Sakaki, Y., ... Hattori, M. (2002). Comparative genomic sequence analysis of the human chromosome 21 Down syndrome critical region. *Genome Research, 12*, 1323–1332.

Yamakawa, K., Huot, Y. K., Haendelt, M. A., Hubert, R., Chen, X. N., Lyons, G. E., & Korenberg, J. R. (1998). DSCAM: a novel member of the immunoglobulin superfamily maps in a Down syndrome region and is involved in the development of the nervous system. *Human Molecular Genetics, 7*, 227–237.

13

The XYY Male

Those working in a psychology department often meet people with mental problems who walk in on their own initiative and ask for help. One such encounter occurred on a Saturday evening in 1973 when I was working late in the office. There was a phone call from a staff member at the library on campus who was trying to help someone find information about the "XYY syndrome." The librarian knew I studied behavioral genetics and thought I might know some facts about XYY, so he asked if he could send a man over to talk with me. I agreed. No more than 30 s later, there was a second phone call from the same librarian, now sounding distressed, who kindly warned me that he thought the man was deranged and might be taking some sort of drug. He suspected the man might be an XYY male. There was no one else in the psychology building at the time, and the thought flashed through my mind that it was time to vacate the premises in a hurry. On the other hand, I had been convinced by numerous research articles that the violence-prone XYY male was a myth, so I got out

the file of reprints on XYY and waited for the visitor. A bundle of scientific articles might prove useful in a genteel academic discussion, but it would offer little defense against an assault, so there was apprehension. It turned out the young man had just been released from prison, where he had been sent because of several assaults. While in prison, he asked a psychiatrist why he seemed to have uncontrollable fits of anger, especially when drunk. The doctor told him that he probably had an extra Y chromosome and that he could never be cured and should be sure to take his medication every day. The man was very upset by this pessimistic diagnosis, and he disliked the drug, a powerful tranquilizer called Largactil (chlorpromazine), because it made him drowsy and slurred his speech. He went to the university to find out about XYY himself. He was polite, and we had a long discussion that I believe offered some hope for a better life.

– D.W.

Prior to 1961, the XYY male was unknown to science. When new karyotyping methods were used, one man with an extra Y chromosome was detected and described in the medical journal *Lancet* (Sandberg, Koepf, Ishihara, & Hauschka, 1961). The man was 44 years old at the time, had never been convicted of a crime, and had no apparent abnormality. He had been karyotyped because one son had Down syndrome. A controversy began in 1965 when a survey of chromosomes of "mentally subnormal male patients with dangerous, violent, or criminal propensities" at the maximum security prison in Carstairs, Scotland, found that 7 of 197 inmates had the XYY karyotype (Jacobs, Brunton, Melville, Brittain, & McClemont, 1965). The next year, a report of a single case of an aggressive male in a mental hospital claimed there was an "XYY syndrome" (Richards & Stewart, 1966). The scientific literature was soon flooded with case studies of violent

offenders, and a comprehensive listing of 115 articles published after the initial 1965 report announced boldly that XYY men "are invariably tall and have criminal record or psychiatric problems" (Borgaonkar et al., 1968). A well-known scientist speculated that "It appears probable that the ordinary quantum of aggressiveness of a normal XY male is derived from his Y chromosome and that the addition of another Y chromosome presents a double dose of those potencies that may under certain conditions facilitate the development of aggressive behavior" (Montagu, 1968). He also suggested that the X chromosome has "a high gentleness component." Thus, the differences in crime rates between females (XX), normal males (XY), and "supermales" (XYY) were viewed primarily as consequences of different chromosome sets determined at conception. In 1968, the widely read magazine *Newsweek* printed a story headlined "Born Bad," and then, an article in 1970 entitled "Chromosomes and crime" appeared in *Newsweek*. Many newspapers ran stories about the "crime chromosome."

SPECKULATION

Hysteria erupted in the mass media in 1968 when it was revealed that a man who murdered a Paris prostitute was XYY. Two years earlier, Richard Speck murdered eight student nurses in Chicago, and in the aftermath of that grisly crime and the subsequent trial, many news stories were published stating that the man was XYY. A medical scientist joined the chorus, proclaiming that "Speck was a likely candidate for the XYY disorder. Independently, a cytogenetic laboratory in Chicago confirmed his hunch" (Telfer, 1968). Other reports claimed Drs. Pergament and Sato did the chromosome tests. A reporter for *The New York Times* contacted the doctors, who said they never examined Speck. Speck's lawyer Gerald Getty stated, "I never knew those doctors existed before I read about them in the paper" (Lyons, 1968). It turned out that Getty had indeed had a chromosome assessment done, hoping there would be something in it to exonerate Speck, but he did not present the results at the trial. Dr. Eric Engel had read about the Speck case and thought the man might be XYY, so he contacted Getty in 1966 and then performed a confidential karyotyping. Amid the furor about XYY in 1968, he finally released a report showing Speck had a normal karyotype (Engel, 1972). Unfortunately, many journalists and textbook writers continued to propagate the false news about Speck for several years.

As soon as exaggerated claims about XYY began to appear in the media, skeptics entered the debate, new facts gave a broader picture of XYY, and the case for the crime chromosome weakened and eventually collapsed. For example, Money (1969) reviewed 13 published reports involving 35 XYY men, less than half of whom were in prison. He cautioned that "From the point of view of XYY babies who will grow up to be law-abiding citizens, it is immoral to spread pseudoscientific rumor that the XYY karyotype inexorably spells crime and jail sentences." It became increasingly clear that many ordinary men with the XYY karyotype were living normal lives undetected by medical science.

FREQUENCIES OF XYY IN LARGE SURVEYS

The next 15 years saw several large surveys of chromosomes of men in prisons or mental institutions published in scholarly journals (Table 13.1). Approximately 1% of men in those institutions had the XYY karyotype. The frequency of XYY in consecutive newborns proved to be much lower. An early survey of karyotypes

TABLE 13.1 Males Karyotyped in Penal Institutions or Previously Convicted[a]

Year	Country	N	XYY	Year	Country	N	XYY
1966	The United Kingdom	942	21	1972	The United States	475	0
1967	The United Kingdom	342	9	1977	France	1171	22
1970	The United States	190	0	1979	Japan	1371	5
1970	The United States	337	4	1980	The United States	3011	30
1972	Denmark	151	3	1981	Finland	1040	9

[a] *Total N in all 10 studies = 9228; total XYY = 110 (1.2%).*
Review by Re, L., & Birkhoff, J. M. (2015). The 47, XYY syndrome, 50 years of certainties and doubts: a systematic review. Aggression and Violent Behavior, 22, 9–17.

of consecutive newborn infants detected three cases of XYY out of a sample of 4500 infants (Lubs & Ruddle, 1970).

At the peak of the debate about XYY in the 1970s, an influential study was published concerning men in Denmark (Witkin et al., 1976). Instead of looking at men already in prison, they started with an entire, unselected population of 31,436 men born in Copenhagen from 1944 to 1947. In Denmark at the time, all males received an extensive examination as part of a military draft program, including a measure of height and a Danish test of mental ability (the Borge Priens Prover or BPP test). Investigators already knew that XYY men tend to be tall, so they decided to study all males at least 184 cm tall at the time of the physical exam, and they were able to obtain blood samples and karyotypes for 4139 of the men who agreed to participate in the study. The records noted all previous criminal convictions and whether each man had passed three separate academic examinations given to all students in the 9th, 10th, and 13th years of school. It is important to note that the karyotypes of the men were determined after all the examinations and arrests had occurred; the parents, teachers, police, and the men themselves did not know who was XY or XYY.

Among the 4139 tall men, there were 12 with the XYY karyotype, about three times the rate in all newborn males, which was expected because the sample was limited to tall men. Among the 4096 XY males, 9.3% had been convicted of at least one crime at some time in their lives, whereas 5 of the 12 XYY males had at least one conviction. Although the conviction rate of the XYY males was higher, the size of the XYY sample was very small. Furthermore, the descriptions of their histories revealed that violence was no more common than among XY convicts. Among the XY men, the average number of school tests passed was 1.6 out of a possible 3, whereas the 12 XYY men passed an average of only 0.6 tests. Average

BPP test score for XY men was 44, whereas XYY men averaged 30 and only 2 of 12 scored above 40 points. Thus, according to data from Denmark, XYY males were often below average in school performance and mental ability test scores.

In more recent surveys, Ratcliffe, Murray, and Teague (1987) reported 18 XYY among 17,522 newborn males, about 1 per 973 males. More recent studies reviewed by Morris, Alberman, Scott, and Jacobs (2008) and Re and Birkhoff (2015) concurred with the XYY incidence at birth of about 0.1%, 10 times less than the frequency observed among institutionalized men.

To judge the impact of the extra Y chromosome on male behavior, it is essential to know how many XYY males eventually are convicted of violent crimes and confined to a prison versus how many are leading reasonably normal lives. A rough approximation of risk of being in prison because of the XYY karyotype can be made from available data without actually doing a chromosome survey of all adult men. Suppose there is a population of 1 million adult males in a particular district (Fig. 13.1). If the population frequency of XYY is 1/1000, then there should be about 1000 XYY men among them. In several jurisdictions, about 0.2% of men are in prison at any one time and could become part of a chromosome survey, as happened for those in Table 13.1. There would then be about 2000 men in prison in the district. Table 13.1 indicates we should expect about 1% of those confined to an institution to have the XYY karyotype, which would amount to 20 XYY males in a prison sample of 2000. Consequently, we expect about $1000 - 20 = 980$ or 98% of XYY males to be living in the district but outside of prison. It is therefore evident that the rate of XYY men in prison could be *10 times higher* than their rate at birth, while at the same time, almost all XYY men are *not* in prison. An elevated risk of being in prison is not a good indicator of the impact of an extra Y chromosome on a man's behavior.

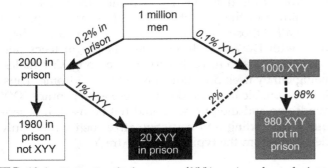

FIG. 13.1 Estimating the frequency of XYY men in and out of prison. Data show that about 1 per 500 men is in prison at any one time in many countries, about 1 per 1000 males has the XYY karyotype, and about 1% of prisoners are XYY. Applying these figures to a hypothetical population of 1 million men, there would be about 2% of XYY men in prison. Thus, although the rate of XYY among prisoners is 10 times higher than in the general population of men, the vast majority of XYY men are not in prison.

Some scientists and especially the mass media interpreted the risk data to mean that a biological factor, the extra Y, impelled men to commit terrible crimes, when in fact, the vast majority of XYY men had done no such thing.

NEW RESEARCH ETHICS AND OLD MYTHS

After the mid-1970s, new studies of XYY in large samples became rare. It was not the balance of evidence that brought a halt to population surveys on this topic, however. A major study of chromosomes of newborn infants in Boston became the focus of intense debate about the ethics of genetic research. Scientists involved in the study were based at Harvard University, and a plea to cancel the study on ethical grounds was made to the university. The university chose to allow the study to continue (Culliton, 1974), but the researchers themselves decided to terminate it (Bauer et al., 1980; Beckwith et al., 1975; Culliton, 1975). The central issue was *informed consent* in medical research. A person who is invited to participate in a medical research project should be given fair and complete information not only about the nature of the procedure to be performed, which for karyotyping was usually a simple taking of a blood sample, but also about any likely harm or benefit that might arise from participation. In a newborn screening study, the parents would be the ones deciding on behalf of their child. Because of the widespread misinformation about the XYY chromosome set, there was likely to be harm to the person's self-image and possibly social discrimination because of the stigma of being XYY. At the same time, because there was no effective way to treat any deficit arising from the extra Y, people would not realize any benefit from taking part in the research, if an anomalous karyotype was detected. It was also held to be a violation of ethics not to inform parents of any peculiarity in the chromosomes of their child that might be found. They had a right to know, and researchers were expected to recognize that right. The large population surveys of the 1960s and early 1970s had not made these risks clear to prospective participants or provided full disclosure of data to them afterward, and the studies were judged to be in violation of the evolving ethical principles of medical research.

By the mid-1970s, the allegation that an extra Y made men more violent and therefore highly likely to commit crimes had been condemned as a myth (Pyeritz, Schreier, Madansky, Miller, & Beckwith, 1977), a judgment upheld by a large review of the evidence, and the history of overblown claims (Bauer et al., 1980). Spreading false news about XYY came to be seen as a serious violation of ethical standards in both science and journalism.

IN SEARCH OF A SYNDROME

The study of the XYY condition has progressed from the time in the 1960s when a few dramatic case reports of tall and violent offenders were enough to establish a clinical syndrome in the minds of many psychiatrists to the present time when the very existence of an XYY syndrome is in doubt. Milunsky (2010) remarked that the extra Y "does not produce a discernible phenotype." A recent review by Gravholt (2013) concluded, "because there does not appear to be a recognizable pattern of neurodevelopmental or behavioral characteristics, the use of the term *syndrome* may be inappropriate." A brief summary of pertinent data shows why specialists now shun the term.

Height

Fig. 13.2 shows data for a diverse sample of 90 XYY males living in Philadelphia, the United States, plotted against norms for XY males (Bardsley et al., 2013). Those XYY males in the teen years were almost all taller than the median or 50th percentile for XY males. In an extensive review of the published literature on XYY, Re and Birkhoff (2015) observed that, "The only phenotypic expression of an extra Y chromosome that can nearly always be found is an increased growth velocity from early childhood, with an average final height approximately 7 cm (3 in) above expected."

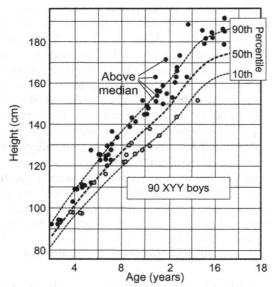

FIG. 13.2 Growth of boys with an XYY karyotype relative to XY boys. The median or 50th percentile is the norm, the value that is equal to or greater than 50% of XY boys. Ten percent of XY boys are at or below the 10th percentile, whereas 90% are below the 90th percentile. There is a wide range for XYY boys at any age, but most of them exceed the 50th percentile (*black dots*), and relatively few are below the median (*gray dots*). *Adapted from Bardsley, M. Z., Kowal, K., Levy, C., Gosek, A., Ayari, N., Tartaglia, N., … Ross, J. L. (2013). 47, XYY syndrome: clinical phenotype and timing of ascertainment. The Journal of Pediatrics, 163, 1085–1094, Fig. 1A, Copyright 2013, with permission from Elsevier.*

Testes and Hormones

The timing of the onset of puberty is usually normal. In XYY boys 14 years and older, about half showed testicular volumes substantially above average for XY boys, but this did not result in generally higher levels of testosterone or other gonadal hormones (Bardsley et al., 2013). It is expected that taller boys would have larger testicles and genitalia simply because the whole body is larger. Prior to birth, hormone levels in XYY fetuses also appeared to be normal (Ratcliffe et al., 1994). Fertility of XYY men was normal, and the extra Y was not transmitted to their sons.

Mental Abilities

The Witkin study done in Copenhagen indicated a reduced level of mental ability on standard tests and in school, and later studies have generally supported this finding. The extensive review of the published literature by Leggett, Jacobs, Nation, Scerif, and Bishop (2010) identified only three studies that met satisfactory criteria for inclusion in their review. The data are still worth examining because they are the only information available, but the biased nature of the samples needs to be kept in mind. Leggett et al. (2010) noted that the XYY males scored close to 100 on standard mental ability tests but their XY controls were 10–15 points higher. Verbal test scores tended to be 7–8 points lower than for nonverbal items for the XYY males. The pattern of lower verbal than performance scores was seen in several recent studies that did not involve XY control groups (Bardsley et al., 2013). This body of weak evidence on XYY males indicates that verbal ability test scores are commonly below the average of XY males, but noteworthy, mental deficiency is uncommon. Almost all XYY males are within the normal range of test scores. Thus, genuine mental disability is not a characteristic of an XYY syndrome.

For a phenotype to serve as a component of a medical syndrome caused by a genetic error, an abnormal form or magnitude of the phenotype should be present in well over half of those with the chromosome anomaly, as happens with Down syndrome. Only height in teens and adults appears to meet this criterion among XYY males. Being tall by itself does not constitute a syndrome. Thus, it is reasonable to conclude that there is no genuine XYY syndrome and the phrase should not be used. It is seriously misleading and can stigmatize men whose only deviation from the typical is one extra Y.

MORTALITY AND TERMINATION

Life expectancy of a newborn infant is a good indicator of the overall health of members of a society. The Danish registries of births, chromosome anomalies, diseases, and deaths revealed that among a sample of 208 XYY males diagnosed between 1968 and 2008, 38 had died by the time of the study (Stochholm, Juul, & Gravholt, 2010). The average life expectancy of XYY men was 67.5 years, considerably less than the 77.9 years for control XY men but nonetheless a rather long life in most societies. Causes of death for XYY men involved several kinds of cancer and heart, lung, and nervous system problems, as well as accident and trauma. There was no evidence connecting the extra Y to any specific medical disease leading to death. Some of the difference might be related to social factors such as lesser education and income of the XYY men.

Whereas a large proportion of Down syndrome fetuses are terminated during pregnancy (Chapter 12), the rates appear to be much lower for XYY fetuses detected during prenatal screening or an assessment of a woman pregnant after the age of 35. A recent report from France (Gruchy et al., 2016) noted that termination rates for XYY fetuses resulting from screening had shifted substantially from about 40% prior to 1997 to about 18% after 1997 when the new Bioethics Act established multidisciplinary committees for prenatal diagnosis (MCPD). The committee consisted of an obstetrician, an ultrasound specialist, a specialist in neonatal medicine, and a geneticist. They were subsequently joined by a psychiatrist, a genetic counselor, and a biologist. When both parents request termination of pregnancy because of a chromosome anomaly, they are required to meet with the team of experts and then obtain authorization in writing from two physicians. Pregnancy termination rates for XYY cases detected fortuitously rather than during a screening program declined greatly from 18% before 1997 to 3% after the new law came into effect (Gruchy et al., 2016).

Data on termination of XYY in other jurisdictions are sparse. The extensive review by Morris et al. (2008) noted a relatively low frequency of termination of XYY but provided no definite numbers. Given what is now known about the absence of clearly adverse symptoms in most cases of XYY, termination rates are likely to be considerably lower than seen for Down syndrome.

CRIME REVISITED

Denmark compiles large amounts of data on many features of its citizens, and the registries can be utilized to address a wide range of social and medical issues. The identities of individuals are carefully protected by the use of code numbers. The Cytogenetic Central Register involves every man in Denmark with a chromosome anomaly detected from 1967 to 2009. There is also a Central Crime Registry with an entry for every criminal conviction since 1978 and a variety of demographic details and social indicators compiled by Statistics Denmark, all of which can be linked to the code numbers in the cytogenetic registry. A recent study examined convictions from 1978 to 2006 (Stochholm, Bojesen, Jensen, Juul, & Gravholt, 2012). Among the 15,356 control cases in the age range 15–70 years that were almost entirely XY men, 6284 (41%) had been convicted of at least one crime, whereas 80 (50%) of 161 XYY males had been convicted. The investigators then converted the results to "hazard ratios" that divided the proportion of XYY men convicted by the proportion of control men convicted. When relatively common traffic offenses were omitted from the analysis, the hazard ratio for XYY men was 2.1, meaning they were twice as likely to have been convicted by a particular age than control men.

The data compiled by Statistics Denmark made it possible to examine the relation between criminal convictions and a variety of social indicators, such as the level of education the man had achieved and whether he was living with a partner, had fathered a child, or had retired after a long history of employment. In the control cases, conviction rates were considerably lower among men with higher education who were married, had a child, and retired from regular work. Among the XYY men, there were more with lower amounts of education who never lived with anyone or fathered a child, things associated with higher conviction rates. The question thus arose as to whether the higher conviction rates for XYY men could have arisen from the less auspicious social aspects of their lives, rather than a propensity for crime per se. Statistical methods were utilized to adjust the XYY data for those four social indicators. Then, the hazard ratio was recomputed. The result was 1.04 when traffic offenses were ignored, which means that the conviction rate for XYY men was nearly the same as for the XY controls. The researchers concluded that "unfavorable socioeconomic conditions may be part of the explanation for the increased rate of convictions" for XYY men.

THE CASE OF MARIO: IMPUTABLE AND DANGEROUS?

The best scientific studies of problem behaviors typically involve large numbers of people chosen without bias and assessed impartially by experts who do not know about their genetic conditions. Nevertheless, knowledge is eventually applied to individuals, not classes of persons, and it is applied by authorities with opinions about what causes what. Examining a single case can teach us much about how genetic anomalies influence real lives. Outside an academic cloister, a man with the XYY karyotype who commits a crime is judged and punished in a legal system with rules and precedents that reflect traditions and beliefs about biological causes.

The Italian case of "Mario" is illuminating (Birkhoff, Re, & Torri, 2015).

Mario attacked members of his family many times in an effort to obtain money to buy drugs to feed his addictions. He was arrested and charged with crimes against his family. In deciding the case, the court had to judge whether he was "imputable" and was socially dangerous. Imputable means he knows the difference between right and wrong and can control his own behavior. Someone socially dangerous cannot be cured and is likely to offend again and again. Mario's case history shows how events unrelated to personal qualities can become woven into the fabric of a life and make it so difficult for a judge or a scientist to discern the causes of despicable conduct. The sequence of formative events leading to his trial is summarized briefly:

- His family was middle class and free from major problems.
- His only remarkable phenotype was "impressive height growth."
- In kindergarten, he began to exhibit problems learning and relating to others.
- When he was 7 years old, his father died, and he developed prolonged grief disorder and was treated with the psychiatric drugs Neuleptil, Haldol, and Depakin.
- He did poorly in school and was held back for two grades, so he was in a class with much younger and smaller students, who frequently mocked him.
- He began to show aggressiveness and could not form close relationships with others.
- He was labeled as different by his peers and became socially isolated.
- Then, he began to use drugs and became addicted.
- He stopped going to school and showed no interest in any kind of work.
- He was diagnosed as having borderline personality disorder.
- A karyotype ordered because of his aggressive conduct revealed the XYY condition.
- He returned to live with his family and began to abuse them in order to get money for drugs.

The court found that he was not imputable because of a mental disorder, but because he committed violent acts under the influence of drugs, he was sent to a forensic psychiatric hospital, a prison for those judged insane.

The authors of his case history explained in their article how "an extra Y seems just to be a genetic substrate which cannot be considered the only cause of a deviant behaviour." Instead, the whole history with its litany of unfortunate experiences needs to be considered. The effect of an extra Y needs to be seen as something indirect, an influence rather than an all-powerful cause. To close the story, they quoted Mario himself: "An extra Y can influence, but I'm the only one who can decide who I am and how to behave, as well as I'm the only one who decided by himself to take drugs and to be a delinquent."

GENES ON THE Y

What is it about the Y chromosome that makes a man a man? Would knowing this perhaps cast some light on the situation of XYY males who have double the dosage of the usual Y chromosome genes? Perusing the Internet, one can find maps of the Y showing genes for typical male behaviors such as channel flipping (FLPN gene), spitting (P2E) gene, and forgetting birthdays (OOPS gene). This tongue-in-cheek exercise invokes the one-gene/one-behavior view, wherein a gene is said to code directly for a specific behavior. Other manifestations of this view appear in the mass media almost every day. Precisely how a chunk of DNA might do this is usually left to the imagination. We can have great fun with this approach by devising a map of X chromosome genes that code for nursing an infant, changing diapers, baking pies, and cleaning the house and by adding the stipulation that they require two gene copies to be active. Those kinds of maps of X and Y could explain so much. For example, women have two copies of the diaper-changing gene, whereas men have only one, so of course, women are better equipped to deal with dirty diapers, and men have one copy of the channel-flipping gene, but women have none, so of course, men must be placed in charge of the TV remote control.

In decades past, there was little that could be done to refute such nonsense decisively, but today, we have a map that shows every gene that is actually on the Y and the X, and databases provide copious details about what many of those genes do. Now, the fun is pretty much over for theories of malevolent genes. If someone avers that a gene on the Y gives a man violent tendencies and disregard for the law, we are entitled to ask exactly which genes do this and how. The recent scientific literature offers no clues about the brain and behavior in relation to the Y. Most of the male-specific genes on the Y seem to be involved in sperm production rather than complex social behaviors.

Let us instead ask a different question, one that could potentially be answered: What goes wrong when a man is XYY? Given that deficits arising from an extra Y are so mild, if there are any at all, it is unlikely that a specific gene on the Y can be pinpointed as the source of problems. It may nevertheless be interesting to review the population of genes on the Y because it is a very interesting chromosome, one quite different from the X and the 22 autosomes. The X is a large chromosome consisting of about 156 million base pairs containing 723 genes that encode proteins, whereas the Y has a substantial 57.2 Mb of DNA but encodes only 48 genes that specify proteins (Vega genome browser). Vast swaths of the Y do not code for any kind of protein that is useful in development or cell function. Much of it consists of evolutionary relics, including 383 pseudogenes that long ago were probably functional but degenerated. There is also a piece of "heterochromatin" 25 Mb long that appears to be functionally inert (Fig. 13.3).

There was no Y chromosome in our remote ancestors about 180 million years ago. Instead, everyone had two X chromosomes that exchanged DNA through meiosis when germ cells were formed. Then, an evolutionary process commenced that led to the loss of many of the genes on one of the X chromosomes. The incredible shrinking X became what we now term the Y, and those with the XY karyotype were males (Wilson Sayres & Makova, 2012). By comparing genomes of a wide range of living species of mammals and marsupials, researchers estimated that the ancestral X housed about 600 protein-encoding genes. Many of those evolved into nonfunctional pseudogenes on the human Y, and another 315 were lost altogether and left no remnants in the human Y. Instead, males today must get that information from the X, of which they have one good copy.

One segment of the p arm of the Y has genes also located on the X and continues to exchange DNA with the X during meiosis. It is termed the "pseudoautosomal region" of the Y (the PAR), and the genes occur in the same order on both the X and Y. There is also a tiny fragment at the far end of the q arm that has just three genes that exchange material with the X. In between, there are a few genes that always occur in only one copy in males but are absent on the X. There are also several genes in bands p11.2 to q11.2 of the Y that have counterparts on the X as well, albeit in slightly different forms. So, the Y has become a hodgepodge of genes, much diminished from its noble origins as a complete chromosome.

Twelve genes that occur on the Y but not the X are involved in the formation of testes and sperm (Table 13.2). One gene of particular interest is SRY, the so-called sex-determining region on the Y (see Fig. 10.5). Only males have this gene. When it is deleted in mice, the animal usually develops into a female. XYY males have two copies of SRY, and this seems to do them no harm. It is the lack of a functional SRY that sometimes changes the sex organs. Several XY males have been observed to have female sex organs or ambiguous genitalia or even totally absent gonads because of a mutation in the SRY gene (OMIM entry 48,000). But sex determination turns out to be a much more complex affair than just SRY telling other genes and cells what to do. The androgen receptor (AR) mutation responsible for AIS (Chapter 10) is a clear example. The GeneCards database lists more than 1000 genes that are sometimes involved in sex reversal. Much of current knowledge about sex determination comes from studies of mice, and it happens that they have many of the same genes as humans, so we presume the molecular events proceed similarly. The WT1 gene at chromosome 11p13 is needed to activate SRY, and the product of SRY along with that from the

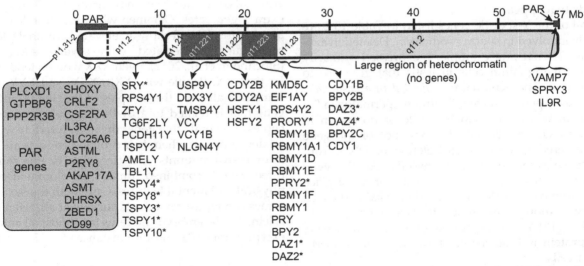

FIG. 13.3 Locations of genes on the human Y chromosome in different bands at various distances in megabases (Mb) of DNA from the far end of the p arm. The pseudoautosomal region (PAR, dark bar) exchanges material with the corresponding portions of the X chromosome during meiosis. Genes shown with * are found only on the Y. Others have counterparts on the X that have very similar DNA sequences, although they do not exchange material with those on the X. The large q11.2 band contains no actively transcribed genes or pseudogenes and has no known function (Quintana-Murci & Fellous, 2001). *GeneCards, OMIM Gene Map for Y and OMIM gene listings, Vega Genome Browser release 66, and Tilford et al. (2001), Fig. 1.*

TABLE 13.2 Genes Only on Y Chromosome, Not on X Chromosome[a]

Symbol	Name	Function	Disorders From Mutation
SRY	Sex-determining region Y	DNA binding, differentiation of embryo gonad into testis	Absent gonads or sex reversal, sterility
TSPY1, TSPY3, TSPY4, TSPY8, TSPY10	Testis-specific protein, Y-linked	Expressed in testis, sperm formation	Male sterility, gonadal cancers
PRORY	Proline-rich, Y-linked	No data, 182 amino acids	No data
PRY2	PTPN13-like, Y-linked	Expressed in testis, sperm	Male infertility
DAZ1, DAZ2, DAZ3, DAZ4	Deleted in azoospermia	Binds to RNA, expressed only in sperm precursors, aids progression to meiosis	No sperm in semen, male infertility

[a] Order on the chromosome is given in Fig. 13.3.
Sources: OMIM, Gene Cards.

NR5A1 gene binds to a testis-specific enhancer of the SOX9 gene located at 17q24.3 that is expressed specifically in gonadal tissue (Kuroki et al., 2013; Sekido & Lovell-Badge, 2008). Defects in several genes (e.g., insulin receptor (INSR), insulin-like growth factor 1 receptor (IGF1R), and insulin receptor-related receptor (INSRR)), many of which are on an autosome, can change the kinds of gonads that appear.

Several of the Y-specific genes listed in Table 13.2 are intimately involved in sperm production. Deleted in azoospermia 1 (DAZ1), for example, is expressed primarily in the spermatogonium cells in testis that give rise to mature sperm. The gene name is *deleted in azoospermia*. In Latin, azoospermia means "without spermatozoa," a condition in which the man has little or no sperm in the semen (Schlegel, 2004). The existence of the gene was discovered when a small deletion of DNA from the Y chromosome led to azoospermia. So, the gene was given a very awkward name because it was a place in the Y chromosome DNA where deletion led to the lack of sperm. Another example, testis-specific protein, Y-linked 1 (TSPY1) is expressed only in testis tissue, and its protein is localized to the cytoplasm of sperm precursor cells.

The gene SHOXY in the pseudoautosomal region, named *short stature homeobox Y*, can reduce adult height when mutated and seems to regulate growth in a manner different from that of human growth hormone (OMIM). Normally, the gene promotes growth, but it is named for the mutant form that reduces growth; the gene itself does not code for short stature. There is a counterpart SHOX on the X chromosome. Girls with Turner's syndrome have just one X and therefore one copy of SHOX, and this may explain why they are often shorter than the typical XX female (Table 13.3). It has been argued that XYY, XXY, and XXX karyotypes all involve three copies of SHOX and/or SHOXY and therefore are taller than average men and women. Conclusive evidence of this plausible claim in humans is lacking, however.

For genes in the pseudoautosomal region, people with the XYY, XXY, and XXX karyotypes will have three copies, and those may generate an overdose of gene-encoded protein, just as occurs in trisomy 21. Three copies could conceivably impair mental development to some extent. At present, this is speculative.

OTHER CHROMOSOME ANOMALIES

XYY is only one of many possible variants of sex chromosome number. Others are mentioned briefly here as a means of comparison with XYY. Extensive information on each of them is available in standard reference works on medical genetics and Internet databases. Table 13.3 lists the better known anomalies. Values for height and IQ are average scores for several cases that usually show a wide range of individual scores.

The commonplace notion that having a Y chromosome confers masculine qualities whereas the X encodes feminine qualities is not easily reconciled with facts from the full range of chromosome anomalies. Although having one extra Y does indeed increase male height and elevates criminal conviction rates, having an extra X instead of an extra Y generates XXY men who are also taller than the average for XY and show an elevated criminal conviction rate that is associated with several social factors (Stochholm et al., 2012). Even XXX women tend to be taller than XX. People with XXX, XXY, and XYY chromosome sets all have three copies of the SHOX or SHOXY gene that is related to growth. They also possess three copies of several other genes that may interfere with mental development when their protein products are present in excessive amounts. It is clear that having more than three X or Y combined impairs development of many physical and mental features. Only with the sex chromosomes can tetrasomy infants survive at all. Even trisomy is lethal to the embryo and fetus for almost all autosomes, except number 21 and rare instances for 13 and 18.

HIGHLIGHTS

• A myth that men with two Y chromosomes are violence prone and often convicted of terrible crimes was propagated widely in the mass media in the 1960s

TABLE 13.3 Sex Chromosome Anomalies

Karyotype	Name	Frequency	Height	Intellectual Disability	Other Common Phenotypes
Female sex organs					
XO	Turner's syndrome	1/3000	Short	None	Infertility, small ovaries, webbed neck, heart
XX	Female typical	>99%	Normal	None	None
XXX	Trisomy X	1/1000	50% tall	Mild	Often none or mild
XXXX	Tetrasomy X	Very rare	No data	Yes	Teeth, eyes
XXXXX	Pentasomy X	Very rare	Short	Yes	Face, hands, feet
Male or ambiguous sex organs					
XY	Male typical	>99%	Normal	None	None
XXY	None	1/650	>179 cm	IQ 89–102	Often none or mild, infertility, some adults show Klinefelter's syndrome—small genitals, breasts, low testosterone
XYY	None	1/1000	>180 cm	IQ 90–100	None or mild
XXYY	None	1/20,000	192 cm	IQ 70–80	Small testes, sparse body hair
XXXY	None	1/50,000	190 cm	IQ 45–75	Genitals, heart, face, 5th finger
XXXXY	None	1/90,000	Very short	IQ 20–60	Very small gonads, face, heart, brain
XXXYY	None	<1/million	No data	Yes	Face, small gonads

Orphanet, GARD; Tartaglia, N., Ayari, N., Howell, S., D'Epagnier, C., & Zeitler, P. (2011). 48,XXYY, 48,XXXY and 49,XXXXY syndromes: not just variants of Klinefelter syndrome. Acta Paediatrica, 100, 851–860.

before scientific research had been done on the topic. It proved to be a false alarm.

- Large population surveys revealed that many XYY men are taller than most XY men, score moderately lower on standard tests of mental ability, and complete fewer grades in school. No association with violence has been reported. There is no consistent phenotypic XYY syndrome.
- Rate of imprisonment is higher for XYY than XY men, but most of the difference seems to be associated with social factors such as the amount of education and employment history.
- Increased height and moderately lower mental ability test scores are common in both XYY and XXY males and XXX females.
- The SHOXY gene on the Y may be responsible for the increased height of XYY men.
- Several genes located on the Y but not the X chromosome are involved in sperm formation.
- Further study of XYY men has become very challenging because of changes in ethical standards of medical research that now require informed consent and some likely benefit or at least no significant harm to the subjects who provide their data.

References

Bardsley, M. Z., Kowal, K., Levy, C., Gosek, A., Ayari, N., Tartaglia, N., ... Ross, J. L. (2013). 47,XYY syndrome: clinical phenotype and timing of ascertainment. *The Journal of Pediatrics*, 163, 1085–1094.

Bauer, D., Bayer, R., Beckwith, J., Bermant, G., Borgaonkar, D. S., Callahan, D., ... Dworkin, G. (1980). Special supplement: the XYY controversy: researching violence and genetics. *Hastings Center Report*, 1–31.

Beckwith, J., Elseviers, D., Gorni, L., Mandansky, C., Csonka, L., & King, J. (1975). Harvard XYY study. *Science*, 187, 298–299.

Birkhoff, J., Re, L., & Torri, D. (2015). 47, XYY karyotype and borderline personality disorder: an Italian judicial case and a review of the literature. *Journal of Child and Adolescent Behaviour*, 3, 220–225.

Borgaonkar, D., Murdoch, J. L., Mckusick, V., Borkowf, S., Money, J., & Robinson, B. (1968). The YY syndrome. *The Lancet*, 292, 461–462.

Culliton, B. J. (1974). The Sloan-Kettering affair: a story without a hero. *Science*, 184, 644–650.

Culliton, B. J. (1975). XYY: harvard researcher under fire stops newborn screening. *Science*, 188, 1284–1285.

Engel, E. (1972). The making of an XYY. *American Journal of Mental Deficiency*, 77, 123–127.

Gravholt, C. H. (2013). 44. Sex-chromosome abnormalities. In D. Rimoin, R. Pyeritz, & B. Korf (Eds.), *Emery and Rimoin's principles and practice of medical genetics* (6th ed.). San Diego, CA: Academic Press.

Gruchy, N., Blondeel, E., Le Meur, N., Joly-Helas, G., Chambon, P., Till, M., ... Vialard, F. (2016). Pregnancy outcomes in prenatally diagnosed 47, XXX and 47, XYY syndromes: a 30-year French, retrospective, multicentre study. *Prenatal Diagnosis*, 36, 523–529.

Jacobs, P. A., Brunton, M., Melville, M. M., Brittain, R. P., & McClemont, W. F. (1965). Aggressive behaviour, mental subnormality and the XYY male. *Nature, 208*, 1351–1352.

Kuroki, S., Matoba, S., Akiyoshi, M., Matsumura, Y., Miyachi, H., Mise, N., ... Tachibana, M. (2013). Epigenetic regulation of mouse sex determination by the histone demethylase Jmjd1a. *Science, 341*, 1106–1109.

Leggett, V., Jacobs, P., Nation, K., Scerif, G., & Bishop, D. V. (2010). Neurocognitive outcomes of individuals with a sex chromosome trisomy: XXX, XYY, or XXY: a systematic review. *Developmental Medicine and Child Neurology, 52*, 119–129.

Lubs, H., & Ruddle, F. (1970). Chromosomal abnormalities in the human population: estimation of rates based on New Haven newborn study. *Science, 169*, 495–497.

Lyons, R. D. (1968). Ultimate speck appeal may cite a genetic defect. *The New York Times, 43*.

Milunsky, J. M. (2010). Prenatal diagnosis of sex chromosome abnormalities. In A. Milunsky, & J. M. Milunsky (Eds.), *Genetic disorders and the fetus: Diagnosis, prevention and treatment* (6th ed., pp.273–312). Oxford: Wiley-Blackwell.

Money, J. (1969). XYY, the law, and forensic moral philosophy. *The Journal of Nervous and Mental Disease, 149*, 309–311.

Montagu, A. (1968). *Man and aggression*. New York: Oxford University Press.

Morris, J. K., Alberman, E., Scott, C., & Jacobs, P. (2008). Is the prevalence of Klinefelter syndrome increasing? *European Journal of Human Genetics, 16*, 163–170.

Pyeritz, R., Schreier, H., Madansky, C., Miller, L., & Beckwith, J. (1977). The XYY male: the making of a myth. In The Ann Arbor Science for the People Editorial Collective (Ed.), *Biology as a social weapon* (pp. 86–100). Minneapolis, MN: Burgess.

Quintana-Murci, L., & Fellous, M. (2001). The human Y chromosome: the biological role of a "functional wasteland" *Journal of Biomedicine and Biotechnology, 1*, 18–24.

Ratcliffe, S. G., Murray, L., & Teague, P. (1987). Edinburgh study of growth and development of children with sex chromosome abnormalities III. *Birth Defects Original Article Series, 23*, 73–118.

Ratcliffe, S. G., Read, G., Pan, H., Fear, C., Lindenbaum, R., & Crossley, J. (1994). Prenatal testosterone levels in XXY and XYY males. *Hormone Research, 42*, 106–109.

Re, L., & Birkhoff, J. M. (2015). The 47, XYY syndrome, 50 years of certainties and doubts: a systematic review. *Aggression and Violent Behavior, 22*, 9–17.

Richards, B., & Stewart, A. (1966). The YY syndrome. *The Lancet, 287*, 984–985.

Sandberg, A. A., Koepf, G. F., Ishihara, T., & Hauschka, T. S. (1961). An XYY human male. *The Lancet, 278*, 488–489.

Schlegel, P. N. (2004). Causes of azoospermia and their management. *Reproduction, Fertility and Development, 16*, 561–572.

Sekido, R., & Lovell-Badge, R. (2008). Sex determination involves synergistic action of SRY and SF1 on a specific Sox9 enhancer. *Nature, 453*, 930–934.

Stochholm, K., Bojesen, A., Jensen, A. S., Juul, S., & Gravholt, C. H. (2012). Criminality in men with Klinefelter's syndrome and XYY syndrome: a cohort study. *BMJ Open, 2*, e000650.

Stochholm, K., Juul, S., & Gravholt, C. H. (2010). Diagnosis and mortality in 47,XYY persons: a registry study. *Orphanet Journal of Rare Diseases, 5*, 15.

Telfer, M. A. (1968). Are some criminals born that way? *Think, 34*, 24–28.

Tilford, C. A., Kuroda-Kawaguchi, T., Skaletsky, H., Rozen, S., Brown, L. G., Rosenberg, M., ... Page, D. C. (2001). A physical map of the human Y chromosome. *Nature, 409*, 943–945.

Wilson Sayres, M. A., & Makova, K. D. (2012). Gene survival and death on the human Y chromosome. *Molecular Biology and Evolution, 30*, 781–787.

Witkin, H. A., Mednick, S. A., Schulsinger, F., Bakkestrøm, E., Christiansen, K. O., Goodenough, D. R., ... Philip, J. (1976). Criminality in XYY and XXY men. *Science, 193*, 547–555.

CHAPTER

14

Complex Traits

The normal functioning of most organelles in cells of the nervous system involves actions of dozens of genes, and an entire cell or bit of tissue involves the expression of more than a thousand genes (Chapters 2–5). The roles of genes in normal bodily and behavioral functions are astonishingly complex. Nevertheless, a major defect in just one important gene or a small portion of an entire chromosome can have major phenotypic effects (Chapters 7–13). Table 14.1 summarizes findings about the seven discrete changes that have been explored here in some depth. Several lessons from these "simple" phenomena need to be kept in mind when considering complex traits influenced by multiple genetic variants:

- The effects of a genetic mutation are highly specific, and they depend on where, when, and how the gene acts.
- Without knowing the specific gene that is involved in some kind of neural or behavioral problem that arises from a transmissible genetic factor, little can be said or done about its consequences. No broad generalizations

are warranted about the inevitability or plasticity of a phenotypic change that arises from an unidentified genetic defect.

- Some rather obvious defects in the genetic material, such as the XYY karyotype, have little or no phenotypic effect at all. Commonplace but ill-founded beliefs about genetic determinants of behavior can have greater consequences than the genetic alteration itself.
- Single-gene defects with large phenotypic effects are in most cases readily identified, and the specific allele carried by a person can often be determined from the DNA sequence. Thus, knowledge of the DNA can be an important aid to the diagnosis of a major gene defect.
- Mutations with large effects on the brain and behavior tend to exist at low frequencies in human populations. Genetic defects that are most readily identified tend to be rare.
- Knowing the nature of a molecular genetic defect usually does not lead directly to the discovery of an effective treatment (e.g., albinism). Treatment for some

TABLE 14.1 Features of Genetic Variants Explored in Chapters 7–13

Phenotype	PKU	HD	AIS	LON	DS	XYY
Gene(s)	PAH	HTT	AR	MT-ND4	Trisomy 21	Y
Location	12q23.2	4p16.3	Xq12	mt11,778	21q22.13	Y
Amino acids	452	3142	920	459	Many genes	48 Genes
Discovery	1939[a]	1993	1985	1988	1959, 1971[b]	1961
Frequency	1/12,000	1/20,000	1/20,000	1/25,000	1/1000	1/1000
Symptom onset	Start of nursing	20–80 years	Embryo	20–50 years	Embryo	Usually none
Preventive treatment	Diet (1955)	None	None	None	None	None
Selection	No[c]	Eight-cell embryo	No	No	Pregnancy termination	Pregnancy termination
Recurrence risk in family	1/4	1/2	1/2 in XY	>50% of males	Very low, maternal age	Very rare

[a] First demonstration of single-gene transmission of a medical disorder in humans.
[b] Shown to be trisomy in 1959, specifically trisomy 21 in 1971.
[c] Possible but usually not done for a treatable disorder.

Genes, Brain Function, and Behavior
https://doi.org/10.1016/B978-0-12-812832-9.00014-2

disorders, such as PKU, was devised long before the human genome was sequenced. In many instances, knowledge of the mode of action of a gene helps to understand why its developmental effects *cannot* be readily prevented or treated.

- Generally speaking, the pathways through which phenotypes are altered by a single-gene defect are themselves quite complex. Those phenotypes can extend beyond the affected individual to involve the family and the larger society.

By definition, a complex trait results from the combined effects of multiple genetic variants and environmental factors. Human intelligence, emotions, and personality are good examples of traits that range widely among people whom we regard as normal. Many complex traits are also *quantitative traits*, things that vary continuously along a scale of measurement and differ between people only in degree. Genetic variants may contribute to differences among people in commonplace characteristics, but the effect of any one gene variant is not likely to be very large. Unlike major gene defects, genetic variants that contribute to deficiencies in complex traits are often quite common. Mental disability, for example, is a complex trait that is identifiable in about 1% of the population. Autism and schizophrenia are complex traits that are far more common than syndromes caused by a major mutation in just one gene that often shows a frequency of less than one in 10,000. Thus, complex traits are things that most of us possess to some degree, and people who have unusually low or high levels of a complex trait are so common in the population that most of us have had experience with them in acquaintances, friends or family members, or ourselves. Because multiple genetic and environmental influences are probably involved, diagnosis of a complex trait can only be done via the phenotype and case history. Knowing someone's DNA sequence is unlikely to shed much light on the nature of a problem, at least not at the present state of knowledge.

TWO APPROACHES TO UNDERSTANDING HEREDITY

In Chapters 7–13, efforts were made to avoid the use of the terms heredity, hereditary, and heritable because they encompass several factors beyond the genes and can lead to confusion. The reasons for this choice were explained in Chapter 3 on gene expression. Now that the discussion has entered the realm of complex traits, phenomena such as parent-offspring and twin resemblance can involve several factors such as epigenetic transmission and the microbiome that extend beyond the scope of Mendelian genes. In this situation, there is little option but to employ

TABLE 14.2 Summary of methods of Mendel and Galton

Researcher	Mendel (1865)	Galton (1869)
Main interest	Laws of heredity, plants	Nature-nurture debate, race, class
Phenotypes	Constant differentiating characters	Continuum, size of seed
Environment (E)	Uniform garden plot, equate E	Wide range of typical gardens
Parentage (H)	Strict control with crossbreeding	Source of pollen uncontrolled
Analysis	Count phenotypic classes, probability	Correlate parent, offspring
Model of heredity	Male and female equal	Male only
Nature of heredity	Discrete factors, separate for different traits (later named "genes")	Blending of substance from parents

imprecise terms such as heredity and heritable. In this book, *heredity* includes all transmissible factors in addition to conventional Mendelian genes that can contribute to parent-offspring resemblance, whereas *genetic* pertains to segments of a DNA molecule in the nucleus that codes for proteins. Even this dichotomy fails to cleave things cleanly into two portions because it takes no account of mitochondria that have DNA that codes for RNA and proteins but does not exhibit Mendelian transmission. Mitochondria can be classified as parts of the broader concept of heredity, but they are usually ignored by theorists who equate heredity with Mendelian genetic influences.

The study of heredity in the past and genetics today is divided into two broad approaches, each with a different focus and scope—relatively large single-gene effects on a well-defined phenotype and relatively small quantitative effects of multiple genes on a complex trait. The former approach was pioneered by Gregor Mendel, whereas the other was pioneered by Francis Galton. The principal differences in their two approaches are summarized in Table 14.2.

Mendel (1865) sought to discern laws of heredity (Platt, 1959). His work was done long before the concept of a gene was formulated. Instead, it contributed greatly to the discovery of genes. He conducted carefully controlled experiments with garden peas and analyzed the results with the aid of probability theory. In many ways, his work serves as a model of good science, although it did not become widely known until 1900 when biologists in several countries discovered it in an obscure journal and noted its great importance (Bateson, 1913). He chose the garden pea because it was amenable to controlled breeding, and he scrutinized the plant for characteristics

TABLE 14.3 Results of F1 and F2 Hybrid Crosses of Mendel's Garden Peas

Characteristic	F1 hybrid	F2 hybrids (F1×F1)		
		Dominant	Recessive	Ratio
Seed color	All yellow	6022 Yellow	2001 Green	3.01:1
Seed shape	All round	5474 Round	1850 Wrinkled	2.96:1
Seed coats	All violet-red	705	224	3.15:1
Pod shape	All inflated	882	299	2.95:1
Pod color	All green	428	152	2.82:1
Flowers	All axial	651	207	3.14:1
Stem height	All tall	787 Tall	277 Short	2.84:1

See Chapter 7 for examples involving humans.
Source: Mendel, G., 1865. Versuche über pflanzenhybriden. *Verhandlungen des Naturforschenden Vereines in Brünn, 4,* 3–47.

that generally appeared in only two forms, those he termed "constant differentiating characters." Many other characteristics were observed but not chosen for analysis. "Constant" meant that the characteristic appeared much the same, generation after generation, and "differentiating" meant that there were just two forms seen among the peas in his garden. Seven such characters were chosen for his research (Table 14.3).

Mendel took great care to raise and nurture all of his plants in the same environment. He did not seek to discern the relative merits of heredity and environment or nature and nurture. As a master gardener, he knew well the importance of proper environment for the health of his plants. He therefore sought to *control* the environment. That experimental control played a crucial role in his ability to perceive patterns of heredity. He also controlled heredity by carefully transferring pollen between plants. Peas are usually self-fertilizing, but Mendel devised methods for cross-pollination between different plants. He also implemented *reciprocal cross-fertilization,* such that some plants that bred true for green seeds were used as the source of seeds and yellow as the source of pollen and vice versa, yellow seeds with green plant pollen. That method led directly to an important discovery: *the contributions to heredity of male and female gametes were equal.* He took great care to count the phenotypic classes that appeared in every kind of hybrid cross, and this too produced a great discovery: *heredities of the parent plants do not blend; instead, there are durable recessive factors that can persist across an entire generation without being expressed* and then reappear in a later generation. Mendel's understanding of the laws of probability helped to interpret the results. With discrete particles of heredity, some of which were recessive and others dominant in an F1 hybrid, *there*

should be a 3:1 ratio of dominant to recessive phenotypes in the F2 offspring (Chapter 7). This indeed happened for all seven of his carefully chosen, constant differentiating characters (Table 14.3), which indicated there was a general principle at work. His discoveries were a tremendous accomplishment by a humble Augustinian monk experimenting in a rural garden at a monastery (Henig, 2000; Orel, 1984).

FINDING GENES AND MUTATIONS

Not long after Mendel's laws and genes themselves were discovered and verified in several species, studies of many kinds of plants and animals suggested that discrete genetic factors were located on the strands of chromosomes. It was shown that the chromosomes are duplicated, and then, their strands separate during meiosis and the formation of germ cells. Under the microscope, it appeared that sometimes two strands crossed over and recombined during meiosis (Janssens, 1909). The next big advance in the understanding of heredity utilized crosses of fruit flies. Morgan and other scientists had investigated the transmission ratios of many kinds of traits such as eye and body color, wing shape, leg shape, and facets in the compound eye. Whenever the traits were present in two distinct forms, results of crosses were similar to what was seen in Mendel's peas. Morgan and his students crossed a large number of flies with different combinations of various traits and obtained what at first looked like a vast and disordered set of numbers. Each trait by itself usually showed a 3:1 ratio in the F2 hybrid, but the results for many of traits did not seem to be independent of each other. Mendel had studied seven traits in detail (Table 14.3), and they were mutually independent. Green and yellow pea seeds occurred in the 3:1 ratio, no matter what was the value of other traits such as pod color or location of flowers. (It was later found that a gene pertinent to each trait in peas was on a different chromosome.) Things seemed to work differently in the flies. Some of the traits seemed to be linked to each other. Morgan (1911) proposed that this might happen if they were on the same chromosome.

Linkage and Chromosome Maps

This idea inspired one of Morgan's graduate students, Alfred Sturtevant, to compile the results for thousands of crosses of flies, and in 1913, he found that certain sets of genes seemed to be transmitted together more often than expected solely by chance. He stayed up all night calculating the percentage of cases when two or three different genes were transmitted together and then had an important insight (Smith, 2013). If two genes are located close to

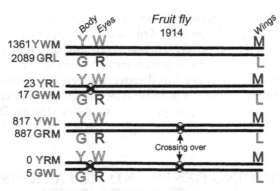

FIG. 14.1 Three genes studied by Morgan in fruit flies occur on the X chromosome of females in a linear order related to their distance along a chromosome. Each gene has two alleles. In that early period, alleles were named for their salient phenotypic effects. Body color is specified by *yellow* (Y) or *gray* (G) alleles, eye color is *white* (W) or *red* (R), and wing shape is miniature (M) or long (L). The initial cross was Y W M by G R L lines. When there is no crossing-over of the chromosome strands during the formation of the germ cells, alleles Y W M remain together on one strand, as do G R L. Crossing-over of strands between the body color and eye color genes gives rise to new kinds of strands Y R L and G W M. The frequency of those strands is low because the two genes are close together, whereas frequencies of Y W L and G R M types are much higher because they are further apart on the chromosome. Numbers 1361–5 are numbers of flies with each allele combination. Source: *Morgan, T. H. (1914). The mechanism of heredity as indicated by the inheritance of linked characters*. Popular Science Monthly, January, 1–16 (Fig. 6).

each other on a chromosome, the probability that they recombine would be much less than when they are far apart. This idea allowed him to arrange genes in a definite order along a chromosome in a kind of *chromosome map*, separated by distances related to how often they recombine in a cross (Fig. 14.1). Genes on different chromosomes always had a 50% chance of recombining. Morgan (1914) developed this finding into a theory of transmission of genes that has withstood the test of time and numerous experiments with diverse species. He coauthored an important book on *The Mechanism of Mendelian Heredity* (Morgan, Sturtevant, Muller, & Bridges, 1915) that explained the linkage of genes and the patterns of transmission of genes on the same chromosome. Later, the unit of distance along the chromosome equivalent to a 1% chance of recombination was named the centimorgan (cM), and in 1933, Morgan himself was awarded the Nobel Prize.

The method of linked genes was employed to construct chromosome maps in many species, such as the fruit fly and lab mouse for which many mutations had been carefully preserved and propagated whenever they were found. The maps showed which genes were located nearest to each other. By 1965, more than 250 Mendelian genes had been identified in mice, and 140 were mapped using Morgan and Sturtevant's methods (Sidman, Green, & Appel, 1965). Many of them had dramatic effects on nervous system structure and gave rise to peculiar defects in motor coordination. The genes were clustered in 20 "linkage groups" that were later shown to correspond to specific chromosomes. Linkage was much more difficult to demonstrate in humans, but evidence slowly accumulated that it worked the same way as in mice. After a substantial number of Mendelian genes had been mapped to the chromosomes, they provided a means to detect and localize a new gene. If a previously unknown gene was located close to a known gene, it would tend to show a similar pattern of transmission to the next generation.

Localizing a newly discovered gene by linkage analysis and chromosome mapping had two significant limitations. First, for some chromosomes, there were large expanses of DNA where no genes were known to occur. Second, there was a real possibility that two genes might be part of the same biochemical pathway or network, which could distort the relative frequencies of the different trait combinations.

Molecular Markers

The situation changed with the advent of molecular analysis that could detect small differences (polymorphisms) in DNA sequence at specific locations along a chromosome. It happens that many of these minor sequence variations are transmitted across generations but do not alter phenotypes. They can serve as neutral markers of a location in the DNA. Chemical analysis can determine which form or allele of the marker is possessed by an individual and relatives, and the markers can be assembled into a chromosome map that has many more locations occupied by markers than by actual genes. Thus, the principal method to detect the location of a Mendelian gene is now to show that a phenotypic variant has almost the same pattern of transmission in a family as a known marker. As the number of markers compiled in catalogs accessible to all scientists grew, the power of the method increased greatly and soon led to clear evidence for many mutations involved in human neurological disorders.

An outstanding example involves Huntington disease. The quest to find that gene engaged dozens of researchers for a decade of intense work. The first breakthrough occurred in 1983 (Fig. 14.2) when a molecular marker was discovered on chromosome 4 that was associated with the HD phenotype in large families (Gusella et al., 1983). The marker name D4S10 indicated it was on chromosome 4 and was the 10th marker to be registered on that chromosome in official catalogs of human genes and markers. Evidence was compelling that the HD disease-causing gene was in band 16.3. The D4S10 marker showed a 4% recombination frequency with HD, but it was not certain on which side of the marker the disease allele was located (Haines et al., 1986). DNA sequencing technology at the time could not deal

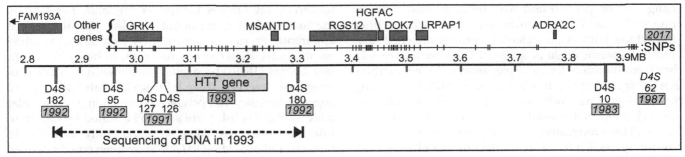

FIG. 14.2 HTT gene on chromosome 4 and nearby DNA markers (D4S) that were important for locating the gene. Year when each marker was discovered is shown in *gray box* (citations in text). The gene was identified when the DNA segment between markers D4S182 and D4S180 was sequenced by The Huntington's Disease Collaborative Research Group (1993). After the full human genome was sequenced in 2001, other nearby genes were also identified. More recently, a large number of single nucleotide polymorphisms (SNPs) have been identified that make mapping mutations much quicker and more precise. Source: *OMIM, GeneCards, dbSNP database, and others cited in the text.*

with such a long segment of a chromosome. More markers were needed. The D4S62 marker appeared to be on the opposite side of D4S10 from HD and showed only 2% recombination with D4S10 (Gilliam et al., 1987; Hayden et al., 1988). A major advance was reported by Allitto et al. (1991) who placed the newly discovered markers D4S125, D4S126, and D4S127 on the growing map of the region. D4S126 and D4S127 proved to be very close to the HD gene. MacDonald et al. (1992) added the markers D4S180 and D4S182, and they concluded that the HD gene was most likely located between those two markers that spanned about 500 kb or 500,000 bases of DNA. That was a big chunk of DNA to decode, but new technologies made it feasible. The immense task was undertaken by The Huntington's Disease Collaborative Research Group (1993), consisting of 58 researchers at six institutions in three countries, and it was crowned with success. They identified and sequenced a new gene termed IT15 (cryptically called "interesting transcript 15," later renamed HTT) that coded for an unknown protein. The gene contained a large series of CAG repeats, as described in Chapter 9, and HD patients had many more repeats than relatives who did not develop HD. The success of the search for the Huntington's disease gene established a paradigm followed by many others looking for different genes—first, identify several markers to localize the gene to a specific band of a chromosome and then sequence the DNA to find it and determine the nature of the protein for which it codes.

SNPs

The slow accumulation of genetic markers soon gave way to a new approach based on the discovery of thousands of seemingly minor genetic variants scattered widely over the genome, often in introns. They made an unexpected appearance after the full genome sequence had been compiled and could be inspected base by base. When two entire genomes of different people were scrutinized, many single-base differences were detected. Termed *single nucleotide polymorphisms* or SNPs, the new genetic variants could be readily identified in an individual because the flanking DNA sequence upstream and downstream from the specific nucleotide could be determined. Then, a molecular probe could be designed to assay the state of that one nucleotide. The full DNA sequence also revealed the presence of several formerly unknown genes lurking near HTT (Fig. 14.2). The new molecular methods unleashed an avalanche of new data that thousands of scientists are still struggling to organize and understand.

An example of just one SNP from the immense dbSNP database is shown here: TTGCAAAGGATATTACCT-GAGTCCCCACAG[**T/C**]CCCCTATGAGGCA-GATGGTGCCATCTGCAG. The SNP is designated **rs34218188**, a number that was assigned when it was added to the database and has no biological meaning. The polymorphism itself is [T/C], while 30 bases from the two flanking regions are shown. Those flanking regions are almost always the same in all people. The rs34218188 SNP is located in an intron on chromosome 4 band 16.3 at nucleotide number 2648176, which is close to the FAM193A gene but not part of it and about 400 kb from the HTT gene. In most human populations, more than 96% of people have nucleotide base T at that position, but a few have C instead of T. Having C seems to create no problems, and the allele is "silent" phenotypically. It is an ideal marker that does not interfere with the trait that a researcher wants to study. Molecular probes applied to a tissue sample from a person can determine whether there is one dose of T and one of C (genotype T/C), double the dose of T (genotype T/T), or double the dose of C (genotype C/C).

The number of cataloged SNPs in a recent version (build 150, 3 February 2017) of dbSNP is staggering. More than 325 million have been assigned an RS code number in the human database, more than half of which are located in a gene. They are distributed widely across every chromosome. Some SNPs result in an amino acid change in the protein and may be associated with an abnormal medical condition. Several manufacturers offer large chips with oligonucleotide probes attached that bind to a specific allele of an SNP. The OmniExpress-24 chip from Illumina, for example, assays 713,599 markers that are spaced at about one SNP every 4000 bases along the entire genome. With so many markers, the chances are very good that several in the vicinity of a particular gene will be informative in any one person and implicate a nearby gene if it has a large effect on the phenotype.

Detecting Mutations

The current state of knowledge and technology in human genetics has brought us to a stage where any mutation in a gene that has a substantial and straightforward effect on a behavioral phenotype in a moderate number of people should be readily detectable. The genome is almost saturated with SNPs and other molecular markers. Molecular and computational methods for searching through such vast quantities of data now exist. The methodology is costly in terms of basic supplies for collecting the data and the time of large teams of researchers devoted to the task.

For example, a recent study identified the effects of two independent genes (neurexin (NXRN1) on chromosome 2 and contactin 6 (CNTN6) on chromosome 3) that could account for about 1% of all cases of a rare neurological condition known as Tourette's syndrome (Huang et al., 2017). The work engaged 64 investigators affiliated with 45 academic and medical institutions in Austria, Canada, France, Germany, Greece, Hungary, Italy, the Netherlands, Poland, the United Kingdom, and the United States, and it left more than 98% of all cases of Tourette's syndrome unexplained.

Tourette's syndrome (TS) is not an instance of a complex trait. Instead, it sometimes arises from a single-gene defect, but different cases arise from defects in different genes. This situation is known as *genetic heterogeneity* of a disorder; different people show essentially the same unusual phenotype for different genetic reasons. A complex trait, on the other hand, typically is influenced by a combination of several relatively common genetic variants involving multiple genes in the same person. When there is heterogeneity, each specific genetic defect, be it a deletion, duplication, or point mutation, acts like a typical single-gene disorder but afflicts only a few people. Examples of genetic heterogeneity are presented in Chapters 15–17.

GALTON AND CORRELATION

Mendel refrained from diving into the murky sea of complex quantitative traits. That plunge was taken by the aristocrat Francis Galton in England. For Galton, the most salient thing about heredity was the way it apparently exerted greater influence on traits, including mental traits, than did environment. In his book *Hereditary Genius* (1869), he argued vehemently that men of eminence deserved their lofty social status because of superior heredity, not privileged upbringing. For similar reasons, the English were said to be suited by nature to rule over Africans. Men were naturally superior to women, and according to Galton, women could not even transmit genius to their sons.

When Galton made a foray into experimentation with heredity, he too chose to study the garden pea, but beyond that, all similarity to Mendel's work vanished (Table 14.2). For his first experiment, he decided to examine a quantitative trait—size of the seed—and made no effort to crossbreed or control the source of the pollen (Galton, 1889). He simply mailed seeds of different sizes to other gardeners and asked them to grow the plants in their own gardens, harvest the next generation of seeds, and mail them back to Galton for measuring (Fig. 14.3). He observed a correlation between size of the parent seed and sizes of its descendants, something he attributed to heredity. But there were a wide range of seed sizes among offspring from a specific parent size that he could not explain. Furthermore, offspring seeds were on average quite a bit smaller than the parent seeds, another unexplained result. Galton's experimental design was clumsy, and his presumptions about heredity were entirely wrong. It was a rough start for the study of quantitative traits.

One of Galton's enduring contributions to the study of individual differences was the use of correlations. His associate Karl Pearson devised a concise statistic termed the correlation coefficient that estimates the strength of association between two measures. Many current studies of genetic aspects of complex traits, especially in psychology, rely on estimates of correlations among relatives. Pearson's *correlation coefficient r* expresses how well one measure (Y) can be predicted by knowing another (X). Three examples are shown in Fig. 14.4, ranging from an almost perfect correlation to a rather weak one. The steps in computing correlation and a straight prediction line are shown in Fig. 14.5. A positive correlation indicates that high values on one measure (X) tend to be *associated* with high measures on the other (Y). Association can arise from many sources. It could happen that X is a cause of Y. If X is a cause and Y a consequence, it makes good sense to predict Y from X. But the correlation coefficient is entirely neutral on this question: Which variable is designated X, and which Y is arbitrary? One could just as well use Y to predict X. The slope of the straight line would change, but the correlation r would remain the same. Many kinds of relationships could give rise to a statistical correlation. A mantra well known to psychology students

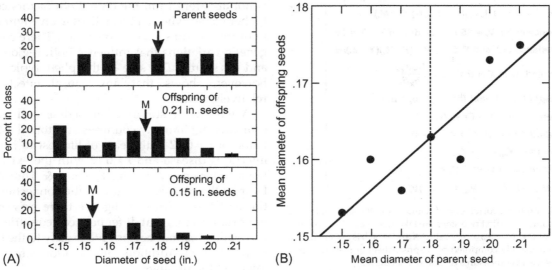

FIG. 14.3 Francis Galton mailed seeds of seven specific diameters to seven other gardeners living from north to south in England for planting, and they mailed the offspring seeds back to him for measurement. (A) There were a wide range of seed sizes among the progeny of a parent with a specific size, and many offspring were far smaller than any parent. On average, seed size was smaller from parents with smaller seeds. (B) There was a clear correlation between average sizes of parent and offspring seeds. *Source: Galton, F. (1889). Natural inheritance. London/New York: Macmillan (Table 2).*

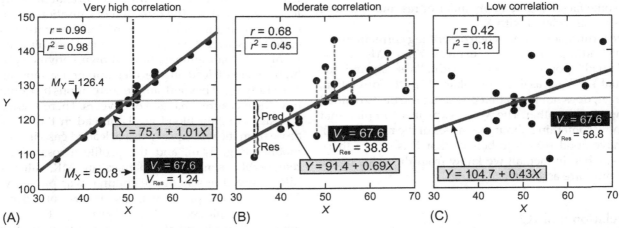

FIG. 14.4 Three XY scatterplots illustrating different magnitudes of correlation (r). (A) very high correlation; (B) moderate correlation; (C) low correlation. The X values are the same in each case. The line for predicting Y from X that gives the best fit to the data is shown as heavy line. The deviation of each data point from the mean of Y is divided into two portions: the predicted part and the residual deviation from the prediction line. The proportion of variance (V) in Y that is accounted for by the prediction is the square of the correlation coefficient (r^2), with the residuals accounting for $(1 - r^2)V_Y$.

is that *correlation does not prove causation*. If X is a cause of Y, they should be correlated. If X and Y are correlated, on the other hand, this by itself does not prove that X causes Y. Correlation is an indicator of *statistical association*, not causation.

The strength of the association is expressed by the coefficient r, but a more readily interpreted indicator of strength is r^2. When we predict Y from X using the best fitting straight line, the deviation of any individual's score Y_i from the group mean of Y, symbolized as $(Y_i - M_Y)$, can be partitioned into two parts—the predicted deviation from the mean, where Y_{Pred} is given by the formula for the

straight line, and the so-called *residual* portion, the deviation of a person's score Y_i from the predicted value. The partition is $Y_i - M_Y = (Y_{Pred} - M_Y) + (Y_i - Y_{Pred})$. The variance of the predicted values computed around the mean of Y is r^2 times the variance of Y itself, whereas the variance of the residual around the prediction line is the fraction $(1 - r^2)$ times the variance of Y. Thus, r^2 is a good indicator of the strength of linear association between X and Y; the larger the value of r^2, the stronger is the association. If the correlation of X and Y is $r = 0.5$, this means that $r^2 = 25\%$ of the variance of Y is associated with X. The other 75% cannot be

Sum of squared dev of Y: $SS_Y = \sum (Y - M_Y)^2$

Variance of Y: $V_Y = SS_Y / N$ St. dev. of $Y = S_Y = \sqrt{V_Y}$

Covariance of X and Y: $Cov_{XY} = \sum (X - M_X)(Y - M_Y)/N$

Correlation of X and Y: $r = \dfrac{Cov_{XY}}{S_X S_Y}$

Straight line of best fit: $Y_{Pred} = b_0 + b_1 X$

Slope of line: $b_1 = Cov_{XY}/V_X$

Predicted portion of $Y = Y_{Pred} - M_Y$

Residual: $Y_{Res} = Y_i - Y_{Pred}$

Partitioned variance: $V_Y = V_{Pred} + V_{Res}$

$V_{Pred} = r^2 V_Y$ $V_{Res} = (1 - r^2) V_Y$

FIG. 14.5   Computing correlation of X and Y and the straight line for predicting Y from X. The first two steps are taken from Fig. 6.3. The algebra itself cares not which measure is called X and which is assigned to be Y. Correlation is usually computed by a spreadsheet or statistical program that uses the raw measures of X and Y as inputs.

accounted for or predicted by knowing X. This statistical approach to analyzing individual differences in phenotypes seeks to *partition phenotypic variance* into two components—a portion of variance in Y that is associated with some factor X and a remainder or residual variance that is not associated with X.

The terminology used when describing correlation can be misleading. Many texts express $Y_{Pred} - M_Y$ as the "explained" portion of an individual's score and the residual $Y_i - Y_{Pred}$ as the "unexplained" part. Correlation, however, does not mean that one thing is truly explained by knowing the other. The word "explained" implies that we know something about the causes of the measure of Y and have good reason to believe that X is one of those causes, when in fact, all we know from the value of r is that things are associated.

Correlation and IQ

Consider some observations about the brain and intelligence. Of course, intelligence depends on brain function, but it would be good to know more specifically what aspects of the brain confer those marvelous properties we call intelligence. A study of the living brains with magnetic resonance imaging (MRI) measured several brain features and correlated them with IQ scores from the same people (Andreasen et al., 1993). The MRI method is able to divide the brain into fractions rich in neurons (gray matter) and myelinated axon tracts that interconnect them (white matter) and fluid filling the ventricles (cerebrospinal fluid, CSF). Correlations between IQ score and volumes of gray matter, white matter, and CSF were 0.35, 0.14, and −0.02, respectively. Thus, about 10% of the variation in IQ scores was associated with variation in the amount of gray matter. Whether the amount of gray matter itself is a direct cause of higher or lower IQ

is not apparent from the data. Both factors could reach higher values when a person lives in an environment rich in nutrition and sensory experience. There could also be genetic variations that influence both regional volumes and IQ scores. The size of the phenotypic correlation, however, shows that any causal effects must be relatively weak.

A worldwide survey examined several correlates of estimated IQ in many countries and found a high correlation of $r = 0.82$ with years of life lost to parasites and infectious disease and $r = 0.72$ with the extent of nutritional deficiencies (Eppig, Fincher, & Thornhill, 2010). Despite the plausible nature of the connections of nutrition and disease with intelligence, there are so many other important factors that differ between countries that those correlations cannot be attributed solely to the two health-related factors. Additional information is needed to help interpret the findings.

This was done in a study from the United States (Nevin, 2009) that revealed a strong correlation between measures of preschool blood levels of lead from the environment and incidence of mental retardation (MR) and later SAT test scores. For MR, the correlation with lead levels was $r = 0.80$. The direction of causation was rather obvious. Being mentally deficient does not cause a young child to seek out toxic environments. Controlled experiments with lab animals have shown convincingly that high levels of lead in the diet impair learning ability. The study went beyond simple correlations by examining how the measures changed over years. There was a major increase in infant blood levels of lead in the 1950s that arose from the widespread use of leaded gasoline in automobiles (Fig. 14.6), and the profile across years was remarkably similar to changes in MR rates when MR observed in school-age children was plotted with a 12-year lag. The profile was also inverse of a decline in SAT scores, plotted with a 17-year lag, that reversed when lead levels fell after leaded gasoline was banned. The time lags showed delayed effects from toxic lead imbibed in early childhood. The case for a causal role of lead exposure was strengthened by changes over years that showed both a rise and then fall in lead levels and psychological outcome measures. Because the data from children were not based on a well-controlled experiment, it cannot be said that every change in MR rate or SAT score was caused only by lead levels, but the weight of the evidence does implicate environmental lead as a source of poor child development.

Correlational data are often the only sources of information about many factors at work in a human population. From a scientific standpoint, they are much inferior to well-controlled experimental data. Lacking a single, decisive indicator of an environmental or even a genetic effect, there often is no recourse, but to assemble all available information, much of it is inconclusive on its own

FIG. 14.6 Blood levels of environmental lead measured in preschool children in the United States in various years plotted against (A) the rate of mental retardation (MR) in public schools and (B) average SAT math and verbal scores. Mental retardation is plotted with a 12-year lag because that is the age when the effect of high lead in early life was most apparent. SAT scores are from children tested 17 years after the early lead exposure. Evidence of a causal effect of lead on mental performance is strengthened because the data show both a rising and then declining severity that corresponds to the use of leaded gasoline in automobiles. *Adapted from Nevin, R. (2009). Trends in preschool lead exposure, mental retardation, and scholastic achievement: association or causation? Environmental Research, 109, 301–310 ((A) Fig. 3 and (B) Fig. 7), Copyright 2009, with permission from Elsevier.*

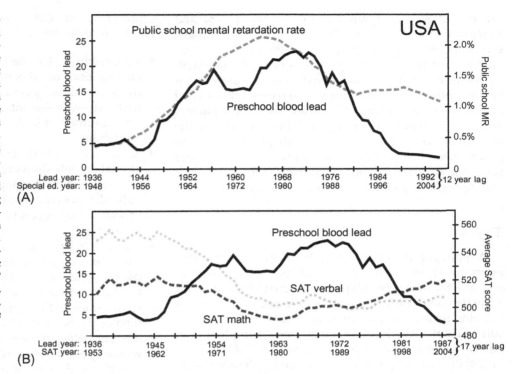

merits and search for patterns that suggest explanations. Any assembly of these kinds of data warrants caution and circumspection when drawing conclusions.

HERITABILITY

Genetic Models of Correlation

An important advance, a conceptual bridge between the Mendelian and Galtonian approaches, was made by Fisher (1918), who showed that quantitative models based on Mendelian genetics could account for Galton's results on correlations among relatives. However, traffic on the bridge coursed exclusively in one direction. Mendelian models could account for Galton's correlations, but the correlations among phenotypes of relatives provided no insights into the nature of genetic transmission. Nevertheless, a quantitative genetic model came to dominate thinking in psychology and other social sciences. Conclusions from the model were mathematical. The quantitative genetic approach did not require any knowledge of how many or which genes might be involved. It required knowledge of algebra. A simplified version is presented here.

The model proposes that the numerical value of a phenotype (Ph), such as a psychological test score, can be partitioned into two parts, G that arises from Mendelian genes and E that is attributable to the individual's environment. It ignores all components of heredity that are transmissible but not Mendelian genetic. Strictly speaking, the transmissible factor should be symbolized as H that includes more than G, but virtually, all theorists in quantitative genetic analysis of human behaviors ignore this complication. The model of an individual's test score presumes that the two components are additive, such that Ph = G + E. If G and E are additive, the variance of their sum can be derived algebraically in a way that partitions phenotypic variance V_{Ph} into two components—variance arising from genetic differences among people (V_G) and variance attributable to environmental differences (V_E). If the two factors G and E are statistically independent or uncorrelated, the model yields the simple relationship $V_{Ph} = V_G + V_E$. Heritability in the broad sense is then defined as $h^2 = V_G/V_{Ph}$, the proportion of phenotypic variance attributable to genetic variance. According to this model, in order to find h^2, one needs only to estimate the phenotypic correlations of different kinds of relatives. There is no need to search for and find evidence of any real gene effects.

Origins of Heritability in Animal Breeding

Prior to 1940, most scientific writings used the term "heritability" in a nonquantitative sense to mean that something was inherited that influenced traits in parent and offspring. It is not clear when the term was first used in the technical, quantitative sense (Bell, 1977), but it became well known in the 1940s in connection with the practical breeding of farm animals. Lush (1945) and Kempthorne and Tandon (1953) showed that comparisons between scores of parents and the average scores of their offspring could be used to estimate the heritability coefficient. Suppose we have data from a large number of families, each consisting of two parents and several offspring. If we symbolize the mean score of the parents as P and mean of the offspring as O, then the slope of the straight line (Fig. 14.5) that predicts offspring mean from parent mean is b_{OP}. If the scores of the two parents are uncorrelated, then the estimate of heritability is $h^2 = b_{OP}$. This formula is commonly used in animal breeding.

Twins

A more popular approach with humans is twins. This offers an advantage when it is difficult or impossible to measure the same phenotype in young children and their adult parents because of the large difference in age. A pair of twins always has the same age. The twin method also suffers from several disadvantages, one being that twin births are only a small fraction (about 33 of 1000 live births) of the total population and are much more likely to have low birth weight than single births. Twins also share a more similar environment because they are conceived at the same time, born at the same time, nurtured in the same womb, and usually reared in the same home and neighborhood. Adjustments for the influence of some factors can be made by comparing the two kinds of twins, monozygotic (MZ) and dizygotic (DZ). MZ twins are derived from the same fertilized egg that then divides into separate embryos, so they have identical genotypes and are of the same sex (Fig. 5.1). They often share a common chorion. DZ twins arise from two eggs fertilized by different sperm, and they always have separate chorions and placentas (Marceau et al., 2016). The method is quite simple. First, the phenotype is measured for each twin, and then, the correlation between scores of twins of each kind is calculated. If the correlations are r_{MZ} and r_{DZ}, the estimate of heritability is $h^2 = 2(r_{MZ} - r_{DZ})$. (The correlation r used with twins is the intraclass correlation.)

A recent review of 2748 published twin studies of a wide range of human traits reported that, when averaged across all kinds of traits including some behavioral or psychological measures, the typical heritability was about 0.49 (Polderman et al., 2015). In the immediate aftermath of the publication of that study, many articles in the mass media proclaimed that genes and environment have equal weights.

Assumptions

A number of questions have been raised about the validity of assumptions that are commonly made to simplify heritability analysis (Goldberger, 1978; Wahlsten, 1990, 1994):

- *All heredity is Mendelian G.* This ignores the contributions of mitochondrial genes, the transmissible epigenetic effects on gene expression, the maternal environment, and the transmissible biome. This assumption is clearly false. It overestimates heritability.
- *No G × E interactions.* This ignores abundant data from animal models showing that the influence of environment on development depends on the individual's genotype. It overestimates heritability to an extent that depends on the specific kind of interaction.
- *No GE correlation.* In human society, children usually experience environments provided by their biological parents, and children from parents having more favorable ancestry tend to be reared in more favorable environments. The presence of positive GE correlation inflates the estimate of heritability.
- *Random mating.* If the mother and father pair randomly with respect to their genotypes, the genetic parts of the equation will be uncorrelated, which greatly simplifies the math. We all know perfectly well, however, that people usually mate with someone more similar to themselves. They choose mates, sometimes with great care, selecting those deemed most suitable on the basis of a wide range of phenotypes. It seems highly likely that some of those traits are to a significant extent related to a person's genetic heritage. Selective mating tends to increase the computed value of heritability.
- *Equal environmental similarity of MZ and DZ twins.* Twin genotypes most likely influence many of their phenotypes to an appreciable extent. MZ twins tend to be markedly more similar phenotypically. This leads them to have more similar experiences for many reasons. The equal environment assumption is not tenable (Joseph, 2015). The MZ twin correlation often exceeds that of DZ twins for reasons that are in part environmental, and this will tend to increase the measured heritability.

Assessing H × E Interaction

The assumption of the additivity of G and E has drawn the most critical attention because it conflicts with a large body of evidence. In order to test for the presence of interaction, a study assesses individuals having the same heredity but reared under different conditions. This is not easily done with people, but lab animals have been evaluated numerous times. Not long after Mendel's laws were shown to apply to most species, experiments were done with standard strains of animals to compare effects of heredity and environment. A study done by Woltereck

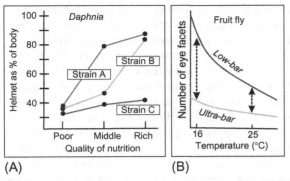

FIG. 14.7 Examples of heredity-environment interaction. (A) Size of the helmet of *Daphnia* water fleas depended strongly on the level of nutrition in two strains but not a third one (Woltereck, 1909). (B) The number of facets in the compound eyes of fruit flies was much greater at cool rearing temperatures in the low-bar strain, and the difference between low bar and ultrabar depended strongly on rearing temperature (Krafka, 1920).

(1909) with the *Daphnia* water flea played an important part in the modern conception of heredity formulated by Johannsen (1911), as discussed briefly in Chapter 2. Woltereck noticed that some lab strains of *Daphnia* had long, pointed head coverings or helmets and others were short and stout, so he reared them with different levels of nutrition. When nutrition was poor, helmets were short for all strains, whereas a food-rich medium boosted the scores of strain B so that they equaled strain A, yet strain C still lagged behind the others (Fig. 14.7A). Strain differences depended strongly on nutrition.

A classical study by Krafka (1920) reared two stocks of fruit flies at different temperatures and measured the number of facets in the compound eyes. The low-bar line generally had more facets than ultrabar but was also far more sensitive to variations in the lab temperature (Fig. 14.7B). Hogben (1933) examined the data and asked how much of the difference between the flies can be attributed to heredity and how much to environment. His answer is as follows: "The question is easily seen to be devoid of any meaning." The magnitude of the difference between stocks depended strongly on the temperature. Facet number was the product of the two factors acting jointly, not their algebraic sum.

Studies of the inbred mouse strains C57BL/6J and DBA/2J revealed that the DBA/2J strain is prone to seizures induced by a loud, high-frequency noise, whereas C57BL/6J is resistant to the seizures. However, a much more intense seizure can be induced if the young animal is "primed" by exposure to a moderate intensity sound during a *critical period* and then tested later. When primed at 14 or 16 days, both strains showed intense seizures (Henry & Bowman, 1970). Thus, the difference between the strains depended on not only whether they were primed but also when they were primed. With priming before 12 days, the data looked like what one would expect for a genetically determined trait, but just 2 days

later, the picture changed dramatically, and the environment appeared to play a critical role, while genotype made little difference.

Another famous example involved two strains of rats selectively bred for high (maze "brights") or low (maze "dulls") scores on a particular kind of maze learned for food reward (Heron, 1941). Cooper and Zubek (1958) reared separate batches of the two strains under three conditions—the usual lab environment, a restricted situation where sensory stimulation and opportunity for exercise were markedly limited, and an enriched sensory and motor condition. Performance of the "bright" line deteriorated when it was reared in restricted conditions, while the "dulls" benefited greatly from enriched rearing. Consequently, there was little difference between the lines in either the impoverished or enriched condition.

In all of those examples, effects of heredity and environment were generally not independent. When there is interaction, phenotypic variance (V_{Ph}) cannot be partitioned into two nonoverlapping and additive portions of variance arising from heredity and environment. Consequently, heritability lacks a mathematical rationale in many circumstances. Once the concept of interaction was understood and demonstrated many times in the lab, the field of behavioral genetics diverged into two more or less isolated camps. Many researchers working with lab animals abandoned heritability analysis and focused on mechanisms of gene action and brain development, whereas the quantitative genetics camp emphasized humans, ignored all those inconvenient interactions, and clung to additive models that simplified the algebra.

Critiques of Heritability

Many criticisms of the methods used to estimate heritability in human populations have been published, and advocates of the methods have offered spirited defenses. The debate began as soon as genetics itself began, and it persists today. Johannsen (1911), author of the genotype/phenotype distinction, commented on the statistical explorations of Galton and Pearson that they are "interesting products of mathematical genius… but they have nothing at all to do with genetics or general biology! Their premises are inadequate for insight into the nature of heredity." Morgan (1929), founder of genetic linkage theory, observed that "As long as a given result is the outcome of different, unknown genetic factors, the statistical treatment may serve as a temporary expedient … but it may be misleading, and admittedly fails to furnish the specific solution to the problem." According to the quantitative geneticist Hogben (1933), "The application of statistical technique in the study of human inheritance is beset with pitfalls. On the one hand the experimental difficulties of the subject-matter necessitate recourse to mathematical refinements which can be dispensed with in animal breeding. On the other there is the danger of concealing assumptions which have no factual basis behind an impressive facade of flawless algebra." The quantitative geneticist Kempthorne (1990) rendered a concise verdict: "… most of the literature on heritability in species that cannot be experimentally manipulated, for example, in mating, should be ignored."

Implications of Heritability

Suppose we are presented with a definite value for the heritability of a phenotype such as intelligence. What does this number tell us about a population of related individuals?

Modifiability

It is sometimes believed that if heritability of a trait is high, its modifiability by changing the environment must be low. Such a belief is not well founded. Heritability analysis is about past environments, conditions experienced by a previous generation, and it provides no basis at all for making predictions about consequences of new environments.

Selective Breeding

As shown by Lush (1945) and Kempthorne and Tandon (1953), when just the additive effects of heredity are considered, the heritability of a trait can predict the response to selective breeding. The basic relationship is quite simple: $R = h^2 S$, where S is the selection differential and R is the response to one generation of selective breeding. S is the difference between the mean score of the original population and the mean score of the parents chosen to propagate the new line of animals. S will be large only when a relatively small fraction, the very-highest-scoring animals, is chosen for breeding. R is the difference between the mean score of the offspring generation and the mean of the original population.

Selective breeding of humans is a core element of the doctrine of eugenics (Chase, 1977; Kevles, 1998; Lombardo, 2008). Heritability is crucial for estimating the expected change in a human trait because of selection. If h^2 is 0.6 instead of 0.3, progress owing to selection is expected to be twice as fast. A detailed example of selective breeding for human intelligence will be presented in the next chapter.

Prospects for Detecting Gene Effects

One of the major claims by those seeking funding for twin studies is that traits with high heritability would be better candidates for studies seeking to detect specific genes related to a phenotype. This expectation has now been evaluated, as discussed in the following chapters, and it has not been well supported (Manolio et al., 2009; Zuk, Hechter, Sunyaev, & Lander, 2012).

Almost 90 years after the criticism by Morgan (1929) that for traits involving "unknown genetic factors, the statistical treatment may serve as a temporary expedient…," that expedient has proved to be not so temporary. The approach pioneered by Galton and Pearson, one that Johannsen (1911) observed more than 100 years ago, has "nothing at all to do with genetics or general biology" and provides no "insight into the nature of heredity"; the method of statistical modeling remains the only recourse when specific genes relevant to a trait cannot be identified. Once the finger can be pointed directly at a bona fide gene with a known DNA sequence, that genetic effect on a trait crosses over to the camp of Mendelian and molecular genetics, where large teams of investigators are striving to apply that new knowledge.

HIGHLIGHTS

- Many complex traits exist on a continuum with no distinct boundary between normal and abnormal.
- The development of a trait in an individual generally reflects the coordinated actions of thousands of genes and numerous features of the environment, whereas a difference in a trait between two individuals may arise from differences at one or relatively few genes.
- A previously unknown single gene can be detected by showing it is linked to genetic marker at a known location on a chromosome, and the DNA sequence in that region of the chromosome can then reveal the identity of the new gene.
- So many genetic markers are now known on every human chromosome that any polymorphic gene with moderate to large effects on a phenotype can be readily detected.
- Some complex neurological disorders exhibit heterogeneity of causes. Different people can show similar symptoms for fundamentally different reasons such as mutations in different genes.
- When the specific genes involved in a complex trait cannot be identified, some researchers choose to study heredity in general, and they often seek to determine the relative strength of the influence of heredity on a trait compared with the strength of environmental influences.
- The study of complex traits in humans relies on correlations of phenotypes among relatives, but correlation by itself cannot reveal the nature of causation.

- If the phenotype is equal to the sum of two components (Ph = G + E) and the variance of the sum is equal to the sum of the variances ($V_{Ph} = V_G + V_E$), heritability can be estimated ($h^2 = V_G/V_{Ph}$).
- Heritability provides a way to predict the response to selective breeding for high levels of a trait.
- Partitioning variance and computing h^2 will be valid only if several assumptions that simplify the analysis are valid. However, most of the principal assumptions of heritability analysis are not valid and have been strongly criticized by many experts in quantitative genetics.

References

Allitto, B. A., MacDonald, M. E., Bucan, M., Richards, J., Romano, D., Whaley, W. L., ... Wasmuth, J. J. (1991). Increased recombination adjacent to the Huntington disease-linked D4S10 marker. *Genomics, 9*, 104–112.

Andreasen, N. C., Flaum, M., Swayze, V., 2nd, O'Leary, D. S., Alliger, R., Cohen, G., ... Yuh, W. T. (1993). Intelligence and brain structure in normal individuals. *The American Journal of Psychiatry, 150*, 130.

Bateson, W. (1913). *Mendel's principles of heredity*. Cambridge: Cambridge University Press.

Bell, A. E. (1977). Heritability in retrospect. *The Journal of Heredity, 68*, 297–300.

Chase, A. (1977). *The legacy of malthus: The social costs of the new scientific racism*. New York: Knopf.

Cooper, R. M., & Zubek, J. P. (1958). Effects of enriched and restricted early environments on the learning ability of bright and dull rats. *Canadian Journal of Psychology, 12*, 159–164.

Eppig, C., Fincher, C. L., & Thornhill, R. (2010). Parasite prevalence and the worldwide distribution of cognitive ability. *Proceedings of the Royal Society B: Biological Sciences, 277*, 3801–3808.

Fisher, R. A. (1918). The correlation between relatives on the supposition of Mendelian inheritance. *Transactions of the Royal Society of Edinburgh, 52*, 399–433.

Galton, F. (1869). *Hereditary genius: An inquiry into its laws and consequences*. London: Macmillan and Co.

Galton, F. (1889). *Natural inheritance*. London/New York: Macmillan.

Gilliam, T. C., Tanzi, R. E., Haines, J. L., Bonner, T. I., Faryniarz, A. G., Hobbs, W. J., ... Conneally, P. M. (1987). Localization of the Huntington's disease gene to a small segment of chromosome 4 flanked by D4S10 and the telomere. *Cell, 50*, 565–571.

Goldberger, A. S. (1978). The nonresolution of IQ inheritance by path analysis. *American Journal of Human Genetics, 30*, 442–445.

Gusella, J. F., Wexler, N. S., Conneally, P. M., Naylor, S. L., Anderson, M. A., Tanzi, R. E., ... Sakaguchi, A. Y. (1983). A polymorphic DNA marker genetically linked to Huntington's disease. *Nature, 306*, 234–238.

Haines, J., Tanzi, R., Wexler, N., Harper, P., Folstein, S., Cassiman, J., ... Conneally, M. (1986). No evidence of linkage heterogeneity between Huntington disease (HD) and G8 (D4S10). *American Journal of Human Genetics, 39*, A156.

Hayden, M., Robbins, C., Allard, D., Haines, J., Fox, S., Wasmuth, J., ... Bloch, M. (1988). Improved predictive testing for Huntington disease by using three linked DNA markers. *American Journal of Human Genetics, 43*, 689–694.

Henig, R. M. (2000). *The monk in the garden*. Boston: Houghton Mifflin.

Henry, K. R., & Bowman, R. E. (1970). Behavior-genetic analysis of the ontogeny of acoustically primed audiogenic seizures in mice. *Journal of Comparative and Physiological Psychology, 70*, 235–241.

Heron, W. T. (1941). The inheritance of brightness and dullness in maze learning ability in the rat. *Journal of Genetic Psychology, 59*, 41–49.

Hogben, L. (1933). *Nature and nurture*. London: Williams & Norgate.

Huang, A. Y., Yu, D., Davis, L. K., Sul, J. H., Tsetsos, F., Ramensky, V., & Chen, J. A. (2017). Rare copy number variants in NRXN1 and CNTN6 increase risk for Tourette syndrome. *Neuron, 94*, 1101–1111.

Janssens, F. A. (1909). Spermatogenese dans les Batraciens V. La theorie de chiasmatype, nouvelle interpretation des cineses de maturation. *Cellule, 25*, 387–411.

Johannsen, W. (1911). The genotype conception of heredity. *The American Naturalist, 45*, 129–159.

Joseph, J. (2015). *The trouble with twin studies: A reassessment of twin research in the social and behavioral sciences*. New York: Routledge.

Kempthorne, O. (1990). How does one apply statistical analysis to our understanding of the development of human relationships? *Behavioral and Brain Sciences, 13*, 138–139.

Kempthorne, O., & Tandon, O. (1953). The estimation of heritability by regression of offspring on parent. *Biometrics, 9*, 90–100.

Kevles, D. J. (1998). Eugenics in North America. In R. A. Peel (Ed.), *Essays in the history of eugenics* (pp. 208–226). London: Galton Institute.

Krafka, J., Jr. (1920). The effect of temperature upon facet number in the bar-eyed mutant of drosophila: part I. *The Journal of General Physiology, 2*, 409–432.

Lombardo, P. A. (2008). *Three generations, no imbeciles: Eugenics, the supreme court, and buck v. bell*. Baltimore: Johns Hopkins University Press.

Lush, J. L. (1945). *Animal breeding plans*. Ames, Iowa: Iowa State College Press.

MacDonald, M. E., Novelletto, A., Lin, C., Tagle, D., Barnes, G., Bates, G., ... Myers, R. (1992). The Huntington's disease candidate region exhibits many different haplotypes. *Nature Genetics, 1*, 99–103.

Manolio, T. A., Collins, F. S., Cox, N. J., Goldstein, D. B., Hindorff, L. A., Hunter, D. J., ... Visscher, P. M. (2009). Finding the missing heritability of complex diseases. *Nature, 461*, 747–753.

Marceau, K., McMaster, M. T., Smith, T. F., Daams, J. G., van Beijsterveldt, C. E., Boomsma, D. I., & Knopik, V. S. (2016). The prenatal environment in twin studies: a review on chorionicity. *Behavior Genetics, 46*, 286–303.

Mendel, G. (1865). Versuche über Pflanzenhybriden. *Verhandlungen des naturforschenden Vereines in Brünn, 4*, 3–47.

Morgan, T. H. (1911). An attempt to analyze the constitution of the chromosomes on the basis of sex-limited inheritance in Drosophila. *Journal of Experimental Zoology Part A: Ecological Genetics and Physiology, 11*, 365–413.

Morgan, T. H. (1914). The mechanism of heredity as indicated by the inheritance of linked characters. *Popular Science*, (Monthly, January), 1–16.

Morgan, T. H. (1929). The mechanism and laws of heredity. In C. Murchison (Ed.), *The foundations of experimental psychology* (pp. 1–44). Worcester, MA: Clark University Press [Reprinted from: Not in File].

Morgan, T., Sturtevant, A., Muller, H., & Bridges, C. (1915). *The mechanism of mendelian heredity*. Holt.

Nevin, R. (2009). Trends in preschool lead exposure, mental retardation, and scholastic achievement: association or causation? *Environmental Research, 109*, 301–310.

Orel, V. (1984). *Mendel*. Oxford, UK: Oxford University Press.

Platt, R. (1959). Darwin, Mendel, and Galton. *Medical History, 3*, 87–99.

Polderman, T. J., Benyamin, B., de Leeuw, C. A., Sullivan, P. F., van, B. A., Visscher, P. M., & Posthuma, D. (2015). Meta-analysis of the heritability of human traits based on fifty years of twin studies. *Nature Genetics, 47*, 702–709.

Sidman, R. L., Green, M., & Appel, S. (1965). *Catalog of the neurological mutants of the mouse*. Cambridge MA: Harvard University Press.

Smith, D. (2013). *The first genetic-linkage map*. Caltech News. Retrieved from: *http://www.caltech.edu/news/first-genetic-linkage-map-38798*.

Wahlsten, D. (1990). Insensitivity of the analysis of variance to heredity-environment interaction. *Behavioral and Brain Sciences, 13*, 109–120.

Wahlsten, D. (1994). The intelligence of heritability. *Canadian Psychology, 35*, 244–258.

Woltereck, R. (1909). Weitere experimentelle Untersuchungen über das Wesen quantitatitiver Artunterschieder bei Daphniden. *Verhandlungen der Deutschen Zoologischen Gesellschaft, 19*, 110–173.

Zuk, O., Hechter, E., Sunyaev, S. R., & Lander, E. S. (2012). The mystery of missing heritability: genetic interactions create phantom heritability. *Proceedings of the National Academy of Sciences of the United States of America, 109*, 1193–1198.

Further Reading

MacDonald, M. E., Ambrose, C. M., Duyao, M. P., Myers, R. H., Lin, C., Srinidhi, L., & Groot, N. (1993). A novel gene containing a trinucleotide repeat that is expanded and unstable on Huntington's disease chromosomes. *Cell, 72*, 971–983.

15

Intelligence

The concept of intelligence and tests to measure it are among the foremost achievements of psychology in the 20th century. Differences of opinion concerning the roles of heredity and environment in determining levels of intelligence existed right at the start of the century, before there were any substantial data on the topic. Now that a large volume of data on the topic is available, differences of opinion still exist but are perhaps less extreme than in the early days of psychology and now more closely bound to facts.

THE RISE OF INTELLIGENCE TESTING

Binet

After the government of the Third French Republic passed a law in 1882 making schooling compulsory for all children, psychiatrists and psychologists offered different perspectives on how to deal with children who did not do well in the schools but nevertheless seemed to be physically healthy (Nicolas, Andrieu, Croizet, Sanitioso, & Burman, 2013). The psychiatrists proposed to identify abnormal children as mentally ill and confine them in special asylums attached to medical facilities, whereas psychologists led by Alfred Binet sought to measure the degree of deficiency in mental development and institute special classes for those who were judged retarded. Those classes were to be part of the normal public schools, where the children would receive special help in a program of "mental orthopedics." Working closely with teachers, Binet and Simon devised a new kind of mental test in 1905. The revised version (Binet & Simon, 1908) was imported to the United States and translated into English, where it was adapted and became the original Stanford-Binet IQ test. In 1909, the government of France passed a new law establishing special classes for "slow" children as part of the elementary schools.

The Binet-Simon method (Binet & Simon, 1905) included several innovations that were widely adopted and are embodied in tests in use today. The test was designed to assess only the "present mental state" of a child and leave out of the evaluation all questions of past history and thoughts about etiology. Of great importance was the idea that intelligence exists on a continuum and a single test can be applied with normal children and those who may be behind in mental development. To achieve this, the test had 30 items arranged in a series of gradually increasing difficulties (Table 15.1). Because young children show rapid mental development, it was deemed essential to compare a child with normal children of the same chronological age. After the test was given to a sample of normal children of different ages, it was then possible to determine the *mental age* of a specific child by comparing her test score with the average scores of children having different chronological ages. The child's mental age was the age at which normal children achieved the same average score. Thus, the test allowed a psychologist to quantify the degree of mental retardation—the number of years by which the child lagged behind normal peers. That lag could then form the basis for an administrative decision about whether to place the child in a special class. A child who lagged behind same-age peers by 2 years might be a candidate for special help in a special class.

The test was carefully constructed so that the items were *heterogeneous* in their sensory and motor requirements and memory and other skills. Some items required vision, some relied on hearing, and some invoked touch to sense weights of objects. Some required that the child simply point at something, while others required spoken words. Thus, intelligence was viewed as an average over a wide range of specific abilities, a "hierarchy of diverse intelligences." Binet did not propose a definition of intelligence per se, but he stressed that intelligence is a higher mental process that involves judgment, comprehension, and reasoning. Those qualities were mainly evident in the more advanced test items.

Pearson

Meanwhile, across the English Channel, a British intellectual was devising his own rating scale of intelligence

TABLE 15.1   Names of Binet Test Items

1. Eye track moving object	2. Response to object in the hand	3. Reaching for object with the hand	4. Recognize food versus wood	5. Unwrap candy in paper
6. Respond to hand gestures	7. Associate object and name	8. Point to named image of objects	9. Name picture of object	10. Compare lines of different lengths
11. Repeat names of three digits	12. Compare two weights	13. Search for named object	14. Verbally define an object	15. Repeat 15-word sentence
16. Compare two objects from memory	17. Visual memory for pictures	18. Draw design from memory	19. Immediate repetition of figures	20. Comparing several objects from memory
21. Compare length of lines	22. Place five weights in order	23. Find missing weight by heft	24. Find rhyming words	25. Replace word in text
26. Use three words to make sentence	27. Answer abstract question	28. Reverse the hands on a clock	29. Draw image of cut paper	30. Define abstract terms

From Binet, A., & Simon, T.H. (1905). New methods for the diagnosis of the intellectual level of subnormals. L'Année Psychologique, 12, 191–244; Kite, E. S. (1915). The Binet-Simon Measuring Scale for Intelligence: What It Is; What It Does; How It Does It; With a Brief Biography of Its Authors, Alfred Binet and Dr. Thomas Simon. Philadelphia, PA: Committee on Provision for the Feeble-Minded.

for a radically different purpose. Pearson (1904), inventor of the Pearson correlation coefficient (r), set out to convince his readers that heredity was by far the most important, even a completely sufficient explanation, for individual differences in mental traits. He attributed the good performance of school children at the top of their class to "innately wise parents." Pearson collected data on several thousand school children using a survey mailed to teachers across the country who responded to ads in educational journals read by teachers. His sample amounted to less than 1% of all teachers, but he nevertheless declared that it was representative of "the nation at large." Everything was based on teachers' ratings of pairs of siblings, children of different ages but the same sex who were known to the same teacher. Teachers were instructed to first read the general directions, which informed them that Pearson was interested in

"collateral heredity" and wanted information on the degree of resemblance of sibs for his work on "quantitative laws of heredity." One section was included "to give some confidence in the scales." He informed the reader that in an earlier study of 150 children rated by three teachers, "agreement in classification was complete in more than 80% of cases." The rating scale for intelligence is shown in Fig. 15.1.

On the basis of a poorly designed, obviously biased methodology and dubious data, Pearson proclaimed that physical and mental characters are inherited "in the same manner and with the same intensity." He argued that to make England more competitive among modern nations, "the remedy lies beyond the reach of revised educational systems" because mental qualities "are not manufactured by home and school and college; they are bred in the bone." His proposed solution was "to alter the relative

Place a cross against the class of each sister under as many headings as possible, except under III and VIII. Please read first the general directions.							

Elder sister. / **Younger sister.**

Name ... ...
Age ... ...
District of home ... ...

I. Physique :

	Very strong.	Strong.	Normally healthy.	Rather delicate.	Very delicate.	Athletic.	Nonathletic.
Elder sister ...							
Younger sister .							

II. Ability : (a) *General scale.*

	Quick intelligent.	Intelligent.	Slow intelligent.	Slow.	Slow dull.	Very dull.	Inaccurate-erratic.
Elder sister ...							
Younger sister ...							

FIG. 15.1   Portion of survey sheet sent by Pearson to school teachers in England who were asked to rate siblings they had known on intelligence and other attributes. There was a different color sheet for girl-girl, boy-boy, and boy-girl pairs. *From Pearson, K. (1904). Mathematical contributions to the theory of evolution: XIII. On the theory of contingency and its relation to association and normal correlation (Vol. 13). London: Dulau and Co., Appendix IB.*

fertility of the good and the bad stocks in the community," a practice known as eugenics, a practice ardently advocated by Francis Galton.

Spearman and Burt

An elaborate theory of intelligence was formulated by Spearman (1904, 1923; see Fancher, 1985) who focused on *"natural innate faculties.* By this definition, we explicitly declare that all such individual circumstances as after birth materially modify the investigated function are irrelevant and must be adequately eliminated" (Spearman, 1904; emphasis in original). He maintained that intelligence resulted from a single factor, general intelligence or "g." Spearman (1923) asserted that the "ultimate basis" for mental abilities was "the influences of heredity and of health." Spearman's ideas were strongly endorsed and applied by Burt (1909), whose expressed interest was in "general, innate endowment, as distinguished from special knowledge and special dexterities, that is to say, from postnatal acquisition." "By intelligence, the psychologist understands inborn, all-round, intellectual ability. It is inherited or at least innate, not due to teaching or training" (Burt, Jones, Miller, & Moodie, 1934). In his final statement on the matter, he claimed that "individual differences in mental ability depend largely on differences in physical structure of the brain" (Burt, 1972).

Terman

Lewis Terman contributed to the translation of the Binet test, and he and his students added items to it and collected extensive normative data on his test given to American children and adults. A revision of the Binet-Simon test entailed 54 items, whereas Terman's version had 90 items. He introduced the concept of the intelligence quotient (IQ) that compared chronological age (CA) with mental age (MA): $IQ = 100(MA/CA)$, and he later replaced it with the version discussed in Chapter 6 that depends on Z scores. The Stanford-Binet became the most widely used intelligence test until the Wechsler tests appeared. Aligning his views with those of Pearson, Spearman, and Burt, Terman (1916) held that intelligence test score represents "original mental endowment." This he achieved without incorporating a single measure of environment into the test protocol. He asked, "Are the inferior races really inferior, or are they merely unfortunate in their lack of opportunity to learn? Only intelligence tests can answer (this) question and grade the raw material with which education works."

Terman's version of the Binet test was welcomed as a major scientific advance in the United States, and many people were keen to be tested to prove just how smart they were. There were some surprises. For example, a headline in the December 29, 1915 edition of the *Chicago Herald* declared, "Experts assail mentality test result as joke. Hear how Binet-Simon method classed mayor and other officials as morons" (cited by Chase, 1977, p. 241). It was decades later that low scores of that generation of older Americans were better understood (Flynn, 1984a, 1987).

The Stanford-Binet test underwent extensive revisions from the original 1916 version to the fifth edition published in 2003 (Becker, 2003). Until 1986, the SB yielded only an index of general intelligence (g). The Fourth Edition (SB4) introduced separate estimates for factors named verbal reasoning, abstract reasoning, quantitative reasoning, and short-term memory. The Fifth Edition (SB5) had a different factor structure consisting of knowledge, fluid reasoning, quantitative reasoning, visual-spatial processing, and working memory. Correlations between SB4 factor scores and SB5 scores ranged from $r = 0.54$ to 0.77, which revealed that the two tests were measuring things somewhat differently.

Without a doubt, there were many good reasons why psychologists with expertise in mental testing made so many changes from one version to the next. Table 6 in the history by Becker (2003) outlined the advantages of each new version and the shortcomings of previous versions that new features were designed to remedy. One thing should be kept in mind, however, when pondering all those revisions. Over a period of many decades, heredity would have changed very little, whereas environments evolved greatly. If Terman's test had been tapping "original mental endowment," why tamper with it? It should be obvious that his test was incapable of distinguishing between things a child had learned and abilities that were supposedly innate. It measured phenotypes alone.

Wechsler

Perhaps, the most successful of all IQ tests (Kaplan & Saccuzzo, 2009) are the versions designed by Wechsler for use with children (WISC) and adults (WAIS). The adult version, named the Wechsler-Bellevue Intelligence Test, was formulated in 1939 when Wechsler was the chief psychologist at the Bellevue Psychiatric Hospital, and a revised version became the WAIS. The WISC (Table 15.2) is structured to provide not only an overall average (full-scale) IQ score but also ratings on several subtests that can help to determine why a particular child may be having difficulties with school work. Five subtests for the verbal part of the test are clearly related to things commonly learned in school, while the five "performance" tests do not require verbal answers, although instructions from the examiner are usually spoken. Each subtest was scaled to have a mean of 10 and limits of 0

TABLE 15.2 Subtests on the Original Version of the Wechsler Intelligence Scale for Children

Verbal intelligence			Performance intelligence		
Kind of subtest	Maximum raw score	Average scaled score	Kind of subtest	Maximum raw score	Average scaled score
Information	30	10	Picture completion	20	10
Comprehension	28	10	Picture arrangement	57	10
Arithmetic	16	10	Block design	51	10
Similarities	28	10	Object assembly	34	10
Vocabulary	80	10	Coding	93	10
Total score		50	Total score		50

Wechsler, D. (1949). Wechsler Intelligence Scale for Children. Manual. *New York: The Psychological Corporation.*

and 20 points derived from the standard deviation of the subtest in the standardization sample. Adding scaled scores on 10 tests, each with a mean of 10, yielded an average total score of 100 with standard deviation of 15, as explained in Chapter 6.

Diversity of Tests

One perplexing fact about intelligence tests today is that the experts do not agree on the nature of intelligence or how many kinds of intelligence exist. Spearman and Burt argued that there was just one important kind— general intelligence. Raymond Cattell claimed there were two, fluid and crystallized intelligence. Robert Sternberg opts for a three-part (triarchic) conceptualization. Leon Thurstone believed that seven different factors combine to make a person intelligent. Howard Gardner sees evidence of perhaps eight and maybe even more kinds of intelligence. Daniel Goleman argues for the existence of emotional and cognitive intelligence. Many of the major theorists devised their own tests to tap the abilities featured in the theories. When the same children take two of these tests, there is usually a substantial positive correlation between the scores, but it never approaches a perfect correlation. Each test seems to weight the various features of the human mind a little differently.

Unlike the profession of psychiatry where there has been a decade-long effort to formulate a common diagnostic terminology and criteria, embodied in the DSM and ICD, psychology has its own professional organizations, usually a main one in each country, but no consensus version of an intelligence test to recommend. Instead, scientific psychology has developed an elaborate technology for constructing, evaluating, and using psychological tests. The American Psychological Association provides a number of standards for professional practice and education that are considered mandatory, but the APA website states emphatically that "APA's Testing Office does not maintain, sell, or endorse any tests." The Buros Center at the University of Nebraska-Lincoln publishes a massive *Mental Measurements Yearbook* (Carlson, Geisinger, & Jonson, 2017) with in-depth reviews of many tests.

After more than 100 years of experience with many kinds of tests, there is at least a consensus that several of them are reasonably good (Table 15.3). They have been revised several times to correct shortcomings and restandardized to reflect continuing changes in the population of people where they will be applied. A full kit including a detailed instruction manual with norms and scoring

TABLE 15.3 Current Intelligence Tests

Test name	Edition	Standardization sample	Ages (years)	Time (min)
Bayley Scales of Infant Development	3	1409	1–42 months	30–90
Kaufman Assessment Battery for Children	2	3025	3–18	25–55
Kaufman Brief Intelligence Test	2	Based on US census	4–90	20
Naglieri Nonverbal Ability Test—Individual	2	1500	5–17	25–30
Raven's Colored Progressive Matrices	2003	No norms	5–11	15–30
Stanford-Binet Intelligence Scales	5	4800	2–85	50
Wechsler Intelligence Scale for Children	5	2200	6–16	60
Wechsler Adult Intelligence Scale	4	2200	16–90	60–90

sheets can be purchased from commercial websites, and several of them are making available much less costly versions that can be administered and scored via the Internet.

EFFECT SIZE

When discussing genetic and environmental effects on intelligence, many people want to know how strong their effects are both in absolute terms and relative to each other. The mean (M) and standard deviation (S) of a sample of N scores are descriptors of results of a study. If the study happens to involve two groups, perhaps a control group and an experimental group that receives a special treatment, each group will have its own M and S. The difference between group means is $M_E - M_C$, and standard deviations are S_E and S_C. If the treatment had a noteworthy effect, $M_E - M_C$ will tend to be fairly large. An indicator of great value in interpreting results for many studies discussed in this chapter and later chapters is *effect size*. This compares the difference *between* groups with the variation *within* groups. If S_C and S_E are similar, we can combine the two estimates to obtain a single estimate S. The effect size indicator is $D = (M_E - M_C)/S$, the fraction of a standard deviation by which group means differ. For the example in Fig. 15.2A with $N = 25$ people in each group, $D = 0.51$.

D provides a good description of results of a two-group study for almost any kind of measurement or test score. Furthermore, there are conventions for what values of D qualify as small, medium, and large effects of a treatment in psychology, neuroscience, and many other fields of research (Fig. 15.2B). For a study of treatment effects on intelligence rated by IQ scores having S close to 15.0, small, medium, and large effects would correspond to treatment effects versus controls of about 3.0, 7.5, and 12.0 IQ points, respectively. A recent review of 3,801 published studies in psychology and neuroscience found that the median effect size for all experiments in the reports was 0.65, a medium to large effect (Szucs & Ioannidis, 2017).

This indicator of effect size can also be used for data on IQ scores of people having different alleles of a gene or different alleles of a SNP. Suppose one SNP allele (T) is very common (94%) and the other (C) occurs in about 6% of the population. About 88.4% of people will have SNP genotype T/T, and 11.3% will be T/C, whereas only 0.4% will be C/C and can safely be ignored for an initial analysis. If the SNP is close to a gene variant that influences intelligence test score by a small amount, there should be a measurable IQ test score difference between people having the genotypes T/T and T/C, and the effect size will be $D = (IQ_{T/T} - IQ_{T/C})/15$ for an IQ test having standard deviation $S = 15$. The same D indicator can be used to compare studies of adoption and other environmental treatments.

ENVIRONMENT

How malleable is intelligence? We need to examine situations where children were reared in special circumstances and later given an IQ test. The major challenge in studies of humans is to find a meaningful comparison group that did *not* receive any special treatment. The difference between the two groups will then indicate an effect of different environments. The question is difficult because there is no single, neat indicator of environment. It is a multidimensional array of influences, most of which are especially effective during specific time periods or in combination with other factors.

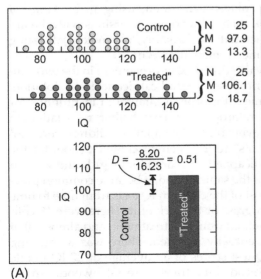

(A)

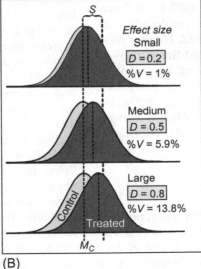

(B)

FIG. 15.2 (A) Simulated IQ scores of two groups of 25 people each sampled randomly from a population with mean = 100 and standard deviation = 15 using Excel function rand(). Effect size is D for the two groups. (B) Standards for small, medium, and large effect sizes showing extent of overlap of the two theoretical distributions and the percentage of total variance attributable to the difference between group means. Note that standard deviation is obtained by pooling sums of squares ($SS = SS_C + SS_E$) and sample sizes N_C and N_E to obtain variance $V = (SS_C + SS_E)/(N_C + N_E)$. Pooled S is the square root of V.

The Flynn Effect

The groups being compared may differ greatly in the year of birth. This can generate an environmental *cohort effect* if conditions in society differed substantially for the cohorts. For mental tests that report the score as IQ, calibrating a test against a standardization sample yields a standard that is not fixed or constant; it estimates the abilities of a population of people in a specific year (Chapter 6). Norms for the test are published in a manual that tells how to convert raw scores to IQ scores based on the original standardization sample, and the same manual is used for many years until a new version of the test is produced. If the psychological entity that the test measures increases in the general population, average raw scores and IQ test scores will rise too, and if this continues for several years, the mean IQ score for children of a specific age may rise to a value well above the value of 100 for original standardization sample. In effect, the original standardization sample serves as the control or comparison group. The painstaking work of Flynn (1984a, 1984b, 1987) has shown that IQ scores have risen on several popular tests in several countries.

It is accepted practice to restandardize a test every few years when several test items are changed, so that content is modernized and the mean IQ score is restored to 100, which, at least in the short term, cloaks the mental growth of the population. In order to insure that the new version of the test given to a new sample is measuring the same thing as the older test, makers of the major IQ tests also give the older version of a test to the new sample of people and check to be sure the correlation of the two sets of test scores is quite high. As discussed in Chapter 6, this has been done several times from 1947 to 1978. The extent of the increase on several major tests over a period of 46 years, about 14 IQ points, amounts to a very large effect size of $D = 0.95$ or about 3 IQ points per decade.

In an updated review of the same tests, Flynn (2007) calculated that the average gain from 1972 to 2002 was 0.31 IQ points per year or 3.1 IQ points per decade, very similar to the findings from 1936 to 1972. He dug more deeply into the data and discovered that the gains were not at all the same for all subtests. Information and arithmetic scores were remarkably stable over more than 50 years, while picture arrangement and similarities showed major improvements. The data offered a strong argument against the notion that IQ test performance reflects mainly a general or "g" factor.

Adoption

When infants are removed from wretched conditions and adopted into good homes, this provides an opportunity to measure effects of a very large, almost lifelong change in a multifaceted environment. Performance on IQ tests can be evaluated before and after adoption, and there can be some kind of nonadopted comparison group. The best known studies of this kind commenced in the 1930s in the State of Iowa in the United States by investigators who did not set out to do an experiment. Unusual circumstances, however, inspired them to look more closely at what was happening in orphanages. In 1934, the Children's Division of the Board of Control of State Institutions established a small psychological services group directed by Harold Skeels with a staff of one, Marie Skodak. The two of them were able to collect a large body of data on children in several institutions (Skodak & Skeels, 1949). Some of the data formed the basis for two separate studies that drew much attention when published and are still cited and debated today.

One remarkable study began serendipitously (Skeels, 1966). The orphans were housed in cramped quarters with very little sensory, motor, or social stimulation and cared for by harried staff who were able to provide only a bare minimum of life-sustaining care. After the age of 2 years, the orphans were moved to similarly crowded and barren cottages and later given an education designed for mentally subnormal children. IQ tests documented a deficit in mental growth. On one occasion, two baby girls from mentally deficient mothers arrived on the ward as "pitiful little creatures" who were emaciated, sad, and inactive. They spent the day rocking and whining. Staff decided these girls would never be suitable for adoption, and the girls were assigned to be sent to an institution for mentally retarded children "at the next available vacancy." The vacancy was most fortuitous. In most instances, children would be housed with those of similar age, but in this case, the girls 15 and 18 months old were placed in a ward housing older and brighter women with ages 18–50 years. In a short time, each infant had been "adopted" by one of the older women who became adoring aunts who showered the girls with love and affection, played with them using many kinds of toys, and took them on various excursions with the aid of sympathetic staff. Remarkably, psychological testing detected significant mental growth.

Inspired by this dramatic improvement in the two children, Skeels proposed "a radical, iconoclastic solution, that is, placement (of orphan infants) in institutions for the mentally retarded…to see whether retardation in infancy was reversible." Eleven more children were sent as "house guests" to the Glenwood State School for the mentally handicapped and cared for by girls and women living there. In the same period, it became common practice to give most of the children IQ tests from time to time to monitor progress or the lack of it. There were 13 children who benefited from this treatment, and almost all of them were eventually adopted. There was a stunning increase in IQ test scores of an average of 27 IQ points during the period after transfer to Glenwood. An ad hoc comparison group was assembled, consisting of children who had remained in the orphanage for several years and had enough IQ test data to allow comparisons at roughly the same ages as the "guest" group. They were a diverse lot and differed from the guest group in several ways. Nevertheless, the overall pattern of results was obvious and dramatic. Many in the guest group were adopted into good homes when their IQ scores and demeanor had improved sufficiently, which ended the experiment. The orphanage group suffered major decline in mental test performance, a decline of about 24 IQ points. Before long, those kinds of orphanages that imposed such severe mental and physical deprivation on young children were closed forever, and the Skeels study will hopefully never be repeated.

Well-controlled adoption studies have provided better estimates of effect size for changed environments (Fig. 15.3). A study done in France compared 20 children who remained with mothers having relatively little education and income versus 32 of their biological siblings from the same mothers who were adopted into homes with well-educated and prosperous parents (Schiff et al., 1978). The group difference amounted to a bit more than one standard deviation on the IQ scale, a very large effect size (Fig. 15.3A). A thorough scrutiny of adoption records in France uncovered several more instances where one sib stayed with a mother low in education and incomes while another sib was adopted into a home

having higher socioeconomic status (SES), as well as instances where one sib was adopted from a high SES family to a low SES family (Fig. 15.3B). IQ differences for the two types of adoption were 11–12 points, a D value of about 0.8 (Capron & Duyme, 1989). The high-low children scored considerably higher than the low-low ones, which could have happened because of differences in the preadoption environment and hereditary factors.

A recent report from Sweden where all males took an IQ test upon reporting for military service involved the largest published data set on matched home-reared and adopted-away children (Kendler, Ohlsson, Sundquist, & Sundquist, 2015). There were 436 pairs where one male was home-reared and a male full sibling (same father) was adopted away into homes with mostly higher income and education and 2341 half-sib pairs (different fathers) with one adopted away. The adopted males scored 3–4 points higher on average on the IQ tests, but their performances were also related to the education level of the adopting parents. Material differences between higher and lower social classes in Sweden are generally not as great as in many countries, especially the United States during the time of Skeels. The IQ difference between adopted and home-reared males with highest and lowest parental education ranged from about 11 IQ points for the full sibs to 6 points for the half-sibs, results that are similar to the findings from the French adoption studies. The authors concluded that "cognitive ability is environmentally malleable and the malleability shows plausible dose-response relations with the magnitude of the environmental difference."

Enrichment

After a series of strident debates and contested claims about heritability and plasticity of IQ sparked by the Jensen (1969) article (see Section "Heritability"), several well-controlled studies of enriched early experience were undertaken in the 1980s and 1990s in the United States with infants at risk for poor postnatal development. Each demonstrated considerable plasticity at least in the shorter term by comparing children randomly assigned to an experimental group that received enriched experiences with a control group from the same population that received only the usual medical care. Improvement in mental test performance was striking, amounting to almost one standard deviation, a very large effect size (Fig. 15.4). The last tests shown in the diagram were given near the time when special treatments ceased, so the longer term impact remained to be determined. A follow-up study of the Infant Health and Development Program (IHDP) study of low-birth-weight children at age 8 (Hill, Brooks-Gunn, & Waldfogel, 2003) showed an

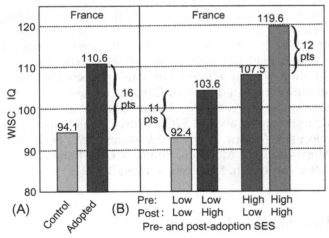

FIG. 15.3 Adoption studies with well-matched control groups. (A) Schiff et al. (1978) compared children who remained with their biological mother with adopted-away siblings placed in homes with well-educated and prosperous parents. (B) Capron and Duyme (1989) identified four groups of siblings that differed in the pre- and postadoption home socioeconomic status (SES). *A: From Schiff, M., Duyme, M., Dumaret, A., Stewart, J., Tomkiewicz, S., & Feingold, J. (1978). Intellectual status of working-class children adopted early into upper-middle-class families.* Science, 200, 1503–1504, *reprinted with permission from AAAS; B: From Capron, C., & Duyme, M. (1989). Assessment of effects of socio-economic status on IQ in a full cross-fostering study.* Nature, 340, 552–554, *Copyright 1989, reprinted by permission from Nature/Springer.*

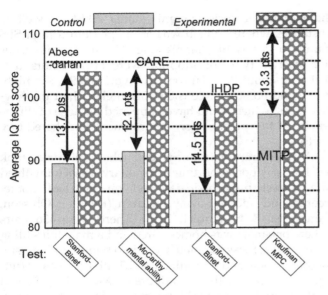

FIG. 15.4 Mean IQ test scores for children randomly assigned to the control and experimental conditions in four studies designed to enhance the development of disadvantaged children in the United States. The abecedarian project represents Stanford-Binet IQ scores at 48 months (Ramey, Yeates, & Short, 1984). Project CARE represents McCarthy Mental Ability scores at 54 months for children in the control with no day care and Child Development Center day care groups (Wasik, Ramey, Bryant, & Sparling, 1990). The Infant Health and Development Program (IHDP) represents Stanford-Binet IQ scores at 36 months for controls versus intervention group that received an average of 500 home visits plus days in day care (Brooks-Gunn, Klebanov, Liaw, & Spiker, 1993; Ramey et al., 1992). The Mother-Infant Transaction Program (MITP) in Vermont represents the Kaufman Mental Processing Composite scores for children at age 9 years that were in the low-birth-weight control and experimental groups (Achenbach, Howell, Aoki, & Rauh, 1993).

average 7–10 IQ point effect that was higher (14 IQ pts) for those with birth weight in the range 2001–2500 g than those born at less than 2001 g (8 IQ pts). Several follow-up studies of the abecedarian study of African-American children living in poverty also found lasting effects. At age 12, average WISC IQ was 93.7 for the enriched group versus 88.4 for the controls that, owing to a standard deviation of about 11 points, amounted to an effect size of $D = 0.5$ (Campbell & Ramey, 1994). When the children entered adulthood, it was noted that the enriched group completed 1.2 more years of schooling, was more likely to graduate from college (23% vs 6%), and earned substantially more (Campbell et al., 2012). It appeared that group differences in terms of effect size were at a maximum at the end of active treatment and then lessened with the passage of time but were substantial years later.

Coaching

Another form of short-term "enrichment" that generates higher test scores is coaching by a teacher who knows the kinds of items that will be on a test and sometimes has access to the test itself. It was known that IQ test scores were generally higher by as much as 7 points if a child took the same test within a few weeks of the first, without any explicit instruction. Familiarity with the test situation and specific test items boosted performance. A study done in 1928 noted a very large improvement when children were coached on the kind of test that would come next (Greene, 1928; cited in Vernon, 1954). The gains gradually faded away but were still evident 1 year later. More recent studies have found somewhat smaller effects (Dempster, 1954; Hausknecht, Halpert, Di Paolo, & Moriarty Gerrard, 2007). A thriving business of coaching students in preparation for taking school entrance exams is now well established.

Schooling

The amount of schooling a child receives is usually correlated strongly with IQ test performance, but little can be learned from this about causes and consequences because so many factors are confounded. Every year, however, a situation occurs that emulates a well-controlled experiment. Most school districts have a strict policy about how old a child must be to start schooling, and this is specified to an exact birth date. The result is that many children just a few days younger than that date must wait 1 year to begin, while others just a few days older make the "cutoff." Because most other factors except age are closely comparable in the two natural groups, comparing their mean scores provides a good estimate of the effects of 1 year of schooling. Of course, the performance on school-related test content will differ greatly, but will IQ show similar effects? For four independent cutoff studies done in Edmonton, Canada, the average of the four studies revealed that 1 year of schooling enhanced IQ test score by about 4.5 IQ points, a small to moderate effect size (Bisanz, Morrison, & Dunn, 1995; Ferreira & Morrison, 1994; Morrison, Smith, & Dow-Ehrensberger, 1995; Varnhagen, Morrison, & Everall, 1994).

How Small Is Too Small?

By common agreement, psychologists regard a group difference of 0.2 standard deviation as small. Perhaps, 0.1 would be very small. Is there a point below which a group difference might be seen as trivially small? No standard has been set yet, but it may be time to ponder this option. A recent report on testing of school children in Denmark prompted parents to request that their children should take standardized mental tests early in the day, rather than later in the afternoon. The study (Sievertsen, Gino, & Piovesan, 2016) compared scores on the national test given to 2,034,964 children of ages 8–15, a test that assessed math abilities and domains related to the sciences. It was not an IQ test as such. Because of a limitation in the number of available computers for test administration, only a fraction of students could take it in a given time slot, so some took it first thing in the school day and others in the last period of the day, making it a "natural experiment" about time of day. Authors found that scores declined by 0.9% of a standard deviation for every hour later in the day, whereas they increased during two scheduled breaks, and they termed the decline a "deterioration" in performance.

One journalist reported the decline as "a nearly 1% drop in scores per hour." A difference between early and late in the day would then be about 3%. This would amount to an equivalent decline in mean on an IQ test of about 3 IQ points, a small effect but something to be concerned about in terms of fairness to all children competing for advancement. In reality, the authors themselves reported an effect of *1% of a standard deviation decline per hour*. That would be $D = 0.01 \times 3$ or effect size $D = 0.03$, equivalent to a drop of 0.45 IQ point over a 3 h period. That is about the size of commonplace rounding errors. IQ of 100.45 might be rounded down to the nearest whole number, 100, while 100.55 might go up to 101, if a school system uses rounding. The time-of-day effect appears to be a trivially small effect. It was detectable at all only because more than 2 million students were tested.

A summary of the studies of IQ and environment reviewed here is shown in Table 15.4. The range of effect sizes is wide indeed. There appears to be a pattern whereby changes in IQ scores are larger when the changes in environment are very large or have a long duration, compared with modest changes of shorter duration.

HERITABILITY

For those studying single-gene effects, the heritability coefficient is at best irrelevant and in many instances downright invalid. For quantitative genetic studies of multiple genes, on the other hand, h^2 is the centerpiece that has occupied many investigators for more than 50 years and continues to inspire new research reports. There has been an ongoing and earnest debate about the best ways to measure and interpret h^2, especially as it pertains to intelligence. As mentioned above, most of the early founders of intelligence testing in the United Kingdom and the United States believed strongly in the genetic determination of that trait and even defined it as innate, beyond the sway of environment. They expected h^2 to be very high, close to 1.0. But within the field of psychology, there were many who were not so sure and wanted to see some data. Thus, the need arose to estimate h^2.

The Difficult Birth of Behavior Genetics

The 1960s was a period of political and intellectual turmoil in the United States. In 1954, the US Supreme Court outlawed racial segregation of public schools, and a civil rights movement arose that demanded action. The Coleman report commissioned by the US government conducted a survey of schools from coast to coast and found that segregation and grossly unequal quality of schools for minority students was prevalent across the country (Coleman et al., 1966). The US congress considered possible ways to ameliorate the many disadvantages experienced by Americans of African ancestry. Many politicians and social scientists advocated a major increase in spending on schools in poorer districts in order to implement "compensatory education."

During that same period, several academics interested in genetic aspects of animal and human behavior decided to establish a new academic discipline—behavior genetics. The University of Colorado approved the founding of the Institute for Behavioral Genetics (IBG), and several professors who had formerly specialized in animal research began to explore how quantitative genetics could be applied to human behavior. They worked hard to found the Behavior Genetics Association that took shape in 1970 and start a new scientific journal named *Behavior Genetics* that published its first volume in 1970. But it was a very difficult birth.

Early in 1969, an inflammatory article by Jensen (1969) was published prominently in the *Harvard Educational*

TABLE 15.4 Effect Sizes of Different Environmental Variations

Effect size D	IQ points	Environmental effect	Comments
From 0.6 to >1.0	10–25	Adoption into superior home environments	Effect is larger when home environments differ greatly and duration is longer
1.0	15	Coaching on the same test	Effects fade when coaching stops
0.8	12	Enriched early environment	Effects largest at end of treatment, then fade over several years
0.5	0.3 pts per year	Flynn effect, increasing IQ of an entire population	Amounts to about 7.5 IQ points in one generation spanning 25 years
0.3	5	Schooling for 1 year	Based on those who barely made "cutoff"
0.03	0.45	Time of day when test is taken	Scores tend to be higher in the morning

Review entitled "How much can we boost IQ and scholastic achievement?" The first sentence of his article declared that "Compensatory education has been tried and apparently it has failed." His main argument was that intelligence is almost entirely determined by genes (heritability of 80%), and therefore, little can be done to improve it through enriched environment. The principal evidence cited to bolster his claim was the twin and family research of Cyril Burt. Jensen claimed that Americans of African ancestry did poorly in school because of poor genes, not poverty and racial discrimination. That unwelcome bombshell landed smack in the middle of IBG and the nascent field of behavior genetics.

Cyril Burt

Burt was one of the most eminent psychologists of his time. As the first psychologist in the world to be appointed to an education authority (London County Council in 1913), he became chair of psychology at the University of London and in 1946 was granted knighthood by the British king. He cofounded the *British Journal of Statistical Psychology* and was editor for many years and a prolific contributor of research articles on twins with his "colleagues" Jane Conway and Margaret Howard. After his death in 1971, fulsome praise was given by Jensen (1972) in an obituary that anointed him "a born nobleman." Jensen noted that Burt bequeathed to the world of science data on twins "consisting of larger, more representative samples than any other investigator."

Meanwhile, a vigorous debate was underway about heritability of intelligence and race differences in America. Leon Kamin of Princeton University decided to take a closer look at Burt's articles because they were often cited as the best available evidence for high heritability of IQ. Kamin soon detected some surprising features about the details of Burt's work. Burt published a series of reports with steadily increasing samples of monozygotic twins reared apart (15, 21, 42, and 53 pairs in 1943, 1955, 1958, and 1966, respectively) in which the monozygotic twin correlations were exactly the same (0.771) in three of the four reports, an impossibility with real data. There were also odd inconsistencies and gaps in Burt's descriptions of his IQ tests and other details. Kamin addressed the Eastern Psychological Association in 1973 and circulated widely a manuscript in which he concluded that Burt's work was scientifically worthless. There were howls of protest from many prominent behavior geneticists. Their intensity increased when Kamin (1974) published *The Science and Politics of IQ* that documented many other odd things about Burt's numbers and other twin studies and placed the issues in their political context. Two British psychologists supported Kamin and noted that data in Burt's 1961 paper on social

class "appear suspiciously perfect" (Clarke & Clarke, 1974), and another scientist reported in *Science* that "beyond reasonable doubt," Burt had "fabricated data on IQ and social class" (Dorfman, 1978).

By that time, the dispute had caught the attention of the British press corps. Journalist Oliver Gillie reported on October 24, 1975 in the *Sunday Times* of London that an extensive search had failed to turn up any trace of "Miss Howard" or "Miss Conway." Another journalist announced in the journal *Science* that "suspicions of fraud becloud" Burt's research (Wade, 1976). Gillie gained access to Burt's daily diaries and reported in *Science* in 1979 that there was no mention of Conway or Howard meeting with Burt. An editorial in *Science* soon appeared, "Tracing Burt's descent to scientific fraud" (Hawkes, 1979). There was outrage on the right, outrage on the left, and confusion down the middle of psychology and behavior genetics.

For a time, the strident debate seemed to be at a standoff, but matters were effectively settled for many of us when the Burt family gave his papers including diaries and extensive correspondence to a trusted family friend, the psychologist L.S. Hearnshaw, so that he could write an informed biography. The shocking results of his investigation were published as *Cyril Burt: Psychologist* (Hearnshaw, 1979). The findings were far more damaging than anyone anticipated. Hearnshaw found that Burt had contact with Miss Howard in 1924 when she was a member of the British Psychological Association as a lay member but was "certainly not in touch with her after 1954," a time period when her name appeared as coauthor on several articles with Burt and letters to the editor. All her "contributions" were to the *British Journal of Statistical Psychology* edited by Burt himself, and they ceased when he stepped down as editor. Burt's diary on April 7, 1962 had him "chiefly doing Howard's reply to Isaacs." The situation with Conway was much the same. Hearnshaw concluded she too was "an 'alter ego' of Burt." The extent of the deception in the journal edited by Burt was breathtaking. More than half of the 40 authors named in the journal could not be identified by Hearnshaw and appeared to be pseudonyms for Burt, who sometimes wrote a reply under a false name to an article authored by himself.

Hearnshaw concluded that Burt made up his coworkers in order to bolster his claim that he continued to collect data on twins in an active research program, a pretense that turned out to be "a complete fabrication." The diaries and letters made it clear that Burt had not collected any data on twins after 1950 and probably after 1939 and had not been visited by any assistant who was collecting data. He had not obtained consent to test school children in that period and had no research funds. Many people had written to ask Burt for copies of his famous data on twins, and he provided a table of IQ and social class numbers for 53 pairs of MZ twins reared apart to Professor Jencks in 1969. This turned out to be a difficult task. Burt recorded in his diary that he spent the entire week of January 2, 1969 "calculating data on twins for Jencks" and on January 11 "finished checking tables for Jencks." Evidently, Burt did not have the data on individual twins, just the correlations published in journals, and he had to devise 53 pairs of scores that fit the published correlation of $r = 0.771$. Burt fabricated the correlations of twins and other relatives so that the numerical values would support his contentions about the overwhelming importance of genes as determinants of intelligence and the irrelevance of environment for individual differences in intelligence.

After Hearnshaw's analysis appeared in print, Burt's defenders assailed Hearnshaw for poor scholarship and sought to rehabilitate the deceased knight (e.g., Joynson, 1989). This prompted a reconsideration of the issue by historians of psychology, including Samelson (1997) who eventually concluded "In this writer's opinion, this accumulated evidence strongly supports the charges of 'scientific misconduct' and—depending on the stringency of one's definition—'fraud'." A review of coverage of heritability and intelligence in genetics textbooks (Paul, 1985) found that several mentioned the Burt affair and suggested his data were not to be relied on, yet they cited or even reprinted a figure showing different kinds of kinship correlations that in fact included Burt's twin data (Erlenmeyer-Kimling & Jarvik, 1963). Paul (1985) concluded, "A majority of genetics students are being taught that intelligence is highly heritable … on the basis of evidence from studies that more properly belong in histories of science or pseudoscience than contemporary textbooks." The dispute persists. A recent text on behavior genetics (Knopik, Neiderhiser, DeFries, & Plomin, 2017) characterized Hearnshaw's scholarly tome

as an "attack" on Burt, stated that "the jury is still out on some of the charges," and cited Joynson's spirited defense of Burt. It did not tell students just what Burt was alleged to have done except to report very high heritability of IQ.

Family and Twin Studies

After the dust had begun to settle, behavior geneticists decided to look more closely at the body of findings untouched by the Burt scandal. Bouchard and McGue (1981) reviewed 111 family and twin studies of IQ and plotted the correlations by the type of relative (Fig. 15.5). Although there were a wide range of values for many kinds of relative, there were also some general patterns in the data that indicated both genetic and environmental variations were important for IQ scores. There was a clear tendency of relatives with more similar genotypes to have more similar IQ scores. Environmental similarities also augmented correlations of relatives. For adoptive relatives where a purely gene-based correlation would be 0, the correlations were well above 0. MZ twins reared apart were less similar than those reared together, also a consequence of environmental similarities. An adopted-away child and biological parent showed far less similarity than expected from a genetic model. From a median correlation of $r_{MZ} = 0.85$ for MZ twins reared together and $r_{DZ} = 0.58$ for DZ twins reared together, a conventional estimate of heritability was $h^2 = 2 (0.85 - 0.58) = 0.54$, considerably lower than the value of 0.8 cited by Jensen mainly on the basis of Burt's correlations. The more recent estimate, of course, assumes that environmental similarities of MZ and DZ twins are the same.

Another team of researchers updated the literature review to include 212 published studies and evaluated the data with a special kind of meta-analysis (Bayesian)

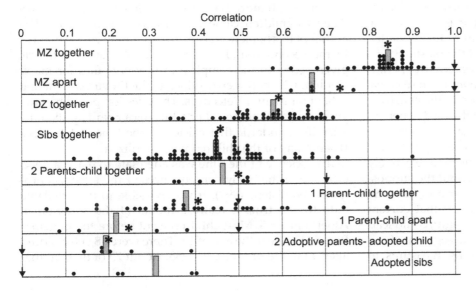

FIG. 15.5 Correlations of IQ scores for different kinds of relatives. Each dot is a value from a previously published study, from Fig. 1 of review by Bouchard and McGue (1981). *Gray bars* are median correlations for each category. *Small arrows* are values expected from a purely genetic model that makes no provisions for environmental effects. Asterisks (✱) are median values from the more recent review of 212 studies by Devlin, Daniels, and Roeder (1997). *From Bouchard, T. J., & McGue, M. (1981). Familial studies of intelligence: a review. Science, 212, 1055–1059, reprinted with permission from AAAS.*

that compared four different statistical models (Devlin et al., 1997). The average correlations of several types of relatives in the expanded sample (shown by * in Fig. 15.5) are very similar to the Bouchard and McGue figures. A model that included effects of maternal environment provided the best account of the observed correlations. Under that model, the estimate of h^2 in the narrow (additive) sense was 0.34, and the broader estimate of h^2 was 0.49.

In recent years, there has been a massive increase in research that utilizes twins to study a wide range of phenotypes. More than 1000 twin studies of various phenotypes have been published since 2004 (Polderman et al., 2015). The sizes of twin studies grew from the period 1950–79 when on average there were fewer than 160 pairs in a single study to the more recent studies after 2003 that averaged more than 1000 twin pairs per study. Vast resources, both human and financial, have been devoted to this work. The Polderman review considered only studies reporting trait correlations for MZ and DZ twins reared together in the same home, and the only observed data were r_{MZ} and r_{DZ}. There were 292 studies that measured some kind of cognitive function. The average value of r_{MZ} was 0.710, and r_{DZ} was 0.441. From these data, an estimate of $h^2 = 2(0.710 - 0.441) = 0.54$ is quite similar to the results for the Bouchard and McGue (1981) and Devlin et al. (1997) reviews. Gathering a very large sample of twin pairs from the literature does not address the fundamental design flaws of the research.

Why Heritability? Selective Breeding

Considering the resources devoted to discerning the heritability of IQ over more than 60 years, we are entitled to ask what it is that behavior geneticists want to learn about intelligence with such studies and why. It is well established that there is a scientific basis for selective breeding and the heritability coefficient is central to it. Eminent geneticists (Feldman & Lewontin, 1975) pointed out that h^2 does not tell anything about the plasticity of a phenotype or how it might change when environment is changed, but it is "an index of the amenability to selective breeding and, as such, is of practical use in the construction of breeding programs." Leaders of the field of behavior genetics usually said little about this kind of application. They were aghast when Jensen (1969) came right out and said heritability could be used to selectively breed humans. Jensen raised concerns about the supposedly higher fertility of Americans of African ancestry who had lower intelligence: "Is there a danger that current welfare policies, unaided by eugenic foresight, could lead to the genetic enslavement of a substantial segment of our population?" He was clearly advocating measures to curtail reproduction of the most disadvantaged sector of American society and cut government social supports for their nutrition, health care, and education. The firestorm of criticism and debate sparked by Jensen's article was still raging when the allegations about fraud by Cyril Burt, whose data were so central to Jensen's argument, fanned the flames. To many inside and outside academia, it appeared that racist academics had deliberately made up data to bolster their schemes to curtail aid for blacks and reinvigorate the eugenics movement.

Eugenic Sterilization

The only practical application of a heritability coefficient is to predict the response to selective breeding for a phenotype. Eugenic sterilization came into common practice in certain states in the United States and the Province of Alberta in Canada in the 1920s and then was implemented on a large scale in Nazi Germany.

The Alberta example reveals much about the inner workings of eugenics. The original purpose of the 1928 Alberta Sexual Sterilization Act was to prevent the reproduction of individuals who were believed to possess defective genes (Christian & Barker, 1974; Wahlsten, 1997; Wallace, 1934). In 1950, a government committee (Alberta Health Survey Committee, 1950) said of the Eugenics Board: "The object of this program is to reduce the level of hereditary mental defects." But no person competent in population genetics was appointed to the Eugenics Board or advised the Board on technical matters until May 27, 1960.

Much information came to light in Alberta when one of the young women ordered sterilized by the government in 1955 and then subjected to salpingectomy (removal of fallopian tubes) later sued for damages and won a decisive court ruling that condemned things done to boys and girls in the name of eugenics (Muir, 2014; Wahlsten, 1997). After that victory, hundreds of other people who had been sterilized as children came forward to seek damages. The author of this book was asked to serve as an expert witness in the second round of sterilization litigation and was given access to the complete institutional files for many children who had been confined to the Provincial Training School for Mental Defectives. The question posed by lawyers was whether the Alberta Eugenics Board applied the scientific standards that existed at the time in the 1950s when many of the children were sterilized.

The science of the time was adequate to make a good estimate of the likely consequences of the eugenics program (Wahlsten, 1999). It was possible to use available data to construct a realistic model of the situation in Alberta in 1955. Most children went before the Eugenics Board at or near the age of 14. There were 18,000 children aged 14 in the province in 1955. About 75 of them at that age were residents of the Provincial Training School for Mental Defectives (PTS), and 37 of them were in fact sterilized. (Those with very low mental functioning who would never be released into the community were not sterilized.) A normal distribution for 18,000 children with mean IQ of 100 and standard deviation of 15 can be constructed. Most of those sterilized had IQ scores of 70 or less, although two with scores of 73 and 77 also fell victim to the surgeon's knife. There would have been about 443 children with IQ < 70 in the province at the time, but the vast majority (about 406) were not in the PTS and instead were living at home with their families. Those in the PTS were mainly children who were not wanted or could not be supported by their families. The sum of the IQ scores for the 37 children was 2195, and the sum for all 18,000 would have been 1,800,000, so the sum for the 17,963 who were not sterilized would be $1,800,000 - 2195 = 1,797,805$. Dividing that figure by the number not sterilized yields a mean IQ of 100.084 and a selection differential of $S = 100.084 - 100 = 0.084$ IQ point. The response to selection ($R = h^2S$) would be half of this when $h^2 = 0.5$, so the expected mean IQ of the next generation of children would be *100.042*. The response would be greater if every one of the 443 children with IQ of 70 or less had been sterilized, which would have yielded a mean IQ of the next generation equal to 100.44. Faster progress in the war against bad genes would require far more people to be forcibly sterilized.

The conclusion from this exercise is that, according to genetic principles that were well understood in the 1950s, the likely effect on the average IQ of Albertans from a program of eugenic sterilization under the Sexual Sterilization Act would have been trivially small, an increase of 0.042 IQ point in 1 year. This is one of the major reasons why many reputable geneticists deserted the eugenics movement in the 1920s and 1930s after the genetics of human populations became better understood. Stoddard (1945) observed wryly that "...there has grown up a certain disjunction between the sober writing of geneticists and the expectation of eugenists."

Facile Dismissal of Environmental Influences

It is noteworthy that IQ testing played a critically important role in the eugenics program. It was the one fact that was relied on by the Alberta Eugenics Board in almost every case. It was routine practice to send a child for an IQ test shortly before the Board hearing when the sterilization order was made. Even if the child scored a little above 70, there was usually an order to sterilize. An exception was made for a girl who scored 116 on the Stanford-Binet; she was not sterilized. In addition to the IQ test report, the Board also had before it the full case

file for each child to be considered. Those files contained extensive information on the home environments of the children before they went to the PTS.

For an expert witness to review those files and then behold the terse orders for sterilization was a very disturbing experience. It appeared that the people in charge of the PTS and the members of the Eugenics Board were so rigidly bound to the genetic determination of intelligence that they dismissed obvious evidence of impoverished environments. Never in any case did a Board member argue against sterilization because of adverse conditions in the upbringing of a child. Just one example will illustrate the matter. Several names have been changed. Quotations and more complete details are provided by Wahlsten (1999).

Dora Anderchuk was born in 1940 in the small village of Weasel Creek, admitted to the PTS in 1951, and sterilized in 1954. Notes by the PTS staff indicated that she was "already an excellent worker and always very dependable." Documents about her home environment offer vivid portrayals of extreme deprivation.

July 8, 1949 (Royal Canadian Mounted Police report on the home). "There are about 5 or 6 small children ... and the parents living in a small one room shack. The shack is filthy and is practically without furniture. The mother and children are filthy and are practically without clothing ... The mother is a mental defective who cannot speak English. The father can talk English but does not appear to be overly bright ... *I have seen plenty of pretty poor homes in this and other districts, but this is one of the very worst and conditions can only be described as deplorable* ..." (emphasis added)

August 24, 1949 (W. Lefkowych, Welfare Inspector). "I don't believe I have ever seen anything like it before. The mother is a mental case. She has been in the Ponoka Mental Institution for quite a while. The husband took her out in 1944 knowing that her condition was not in any way bettered. I spoke to her for a few minutes, and she is definitely not in her right mind... She was only scantily dressed... She sometimes wanders away from home ... She does not wash or feed the children ... There have been four children since the mental condition set in ... These children should either be placed in a children's home or perhaps be looked after by relatives. *It is hard to believe that in the twentieth century, conditions such as these exist. There is evidence of severe neglect on part of both parents.*" (emphasis added)

November 10, 1949 (RCMP report on home). "The home was found to be in the filthiest condition imaginable. The children were only partially clothed... Two basins of slops sat on the stove that had been there for 7 days at least ... Mrs. ANDERCHUK's major problem has undoubtedly been the raising of her children, and their ages will give an indication of the strain the woman has undergone, and which no one can gainsay is the reason for her present mental condition. The children are as follows:

Staci   1 year
Ida   2 years
Hektor   4 years
Vonolia   5 years
Natan   8 years
Dora   9 years."

So, Dora was the oldest of six children and must have had the main responsibility for taking care of her younger siblings because her mother spent most of the time in a mental hospital. Dora would have missed many days of school because of this.

At the time of her Stanford-Binet test by Mr. Taylor, he commented, "Mid-grade moron. Rather tense—short of breath, 'sob breathing.' Poor in attention and memory." Results of any IQ test given under those conditions to a child overcome by anxiety would be unworthy of consideration. But for an inmate of the PTS who was already scheduled for surgical sterilization, test conditions did not matter.

Evidently, the eugenics creed blinded the Eugenics Board to the reality and significance of impoverished conditions in the homes of many children. It can be argued that an added practical application of genetics in the eugenics movement was to minimize or dismiss outright the role of environment in shaping the minds of children. That is why a single IQ score could be taken as a sign of a genetic defect, an inborn imperfection that some believed could never be alleviated.

STATISTICAL INTERLUDE: INFERENTIAL STATISTICS

Many of the tests of genetic effects to be presented here involve a decision rule that is used in *statistical inference*. In order to be taken seriously by other scientists, the results of a study should be able to pass a *significance test*. This is a major issue in the search for single-gene effects on behavior and needs some explanation. Scientific studies often compare two hypotheses about the data for two groups of people. The *null hypothesis* is that there was no treatment or genetic effect and group differences arose merely by chance. The *alternative* to the null is that there was a real effect. According to the practice of *null hypothesis significance testing*, an inference in favor of the alternative is made when the data warrant rejection of the null. The options in making an inference are portrayed in Table 15.5. An important feature of this approach is that the hypotheses and decision criteria are set *before the data are collected*. It is unwise to look at the data and then set the criteria for rejecting the null. This would open the door wide to many kinds of unruly biases and wishes.

TABLE 15.5 The Logic of Null Hypothesis Testing[a]

Null	Decision about null	Status of that decision	Probability of decision
True	Retain	Correct	$1 - \alpha$
True	Reject	False positive	Type I error Probability $= \alpha$
False	Retain	False negative	Type II error Probability $= \beta$
False	Reject	Correct	Power $= 1 - \beta$

[a] *Null hypothesis asserts that treatment had no real effect and difference between controls and treated groups arose merely from chance.*

Scientists in many fields of study have adopted a decision rule for judging the null hypothesis that sets the probability of a type I error equal to $\alpha = 0.05$, one chance in 20. The belief is that this criterion will render a correct decision in 95% of the instances when there really is no treatment effect. False positives should be rare. The statistical test compares the observed difference in group means to the hypothetical value of 0.

Multiple Hypothesis Tests

Only rarely will a research study involve just one hypothesis test. Often, it entails several experiments with different samples of subjects and several tests of a null hypothesis. This can inflate the probability of making one or more false claims of a significant effect. If several research teams are studying the same problem, false reports may proliferate. Consider a simple game played with a 20-sided die, the dodecahedron. When tossed, each face has the same chance of being at the top when the die comes to rest: 1 in 20. Let us define the event "20" as a significant "discovery" and events "1" through "19" as nondiscoveries. The discovery will of course be a false discovery because the outcomes are purely a matter of random bounces of the die. Nevertheless, this simple device simulates quite well the perils of null hypothesis testing. Now and then, there will be apparent discoveries that have no real basis apart from chance outcomes. This is implicit in the practice of setting a criterion for rejecting the null such that probability of type I error is $\alpha = 0.05$.

Next, let us toss two such dice and let them come to rest. We would like to know the chances of making one or more "discoveries." This will be equal to one minus the probability of making no discovery with either die. For one die, the probability of no discovery is 19/20. When the two dice are independent, probabilities multiply, and the probability of neither coming up "20" is $(19/20)(19/20) = 361/400 = 0.9025$. Therefore, the probability of one or two "20s" is $1 - 0.9025 = 0.0975$, which is almost double $\alpha = 0.05$. In general, if we conduct J independent tests of the null, the probability of one or more false discoveries will be $1-(19/20)^J$. Suppose one study involves $J=10$ such tests, a plausible quantity that is often seen in published work. Then, $1-(19/20)^{10}=0.598$. Even within the confines of a single study, there would likely be at least one false-positive rejection of a true null hypothesis.

For a popular topic of research such as genes and behavior, there are often several labs in the hunt for a new gene that changes behavior. Their work will certainly be done independently of each other, so if there are K such labs, the probability that one or more experiments result in a false discovery will be $1-(19/20)^K$ if each does just one test and $1-(19/20)^{JK}$ if each lab does J tests. Suppose that each lab does $J=5$ tests of the null and there are $K=5$ labs working on the same topic. Probability of one or more false positives will be $1-0.277=0.723$.

This is a major problem in all kinds of research but especially in genetic research where one study involves tests of possible differences in SNPs near many genes on different chromosomes, and labs sometimes race to be the first to make a discovery. It is essential that this problem be addressed by using a more stringent criterion for type I error probability. One common recourse is the Bonferroni adjustment that divides α by J when there are J independent tests and uses $\alpha/J=0.05/J$ as the criterion for setting a critical value of Z or other test statistics. For example, if one plans to do 10 tests of the null in a single study, it is wise to use $\alpha=0.005$. In the case of 10 tosses of the dodecahedron using $\alpha=0.005$, the probability of one or more false positives would then be $1-(0.995)^{10}=1-0.951=0.049$, which is very close to 0.05. A recent consideration of these issues observed that the false-positive rate in psychological research is so high that the criterion for significance should routinely be $\alpha=0.005$ even within a single study that does just one significance test (Benjamin et al., 2018).

Thus, null hypothesis significance testing is an error-prone enterprise, but the frequency of errors of inference can be reduced considerably by employing the Bonferroni adjustment to the significance criterion. Even then, any one study can lead us to a wrong conclusion, an exciting "discovery" that makes a short-lived splash in news media. The more stringent test of truth is whether a lovely result can be replicated in the same lab and, even more importantly, by other researchers working independently. No single experiment by even the best funded and most brilliant researcher can establish the truth or falsity of a scientific hypothesis.

SINGLE GENES AND IQ

After the successes of mapping and characterizing the HTT gene in 1993 (Fig. 14.2) and many other major neurological disorders, including the NXRN1 and CNTN6

genes that can generate Tourette syndrome, many researchers in behavior genetics sensed that there were now enough DNA markers to warrant an intensive search for genes having effects on intelligence test scores in the normal range of variation. The main problem was that the effect size of such genes would likely be considerably smaller than those known to generate highly abnormal traits. How much smaller was unknown, and there was no way to find out but to start the search.

Quest for Major Genes

The first apparent success in the quest (Chorney et al., 1998) was published in *Psychological Science*, a journal directed primarily to psychologists who know little of the intricacies of genetics. The authors examined 37 DNA markers on the q arm of chromosome 6 and found one in the vicinity of the insulin-like growth factor 2 receptor (IGF2R gene) where alleles were associated with IQ scores. Allele 4 of the marker occurred in 66% of people with very high IQ and 81% of people in a control group. The observed effect size was 4 IQ points or about $D=0.25$, just barely large enough to cause the authors to reject the null that IGF2R alleles did not alter IQ. They used the uncorrected $\alpha=0.05$ criterion for significance. Two nearby markers (D6S1550 and D6S411) showed no association with IQ. Looking more closely at that region on chromosome 6, it was seen that the marker associated with IQ was not actually in the IGF2R gene; it was in a region that was not translated into protein. The association was with a marker, not an actual gene and certainly not a gene "for" IQ. The authors stressed the need to replicate the finding but speculated that this kind of work "will revolutionize research on cognition" and worried about possible implications of genetic screening that could affect insurance and employment.

The mass media celebrated the report as "the first gene to be linked with high intelligence" (Wade, 1998), a finding that is "a tender green shoot arising from the ashes of long-smoldering debate about whether intelligence is determined by people's genes or by the circumstances of their upbringing." Robert Plomin, lead author of the study, told the press that this discovery should end years of argument over whether genes can affect intelligence because "It is harder to argue with a piece of DNA" (Highfield, 1998). A review article (Lubbock, 2000) cited this study as showing it is "well established that people with this variant of the normal gene are far more intelligent than average."

The celebration was cut short after Plomin and his team sought to replicate the finding 4 years later (Hill, Chorney, Lubinski, Thompson, & Plomin, 2002). This time, for reasons that are not at all clear, they decided to focus on allele 5, not 4 that was featured in 1998.

In 1998, 32% of the high IQ group had allele 5 versus 16% of controls, whereas in the new sample, allele 5 appeared in 19% of high IQ people and 24% of controls, a clearly nonsignificant difference. The IGF2R effect was not replicated, and the earlier claim was withdrawn.

This pattern of a claim of an association of a gene with IQ followed by celebration and later by retraction was repeated over and over. For example, a report of IQ related to alleles of the CHRM2 gene (cholinergic muscarinic 2 receptor) also found an effect size of about 4 IQ points in a sample of 828 people (Comings et al., 2003). This and other similar findings were not replicated. Reviews of the burgeoning literature on single genes related to IQ sang the same old dirge:

Gray and Thompson (2004): "none of these associations has yet been replicated."

Posthuma and de Geus (2006): "poor replication … is a common concern."

Meanwhile, the number of genetic markers distributed widely across the genome grew rapidly, and amazing chips with thousands of DNA marker sequences embedded in them made it possible to scan broad regions of all chromosomes for alleles related to quantitative phenotypes such as IQ.

The immediate impact of the new technology, however, was to demonstrate that, as one group of researchers put matters, "most reported genetic associations with general intelligence are probably false positives" (Chabris et al., 2012). Previous claims could not be verified by the newer and more powerful technology. To emphasize this point, one reviewer placed two charts of the chromosomes side by side, one showing a plot of 12 prior claims of single-gene effects on IQ and the other showing the situation after the new chromosome scan technology was applied (Wahlsten, 2012). The more recent chart was blank. None of the earlier claims had been replicated.

GWAS

Large contingents of researchers are now engaged in an intense hunt for specific genes that they believe give rise to differences in intelligence. The principal method for this work is the genome-wide association study (GWAS) that scans all the chromosomes for possible relationships of a SNP to a gene that somehow influences intelligence. Geneticists pronounce the abbreviviation SNP as "a snip" rather than "an es-en-pee." The quest begins with measures of IQ in a large sample of people in a well-defined population. It is believed that the IQ score is influenced by many genes and environmental factors that combine in unknown ways to change the phenotype.

The method presumes that here and there in exons across all the chromosomes, there are genes with two or more alleles that code for proteins involved in brain development or other functions that pertain to mental abilities, perhaps the regulation of the expression of important nervous system genes. When there are two alleles of such a gene, one would increase the average IQ score by a small amount, while the other would tend to reduce it a little in comparison with the overall average test score. If a SNP allele is associated with IQ score, an inference is made that a real gene influencing intelligence is located close to the SNP. In conventional linkage analysis, patterns of the SNPs and the phenotype of interest are compared in close relatives. In GWAS studies, on the other hand, a sample of people from the entire population in one generation is scanned for statistical associations. The GWAS method requires very large samples of SNPs and people who have been phenotyped, and it involves sophisticated data analysis that is usually performed by teams of mathematical biologists. The history of this approach has been lucidly outlined by Flint, Greenspan, and Kendler (2010). The method has had success in identifying a number of genes with small to moderate effects that contribute to biomedical diseases such as cystic fibrosis and types 1 and 2 diabetes.

One exceedingly important issue arises when so many statistical associations of IQ with millions of SNPs are assessed. If $\alpha = 0.005$ was used as the type I error criterion in a GWAS study, thousands of false-positive "genes" would be detected. Accordingly, statisticians determined through math and data simulation that a criterion for any one null hypothesis test in a GWAS study should be $\alpha = 5 \times 10^{-8}$ or $\alpha = 0.00000005$, one chance in 50 million. A false-positive "discovery" of a gene related to IQ can still occur, but there should not be very many of them.

Table 15.6 summarizes several recent studies that have used increasingly larger sample sizes. One major challenge is immediately apparent. In order to assemble enough people for the genome scan, it was necessary to combine data from separate samples that were collected in different places, often by different research teams using different mental tests. Furthermore, in some instances, samples involved in an earlier study were combined with newer data in a subsequent study in order to boost the sample size, so results of the studies are cumulative, not independent.

Results of the first two genome scans were disappointing. Even with more than 17,000 people in the study, not even one SNP out of more than 2 million tested met the preset criterion for rejecting the null. Authors then conducted a second analysis (genome-wide gene association study (GWGAS)) of the same set of SNPs but limiting attention to just SNPs located within regions of known genes. Because that scan involved far fewer tests of the null, the criterion for a significant association with IQ

TABLE 15.6 Summaries of Recent GWAS Studies of IQ

First author	Davies*	Benyamin*	Davies*	Davies	Sniekers*	Zabaneh
Year	2011	2014	2015	2016	2017	2017
Number of people	3511	17,989	53,949	112,151	78,308	8185
Ethnicity	Caucasian England Scotland	European England Australia	European	White British	White European	White Caucasian
Number of cohorts pooled	5	6	31	UK Biobank	13	2 (the United States)
Number of IQ indicators	9	9	Different test batteries	None	"Various tests"	SAT
Number of SNPs	549,692	2,611,179	2,478,500	?	12,104,294	6,773,587
Significant SNPs in GWAS	0	0	13	149	336	1
Genes/SNP clusters	0	0	3	10/3	18/13	0
GWGAS genes	1 or 0	1 FNBP1L	1 HMGN1	17	46	0
Polygenic prediction	$R^2 = 1\%$	0.5%–3.5%	1.3%	2.8%	4.8%	1.6%–2.4%

Certain studies added new samples to people who had been studied previously, so that the results for studies marked with the asterisk (*) are not independent and also shared some of the same authors.

was relaxed to $P = 3 \times 10^{-6}$. There was one significant "hit" in the FNBP1L gene that was not replicated in future studies.

Only when the total sample size exceeded 50,000 people were several significant SNPs detected (Davies et al., 2015). The fact that they occurred in three clusters supported the hypothesis that there was a real gene somewhere in the cluster that affected IQ. But the so-called cluster on chromosome 19 was not a cluster at all. Just one SNP met the criterion for significance, while others nearby were not even close to that criterion. Furthermore, that one SNP was close to three genes—TOMM40, APOE, and APOC1. There was no evidence to favor one over the other. If a researcher is faced with a choice of three or more genes that might be responsible for the SNP effect, she certainly cannot claim discovery of a gene influencing IQ. Much more work needs to be done in support of that claim. On chromosome 6, the rs10457441 marker was not in any gene. It was about 100 kb away from an obscure RNA encoding gene, and there were SNPs much closer to that gene that did not show a significant association with IQ. So, there was no basis for claiming the MIR2113 gene is an IQ-related gene. Mercifully, there was a definite cluster in chromosome band 14q12 that appeared to be located in the AKAP6 gene (A-kinase anchoring protein 6). No alleles of the gene itself were identified and correlated with IQ, so conclusive proof that AKAP6 harbored a polymorphism that altered IQ was lacking. In summary, the Davies et al. (2015) study provided suggestive evidence for one gene that might affect IQ score. One valuable fact about effect size was mentioned in the report. Significant SNPs appeared to be responsible for a difference of $D = 0.05$ standard deviation or about 0.75 IQ point.

When the sample size was boosted by adding more than 24,000 people to previous samples (Davies et al., 2015), there were many more SNPs showing statistical significance (Sniekers et al., 2017), and several occurred in tightly packed clusters that left little room for doubt that the association with IQ was real. Study authors chose to place each of them in the tally of genes influencing IQ, but a closer look reveals some concerns. For certain SNPs, the cluster was not in the range of any known gene and instead occupied the so-called intergenic region between genes. For example, the cluster around SNP rs2251499 was about 200 kb away from the closest gene, whereas there were no SNPs that were even suggestive within the limits of the nearest gene itself. It is conceivable that the SNP cluster marked some kind of regulatory region that affected expression levels of another gene further away. Which gene that might be was unknown. Five of the 18 SNP clusters were intergenic and implicated no specific gene.

Table 15.7 cites genes implicated by the two largest studies using the GWAS and GWGAS methods. The harvest of genes was much richer for the GWGAS method that considered only SNPs within the limits of a known gene and used a much more lenient criterion for significance (2.7×10^{-6}), 200 times larger than the criterion of 5×10^{-8} for GWAS that assessed all SNPs. The immense sample of more than 100,000 people amassed by Davies et al. (2016) was largely independent from the Sniekers et al. (2017) sample, and the GWAS uncovered evidence of significant SNPs involving nine genes on three

TABLE 15.7 Genes Located Near Significant SNPs in Two Large GWAS Studies[a]

Chr	Davies et al. (2016) N = 112,151		Sniekers et al. (2017) N = 78,308	
	GWAS	GWGAS	GWAS	GWGAS
1				COL16A1, HCRTR1, PEF1, WNT4
2		ZNF638[b]	ARHGAP15, ZNF638	ARHGAP15 ZNF638
3				LPGK1, NK1RAS1, RNF123, RPL15
4		CCDC149		YIPF7
5			MEF2C	GRK6, PRR7
6			FOXO3	FOXO3
7	PDE1C		EXOC4, PDE1C	EXOC4, PDE1C
9		APBA1,[b] CRAT, PPP2R4	APBA1	APBA1
10		GBF1[b]		GBF1, JMJD1C
12				DDN, EEA1
14	FUT8			DCAF5
15				EFTUD1
16		ATXN2L,[b] SH2B1,[b] NPIPB8, XPO6	ATXN2L, SH2B1, ATP2A1, TUFM	ATXN2L, SH2B1, ATP2A1, TUFM, APOBR, CCDC101, ZFEX3
17				SEPT4
18				DCC, ZNF407
20			CSEIL	CSEIL, ARF6EF2, STAO1
22	CYP2D6,[b] NAGA,[b] NDUFA6,[b] SEPT3,[b] TCF20,[b] WBP2NL,[b] FAM109B[b]	CYP2D6, NAGA, NDUFA6, SEPT3, TCF20, WBP2NL, FAM109B	CYP2D6, NAGA, NDUFA6, SEPT3, TCF20, WBP2NL	CYP2D6, NAGA, NDUFA6, SEPT3, TCF20, WBP2NL, DRG1, PIK31P1, RNF185, SHANK3

[a] Full gene names are provided in OMIM and GeneCards websites.
[b] Denotes gene detected in both studies.

chromosomes. One gene (FUT8) on chromosome 14 was novel, while seven others were also flagged by Sniekers et al. Six of them were in the cluster of seven genes on chromosome 22. This is not proof that they are indeed functional genes influencing IQ but it does make them good bets. Also promising was phosphodiesterase 1C (PDE1C) on chromosome 7.

The more lenient GWGAS criterion resulted in a harvest of 46 genes in the Sniekers sample, several of which were also detected using GWAS or also seen by Davies et al. (2016), which bolstered the case for their relevance to IQ. Among the 52 genes in at least one of the four categories in Table 15.7, 34 were identified in only one category and were unsupported by other evidence in the studies. This does not necessarily prove that the isolated cases were false positives, but some could well have been. The authors tallied any gene that was flagged in any analysis as a hit, a gene implicated in human intelligence. The figure of 52 genes was the one conveyed to the media and celebrated by NBC News as a major breakthrough (Fox, 2017).

Given the history of false-positive results that have littered the scientific journals in this field of study, greater rigor and caution are warranted. It is reasonable to ask that three basic criteria be met before declaring a specific gene is related to IQ in a way that indicates causation. First, it should be closely associated in a clearly significant way with a cluster of SNPs in the same region as the gene. Second, that association should be replicated at least once in separate samples of people. Third, there should be clear evidence that the gene has two or more alleles that alter the amino acid sequence of a protein that might reasonably be involved in brain function or some other processes that could alter intelligence. Researchers would need to be somewhat flexible on this third stipulation because so many things could alter the nervous system in a surprising way (e.g., PKU). The first criterion has been achieved for 52 SNP clusters in Table 15.7. The second has been met for 11 genes that scored hits in both samples. The third has not yet been met at all. Thus, no bona fide gene having two or more functional alleles associated with IQ score has yet been identified through GWAS or GWGAS.

We must resist the temptation to relax the criteria for gene discovery. The criteria have been well established in GWAS research since an influential review in 2008 (McCarthy et al., 2008). They viewed SNPs that meet a criterion for significance as "association signals," not proof of a causal connection with a phenotype. They strongly emphasized "the preeminent role of replication" and noted "it is important to use independent replication samples" involving the same alleles of an SNP and the same phenotype, not some loosely specified proxy indicator. The isolated "hits" in Table 15.7 certainly did not meet criteria for discovery of a new "IQ gene."

Effect Sizes

Statistical significance of a test of a SNP tells only whether it may be relevant to IQ. The far more important question is the size of the effect expressed either as D or IQ points. The Davies et al. (2015) study suggested that individual SNP effect sizes are sometimes around $D = 0.05$ standard deviation, equivalent to a difference of about 0.75 IQ point. In supplementary Table S4 of Sniekers et al. (2017), many effect sizes of individual SNPs were in the range of 0.05%–0.07% of phenotypic variances, equivalent to about 0.7–0.8 IQ point. By the same standards applied to studies of environmental effects on IQ, these would be regarded as trivially small. Only when the combined effects of many such genes are considered could there be an effect worthy of note.

Polygenic Prediction

Another approach to exploring effect size is to see how much of the variation in IQ of one sample of people is associated with predictions from a linear equation fit to data for another sample. The multiple regression equation would have the form $Y = B_0 + B_1 X_1 + B_2 X_2 + B_3 X_3 \ldots$, where each X is either 0, 1, or 2, depending on how many SNP alleles are possessed by a person for each locus, and B is the regression coefficient for predicting IQ for that one gene from the data. The common practice in SNP studies is to fit a model with hundreds or even thousands of X terms. The investigators would have no idea what genes are involved, just what IQ effect size might be associated with the marker alleles. Table 15.6 shows percentages of variance in IQ associated with the polygenic prediction equation. These are far less than the typical value of heritability of $h^2 = 50\%$ and by any criterion are trivially small.

At this stage in research on single-gene influences on intelligence, it is evident that the genetic pieces of the puzzle arise from large numbers of genes, each with very small effect and a very small combined effect as well. At the present time, there is not even one gene where alleles causing differences in gene function in the normal range of IQ variation are well documented and widely accepted among expert geneticists.

INTELLECTUAL DISABILITY

When intelligence test scores occur below a threshold of about 70 IQ points, this may be viewed as an abnormal state (Vissers, Gilissen, & Veltman, 2016), even though there are no external physical signs of malformation. Three diagnostic categories (idiot, imbecile, and moron) proposed by Doll (1936) and once used in some public institutions for the mentally disabled had already become part of the popular lexicon in North America as vulgar terms of abuse. Terminology gradually evolved to be less pejorative, moving from mentally defective to retarded, to deficient, to disabled, most recently in DSM-5 to the preferred term "intellectual disability." In many research projects, intellectual disability (ID) is defined as IQ < 70, and severe ID is IQ < 50 or 35. About 0.5% of people in many countries are judged severely intellectually disabled, and from 1% to 3% are termed simply ID (Maulik, Mascarenhas, Mathers, Dua, & Saxena, 2011; Vissers et al., 2016). Thus, intellectual disability in the absence of significant physical disability is far more common than Down syndrome or sex chromosome anomalies.

There appears to be an important distinction between cases of ID not far below the IQ criterion of 70 and those with IQ < 35. In Sweden and Israel where all young men took a mental ability test as part of universal military service, researchers were able to identify more than 400,000 nontwin brothers and 8700 male twins where at least 1 had intellectual disability (Reichenberg et al., 2016). If the genetic part of the process works in the same way for borderline cases of IQ and severe cases, average test scores of the brothers should be much lower if the index brother showed severe ID. The actual pattern was just the opposite. For a brother not too far below IQ = 70, the other brothers averaged well below IQ = 100 but above 70, whereas for a pair where one brother was IQ = 35 or less, the other brother commonly had a normal IQ score. This suggests that severe ID has different causes than mild ID. There could have been a major gene mutation or serious difficulties before or during birth for those with IQ < 35, whereas those closer to IQ = 70 may represent polygenic influences of many genes, each with very small effect combined with many environmental influences. Prospects for detecting a major gene effect when IQ of a child is less than 35 involve an effect size of the gene of more than $D = 3.0$ or 45 IQ points, a dramatically large effect compared with polygenic effects that usually are 0.75 IQ point or less for one SNP.

The search for genes pertinent to severe ID has been very productive. The design of the research usually involves parent-child trios—two parents and one child, where the child shows severe ID. In almost all cases, the parents appear quite normal but could be carrying a defective gene. If they both carry a recessive mutation, for example, the child might be homozygous for a recessive mutation and suffer impaired development. Alternatively, neither parent may carry any significant genetic defect, and the situation in the ID child may arise *de novo* through a new mutation. Given the likelihood of a mutation having a large effect in ID, the search is not done with SNPs that so often reside in an intron or in the vast regions between genes. Instead, researchers opt for *whole-exome sequencing* that determines the entire nucleotide base sequence of just the exons for thousands of genes. Because the sequence of the entire human genome is known, computerized comparisons of sequences for one child can quickly highlight places where the code departs from typical. The computer can look up the amino acid encoded by the altered sequence (Table 2.2) and then predict whether the change would likely alter function of the protein. If the change is *synonymous*, the DNA will code for the same protein, and there will be no change in the amino acid sequence. Table 2.2 indicates there are usually several codons that specify the same amino acid. A *nonsynonymous* change would alter an amino acid at a specific position in the protein. Because there are long portions of some proteins that are not part of the active site where an enzyme binds to a substrate, there are instances where a change in one amino acid may be nonsynonymous but is not likely to generate pathology. The computer can make this judgment, but it needs to be double-checked by a scientist. If the change is likely to be pathogenic, then the mutation is compared with tabulations of known mutations in that gene. If it is not on the official list, it is a novel mutation.

One of the first exon sequencing studies of ID identified 87 *de novo* variants among 51 cases of severe ID and concluded 6 of them represented novel pathogenic ID genes and 16 known ID genes (Rauch et al., 2012).

A larger sample of 100 ID cases yielded 79 *de novo* mutations and about 22 novel candidate genes (de Ligt et al., 2012). Fewer than half of the ID cases in that study yielded a diagnosis based on a genetic defect, so another group (Gilissen et al., 2014) used improved technology to do a follow-up study of 50 people not diagnosed with a genetic defect in the de Ligt et al. (2012) study. Among those 50, they found 84 *de novo* variants, 65 of which were missed in the original study, and they identified another 8 candidate ID genes. Combining the data from the two studies, more than half of the cases received some kind of genetic diagnosis pertinent to their ID. All of them involved a genuine difference in DNA sequence in an exon that altered protein structure (Table 15.8). Reviewing these and other exome sequencing studies, Vissers et al. (2016) commented that "establishing that a candidate ID gene is an ID-causing gene when mutated is still complicated."

Studies of ID published in the 1970s were able to identify a genetic diagnosis in only about 4% of cases, whereas by 2015, the newer methods were able to diagnose more than 60% of severe ID cases. The list of known genes giving rise to ID or likely to cause it has grown to more than 700, including more than 100 X-linked recessive genes, more than 300 autosomal recessive genes, and more than 300 autosomal dominant genes (Vissers et al., 2016). Despite the large number of pathogenic genes, their frequencies in the population are very low because children with severe ID are rarely able to become adults capable of reproduction. In recent studies, the vast majority of genes detected that caused ID arose *de novo* and were absent in both parents. For the majority of those genes, the pathogenic allele was detected in *just one case* and most likely would not be passed to the next generation. ID is highly heterogeneous genetically. Consequently, a genetic

TABLE 15.8   ID-Related Genes Detected in Parent-Child Trios Using Exome Sequencing[a]

Type of mutation	Known genes	Novel ID genes	Candidate ID genes
de Ligt et al. (2012); $N=100$ patients with IQ $<$ 50 plus their parents			
Missense	ARFGEF2, GRIN2A, TCF4, TUSC3	DYNC1H1	ASH1L, CAMKIIG, COL4A3BP, EEF1A2, GRIA1, KIF2C, LRP1, MIB1, PHACTR1, PPP2R5D, PROX2, PSMA7, RAPGEF1, TANC2, TNPO2, TRIO
Nonsense	SCN2A[b]	GATAD2B	PHIP, WAC
Frameshift	LRP2, PDHA1, SLC6A8, TUBA1A	CTNNB1	MTF1, ZMYM6
Rauch et al. (2012); $N=51$ patients			
Missense	NAA10, SATB2, SCN2A[b], SCN8A, SLC2A1, STXBP1, TCF4		CUX2, DEAF1, EIF2C1, KCNQ3, STAG1, ZNF238
Nonsense	IQSEC2, SETBP1		SETD5
Frameshift	MECP2, SCN2A[b], SYNGAP1		ARIH1, CHD2, HIVEP2, SLC6A1, SYNCRIP

[a] *Full gene names are provided in OMIM and GeneCards websites.*
[b] *Gene detected in both studies.*

screening program for genes related to ID would likely be of little value. A mutation observed only once becomes a candidate gene that needs to be confirmed as the cause of other ID cases before it can be regarded as a genuine ID gene. Many of the gene effects included on the long list of *de novo* mutations have not yet been replicated.

A few genes pertinent to ID have been explored in greater depth. A recent study (Dias et al., 2016) documented 11 cases of ID that arose from a mutation in the BCL11A (B-cell lymphoma/leukemia 11A) gene. Each of the mutations involved a different part of the same gene, and apart from the ID, phenotypes were highly variable. Five mutations were small deletions or insertions of a bit of DNA that shifted the frame for transcribing the RNA and caused a total loss of function (LOF) of the protein. Three involved a change of a codon to a nonsense code that stopped the synthesis of the protein and also caused LOF. Three others were missense mutations that changed one of the amino acids in an important part of the protein that greatly impaired its function. In each instance, the protein from the mutant allele was nonfunctional, and the child's one good copy could not compensate for the one botched allele.

Contrasting IQ Genes and ID Genes

Neither of these terms is valid. There is no gene that codes for IQ or even a tiny fraction of it, nor is there a gene coding for ID. The terms "ID gene" and "IQ gene" are just shorthand ways of saying that different alleles of a gene have been shown to cause a difference in the phenotype under certain conditions. For the so-called IQ genes, even this has not yet been demonstrated. Only statistical association has been shown. There are many genes that can alter mental functioning through a diversity of mechanisms. Most of them also influence other phenotypes. Those pertinent to severe ID typically have very large effects and may disable a gene of vital importance for brain or bodily function, while those pertinent to IQ variations in the normal range have exceedingly small effects that leave much room for doubt about their roles in many instances. Under the criteria applied when considering ID genes, no so-called IQ gene identified in recent GWAS studies would even make it to the first stage where a base change in an exon has been detected.

HIGHLIGHTS

- Psychologists have invented several kinds of intelligence tests that are in common use today, but they have not been able to agree of a definition of intelligence or a consensus version of a test.
- The inventors of most early tests believed they were measuring an innate, genetically determined ability.
- Intelligence is a phenotype, and factors that influence test scores need to be discovered through research, not asserted on the basis of ideology.
- In many countries, the level of intelligence in the entire population has been gradually increasing for several decades at a rate of about 3 IQ points per decade.
- Adoption of a child into a better home environment can increase IQ score by 8–15 points.
- Several reviews of twin studies have noted heritabilities of IQ close to $h^2 = 0.5$.
- Much higher values of h^2 reported by Cyril Burt were products of scientific misconduct and outright deception.
- Heritability can be used to anticipate the change of a phenotype affected by selective breeding. The size of the change in certain eugenic programs turned out to be trivially small (e.g., 0.04 IQ point in Alberta, Canada).
- Among the 12 genes implicated in IQ differences between 1998 and 2010, not 1 could be replicated.
- Genome-wide scans using millions of SNPs have detected a few dozen markers associated with IQ, several of which have been replicated, but their effect sizes are very small, about 0.75 IQ point.
- Major intellectual disability involving IQ scores below 40 has been traced to a major genetic mutation in several dozen genes using exome sequencing. In most cases, neither parent carries the mutation, and there is just one child who shows a particular kind of mutation.
- Most cases of intellectual disability are generated anew each generation through *de novo* mutations involving a wide diversity of genes.

References

Achenbach, T. M., Howell, C. T., Aoki, M. F., & Rauh, V. A. (1993). Nine-year outcome of the Vermont intervention program for low birth weight infants. *Pediatrics, 91,* 45–55.

Alberta Health Survey Committee (1950). *A survey of Alberta's health.* Edmonton, AB, Canada: Alberta Health Survey Committee.

Becker, K. A. (2003). *History of the stanford-binet intelligence scales: Content and psychometrics* (5th ed.). Itasca, IL: Stanford-Binet Intelligence Scales. Assessment service bulletin number 1.

Benjamin, D. J., Berger, J. O., Johannesson, M., Nosek, B. A., Wagenmakers, E.-J., Berk, R., … Camerer, C. (2018). Redefine statistical significance. *Nature Human Behaviour, 2,* 6–10.

Binet, A., & Simon, T. (1908). Le dévelopment de lintelligence chez les enfants. *L'Année Psychologique, 14,* 1–94.

Binet, A., & Simon, T. H. (1905). New methods for the diagnosis of the intellectual level of subnormals. *L'Année Psychologique, 12,* 191–244.

Bisanz, J., Morrison, F. J., & Dunn, M. (1995). Effects of age and schooling on the acquisition of elementary quantitative skills. *Developmental Psychology, 31,* 221–236.

Bouchard, T. J., & McGue, M. (1981). Familial studies of intelligence: a review. *Science, 212,* 1055–1059.

Brooks-Gunn, J., Klebanov, P. K., Liaw, F. r., & Spiker, D. (1993). Enhancing the development of low-birthweight, premature infants: changes in cognition and behavior over the first three years. *Child Development, 64*, 736–753.

Burt, C. (1909). Experimental tests of general intelligence. *British Journal of Psychology, 3*, 94–177.

Burt, C. (1972). Inheritance of general intelligence. *American Psychologist, 27*, 175–190.

Burt, C., Jones, E., Miller, E., & Moodie, W. (1934). *How the mind works.* New York: D. Appleton-Century Company.

Campbell, F. A., Pungello, E. P., Burchinal, M., Kainz, K., Pan, Y., Wasik, B. H., … Ramey, C. T. (2012). Adult outcomes as a function of an early childhood educational program: an Abecedarian Project follow-up. *Developmental Psychology, 48*, 1033–1043.

Campbell, F. A., & Ramey, C. T. (1994). Effects of early intervention on intellectual and academic achievement: a follow-up study of children from low-income families. *Child Development, 65*, 684–698.

Capron, C., & Duyme, M. (1989). Assessment of effects of socio-economic status on IQ in a full cross-fostering study. *Nature, 340*, 552–554.

Carlson, J. F., Geisinger, K. F., & Jonson, J. L. (Eds.), (2017). *The twentieth mental measurements yearbook.* Lincoln, NE: Buros Center for Testing.

Chabris, C. F., Hebert, B. M., Benjamin, D. J., Beauchamp, J., Cesarini, D., van der Loos, M., … Laibson, D. (2012). Most reported genetic associations with general intelligence are probably false positives. *Psychological Science, 23*, 1314–1323.

Chase, A. (1977). *The legacy of malthus: The social costs of the new scientific racism.* New York: Knopf.

Chorney, M. J., Chorney, K., Seese, N., Owen, M. J., McGuffin, P., Daniels, J., … Plomin, R. (1998). A quantitative trait locus associated with cognitive ability in children. *Psychological Science, 9*, 159–166.

Christian, T., & Barker, B. M. (1974). *The mentally Ill and human rights in Alberta: A study of the Alberta sexual sterilization act.* Edmonton, AB, Canada: Faculty of Law, University of Alberta.

Clarke, A., & Clarke, A. (1974). Experimental studies: an overview. In A. Clarke, & A. Clarke (Eds.), *Mental deficiency: The changing outlook* (3rd ed., pp. 259–329) London: Methuen.

Coleman, J. S., Campbell, E. Q., Hobson, C. J., McPartland, J., Mood, A. M., Weinfeld, F. D., & York, R. L. (1966). *Equality of educational opportunity.* Washington, DC: National Center for Educational Statistics.

Comings, D., Wu, S., Rostamkhani, M., McGue, M., Cheng, L. S., & Macmurray, J. (2003). Role of the cholinergic muscarinic 2 receptor (CHRM2) gene in cognition. *Molecular Psychiatry, 8*, 10–11.

Davies, G., Armstrong, N., Bis, J. C., Bressler, J., Chouraki, V., Giddaluru, S., … Deary, I. J. (2015). Genetic contributions to variation in general cognitive function: a meta-analysis of genome-wide association studies in the CHARGE consortium (N=53949). *Molecular Psychiatry, 20*, 183–192.

Davies, G., Marioni, R. E., Liewald, D. C., Hill, W. D., Hagenaars, S. P., Harris, S. E., … Lyall, D. (2016). Genome-wide association study of cognitive functions and educational attainment in UK Biobank (N= 112 151). *Molecular Psychiatry, 21*, 758–767.

de Ligt, J., Willemsen, M. H., van Bon, B. W., Kleefstra, T., Yntema, H. G., Kroes, T., … Vissers, L. E. (2012). Diagnostic exome sequencing in persons with severe intellectual disability. *New England Journal of Medicine, 367*, 1921–1929.

Dempster, J. J. (1954). Symposium on the effects of coaching and practice in intelligence tests. *British Journal of Educational Psychology, 24*, 1–4.

Devlin, B., Daniels, M., & Roeder, K. (1997). The heritability of IQ. *Nature, 388*, 468–471.

Dias, C., Estruch, S. B., Graham, S. A., McRae, J., Sawiak, S. J., Hurst, J. A., … Turner, C. (2016). BCL11A haploinsufficiency causes an intellectual disability syndrome and dysregulates transcription. *The American Journal of Human Genetics, 99*, 253–274.

Doll, E. A. (1936). Idiot, imbecile, and moron. *Journal of Applied Psychology, 20*, 427–437.

Dorfman, D. D. (1978). The Cyril Burt question: new findings. *Science, 201*, 1177–1186.

Erlenmeyer-Kimling, L., & Jarvik, L. F. (1963). Genetics and intelligence: a review. *Science, 142*, 1477–1479.

Fancher, R. E. (1985). *The intelligence men: Makers of the IQ controversy.* New York: Norton.

Feldman, M. W., & Lewontin, R. C. (1975). The heritability hang-up. *Science, 190*, 1163–1168.

Ferreira, F., & Morrison, F. J. (1994). Children's metalinguistic knowledge of syntactic constituents: effects of age and schooling. *Developmental Psychology, 30*, 663–678.

Flint, J., Greenspan, R. J., & Kendler, K. S. (2010). *How genes influence behavior.* Oxford, UK: Oxford University Press.

Flynn, J. R. (1984a). IQ gains and the Binet decrements. *Journal of Educational Measurement, 21*, 283–290.

Flynn, J. R. (1984b). The mean IQ of Americans: massive gains 1932 to 1978. *Psychological Bulletin, 95*, 29–51.

Flynn, J. R. (1987). Massive IQ gains in 14 nations: what IQ tests really measure. *Psychological Bulletin, 101*, 171–191.

Flynn, J. R. (2007). *What is intelligence?: Beyond the flynn effect.* New York: Cambridge University Press.

Fox, M. (2017). Forty more genes for intelligence discovered. *NBC News.* May 22.

Gilissen, C., Hehir-Kwa, J. Y., Thung, D. T., van de Vorst, M., van Bon, B. W., Willemsen, M. H., … Schenck, A. (2014). Genome sequencing identifies major causes of severe intellectual disability. *Nature, 511*, 344–347.

Gray, J. R., & Thompson, P. M. (2004). Neurobiology of intelligence: science and ethics. *Nature Reviews Neuroscience, 5*, 471–482.

Greene, K. B. (1928). The influence of specialized training on tests of general intelligence. In: *Twenty-seventh yearbook of the national society for the study of education*Vol. 1, (pp. 421–428). , pp. 421–428.

Hausknecht, J. P., Halpert, J. A., Di Paolo, N. T., & Moriarty Gerrard, M. O. (2007). Retesting in selection: a meta-analysis of coaching and practice effects for tests of cognitive ability. *Journal of Applied Psychology, 92*, 373–385.

Hawkes, N. (1979). Tracing Burt's descent to scientific fraud. *Science, 205*, 673–675.

Hearnshaw, L. S. (1979). *Cyril burt psychologist.* New York: Vintage Books.

Highfield, R. (1998). Scientists discover gene that creates human intelligence. *Daily Telegraph*, 13. October 31.

Hill, J. L., Brooks-Gunn, J., & Waldfogel, J. (2003). Sustained effects of high participation in an early intervention for low-birth-weight premature infants. *Developmental Psychology, 39*, 730–744.

Hill, L., Chorney, M. J., Lubinski, D., Thompson, L. A., & Plomin, R. (2002). A quantitative trait locus not associated with cognitive ability in children: a failure to replicate. *Psychological Science, 13*, 561–562.

Jensen, A. (1969). How much can we boost IQ and scholastic achievement? *Harvard Educational Review, 39*, 1–123.

Jensen, A. R. (1972). Sir Cyril Burt (1883–1971). *Psychometrika, 37*, 115–117.

Joynson, R. B. (1989). *The Burt affair.* Florence, KY: Taylor & Frances/Routledge.

Kamin, L. J. (1974). *The science & politics of I.Q.* Mahwah, NJ: Earlbaum.

Kaplan, R. M., & Saccuzzo, D. P. (2009). *Psychological testing: Principles, applications and issues* (7th ed.). Belmont, CA: Wadsworth.

Kendler, K. S., Ohlsson, H., Sundquist, J., & Sundquist, K. (2015). IQ and schizophrenia in a Swedish national sample: their causal relationship and the interaction of IQ with genetic risk. *American Journal of Psychiatry, 172*, 259–265.

Knopik, V. S., Neiderhiser, J. M., DeFries, J. C., & Plomin, R. (2017). *Behavioral genetics* (7th ed.). New York: Worth Publishers.

Lubbock, R. (2000). *A helpful tour of the gnomic genome.* The Globe and Mail. March 11.

Maulik, P. K., Mascarenhas, M. N., Mathers, C. D., Dua, T., & Saxena, S. (2011). Prevalence of intellectual disability: a meta-analysis of population-based studies. *Research in Developmental Disabilities, 32*, 419–436.

McCarthy, M. I., Abecasis, G. R., Cardon, L. R., Goldstein, D. B., Little, J., Ioannidis, J. P., & Hirschhorn, J. N. (2008). Genome-wide association studies for complex traits: consensus, uncertainty and challenges. *Nature Reviews Genetics, 9*, 356–369.

Morrison, F. J., Smith, L., & Dow-Ehrensberger, M. (1995). Education and cognitive development: a natural experiment. *Developmental Psychology, 31*, 789–799.

Muir, L. (2014). *A whisper past: Childless after eugenic sterilization in Alberta*. Victoria, BC, Canada: Friesen Press.

Nicolas, S., Andrieu, B., Croizet, J.-C., Sanitioso, R. B., & Burman, J. T. (2013). Sick? or slow? on the origins of intelligence as a psychological object. *Intelligence, 41*, 699–711.

Paul, D. B. (1985). Textbook treatments of the genetics of intelligence. *The Quarterly Review of Biology, 60*, 317–326.

Pearson, K. (1904). *Mathematical contributions to the theory of evolution: XIII. On the theory of contingency and its relation to association and normal correlation. 13*. London: Dulau and Co.

Polderman, T. J., Benyamin, B., de Leeuw, C. A., Sullivan, P. F., van, B. A., Visscher, P. M., & Posthuma, D. (2015). Meta-analysis of the heritability of human traits based on fifty years of twin studies. *Nature Genetics, 47*, 702–709.

Posthuma, D., & de Geus, E. J. C. (2006). Progress in the molecular-genetic study of intelligence. *Current Directions in Psychological Science, 15*, 151–155.

Ramey, C. T., Bryant, D. M., Wasik, B. H., Sparling, J. J., Fendt, K. H., & La Vange, L. M. (1992). Infant health and development program for low birth weight, premature infants: program elements, family participation, and child intelligence. *Pediatrics, 89*, 454–465.

Ramey, C. T., Yeates, K. O., & Short, E. J. (1984). The plasticity of intellectual development: insights from preventive intervention. *Child Development, 55*, 1913–1925.

Rauch, A., Wieczorek, D., Graf, E., Wieland, T., Endele, S., Schwarzmayr, T., … Di Donato, N. (2012). Range of genetic mutations associated with severe non-syndromic sporadic intellectual disability: an exome sequencing study. *The Lancet, 380*, 1674–1682.

Reichenberg, A., Cederlöf, M., McMillan, A., Trzaskowski, M., Kapra, O., Fruchter, E., … Larsson, H. (2016). Discontinuity in the genetic and environmental causes of the intellectual disability spectrum. *Proceedings of the National Academy of Sciences of the United States of America, 113*, 1098–1103.

Samelson, F. (1997). What to do about fraud charges in science; or, will the Burt affair ever end? *Genetica, 99*, 145–151.

Schiff, M., Duyme, M., Dumaret, A., Stewart, J., Tomkiewicz, S., & Feingold, J. (1978). Intellectual status of working-class children adopted early into upper-middle-class families. *Science, 200*, 1503–1504.

Sievertsen, H. H., Gino, F., & Piovesan, M. (2016). Cognitive fatigue influences students' performance on standardized tests. *Proceedings of the National Academy of Sciences of the United States of America, 113*, 2621–2624.

Skeels, H. M. (1966). Adult status of children with contrasting early life experiences: a follow-up study. *Monographs of the Society for Research in Child Development, 31*, 1–65.

Skodak, M., & Skeels, H. M. (1949). A final follow-up study of one hundred adopted children. *Journal of Genetic Psychology, 75*, 85–125.

Sniekers, S., Stringer, S., Watanabe, K., Jansen, P. R., Coleman, J. R., Krapohl, E., … Zabaneh, D. (2017). Genome-wide association meta-analysis of 78,308 individuals identifies new loci and genes influencing human intelligence. *Nature Genetics, 49*, 1107–1112.

Spearman, C. (1904). "General intelligence," objectively determined and measured. *American Journal of Psychology, 15*, 201–293.

Spearman, C. (1923). *The nature of "intelligence" and the principles of cognition*. London: Macmillan.

Stoddard, G. D. (1945). *The meaning of intelligence*. New York: Macmillan.

Szucs, D., & Ioannidis, J. P. (2017). Empirical assessment of published effect sizes and power in the recent cognitive neuroscience and psychology literature. *PLoS Biology, 15*, e2000797.

Terman, L. M. (1916). *The measurement of intelligence: An explanation of and a complete guide for the use of the stanford revision and extension of the binet-simon intelligence scale*. Boston, MA: Houghton, Mifflin and Company.

Varnhagen, C. K., Morrison, F. J., & Everall, R. (1994). Age and schooling effects in story recall and story production. *Developmental Psychology, 30*, 969–979.

Vernon, P. (1954). Practice and coaching effects in intelligence tests. *The Educational Forum, 18*, 269–280.

Vissers, L. E., Gilissen, C., & Veltman, J. A. (2016). Genetic studies in intellectual disability and related disorders. *Nature Reviews Genetics, 17*, 9–18.

Wade, N. (1976). IQ and heredity: suspicion of fraud beclouds classic experiment. *Science, 194*, 916–919.

Wade, N. (1998). First gene to be linked with high intelligence is reported found. *The New York Times*, A16. May 14.

Wahlsten, D. (1997). Leilani Muir versus the philosopher king: eugenics on trial in Alberta. *Genetica, 99*, 185–198.

Wahlsten, D. (1999). *Report on eugenic sterilization in Alberta from 1950 to 1968*. Edmonton, AB, Canada: University of Alberta, Department of Psycology. June 15.

Wahlsten, D. (2012). The hunt for gene effects pertinent to behavioral traits and psychiatric disorders: from mouse to human. *Developmental Psychobiology, 54*, 475–492.

Wallace, R. C. (1934). The quality of the human stock. *Canadian Medical Association Journal, 31*, 427–430.

Wasik, B. H., Ramey, C. T., Bryant, D. M., & Sparling, J. J. (1990). A longitudinal study of two early intervention strategies: project CARE. *Child Development, 61*, 1682–1696.

Further Reading

Benyamin, B., Pourcain, B., Davis, O. S., Davies, G., Hansell, N. K., Brion, M.-J., … Miller, M. (2014). Childhood intelligence is heritable, highly polygenic and associated with FNBP1L. *Molecular Psychiatry, 19*, 253–258.

Davies, G., Tenesa, A., Payton, A., Yang, J., Harris, S. E., Liewald, D., … Deary, I. J. (2011). Genome-wide association studies establish that human intelligence is highly heritable and polygenic. *Molecular Psychiatry, 16*, 996–1005.

Department of Public Health, & Province of Alberta (1959). *Annual report of the bureau of vital statistics*. Edmonton, AB, Canada: Department of Public Health, & Province of Alberta.

Kite, E. S. (1915). *The Binet-Simon measuring scale for intelligence: What it is; what it does; how it does it; with a brief biography of its authors, Alfred Binet and Dr. Thomas Simon*. Philadelphia, PA: Committee on Provision for the Feeble-Minded.

Wechsler, D. (1949). *Wechsler intelligence scale for children. Manual*. New York: The Psychological Corporation.

Zabaneh, D., Krapohl, E., Gaspar, H., Curtis, C., Lee, S., Patel, H., … Breen, G. (2017). A genome-wide association study for extremely high intelligence. *Molecular Psychiatry, 23*, 1226–1232.

16

Autism

On Mother's Day of 2013, a young man went to the market with his helper and wanted to buy a flower for his mother. This was not an easy task because he was struggling with nonverbal autism and ordinarily could not speak even one sentence. In his book *Fall Down 7 Times Get up 8*, Naoki Higashida describes the difficult thought process that enabled him to say "carnation" and "buy" to his helper. For him, this was a major victory, and his mother was elated when he proudly presented that flower to her. Later he used a 40-character form board to write the story of how he did it, pointing to one character at a time so that each could be recorded by his helper.

For years, his family and teachers had tried to help him to speak a few simple phrases. Eventually, it became apparent that he could hear and see quite well and understand many things. Using the form board, at the age of 13, he completed his first book, *The Reason I Jump*, and in 2017, his collection of essays *Fall Down 7 Times Get Up 8* became a best seller. Those works provide insights into the mind of a remarkable boy with autism from his own point of view.

Autism is a common behavior pattern that was elevated to cultural icon status in the movie *Rain Man* modeled on autistic savant Kim Peek. Savants like him often have extraordinary memories in a narrow domain and poor social skills. The term "idiot savant" was coined by Langdon Down, author of Mongolian idiocy (Chapter 12), in 1887, but this was a gross misnomer. Almost all known savants have IQ scores above 50, and many are in the range for typical children above 70 and often much higher. Today, the more appropriate term is "savant syndrome." About half of the cases of savant syndrome also expresses autism spectrum disorder, whereas about 10% of people found to be autistic show some degree of savant syndrome and demonstrate a "jarring juxtaposition of ability and disability" (Treffert, 2009). In some samples of autistic adults, the frequency of savant skills exceeds 25% (Howlin, Magiati, & Charman, 2009). Savant syndrome is not a category recognized in DSM-5.

People diagnosed as autistic express a wide range of IQ scores, from below 30 to a few cases as high as 160. Intellectual disability is not a requisite part of the syndrome, although it occurs in about 30% of cases. Some autistic adults perform at a high level in society despite their disability. An outstanding example is the veterinarian and professor of animal science Dr. Temple Grandin whose 2013 book with Richard Panek, *The Autistic Brain*, provides not only insights into her own experiences as an autistic person and those of many others but also an in-depth review of the scientific knowledge about the syndrome.

DEFINITION AND DIAGNOSIS

Autism has been recognized as a distinct psychiatric disorder since Kanner published the first article on the topic in 1943. In the next year, a similar disorder was described in German by Asperger (1944) in an article that drew little attention in the English speaking world of psychiatry when it appeared in the midst of WWII. The official diagnostic criteria for autism have changed considerably since that time. The most recent version in DSM-5 was a sweeping change that occurred at the same time when criteria for mental deficiency were also changed in a major way, more than just a name change from mental retardation to intellectual disability (ID). The criteria for ID have relevance for autism because the two disorders often show *comorbidity* in which the same person can express two different DSM categories. Many children with autism also meet the criteria for ID. Researchers are hoping that genetic studies will help to clarify the distinctions between the two.

Changes in diagnostic criteria and names have been implemented by professional psychiatrists and psychologists, but they now share the field with the growing ranks of disability activists and their families who are exerting leadership and demanding greater respect and better treatments for their children. The kinds of objective criteria that are important to researchers when combining patient samples from many sources for a GWAS study, for example, are not necessarily the best guides for clinicians working with disabled people on a daily basis. The recent transformation of criteria for ID illustrates this point.

Intellectual Disability

DSM-3 and DSM-4 used the category "mental retardation" that was perceived as offensive by many disabled people and their families. That relatively benign phrase was sometimes shortened to the abusive epithet "retard," and the children so labeled by their peers felt the sting of disrespect. One of the earliest signs of dissatisfaction appeared in 1992 when the Association for Retarded Citizens in the United States removed the word "retarded" from its name and called itself simply "The Arc" (Ford, Acosta, & Sutcliffe, 2013). In 2003, President George W. Bush changed the name of an advisory committee to the President's Committee on People with Intellectual Disabilities, and in 2004, the Special Olympics adopted ID for its many activities. A major step was taken by Rosa Marcellino, a girl with Down syndrome who competed in the 100m sprint in the Special Olympics. She objected to being called "retarded" at her school, and her family, who had already banned the term at home, began a campaign in 2009 to have the official term changed. This soon led to a law passed by the State of Maryland adopting the phrase "intellectual disability." Support then grew rapidly across the United States, culminating on October 5, 2010, when "Rosa's law" was signed by President Barack Obama. It directed that "mentally retarded" be replaced by "intellectual disability" in almost all federal statutes (Rosa's Law, 2010).

Terminology was modernized in 2013 when the new DSM-5 was published, but the criteria for intellectual disability were also changed radically. As shown in Table 16.1, the four degrees of retardation were defined succinctly in DSM-3 by ranges of IQ scores with sharp cutoffs. That definition had some utility for researchers comparing data from different samples of children. At the same time, psychologists were aware of how test scores could change from one occasion to another and be influenced by many other factors, and this awareness resulted in the criteria in DSM-4 having fuzzy boundaries. DSM categories 317 and 318 had always included deficits in adaptive behavior, but degrees of impairment in that domain were not specified in DSM-3, whereas DSM-4 added brief verbal descriptions of the amount of support a child needed under each degree of impairment. DSM-5 changed the fundamental structure of intellectual disability by dispensing with the reliance on IQ scores altogether and relying almost entirely on descriptions of adaptive functioning. Features of speech and social interactions specified in ID had clear counterparts in parallel definitions of autism.

Autism Spectrum Disorder and Asperger's Syndrome

Infantile autism first appeared as a distinct category in DSM-3 in 1980. Four criteria were supposed to be satisfied to warrant a diagnosis, and cases meeting just two or three criteria were not considered sufficiently serious to receive a DSM code number. For children whose symptoms were similar but appeared later than 30 months, pervasive developmental disorder (PDD) was the preferred diagnosis, especially if it involved several bizarre behaviors. Asperger's syndrome was recognized by some psychiatrists and was thought to be fitting for children who clearly did not show mental deficiency, but it was

TABLE 16.1 Criteria for Mental Retardation (DSM-3 and DSM-4) and Intellecutal Disability (DSM-5)

	Criterion of impaired intellect			
DSM	**Mild**	**Moderate**	**Severe**	**Profound**
3	IQ 50–70	IQ 35–49	IQ 20–34	IQ < 20
4	From 50–55 to about 70	From 35–40 to 50–55	From 20–25 to 35–40	<20 or 25
5	Severity is assigned by adaptive functioning, not intellectual capabilities			
	Criterion of impairments in adaptive behavior			
3	Same description for all severities			
4	May need supervision and guidance	Difficulty recognizing social conventions	Needs closely supervised settings	Needs constant supervision
5	Immature in social interactions; needs some support with daily living tasks	Much teaching needed to become independent in self-care, household tasks	Speech is single word, phrases; needs supervision at all times	Uses nonverbal communication; dependent on others for all aspects of life

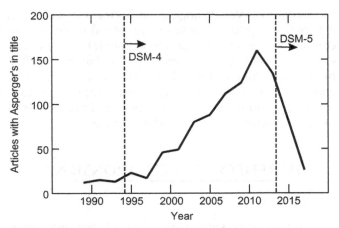

FIG. 16.1 Scholarly articles with the term "Asperger's" in the title according to Google Scholar. *Dashed lines* show when new versions of the Diagnostic and Statistical Manual (DSM) appeared. Asperger's syndrome was included for the first time in DSM-4 and then removed from DSM-5.

not in DSM-3. Prior to 1995, there was a slow but steady stream of publications about Asperger's syndrome (Fig. 16.1), but it had not yet received official sanction as a distinct disorder, and some psychiatrists regarded it simply as high-functioning autism.

The vagueness of PDD was not ameliorated when the suffix "not otherwise specified" (NOS) was added. In reviewing the rationale for revising DSM-3, Szatmari (2000) noted that there were parents "who did not appreciate being told their child had a "nonspecific disorder." They wanted something specific to explain their child's developmental difficulties and to give them a concrete plan for intervention." Of course, giving something a name does not explain its origins or prescribe treatment, but it does imply that the medical profession understands it well enough to differentiate it from similar syndromes. Several publications in the period leading up to DSM-4 supported Asperger's syndrome as something distinct from infantile autism (Bishop, 1989; Green, 1990; Szatmari, 1991).

DSM-4 altered the criteria for early onset autism, added Asperger's syndrome, and retained PDD-NOS as a residual category for milder cases. The category for Asperger's syndrome (AS) proved to be very popular with frontline clinicians, and the frequency of diagnosis of AS "increased dramatically" (Szatmari, 2000). A survey of 14 autism centers in the United States found that the prevalence of autism and closely associated disorders almost doubled from 2000 to 2008 (Anagnostou et al., 2014). Publications in the scientific literature about Asperger's syndrome increased greatly in just 10 years (Fig. 16.1). An "Asperger's community" of people sharing the new diagnosis emerged (Grens, 2013) and lobbied for better provision of services. Because AS was a bona fide psychiatric disorder legitimized in DSM-4, there

was an expectation that expensive support services would be covered by insurance plans, just as they would for autism.

There were suggestions that clinicians were using criteria for AS that were not precisely those in DSM-4, and it became increasingly difficult to demonstrate any clear distinction between children diagnosed as autistic and those designated Asperger's syndrome. As expressed by Szatmari (2000), "Is AS different from high-functioning autism? Yes, no, and it depends." He noted that using distinct categories that included PDD-NOS made things "hopelessly confusing for parents and front-line clinicians." Placing them on the same dimension where differences were only in severity began to have greater appeal. After several years' experience with DSM-4, opinion among psychiatrists and psychologists gradually shifted toward a single autism spectrum disorder (ASD) that would give no special status to Asperger's syndrome or PDD, and this was formalized in DSM-5. Concern arose that eliminating Asperger's syndrome would cause many children and adults to lose their insurance benefits and access to clinical services. A large study of 657 cases included in the autism spectrum under DSM-4 found that only 61% of them would qualify as ASD under the new DSM-5 criteria (McPartland, Reichow, & Volkmar, 2012). About 76% of cases classified as 299.0 autistic disorder under DSM-4 would be included in the new ASD, whereas only 25% of the former Asperger's syndrome cases and 28% of the former PDD-NOS would be in the new ASD. The authors observed that "A more stringent diagnostic rubric holds significant public health ramifications regarding service eligibility and compatibility of historical and future research." At the present time, it is not clear how the frequency of diagnoses of autism has changed as a result of DSM-5 criteria (Lyall et al., 2017).

With the advent of DSM-5, articles about Asperger's syndrome declined rapidly (Fig. 16.1). It appears that there was a trend toward reporting more cases of both autism and Asperger's syndrome before the appearance of DSM-5, partly from pressure exerted by parents who were becoming better informed about psychiatric diagnosis and available treatments.

Autism Rating Scales

Alternative diagnostic systems have been devised. The Autism Diagnostic Observation Schedule, (ADOS-2) involves observation by a trained professional of a child in a series of standard situations appropriate for the child's level of language functioning (Gotham et al., 2008; Gotham, Risi, Pickles, & Lord, 2007). In each situation, a score is assigned from 0 (no abnormality) to 3 (moderate to severe abnormality). Scores are then added to yield totals for social interaction, communication, restricted repetitive behavior, and social affect domains. The raw score totals can be used for studies of test validity (Gotham et al., 2008), or they can be compared with cutoff scores to determine a clinical diagnosis. For example, in module 1 intended for children who cannot speak any words, a score of 10 or less in a domain is considered outside the autism spectrum, 11–15 is in the spectrum, and 16 or more is considered autism.

The Autism Diagnostic Interview-Revised (ADI-R) consists of 93 standardized questions addressed to a parent or caregiver about a child (Le Couteur, Lord, & Rutter, 2003; Lord, Rutter, & Le Couteur, 1994). The questions are divided into items to assess social interaction; communication and language; and repetitive, restricted, and stereotyped behaviors. Cutoff scores must be exceeded for all domains in order to support a diagnosis of autism. Using results from both the ADOS-2 and ADI-R to decide on the autism diagnosis is considered the "gold standard" in child psychiatry (e.g., Reaven, Hepburn, & Ross, 2008) and is commonly employed in studies of genetic involvement.

Another useful instrument is the Childhood Autism Rating Scale (CARS; Chlebowski, Green, Barton, & Fein, 2010) that involves 15 items, each rated from 1 (low) to 4 (high impairment). The authors compared test sensitivity and specificity when different cutoff scores were used. *Sensitivity* is the proportion of cases that have bona fide autism as assessed by a clinician that are also classed as autistic by the CARS, while *specificity* is the proportion of children who are not clinically judged autistic who are also not regarded as autistic by CARS. The authors found that a CARS cutoff score of 32 most effectively distinguished those with autistic disorder from PDD-NOS and a cutoff of 25.5 best distinguished cases of PDD-NOS from nonspectrum children. Agreement of CARS with DSM-4 and ADOS was generally good but imperfect (about 70%).

Gender Differences

Many more boys are judged autistic than girls in most samples, with male-female ratios often being 4:1 or higher. There are probably many reasons for this striking imbalance (Schaafsma & Pfaff, 2014), one being gender-based socialization practices and gender-biased criteria for diagnosing ASD (Goldman, 2013). In a large UK population survey of more than 14,000 children, 71 cases of diagnosed ASD were detected (with a 9:1 boy-girl ratio), but additional assessments of all children using 27 measures of social interaction, communication, and repetitive behaviors identified another 142 who met the criteria for autism but had not been diagnosed (Russell, Steer, & Golding, 2011). A higher proportion of girls was undiagnosed (81%) than boys (63%). A literature review and meta-analysis (Van Wijngaarden-Cremers et al., 2014) reported no gender differences on social behavior and communication ratings but a moderate ($D = 0.51$) difference in repetitive and stereotyped behaviors with a higher frequency in boys than girls. Because repetitive and stereotyped behaviors are a core symptom in ASD, a more common occurrence in boys would probably lead to more diagnosis of boys as showing ASD.

HEREDITY AND ENVIRONMENT

Prevalence

If autism were simply the result of defective genes passed from ancestors, one would expect the population frequency to be fairly stable. One contrary and worrisome clue about causes, however, is the large increase in prevalence of the diagnosis of autism in recent years. The rate was estimated at less than 0.1% in the 1960s, before autism was listed in the official DSM, and then rose to about 0.67% only 9 years ago (Geschwind, 2008) and at least 1.5% recently (Lyall, Croen, Daniels, et al., 2017). A secular change of that magnitude could not arise from recent changes in gene frequencies. Less stringent diagnostic criteria could play a part along with increasing awareness of autism among family physicians, psychiatrists, and the general population, as well as an increasing availability of support services. The trend also suggests an upswing in possible environmental factors that may harm fetal and infant brain development.

Familial Risk and Heritability

Quantitative behavior geneticists compute heritabilities using twin data for this purpose (Chapter 15). Large samples of twins can be recruited when assessing a quantitative trait that varies in degree in the population, for example, IQ, but it is more difficult to find enough twin pairs when examining an infrequent psychiatric diagnosis. Prior to 2010, three small studies had located only 36 monozygotic twin pairs wherein at least one member of a pair was autistic. All were part of clinical samples, not population surveys, and had not been rated with the same assessment scale. A much larger sample was obtained from a registry in the State of California of all twins born between 1987 and 2004 in which at least one had been diagnosed as autistic (Hallmayer et al., 2011). Standard ratings with the ADOS and ADI-R were obtained for 54 MZ and 138 DZ pairs. Among MZ twins, 58% of co-twins both showed autism, whereas among DZ, the rate was about 24%. Statistical analysis with a multivariable model yielded an estimate of 37% of variance attributable to genetic factors, 55% for environmental factors common to a twin pair, and about 8% unique to a twin. This result implicates features of the prenatal and early postnatal environments.

Early Environments

Many studies of possible environmental influences on ASD have been published, but the main bulk of this evidence has been inconsistent, neither condemning nor exonerating a factor. Some evidence supports the notion that fetuses or infants who are exposed to multiple risk factors will indeed show higher frequencies of ASD, but there have been no large-scale studies that have put this hypothesis to a serious test. Generally speaking, there does not yet seem to be systematic, well-funded research on environmental influences involving multiple universities and medical institutions of the scope seen in research on genetics. Prior to 2010, before the large Hallmayer twin study, many authors cited high heritability estimates for ASD as grounds to focus on genetics and largely ignore environmental factors. That is not a valid application of the h^2 concept.

Several environmental factors reviewed by Lyall et al. (2017) are summarized in Table 16.2. There is clear and consistent evidence for an increased risk of ASD among infants conceived less than 12 months after a previous birth or those born prematurely. Maternal infection during pregnancy is also a well-established risk factor on the basis of a large study of more than 2 million live births. Antiepileptic drugs have shown consistently increased risk for ASD in infants who were in utero during maternal drug use. For a large number of other factors, results of several small studies have been inconsistent. Hazardous air pollution shows fairly consistent increases in ASD,

but the studies entail confounds with social factors such as parental education and income. Maternal smoking and alcohol consumption do not reveal any consistent relation with ASD.

Vaccinations, Thimerosal, and Autism

A report in the medical journal *The Lancet* in 1998 sparked a major controversy about causes of autism, one that has not yet been extinguished. Wakefield et al. (1998) claimed to have studied 10 children who presented with symptoms of autism or pervasive developmental disorder soon after being vaccinated for measles, mumps, and rubella viruses (MMR). The vaccine contained diethylmercury in the form of thimerosal as a preservative. Authors suggested a causal link, and the media were quick to sound the alarm. Activist groups of parents contributed to the uproar. The news spread so widely so fast and generated such worries among parents that the actual rate of child vaccination in the United Kingdom and the United States fell to about 80%, well below the rate of 95% needed to effectively protect a population via "herd immunity." In 2008, measles was once again declared endemic in England and Wales (Burns, 2010).

Reports to the contrary soon appeared. Taylor et al. (1999) pointed out that ASD rates had been rising rapidly during a period when rate of MMR vaccination remained steady, and they reported data on 29 children under their care who were diagnosed with autism, all of whom had shown symptoms *before* receiving the MMR shot. Wakefield clung to his story as the weight of evidence grew. Investigative journalists began to dig into facts behind the scene, and officials at the General Medical Council in the United Kingdom where the original study had been done started an investigation. Things got messy. The editor of *The Lancet* revealed that the children in the 1998 study had been referred to Wakefield by a society working with autistic children and they were evaluated for the Legal Aid Board well before the Lancet article was published (Horton, 2004). The Legal Aid Board was considering a lawsuit. In 2008, Paul Offit published *Autism's False Prophets* that exposed the forces pushing the autism-vaccine link.

Meanwhile, the scientific literature on the topic accumulated and pointed to no causal link at all. Rarely has the case against a hypothesis been so clear. Plotkin, Gerber, and Offit (2009) reviewed 20 studies in several countries, some involving more than 500,000 children and concluded the following:

1. Considering that about 50,000 British children were receiving the MMR vaccine each month and knowing the rate of autism at the time, they expected that about 25 children per month would first show symptoms of autism within a month of receiving the vaccine, purely by chance.

TABLE 16.2 Environmental Factors That May Increase Risk of Autism[a]

Factor	Evidence	Factor	Evidence
Maternal infection during pregnancy	Strong; study of >2 million births	Caesarian delivery Antidepressant drugs Folic acid supplementation Pesticides Endocrine disruptors PCBs Mercury, lead	Inconsistent
Interpregnancy interval <12 months Preterm birth Antiepileptic drug exposure during pregnancy	Consistent	Maternal alcohol use Maternal smoking Vaccines	No effect

[a] Based on review by Lyall, Croen, Daniels et al. (2017).

2. No change in the rate of autism diagnoses was seen shortly after MMR vaccinations began in 1987, and autism rates were similar for children who got MMR and those who did not.

3. For very large samples in several countries, cases of autism did not cluster relative to timing of a child's MMR vaccination.

4. When thimerosal was removed from the MMR vaccine in 1992 in the United Kingdom and 2001 in the United States, there was no decline in autism rates in either country, where a gradual increase in rates of autism had commenced in 1990. This result was confirmed by Hurley, Tadrous, and Miller (2010).

Finally, the General Medical Council in 2010 concluded its investigation, the longest in its history, and banned Wakefield from practicing medicine in the United Kingdom. Unbowed, he moved to Texas and continued his work against vaccinations (Burns, 2010). The staid *British Medical Journal* published an editorial accusing him of outright fraud, unethical treatment of children, and conflict of interest (Godlee, Smith, & Marcovitch, 2011). The *BMJ* editorial lamented "the energy, emotion, and money that have been diverted away from efforts to understand the real causes of autism and how to help children and families who live with it."

TREATMENT

For many children with the diagnosis of ASD, progress in the typical classroom with conventional instruction is painfully slow or absent. Those who do not receive special help have diminished prospects for later employment, independent living, and raising a family. Accordingly, therapies have been devised specially to treat autism, and there is clear evidence that they can improve social and language functions. The treatments sometimes do not generate an outright cure where the child is no longer diagnosed as autistic and enters the mainstream with no further assistance. Nevertheless, longer-term prospects can be enhanced by intensive therapy in early childhood.

The most successful approach is based on the psychology of learning, as pioneered by Skinner and others working with lab animals. The basic idea is quite simple: behaviors that are followed closely in time by a desirable event such as giving a small bit of food or, in the case of a pet dog or a child, praise are likely to increase in frequency, while those not reinforced or perhaps subjected to mild punishment (a gentle "no") decrease in frequency. The big challenge is to analyze behaviors in detail and devise a plan of action to change things in the desired direction. This general approach is termed "applied behavior analysis" or ABA. The first implementation of

TABLE 16.3 Therapies to Treat Autism Spectrum Disorder[a]

Applied behavior analysis (ABA-EIBI)	Pivotal response training
Developmental social-pragmatic intervention	Prompt for restructuring phonetic targets
Early Start Denver Model	Reciprocal imitation training
Enhanced milieu training	Stimulus-stimulus pairing
Joint Attention-Mediated Learning	Video modeling
Natural language paradigm	Within-stimulus prompting
Picture exchange communication system	

[a] *Therapies included in review by Roane et al. (2016) and analysis by Debodinance, Maljaars, Noens, and Van den Noortgate (2017).*

the technique was "early and intensive behavioral intervention" (EIBI) developed by Lovaas (1981). A recent primer designed for pediatricians provides a concise introduction to the methods (Roane, Fisher, & Carr, 2016). Most therapies are applied one-on-one by a highly trained psychologist, beginning with a child 2 or 3 years old who has recently been diagnosed as ASD, and extending for several hours each day, 5 days each week, for several months or years. Some therapies can be administered at home by a parent who receives instruction on the technique. Table 16.3 lists some of the currently available therapies. A more complete list of 27 ABA-based intervention programs, both comprehensive and more narrowly focussed, has been compiled (Wong et al., 2014). The ABA approach now involves a formal mechanism for certifying the competence of professionals trained in the method by the Behavior Analyst Certification Board. The method has earned widespread acceptance, and health-related branches of the US government "all recognize applied behavior analysis as an efficacious or accepted approach for treating the core symptoms of autism" (Greer & Kodak, 2017).

Efficacy and Meta-Analysis

Several meta-analyses have determined the mean and range of effect sizes for ABA and its variants. One study reviewed five different meta-analyses (Reichow, 2012), but close inspection reveals that the different meta-analyses included several of the same studies and therefore were not independent. For example, the meta-analysis by Peters-Scheffer, Didden, Korzilius, and Sturmey (2011) involved 10 studies, eight of which were included in a much larger analysis by Virués-Ortega (2010). Results of the larger meta-analysis are described here. Studies involved children diagnosed using several criteria, usually within the spectrum of autism,

Asperger's syndrome, and PDD-NOS. The studies also applied different treatments and measures of outcomes. Although IQ is not a defining characteristic of autism, many children expressing autistic symptoms also show ID, and most studies collected data on IQ using a variety of tests. The meta-analysis was performed using the effect size indicator D (Chapter 14), number of standard deviations separating the means for treated and control groups. Effect sizes for the most commonly reported measure, full-scale IQ, are shown in Fig. 16.2. The average effect of $D = 1.2$ amounts to a very large effect of treatment, but the plot of individual study results shows a very wide range of outcomes. Roane et al. (2016) highlighted a typical study done with the Early Start Denver Model that boosted IQ by 19 points after 2 years of treatment for 25 h/week. Some of the differences between studies appeared to arise from intensity (h/week) and duration of treatments (total weeks). Measures of receptive and expressive language, core indicators of autism, revealed very large effects of treatment ($D = 1.66$) on average for nine studies reporting that measure, but the range of outcomes was remarkably wide.

A recent assessment of 35 studies involving 82 children also found very large effects of treatments on symptoms of autism (Debodinance et al., 2017). Sample sizes within a study were small, and there were many features of the treatments that varied among studies, so the authors used a sophisticated multivariate analysis to test effects of different kinds of treatments. The average effect size across all studies was a massive improvement in performance on the measures of autism-relevant behaviors of 2.14 standard deviations. Effect sizes were generally large and positive for studies done in the home, a clinic, a class,

or the community, as well as training given by a parent, a researcher, a teacher, or a professional therapist.

How long are gains from ABA likely to last when the special training comes to an end? How many of the children who started in the programs early and continued for several years are then able to enter the mainstream classroom and thrive? Further study is needed to answer these important questions.

Drug Therapy

There are no chemical treatments that directly address the core deficiencies in autism (Anagnostou et al., 2014). This is not at all surprising because, as shown in the section on genetics, causes are heterogeneous across many cases. Instead, drug therapy is often directed to specific neurological or behavioral symptoms that are causing difficulties for a specific child. Some of the drug treatments are considered "off label," indicating the seller makes no claim of efficacy for symptoms specifically related to autism. Aripiprazole and risperidone, for example, are sometimes given to control serious problems with irritability, aggression, and repetitive behavior. Methylphenidate and atomoxetine may be given to control symptoms common in attention-deficit hyperactivity disorder, and melatonin may be given to treat insomnia.

SINGLE GENE MUTATIONS

The search for genes pertinent to childhood autism has followed a trail similar to that for intelligence and intellectual disability, but geneticists have been hunting for a quarry with much larger effects of any one gene. The result is that several gene mutations that can cause autism have been identified and replicated, and in some instances, there are indications of how brain development and function are impaired.

Neuroligin and Others

Until the era of SNPs and GWAS studies, the search was more or less hit and miss. Now and then, a family with enough cases made it possible to conduct a traditional linkage study using DNA markers. Several extended families in France were identified that involved multiple instances of intellectual disability in conjunction with autism diagnosed using DSM-4 and ADI-R. Family number 118 had 13 males diagnosed as autism spectrum disorder or intellectual disability (Laumonnier et al., 2004). Researchers proceeded on the assumption that their mothers would all have been unaffected carriers of a mutant gene on the X chromosome, and they focussed on a few genes that had been found to reside

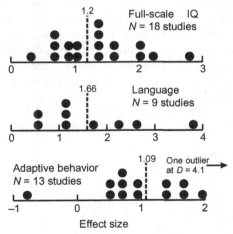

FIG. 16.2 Meta-analysis of the effects of applied behavior analysis therapy on three measures of children with ASD. Numbers at *dashed lines* are means of studies reporting that measure. *Reprinted from* Clinical Psychology Review, 30, *Virués-Ortega, Applied behavior analytic intervention for autism in early childhood: Meta-analysis, meta-regression and dose–response meta-analysis of multiple outcomes, 387–399, Copyright 2010, with permission from Elsevier.*

on the X. Using 41 DNA markers spanning the X, they obtained data indicating a pathogenic gene close to a marker designated DXS996 near chromosome band Xp21.2. A particularly intriguing suspect was the recently discovered gene *Neuroligin 4* (NLGN4) that was very closely linked to DXS996. Investigators decided to sequence the exons of that gene in family members who could provide DNA samples. With the technology of the time, it was a tedious task, reading off one nucleotide base and then the next and then the next, working their way along the DNA of just that one gene. In exon 5, something clearly was wrong with the gene of affected males: a two-base sequence AG had been deleted from NLGN4 in every affected person. This shifted the translation of RNA to protein over by two bases, and the RNA polymerase then encountered a STOP codon about half way through the gene, which yielded a greatly shortened neuroligin protein having 429 amino acids instead of the normal 816 (Fig. 16.3). Because males have only one X chromosome, one copy of the mutation was sufficient to cause a significant defect in a protein involved in synapse formation. The unaffected mothers had one copy of the mutation and one normal copy, which was enough to get the job done. A survey of other French families harboring autism failed to find any other with the neuroligin defect. Thus, the discovery was a noteworthy scientific achievement but explained only a very small proportion of total autism cases. This is typical of many genetically heterogenous disorders, wherein an apparently similar phenotype can be generated by a great diversity of different genetic and/or environmental factors in different families. Gradually, more instances of single-gene defects apparently leading to autism were identified; until in 2012, the GeneCards database listed 25 genes that appeared to be connected with one form of autism or another, often in conjunction with intellectual disability. Table 16.4 lists those genes in alphabetical order to make comparisons with other studies easier.

GWAS

The next major improvement in technology was GWAS scanning for SNPs located close to genes that generate phenotypic changes. Because relatively large effect sizes were expected when children had a diagnosis of autism, it was hoped that a few thousand cases would suffice to find a few genes. A typical result was the GWAS study by Anney et al. (2010) that assessed 2394 ASD cases with a microarray capable of detecting variants at 1,072,820 SNP markers. That was far more people and markers than were typically studied in pre-GWAS research projects, but only one gene (MACROD2 at chromosome location 20p12.1) met the criterion for significance ($P < 5 \times 10^{-8}$). Two other GWAS studies found evidence for the SEMA5A gene at 5p15 and MSNP1 at 5p14.1. Those were meager harvests for such great effort and expense, and none of the effects was replicated in a recent study focused on just those three genes (Torrico et al., 2017).

Exon Sequencing

Instead of employing new arrays that could detect millions of SNPs in huge samples of subjects, as was done in the hunt for IQ genes of small effects, investigators took advantage of new technology that made it convenient to sequence all the exons of numerous genes across the genome, as was done for ID (Chapter 15). That approach no longer relied on relatively insensitive linkage data with distant markers. It searched directly for mutations in protein-encoding DNA sequences, reasoning that a mutation having a major effect on a protein sequence might very well cause a substantial deficit in brain development that could produce autism. This surmise proved to be correct (Table 16.4). A whole-exome sequencing study of 2508 ASD children, 1911 non-ASD siblings, and their parents found mutations in 353 genes that were judged to be "likely gene disrupting" on the basis of the change in protein structure (Iossifov et al., 2014). All of them were de novo mutations, occurring in the ASD cases but not the parents. Most mutations occurred in only one ASD case, but 27 were recurrent, appearing in more than one unrelated family, which strengthened the evidence that the gene was indeed related to autism. No instances of a recessive mutation that produced autism when

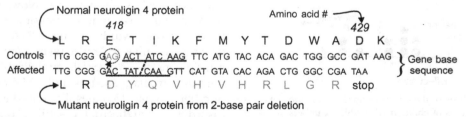

FIG. 16.3 Nucleotide base sequence in control genome and affected male family members with a two-base (AG) deletion in the neuroligin 4 X-linked (NLGN4X) gene and amino acid sequences of a portion of the neuroligin protein encoded by the two genes. In those with the mutation, every amino acid after #417 is altered and then translation stops at #429, which causes a complete loss of function of the protein and then autism. *Reprinted from* The American Journal of Human Genetics, 74, *Laumonnier et al., X-linked mental retardation and autism are associated with a mutation in the nlgn4 gene, a member of the neuroligin family, 552–557, Fig. 2B, Copyright 2004, with permission from Elsevier.*

homozygous were detected. Whenever a mutation was detected in one chromosome strand, the allele in the other chromosome strand was normal. It was worrisome that the exome assessment did not detect even one of the genes flagged in the 2012 version of the GeneCards database (Table 16.4).

Whole Genome Sequencing

Further advances in DNA sequencing technology made it possible to determine almost the entire genome sequence of 3 billion nucleotide bases for a single individual. Whole-genome sequencing (WGS) offered the possibility of detecting not only many de novo mutations in exons, gene regulation sequences outside the exons, and copy number variations (insertions or deletions) involving large chunks of a chromosome but also very small sequences. A large study using WGS assessed 2066 families with at least one ASD child (Yuen et al., 2017). In order to assemble a sufficiently large sample, nine different data sets were pooled. Diagnosis was "research quality" using ADI-R and ADOS for most children but DSM-4 or DSM-5 for others.

The total harvest of genetic variants was immense and revealed what appeared to be a high rate of mutations in all people in the study. The average number of "events" detected per individual genome was 86.4, 73.8 of which were single-nucleotide variants (SNVs) and 12.6 were insertions or deletions. The vast majority of those events, however, were in introns, whereas there was an average of only 1.3 de novo SNVs in an exon. One or two mutations in an exon of a person having about 20,000 genes, each with many exons, can be seen as remarkable stability of the protein-encoding genome. Nevertheless, there were 112 mutations among 2620 ASD children that were expected to yield the loss of function for a protein, about 4% of the total sample of children. Among them, 54 were considered putative ASD risk genes plus another seven risk genes on the X chromosome of males for a total of 61 different genes in which a mutation would cause the loss of function of the protein and increase risk of ASD. Forty-three of them had been reported previously in various studies, whereas 18 were new (Table 16.4). It is noteworthy that 20 of the genes detected through exome sequencing by Iossifov et al. (2014) were also detected by Yuen et al. (2017) using WGS. There were also six genes flagged in the GeneCards list in 2012 that were detected in the WGS study. Almost all of the putative risk genes in the 2017 study were de novo mutations that were not present in either parent.

The WGS method also detected many copy number variations in which portions of a chromosome were present in one or three copies instead of the normal two. There were on average 401 copy number variants (CNVs) per genome, most residing in introns, but among the 2620 ASD children, 189 copy number variants were likely to be pathogenic and were probably causes of the children's ASD. Altogether, among the 2620 ASD cases, there were 112 single-nucleotide variants (SNVs) and 189 copy number variants (CNVs) for a total of 11.2% of children who showed some genetic defect that was deemed likely to contribute to their autism. At the same time, this highly sensitive and comprehensive assessment of the entire genome of each child did not detect anything of relevance in almost 90% of the cases. It appears that the overwhelming majority of genetic variation pertinent to autism is generated anew each generation through de novo loss of function mutations that are not hereditary. This comes as a surprise to the more traditional Mendelian genetics that expects many hereditary rare recessive or dominant genes in a population that contribute to parent-offspring, sibling, and twin resemblance.

Gene Function and Networks

As more genes pertinent to ASD are detected from de novo mutations of large effect, it should be possible to assemble a picture of processes in the brain such as synapse formation that often go awry in autism. Fig. 16.4 shows an interaction network of genes identified in the WGS study by Yuen et al. (2017) using a database-graphing program called Cytoscape. Information about which genes are most similar in terms of their neural functions was compiled from many sources, mainly relying on biochemical data from mice, fruit flies, zebra fish, and nematode worms that also possess versions of the genes listed in Table 16.4. Most of the genes had important roles in prenatal gene expression or the formation of neurons and synapses. In many instances, the actions of just one gene had an impact on the functions of five or more other genes in the same networks, and a few genes extended their influences to other functional clusters. Thus, genes that are important for synapse formation are not walled off or isolated from other functions. RNA processing and transcription are important for the proper functioning of large numbers of other genes, and a debilitating mutation of one of them can have widespread ramifications. Future research on autism will undoubtedly add many other genes to this already intricate portrayal of gene interactions.

ASD, IQ, and ID Genes

Comorbidity of autism and intellectual disability should be evident in the overlap of the sets of genes identified as important for each diagnostic category. For the

TABLE 16.4 Genes Associated With Autism Spectrum Disorder[a]

Study	GeneCards (2012)	Iossifov et al. (2014)	Yuen et al. (2017)	Yuen et al. (2017)
Method	Mostly linkage; single studies	Whole-exome sequencing	Whole-genome sequencing	Whole-genome sequencing
ASD cases	Variable; small N	2508	2620	2620
Gene criteria	Variable	Recurrent de novo	Recurrent known	New recurrent
Genes in alphabetical order	ADSL	**ADNP**	**ADNP**	ADCY3
	CADPS2	**ANK2**	AFF2 (X)	AGAP2
	CNTN4	**ANKRD11**	**ANK2**	CIC
	CNTNAP2	**ARID1B**	**ANKRD11**	CLASP1
	EN2	**CHD2**	**ARID1B**	CNOT3
	FRA16A	**CHD8**	ASH1L	DIP2C
	GABRA4	DIP2A	ASXL3	DYNC1H1
	GABRB3	**DSCAM**	CACNA2D3	FAM47A
	GRPR	**DYRK1A**	**CHD2**	(X)
	MECP2 (X)	**FOXP1**	**CHD8**	KIAA2022
	MET	**GIGYF1**	CUL3	(X)
	NBEA	**GRIN2B**	DDX3X	MED13
	NLGN3 (X)	KATNAL2	DNMT3A	MYO5A
	NLGN4X (X)	KDM5B	**DSCAM**	PAX5
	NRCAM	**KDM6B**	**DYRK1A**	PCDH11X
	PTEN	KMT2E	**FOXP1**	(X)
	RELN	MED13L	**GIGYF1**	PHF3
	SCTR	NCKAP1	**GRIN2B**	SMARCC2
	SEMA5A	**PHF2**	KDM6A	SRSF11
	SHANK2	**POGZ**	**KDM6B**	TAF6
	SHANK3	RIMS1	KMT2A	UBN2
	SLC9A9	**SCN2A**	KMT2C	
	SLC25A12	**TBR1**	KMT5B	
	ST7	**TCF7L2**	**MECP2** (X)	
	ZNF778	**TNRC6B**	MYT1L	
		WAC	NAA15	
		WDFY3	**NLGN3** (X)	
			NLGN4X (X)	
			PHF2	
			POGZ	
			PTEN	
			SCN2A	
			SHANK2	
			SHANK3	
			SLC6A1	
			SPAST	
			SYNGAP1	
			TBR1	
			TCF7L2	
			TNRC6B	
			UPF3B	
			WAC	
			WDFY3	

[a] *Times-Roman bold designates genes listed in two or more lists of genes in this table.* (X) *is on X chromosome.*

genes listed in Table 15.7 that were identified in GWAS studies of IQ in the normal range, only one (SHANK3) appears once there and also in Table 16.4. For ID, on

the other hand, the de Ligt et al. (2012) study (Table 15.8) detected four using exome sequencing (DYNC1H1, ASH1L, SCN2A, WAC) that are also on the list for ASD, and Rauch et al. (2012), also using exome sequencing, took note of five that overlap with ASD (SCN2A, MECP2, SYNGAP1, CHD2, SLC6A1). Knowing these genes are common to both phenotypes does not aid in diagnosis, which is nevertheless based on behavioral phenotypes, but it may advance our understanding of cellular and neural functions that are disturbed in both kinds of disability.

SNP Prediction of ASD

If sites in the genome can be identified that reliably are associated with occurrence of autism, it might be possible to predict which children are most likely to develop autism from the knowledge of their genotypes at those sites. This kind of prediction was reported by a team of researchers in Australia (Skafidas et al., 2014) who used a set of 237 SNPs to predict autism. The original report gave data for 30 of the SNPs that were most closely correlated with ASD and suggested good predictions could be made from relatively small samples of SNPs. To several other groups of geneticists, however, this seemed unlikely to be true. Robinson, Lichtenstein, Anckarsäter, Happé, and Ronald (2013) applied the data from the 30 SNPs to another sample of more than 5400 ASD cases and a similar number of controls, and they found that the 30 SNPs could not predict autism above a chance level.

Viewing this dispute from a distance, the question arises as to why the scientists would even care to predict ASD from a set of SNPs. Why not just wait until the child is old enough and see if real behavioral symptoms emerge? The original report (Skafidas et al., 2014) hinted that the SNPs might provide "biomarkers to aid diagnosis." But recent studies find that many cases of ASD are caused by de novo copy number variants or SNV mutations that would bear no relation to prior SNP data.

Genes and Sex Differences

A portion of the 4:1 boy-girl difference in rates of autism may arise from genetic sources, although precisely how much has not been determined. Seven of the genes identified in the recent whole-genome sequencing study (Yuen et al., 2017; Table 16.4) are located on the X and may contribute to sex differences in ASD. A recent review by Schaafsma and Pfaff (2014) presents several possibilities. The origins of sex differences in ASD-related behaviors seem to be complex.

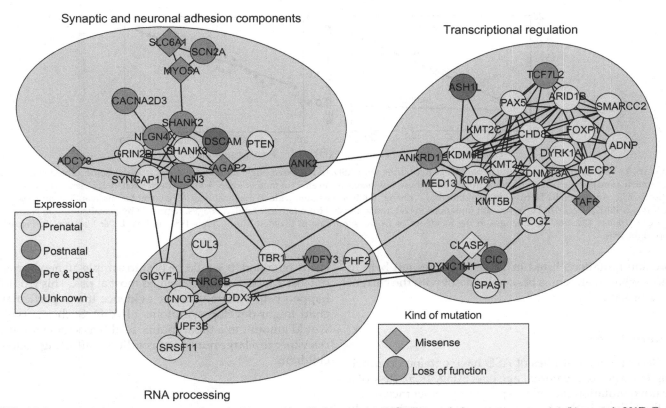

FIG. 16.4 Interrelations among gene products for 61 genes identified in a study of ASD using whole-genome sequencing (Yuen et al., 2017). Each gene was found to have abnormal DNA sequences in an exome in several children with ASD that would lead to the loss of function of the associated protein. The genes were part of three clusters with different biochemical and cellular functions. The clusters had been identified previously and connections within and between them documented in genetically normal samples. In a loss of function mutation, certain of the connections with a specific gene would no longer function normally. *Reprinted by permission from Nature/Springer, Nature Neuroscience, 20, Whole genome sequencing resource identifies 18 new candidate genes for autism spectrum disorder, Yuen et al., Copyright 2017.*

Interactions

The exome and whole-genome sequencing studies have treated each gene as though it is an independent entity, estimating the effects of different alleles by averaging over all other sources of variation. This approach is consistent with the additive model underlying heritability analysis, but it does not accord with what is known about gene actions at the molecular level. The consequences of having a particular pair of alleles for one gene often depend on the individual's genotype at other genes, a phenomenon termed *gene-gene interaction* or *epistasis*. The interaction can even extend across generations, as when effects in mother and child are interdependent (Tsang et al., 2013). Effects of early environment can also depend on the genotype, as happens for vulnerability to prenatal stress where response in the child depends on the mother's genotype at the serotonin transporter 5-HTTLPR gene (Hecht et al., 2016). Many instances of gene-environment interactions are well known from controlled studies with lab animals (Chapter 14). It is very difficult to assemble enough evidence of specific interactions involving children who develop autism because there are so few instances of children that share the same genetic anomaly. Risk in autism appears to be widely dispersed across many genes in the population, each with its own unique function and environmental sensitivities.

REDUCING THE PREVALENCE OF AUTISM

Genetic Screening

The tripling of the prevalence of ASD in just a few years invites discussion of things that might be done to reduce the frequency. One option is clearly not viable—prenatal screening for DNA anomalies. Because so many new cases of ASD arise from de novo mutations, each of them quite rare, it would not be sufficient to screen just relatives of existing ASD cases. Instead, it would be necessary to screen every new pregnancy using exome or whole-genome sequencing. Obtaining informed consent from all those expectant parents would be a formidable challenge. This does not seem like a realistic option. Furthermore, knowing that a child carries one of the losses of

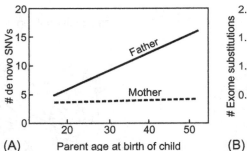
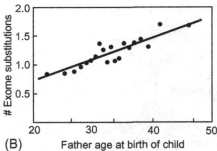

(A)   Parent age at birth of child   (B)   Father age at birth of child

FIG. 16.5 Mutation rates in relation to parent age in children with ASD. (A) De novo single-nucleotide variants (SNVs), some located in introns, were much more likely when sperm came from older fathers, but there was no effect of mother's age. Adapted from Yuen, R. K., Merico, D., Cao, H., Pellecchia, G., Alipanahi, B., Thiruvahindrapuram, B., . . . Zhang, T. (2016). Genome-wide characteristics of de novo mutations in autism. *npj Genomic Medicine, 1*, 16027, Fig. 2. (B) De novo exon mutations in children with ASD from fathers of different ages. Reprinted by permission from Nature/Springer, *Human Genetics, 72*, Changes in the incidence of Down syndrome in Sweden during 1968–1982, Iselius, L., & Lindsten, J., Fig. e4, Copyright 1986.

function mutations listed in Table 16.4 does not help to know what treatment is best, at least not with the current state of knowledge.

Paternal Age

Recent genetic studies of ASD have obtained compelling evidence of a paternal age effect (Fig. 16.5). Risk of a de novo mutation carried by sperm from older men is two to three times higher than from young men in their early 20s. This could be addressed at a population level by routinely freezing sperm from a young man and then thawing it when the time comes to be a father.

Better Spacing of Pregnancies

There is a large effect of very short intervals between birth of one child and conception of the next, about a doubling of risk of an ASD child the second time. There would also be advantages to the overall health of the mother if she had time to recover from the stresses of childbirth and the demands of breast feeding and changing all those diapers before the next pregnancy. Other measures to enhance the health of expectant mothers would also be likely to reduce ASD frequency by reducing preterm births and maternal infections during pregnancy (Table 16.2).

Toxins in the Environment

There is some evidence that a number of environmental toxins can contribute to ASD (Table 16.2). A direct test of this notion could be achieved by taking samples of umbilical cord blood at birth in a large nationwide survey. That blood is almost entirely from the fetus, and it could be analyzed for many kinds of pollutants and nutrients to obtain a snapshot of infant exposure to adverse conditions during gestation. Parental consent should be

easily obtained. It would be best to sample all newborns, not just a subset of those deemed most at risk. This would happen before there was any evidence that a particular child might develop symptoms of ASD. Such a survey would amount to a broad public health assessment with relevance to a large number of conditions afflicting young children.

HIGHLIGHTS

- Autism spectrum disorder (ASD) is a disorder of social communication that shows extensive comorbidity with intellectual disability (ID). About 30% of cases of ASD also qualify as having ID.
- The prevalence of a diagnosis of childhood autism increased more than 10 times in the United States since the 1960s and is now close to 1.5% of all children.
- Criteria for autism were broadened considerably when Asperger's syndrome that does not involve ID was included in the autism spectrum in DSM-4. That change contributed to a doubling of cases of ASD in a 10-year period. Asperger's syndrome was then removed from DSM-5.
- Autism spectrum disorder (ASD) exists on a continuum of severity, and the cutoff score for a formal diagnosis of ASD is somewhat arbitrary.
- Several environmental factors are clearly and strongly related to the frequency of ASD. They include a short interval between successive births of a child, premature birth, and maternal infection during pregnancy.
- A few environmental factors clearly have no influence on the frequency of ASD, including maternal smoking and alcohol use and vaccination for measles, mumps, and rubella.
- A psychological treatment—applied behavior analysis—has had good success in reducing autism

symptoms in children but must be implemented intensively from an early age.

- Exome sequencing of the two parents and a child with ASD has detected several dozen major gene mutations in the children. Virtually, all of them appeared de novo and were not seen in either parent.
- Any one specific mutation is usually seen in just one or a very few cases. Thus, autism shows pronounced genetic heterogeneity with many different molecular and environmental causes in different children.
- The molecular pathways affected by the mutations relevant to ASD involve a wide range of processes, including gene transcription, RNA processing, synapse formation, and neural cell adhesion.
- The great variety of mutations contributing to ASD is generated anew each generation and is usually not transmitted to future generations.
- The rate of mutations contributing to ASD is much higher in fathers older than 35 years.

References

Anagnostou, E., Zwaigenbaum, L., Szatmari, P., Fombonne, E., Fernandez, B. A., Woodbury-Smith, M., … Drmic, I. (2014). Autism spectrum disorder: advances in evidence-based practice. *Canadian Medical Association Journal*, 186, 509–519.

Anney, R., Klei, L., Pinto, D., Regan, R., Conroy, J., Magalhaes, T. R., … Hallmayer, J. (2010). A genome-wide scan for common alleles affecting risk for autism. *Human Molecular Genetics*, 19, 4072–4082.

Asperger, H. (1944). Die "Autistischen Psychopathen" im Kindesalter. *European Archives of Psychiatry and Clinical Neuroscience*, 117, 76–136.

Bishop, D. V. (1989). Autism, Asperger's syndrome and semantic-pragmatic disorder: where are the boundaries? *British Journal of Disorders of Communication*, 24, 107–121.

Burns, J. F. (2010). British Medical Council bars doctor who linked vaccine with autism. *The New York Times*, May 24 Retrieved from (2010). https://www.nytimes.com/2010/05/25/health/policy/25autism.html.

Chlebowski, C., Green, J. A., Barton, M. L., & Fein, D. (2010). Using the childhood autism rating scale to diagnose autism spectrum disorders. *Journal of Autism and Developmental Disorders*, 40, 787–799.

de Ligt, J., Willemsen, M. H., van Bon, B. W., Kleefstra, T., Yntema, H. G., Kroes, T., … Vissers, L. E. (2012). Diagnostic exome sequencing in persons with severe intellectual disability. *New England Journal of Medicine*, 367, 1921–1929.

Debodinance, E., Maljaars, J., Noens, I., & Van den Noortgate, W. (2017). Interventions for toddlers with autism spectrum disorder: a meta-analysis of single-subject experimental studies. *Research in Autism Spectrum Disorders*, 36, 79–92.

Ford, M., Acosta, A., & Sutcliffe, T. (2013). Beyond terminology: the policy impact of a grassroots movement. *Intellectual and Developmental Disabilities*, 51, 108–112.

Geschwind, D. H. (2008). Autism: many genes, common pathways? *Cell*, 135, 391–395.

Godlee, F., Smith, J., & Marcovitch, H. (2011). Wakefield's article linking MMR vaccine and autism was fraudulent. *British Medical Journal*, 342, c7542.

Goldman, S. (2013). Opinion: sex, gender and the diagnosis of autism—a biosocial view of the male preponderance. *Research in Autism Spectrum Disorders*, 7, 675–679.

Gotham, K., Risi, S., Dawson, G., Tager-Flusberg, H., Joseph, R., Carter, A., … Hyman, S. L. (2008). A replication of the Autism Diagnostic Observation Schedule (ADOS) revised algorithms. *Journal of the American Academy of Child and Adolescent Psychiatry*, 47, 642–651.

Gotham, K., Risi, S., Pickles, A., & Lord, C. (2007). The Autism Diagnostic Observation Schedule: revised algorithms for improved diagnostic validity. *Journal of Autism and Developmental Disorders*, 37, 613–627.

Green, J. (1990). Is Asperger's a syndrome? *Developmental Medicine and Child Neurology*, 32, 743–747.

Greer, B. D., & Kodak, T. (2017). Introduction to the special issue on applied behavior analysis. *Learning and Motivation*,(2017). https://doi.org/10.1016/j.lmot.2017.03.005.

Grens, K. (2013). Disorder no more: researchers hunt for biomarkers of Asperger's syndrome a condition that officially no longer exists. *The Scientist*,(December 1) Retrieved from (2013). https://www.the-scientist.com/?articles.view/articleNo/38378/title/Disorder-No-More/.

Hallmayer, J., Cleveland, S., Torres, A., Phillips, J., Cohen, B., Torigoe, T., … Smith, K. (2011). Genetic heritability and shared environmental factors among twin pairs with autism. *Archives of General Psychiatry*, 68, 1095–1102.

Hecht, P. M., Hudson, M., Connors, S. L., Tilley, M. R., Liu, X., & Beversdorf, D. Q. (2016). Maternal serotonin transporter genotype affects risk for ASD with exposure to prenatal stress. *Autism Research*, 9, 1151–1160.

Horton, R. (2004). A statement by the editors of The Lancet. *The Lancet*, 363, 820–821.

Howlin, P., Magiati, I., & Charman, T. (2009). Systematic review of early intensive behavioral interventions for children with autism. *American Journal on Intellectual and Developmental Disabilities*, 114, 23–41.

Hurley, A. M., Tadrous, M., & Miller, E. S. (2010). Thimerosal-containing vaccines and autism: a review of recent epidemiologic studies. *Journal of Pediatric Pharmacology and Therapeutics*, 15, 173–181.

Iossifov, I., O'roak, B. J., Sanders, S. J., Ronemus, M., Krumm, N., Levy, D., & Patterson, K. E. (2014). The contribution of de novo coding mutations to autism spectrum disorder. *Nature*, 515, 216–221.

Laumonnier, F., Bonnet-Brilhault, F., Gomot, M., Blanc, R., David, A., Moizard, M.-P., … Calvas, P. (2004). X-linked mental retardation and autism are associated with a mutation in the NLGN4 gene, a member of the neuroligin family. *American Journal of Human Genetics*, 74, 552–557.

Le Couteur, A., Lord, C., & Rutter, M. (2003). *Autism Diagnostic Interview-Revised (ADI-R)*. Los Angeles, CA: Autism Genetic Resource Exchange.

Lord, C., Rutter, M., & Le Couteur, A. (1994). Autism Diagnostic Interview-Revised: a revised version of a diagnostic interview for caregivers of individuals with possible pervasive developmental disorders. *Journal of Autism and Developmental Disorders*, 24, 659–685.

Lovaas, O. I. (1981). *Teaching developmentally disabled children: the me book*. Austin, TX: Pro-Ed.

Lyall, K., Croen, L., Daniels, J., Fallin, M. D., Ladd-Acosta, C., Lee, B. K., … Volk, H. (2017). The changing epidemiology of autism spectrum disorders. *Annual Review of Public Health*, 38, 81–102.

Lyall, K., Croen, L. A., Sjödin, A., Yoshida, C. K., Zerbo, O., Kharrazi, M., & Windham, G. C. (2017). Polychlorinated biphenyl and organochlorine pesticide concentrations in maternal mid-pregnancy serum samples: association with autism spectrum disorder and intellectual disability. *Environmental Health Perspectives*, 125, 474–480.

McPartland, J. C., Reichow, B., & Volkmar, F. R. (2012). Sensitivity and specificity of proposed DSM-5 diagnostic criteria for autism spectrum disorder. *Journal of the American Academy of Child and Adolescent Psychiatry*, 51, 368–383.

Peters-Scheffer, N., Didden, R., Korzilius, H., & Sturmey, P. (2011). A meta-analytic study on the effectiveness of comprehensive

ABA-based early intervention programs for children with autism spectrum disorders. *Research in Autism Spectrum Disorders, 5*, 60–69.

Plotkin, S., Gerber, J. S., & Offit, P. A. (2009). Vaccines and autism: a tale of shifting hypotheses. *Clinical Infectious Diseases, 48*, 456–461.

Rauch, A., Wieczorek, D., Graf, E., Wieland, T., Endele, S., Schwarzmayr, T., … Di Donato, N. (2012). Range of genetic mutations associated with severe non-syndromic sporadic intellectual disability: an exome sequencing study. *The Lancet, 380*, 1674–1682.

Reaven, J. A., Hepburn, S. L., & Ross, R. G. (2008). Use of the ADOS and ADI-R in children with psychosis: importance of clinical judgment. *Clinical Child Psychology and Psychiatry, 13*, 81–94.

Reichow, B. (2012). Overview of meta-analyses on early intensive behavioral intervention for young children with autism spectrum disorders. *Journal of Autism and Developmental Disorders, 42*, 512–520.

Roane, H. S., Fisher, W. W., & Carr, J. E. (2016). Applied behavior analysis as treatment for autism spectrum disorder. *The Journal of Pediatrics, 175*, 27–32.

Robinson, E. B., Lichtenstein, P., Anckarsäter, H., Happé, F., & Ronald, A. (2013). Examining and interpreting the female protective effect against autistic behavior. *Proceedings of the National Academy of Sciences of the United States of America, 110*, 5258–5262.

Rosa's Law, 111th Congress, Pub. L. No. 111-256, 124 Stat. 2643-2645 (2010).

Russell, G., Steer, C., & Golding, J. (2011). Social and demographic factors that influence the diagnosis of autistic spectrum disorders. *Social Psychiatry and Psychiatric Epidemiology, 46*, 1283–1293.

Schaafsma, S. M., & Pfaff, D. W. (2014). Etiologies underlying sex differences in autism spectrum disorders. *Frontiers in Neuroendocrinology, 35*, 255–271.

Skafidas, E., Testa, R., Zantomio, D., Chana, G., Everall, I. P., & Pantelis, C. (2014). Predicting the diagnosis of autism spectrum disorder using gene pathway analysis. *Molecular Psychiatry, 19*, 504–510.

Szatmari, P. (1991). Asperger's syndrome: diagnosis, treatment, and outcome. *Psychiatric Clinics of North America, 14*, 81–93.

Szatmari, P. (2000). The classification of autism, Asperger's syndrome, and pervasive developmental disorder. *Canadian Journal of Psychiatry, 45*, 731–738.

Taylor, B., Miller, E., Farrington, C., Petropoulos, M.-C., Favot-Mayaud, I., Li, J., & Waight, P. A. (1999). Autism and measles, mumps, and rubella vaccine: no epidemiological evidence for a causal association. *The Lancet, 353*, 2026–2029.

Torrico, B., Chiocchetti, A. G., Bacchelli, E., Trabetti, E., Hervás, A., Franke, B., … Duketis, E. (2017). Lack of replication of previous autism spectrum disorder GWAS hits in European populations. *Autism Research, 10*, 202–211.

Treffert, D. A. (2009). The savant syndrome: an extraordinary condition. A synopsis: past, present, future. *Philosophical Transactions of the Royal Society of London B: Biological Sciences, 364*, 1351–1357.

Tsang, K. M., Croen, L. A., Torres, A. R., Kharrazi, M., Delorenze, G. N., Windham, G. C., … Weiss, L. A. (2013). A genome-wide survey of transgenerational genetic effects in autism. *PloS One, 8*, e76978.

Van Wijngaarden-Cremers, P. J., van Eeten, E., Groen, W. B., Van Deurzen, P. A., Oosterling, I. J., & Van der Gaag, R. J. (2014). Gender and age differences in the core triad of impairments in autism spectrum disorders: a systematic review and meta-analysis. *Journal of Autism and Developmental Disorders, 44*, 627–635.

Virués-Ortega, J. (2010). Applied behavior analytic intervention for autism in early childhood: meta-analysis, meta-regression and dose–response meta-analysis of multiple outcomes. *Clinical Psychology Review, 30*, 387–399.

Wakefield, A., Murch, S., Anthony, A., Linnell, J., Casson, D., Malik, M., … Walker-Smith, J. (1998). RETRACTED: ileal-lymphoid-nodular hyperplasia, non-specific colitis, and pervasive developmental disorder in children. *The Lancet, 351*, 637–641.

Wong, C., Odom, S. L., Hume, K., Cox, A. W., Fettig, A., Kucharczyk, S., … Schultz, T. R. (2014). *Evidence-based practices for children, youth, and young adults with autism spectrum disorder*. Chapel Hill: The University of North Carolina, Frank Porter Graham Child Development Institute, Autism Evidence-Based Practice Review Group.

Yuen, R. K., Merico, D., Bookman, M., Howe, J. L., Thiruvahindrapuram, B., Patel, R. V., … Wang, Z. (2017). Whole genome sequencing resource identifies 18 new candidate genes for autism spectrum disorder. *Nature Neuroscience, 20*, 602–611.

Further Reading

Yuen, R. K., Merico, D., Cao, H., Pellecchia, G., Alipanahi, B., Thiruvahindrapuram, B., … Zhang, T. (2016). Genome-wide characteristics of de novo mutations in autism. *npj Genomic Medicine, 1*, 16027.

17

Schizophrenia

Today, schizophrenia has become part of the vocabulary of most adults through exposure to films, television, and newspapers, but how many people can tell us precisely who is schizophrenic and who is not? Psychiatrists themselves do not find it easy to make this distinction. Is schizophrenia a distinct disease entity, a mental illness, or a mode of thinking and behaving that some people use to cope with confusing or seemingly hopeless circumstances? There is no biochemical or physiological test to identify schizophrenia. Chromosome karyotyping, DNA microarrays, and MRI brain scans do not provide unambiguous answers, either. At the present time, schizophrenia is diagnosed solely on the basis of patterns of thought and behavior (Box. 17.1).

Although specialists may disagree on who has what mental problem, there is usually one official definition in every country with a strong central government. One stimulus toward uniformity is the desire to classify and count people in a census or a survey of medical services. In the United States, for example, the census of 1840 recognized only "idiocy," whereas by 1880, there was a list of seven maladies including mania, dementia, and dipsomania. In 1889, the International Congress of Medical Science met in Paris to discuss and decide on the kinds of mental problems recognized by physicians, and it issued a list of 11 mental disorders, but the categories were not adopted by all the countries represented in Congress (Spitzer & Williams, 1985).

At the 29th Congress of Southwestern German Psychiatry in 1898, Kraepelin (1898) presented a paper on "dementia praecox," a syndrome of insanity with early onset (usually in adolescence) and inevitable deterioration. The Swiss psychiatrist Bleuler then proposed the term "schizophrenia" in 1911 to identify a broad category of people who suffered from "breaking of associative threads" of thought because of an unseen disease process (Neale & Oltmanns, 1980). There was competition between those men and their followers for recognition

of their syndromes. As we now know, schizophrenia was the survivor.

THE ERA OF ICD AND DSM

After World War II, the World Health Organization, an agency of the newly founded United Nations, undertook a revision of the International List of Causes of Death and in 1948 included certain mental disorders for the first time (ICD-6). There was widespread unhappiness with its treatment of psychiatric categories, and it was officially adopted in only five countries, including the United Kingdom. In the United States, the American Psychiatric Association (APA) published its own classification, the Diagnostic and Statistical Manual of Mental Disorders (DSM-1) in 1952. The system reflected the prevailing views of the American psychiatrist Adolph Meyer, who spoke of a "schizophrenic reaction" to adverse environment. Besides favoring an environmental explanation of schizophrenia, American psychiatrists further broadened the diagnostic criteria, to the extent that 80% of mental patients in New York City were labeled "schizophrenic" in 1952 (Neale & Oltmanns, 1980).

In 1968, the APA published a sweeping revision of its classification scheme, DSM-2. The ICD was then in its eighth revision, and international efforts had been made to adopt similar terms in ICD and DSM. Schizophrenia had been labeled "schizophrenic reaction" in DSM-1. The foreword to DSM-2 stated as follows: "This change of label has not changed the nature of the disease, nor will it discourage continuing debate about its nature or causes. Even if it had tried, the Committee could not establish agreement about what this disorder is; it could only agree on what to call it." (Gruenberg, 1968)

An obvious difference was seen between standards of British and American psychiatrists that led to the establishment in 1972 of the US-UK Cross-National Project

211

BOX 17.1

"After that year at college, I returned to school teaching for a year ... During those months my mind convinced me completely of intense feelings of other people towards me - feelings of love, hate, indifference, spite, friendship. These certainties were groundless and led me into dreadful relationships with people... Not knowing that I was ill, I made no attempt to understand what was happening, but felt that there was some overwhelming significance in all this, produced either by God or Satan, and I felt that I was duty-bound to ponder on each of these new interests, and the more I pondered the worse it became. The walk of a stranger on the street could be a 'sign' to me which I must interpret. Every face in the windows of a passing streetcar would be engraved on my mind, all of them concentrating on me and trying to pass me some sort of message. By the time I was admitted to hospital I had reached a state of 'wakefulness' when the brilliance of light on a window sill or the colour of blue in the sky would be so important it could make me cry... Completely unrelated events became intricately connected in my mind." (MacDonald, 1960)

---

to study the problem. Staff psychiatrists working on the project carefully diagnosed patients in both New York and London using the same ICD-8 criteria, and in both cities, they called about 30% of the patients "schizophrenic." Nevertheless, hospital psychiatrists in New York who were not part of the project team diagnosed about 118 (60%) of their 192 patients schizophrenic, a rate twice as high as their London hospital counterparts who diagnosed 59 (34%) of their 174 patients as schizophrenic. Many patients considered depressive, manic, neurotic, or disordered personality under ICD-8 criteria were regarded as schizophrenic in New York (Kendell, Cooper, & Gourlay, 1971).

SANE IN INSANE PLACES

The dispute over diagnosis reached a crisis level in 1973 with the publication in *Science* of results of a study by D. L. Rosenhan, a Stanford University professor interested in the behavior of psychiatrists toward patients in hospital wards (Rosenhan, 1973). To gather his data, Rosenhan recruited seven mentally sound adult volunteers and instructed them to seek admission to 12 different psychiatric hospitals in five states. He himself also sought admission. Each of the eight pseudopatients phoned for an appointment and during the hospital interview complained of hearing unfamiliar voices saying "empty," "hollow," and "thud." Apart from faking that one symptom and their names, the pseudopatients reported other details of their past accurately. If admitted, they were instructed to cease faking any symptoms. Results of the ruse were shocking. On all 12 occasions, the pseudopatients were admitted, 11 with a diagnosis of schizophrenia and one as a manic-depressive. It took an average of 19 days and as many as 52 days for staff psychiatrists to conclude the people were well enough to return home. Not one staff member showed suspicion that any one of the eight was actually healthy, but real patients frequently did notice this. One inmate addressed a pseudopatient, who had been taking notes about life in the hospital: "You're not crazy. You're a journalist or a professor. You're checking up on the hospital."

Rosenhan then presented a new challenge to the psychiatric profession. He told the results of his study to the staff at a research hospital and forewarned them that in the next 3 months, he would send one or more pseudopatients there to seek admission. Now on the alert, staff members from chief psychiatrist to ward attendants scrutinized 193 newly admitted patients. This time, they had serious doubts about several admissions. At least one psychiatrist voiced suspicion about 23 of the patients. On 41 occasions, at least one staff member claimed with virtual certainty that a pseudopatient had been admitted. Imagine the chagrin of these high and mighty judges of sanity when it was finally revealed that the wily Rosenhan had not sent even one pseudopatient in that period! The report prompted an indignant reply from R. L. Spitzer, a professor at Columbia University and a member of the committee that drafted DSM-2, who denounced Rosenhan's work as "pseudoscience" (Spitzer, 1976).

DSM-3 TO -5

Another vexing difficulty in diagnosis was evident from disagreements among psychiatrists at a single institution when presented with the same patients. The index of agreement is termed "kappa" or κ, a coefficient that is 1.0 if there is perfect agreement and 0 if agreement is at the level of pure chance or guessing. A review of seven studies that used DSM-1 or DSM-2 criteria reported kappa values with an average of 0.56 and a range from 0.27 to 0.77 (Spitzer & Fleiss, 1974). In view of these and other practical problems of diagnosis, the APA in 1974 set up a new Task Force on Nomenclature and Statistics to revise the DSM once more. Spitzer was appointed the chairman of the task force, and he selected the initial members of the committee. Eventually, over 100 psychiatrists were involved in writing the draft of DSM-3. The first printing of the draft was sent to thousands of APA members and other professionals in January of 1978, asking for their criticisms and suggestions.

The debate over terms and criteria in DSM-3 revealed much about professional and cultural influences on definitions of mental disorders. Psychoanalysts were angered by the omission of the term "neurosis," which was still widely used by followers of Freud. Another contentious point arose in the definition of schizophrenia, which was identified with 10 different "characteristic delusions," at least one of which must be present during the active phase of the disorder (The Task Force on Nomenclature and Statistics of the American Psychiatric Association, 1978). Item 3 on the list on page C-10 was as follows: "Thought insertion: Experiences thoughts, which are not his own, being inserted into his mind (other than by God)." Psychiatrists had always found it difficult to distinguish between socially acceptable religious belief and outright insanity with delusions that often feature strong religious content, such as the belief one is Jesus Christ returned to earth. Some wanted an explicit statement in the DSM that people who believed God could insert thoughts were sane, whereas those who believed living humans or dead relatives could do this might be schizophrenic. The APA Committee, in its infinite wisdom, decided to delete the reference to God and leave it up to the discretion of the individual psychiatrist to separate sincere believers from the mentally ill.

Finally, DSM-3 was published in 1980. It embodied stricter criteria that would exclude Rosenhan and his confederates from mental hospitals. The new version included a mandatory active phase with at least one significant symptom from a list of six and a deterioration from previous functioning, plus a minimum duration of 6 months and onset before age 45. DSM-2, on the other hand, had not required an active phase or deterioration and set no minimum duration or age limit.

Many people who would have been categorized as schizophrenic under DSM-2 would not be classified that way under the new regime. A study in Illinois used taped interviews and written case histories of 252 patients who had originally been diagnosed under DSM-2 and then were diagnosed by a clinical research team using DSM-3. Of the 103 classed as schizophrenic under DSM-2, only 57 qualified under DSM-3 (Silverstein, Warren, Harrow, Grinker, & Pawelski, 1982). Another study evaluated the records of 283 patients discharged from Johns Hopkins Hospital between 1948 and 1960. The original diagnoses done with DSM-1 considered 236 of them schizophrenic, whereas DSM-3 classified only 105 as schizophrenic (Stephens et al., 1982).

The reduced rate would not become evident from hospital admissions immediately, because there is no legal requirement that a psychiatrist use DSM or any other system for making a diagnosis. A survey of practicing psychiatrists in the United States found that criteria similar to DSM-3 were used mainly by younger doctors, whereas the more experienced ones tended to stick with their previous systems (Lipkowitz & Idupuganti, 1983). Those authors remarked that "A high number of psychiatrists continued to approach the diagnosis of schizophrenia in an individualistic, unsystematic way."

The DSM was revised again and then again, but for schizophrenia, changes in criteria were minimal. The principal change was that DSM-4 and DSM-5 required that at least two core symptoms in group A be expressed, whereas DSM-3 was satisfied with just one. The current requirements under DSM-5 indicate that, for milder cases that do not last more than 6 months, schizophreniform disorder (DSM 295.40, ICD F20.81) has a set of criteria much like that of schizophrenia and, for cases wherein symptoms last more than 1 day but less than 1 month, brief psychotic disorder (DSM 298.8, ICD F23) may be an appropriate diagnosis. Otherwise, it appears that, after many years of development, a stable set of criteria for diagnosis of schizophrenia and closely related disorders has been achieved.

RESEARCH DOMAIN CRITERIA (RDoC)

The apparent stability did not last long. The National Institute of Mental Health in the United States has grown increasingly dissatisfied with the current system of diagnosis. Just as DSM-5 was being proclaimed, the NIMH director (Insel, 2013) decried a system based on a "consensus about clusters of clinical symptoms" that cannot rely on "any objective laboratory measure." Insel (2014) remarked, "So far, we don't have rigorously tested, reproducible, clinically actionable biomarkers for any psychiatric disorder." NIMH therefore initiated a new approach that is to be based on new research data that may cluster in new categories. Funding of research by NIMH is to be organized around five major research domains or systems (RDoC): negative valence, positive valence, cognitive, social processes, and regulatory (Simmons & Quinn, 2014). Each domain encompasses several psychological/behavioral constructs. Research will then be done to explore each construct at multiple levels: genes, molecules, cells, circuits, physiology, behavior, self-reports, and paradigms. The longer-term goal is to fill in the cells in the grid of domains by levels with up-to-date knowledge, but the grid does not include everything; Insel (2014) noted it has no place for information on development or environmental exposures. He earlier emphasized that "mental disorders are biological disorders" and diagnosis should be based on biology. Placing biology at the center of the undertaking and excluding environment appears rather one-sided, and it contradicts the declared desire to base new categories on new data rather than old ideas.

ENVIRONMENT

To assess the possible role of experience in the etiology of schizophrenia, there must be a good way to measure experience. This is extraordinarily difficult. A person can travel through a richly varied environment and yet be affected by only those few features that draw the attention, while myriad other features of the surroundings register no lasting change in the brain. A home may be full of unread books. Environment involves things in the surroundings that might impinge on the individual, whereas experience includes only those things that actually do effect a change in the individual. An indicator of environment is indirect and may have no relevance at all for a specific person. For example, a concussion appears to increase the later risk of developing schizophrenia (Molloy, Conroy, Cotter, & Cannon, 2011). A blow to the head is certainly a kind of experience, whereas playing hockey, rugby, or football is a feature of the environment that may be a risk factor for later cognitive decline but on its own does not damage the brain. Researchers often opt for recording aspects of the environment that can be easily tallied, such as parents' past education, current income, or type of neighborhood, even though their relations with experiences of the child may be tenuous.

Large Surveys and Meta-Analyses

Several factors that increase the frequency of schizophrenia have been identified through large population studies, extensive literature reviews, and formal meta-analyses of published studies. Every study involved a different array of adverse childhood experiences or family environmental indicators, but all of them detected significant increases in cases of schizophrenia. The National Comorbidity Survey in the United States questioned a nationally representative sample of 5782 adults about childhood and later adverse experiences (Table 17.1) and also assessed the frequency of psychosis that was mainly schizophrenia. About 3%–8% of the sample experienced various events, and about 0.7% of adults were judged psychotic. In the British psychiatric morbidity survey, the frequency of experiencing bullying or serious illness was quite high, but the frequency of psychosis was also about 0.7%. The surveys described by Shevlin, Houston, Dorahy, and Adamson (2008) found that

TABLE 17.1 Population Surveys of Adverse Childhood Experiences[a]

National Comorbidity Survey (NCS)—the United States $N = 5782$ adults (41 cases of psychosis)		British Psychiatric Morbidity Survey (BPMS) $N = 8580$ adults (60 cases of psychosis)	
Adversity	**%**	**Adversity**	**%**
Neglect	2.8	Bullying	17.3
Physical abuse	4.2	Serious illness	27.9
Physical assault	8.4	Sexual abuse	4.0
Rape	4.4	Violence in home	8.3
Sexual touching	7.1	Violence at work	3.8

[a] *Percentages denote the proportion of the entire adult sample that reported that kind of experience.*
Adapted from Shevlin, M., Houston, J. E., Dorahy, M. J., & Adamson, G. (2008). Cumulative traumas and psychosis: an analysis of the National Comorbidity Survey and the British Psychiatric Morbidity Survey. Schizophrenia Bulletin, 34, *193–199, by permission of Oxford University Press.*

incidence of psychosis rose with the number of adverse experiences (Fig. 17.1). Rate of psychosis was quite low (1%) when there were no adverse events queried by the survey and only slightly higher when there was one such event, but incidence increased dramatically when a child experienced multiple adverse events. Four adverse experiences were associated with a 5% rate of psychosis, and five adverse events had a rate of schizophrenia over 15%.

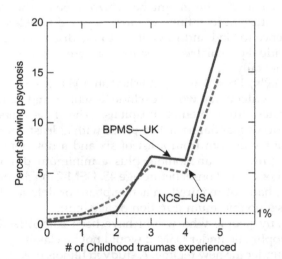

FIG. 17.1 Percent of adults exhibiting psychosis in relation to number of childhood traumas or adverse experiences reported in nationwide surveys of adults. Survey items were somewhat different in the two countries (Table 17.1). Data are from the National Comorbidity Survey (NCS) in the United States and the British Psychiatric Morbidity Survey in the United Kingdom. The dotted line at 1% is the approximate population risk for psychosis. *Reprinted from Shevlin, M., Houston, J. E., Dorahy, M. J., & Adamson, G. (2008). Cumulative traumas and psychosis: an analysis of the National Comorbidity Survey and the British Psychiatric Morbidity Survey.* Schizophrenia Bulletin, 34, *193–199, by permission of Oxford University Press.*

TABLE 17.2 Literature Review and Meta-Analyses of Early Adversity and Later Schizophrenia

van Os et al. (2010)		Varese et al. (2012)		Matheson, Shepherd, Pinchbeck, Laurens, and Carr (2013)	
Review of >30 studies		Meta-analysis of 36 studies[a]		Meta-analysis of 7 studies	
Childhood indicator	Studies	Experience	Odds ratio	Experience	Odds ratio
Cannabis use[b]	8	Bullying	2.39	Incest	3.60[c]
Childhood trauma[b]	7	Emotional abuse	3.40	Neglect	
Minority group status[b]	7	Neglect	2.90	Physical abuse	
Urban living[b]	6	Parent death	1.70	Sexual abuse	
Several studies showed inconsistent effects of prenatal environments		Physical abuse	2.95	Trauma	
		Sexual abuse	2.38	Victimization	

[a] Fewer than 10 studies reported data for certain experiences.
[b] Showed consistent increase in frequency of schizophrenia.
[c] The value applies to the combined effects of the six experiences.

A review of more than 30 published studies by van Os, Kenis, and Rutten (2010) found that four indicators of adverse environment were consistently reported to increase the risk of later showing schizophrenia (Table 17.2). Several studies also assessed a wider range of environments, but there were too few that examined the same risk factors to warrant a firm conclusion. The inconsistency of environmental factors assessed by different research teams is a major weakness in this field of research. What to measure and what scale is used to rate a factor have not been standardized at all.

A large meta-analysis of 36 studies (Varese et al., 2012) found sufficient data pertaining to six kinds of childhood experiences (Table 17.2). Results were quantified in terms of odds ratio (OR) that compares the risk of psychosis when an adverse experience is present to risk when it is absent. For an experience that has no impact on schizophrenia at all, the ratio would be close to 1.0, whereas a factor that doubles the risk would be rated at OR = 2.0. Several factors in the meta-analysis yielded OR close to

3, which is considered to be a moderate to large effect. Another meta-analysis involving seven studies found an odds ratio of 3.60 for experiences of one or more adverse events (Matheson et al., 2013).

Another approach to assessing environmental influences is to note the timing of onset of symptoms. It is commonly observed that an episode of outright psychotic symptoms is usually preceded by a *prodromal phase* of cognitive decline, social withdrawal, and sometimes depression that do not themselves warrant a diagnosis under DSM but are noteworthy and often noticed by family and friends. The age at onset of prodrome was assessed in 750 male schizophrenic patients in the Göttingen Research Association for Schizophrenia data set in Germany, and several environmental risk factors were also tabulated (Stepniak et al., 2014). For some factors, the onset of prodrome was hastened by about 2 years in comparison with men who did not share that experience (Fig. 17.2). Even more compelling was the clear pattern of earlier onset of prodrome when there was more

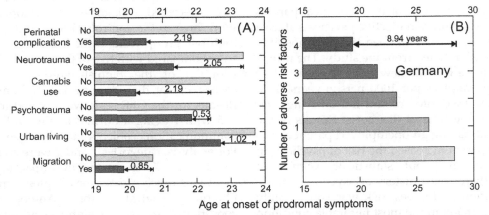

FIG 17.2 (A) Average age at onset of prodromal symptoms of schizophrenia in patients in Germany who had experienced several kinds of risk factors versus those who lacked them. (B) Average age of the onset of prodromal symptoms in relation to the number of risk factors the person experienced. *Reprinted from Stepniak, B., Papiol, S., Hammer, C., Ramin, A., Everts, S., Hennig, L., ... Ehrenreich, H. (2014). Accumulated environmental risk determining age at schizophrenia onset: a deep phenotyping-based study. The Lancet Psychiatry, 1, 444–453, Copyright 2014, with permission from Elsevier.*

than one adverse experience, amounting to 9 years when four adverse risk factors were present.

In all of the above studies, the higher rates of psychosis are not necessarily attributable solely to environmental factors because there could have been hereditary differences between families that contributed to an abysmal home environment for the child and also elevated frequency of adult psychosis. It is fair to conclude that the data strongly support the idea that adverse childhood environment contributes to later psychosis but do not offer unequivocal proof.

Adoption

Heredity and environment can be disentangled to some extent when a child is adopted into a better home and reared by nonrelatives. A large study in Sweden searched registries of all adoptions and all hospital admissions for different kinds of mental disorders (Wicks, Hjern, & Dalman, 2010) and indicators of socioeconomic status (employment, single parent, and apartment living). There were 13,116 adoptees identified in the register. Many dated from the 1950s when the rate of adoption was very high (869 children per year) because unwed teenage girls often gave up a child for adoption, especially when the father was unknown. By 1984, the rate of adoption had fallen to only 35 children per year, which suggests that large-scale adoption studies will not be feasible in future years. There were 230 (1.7%) adoptees hospitalized for nonaffective psychosis, largely schizophrenia but not major depression or bipolar disorder. The comparable rate among 2.9 million nonadoptees in Sweden was 0.85%. The only data on environments of the adoptees came from a national census taken every 5 years. Only adoptees whose biological mother was known were included in the study. In 41% of cases, the biological father was unknown. Furthermore, adopting parents were identified in registries. Researchers defined genetic liability as having one or two biological parents who were hospitalized for nonaffective psychosis at least once. If an adopting parent had been similarly hospitalized, the case was eliminated from the study. The principal outcome measure was the new cases of nonaffective psychosis among adoptees per 1000 person-years. That measure sought to take into account the differing ages of the adoptees at the time of diagnosis. Results of the study showed that having an unemployed parent or living with a single parent increased the risk of psychosis in the adoptees by small but statistically significant amounts. When there was a history of psychosis among the biological parents, adoptees' risk was considerably higher even when living in the most favorable condition with two employed parents, which supported the notion that genetic factors passed from parent to child could

influence psychosis, but the most striking finding was the very large increase in incidence of psychosis among adoptees having a psychotic biological relative *and* living in more difficult circumstances involving rearing by a single mother or having an unemployed parent.

A similar adoption study done in Finland involved 19,447 adult women who had been hospitalized between 1960 and 1979 with a diagnosis of schizophrenia or paranoid psychosis (Tienari et al., 2004). A search of population registers detected several hundred cases in which a child of one of those women had been given up for adoption. A control group of adoptees whose mothers were not psychotic was also studied. Biological mothers, adoptees, and adopting parents were evaluated with DSM-3 revision (DSM-3R) criteria, and the entire adoptive family was assessed with interviews and standard psychological tests. Features of the adoptive family environment were also assessed with the global family rating (GFR) scale. Adoptees were assessed 12 years after adoption and then again in a 21-year follow-up by searching registers of patients admitted for mental disorders. Individual psychiatric interviews and diagnosis were done for 389 adopting mothers and 368 adopting fathers. There were 333 families with complete data (Wynne et al., 2006).

High-risk adoptees were those with a biological mother diagnosed as schizophrenic or in the schizophrenic "spectrum," whereas low-risk adoptees had a mother with no diagnosis or a diagnosis not in the spectrum. Among the 166 high-risk adoptees, 123 had a mother who met narrowly defined criteria for schizophrenia, and another 43 had a mother in the broader "spectrum." There were 167 control adoptees. The "spectrum" in this study was wide indeed. It included DSM-3R diagnoses of typical schizophrenia, schizoaffective psychosis, schizophreniform psychosis, delusional disorder, psychosis not otherwise specified, "odd cluster" personality disorders, and bipolar and depressive disorders with psychotic features.

For the adoptees themselves, 12 in the high-risk group (7.2%) received a diagnosis of schizophrenia compared with three in the controls (1.8%), and another 12 in the high-risk group were in the less seriously affected spectrum compared with five of the controls. The unique contribution of the Finnish study was the thorough assessment of the adopting family home environment. Results for the GFR scale were scored in five categories 1–5 with 1 being a normal family with no noteworthy communication problems and 5 being severely abnormal. For analysis, categories 1 and 2 were combined, as were 4 and 5. Fig. 17.3 shows that the impact of family environment was much greater for adoptees who had a schizophrenic biological mother. Among those in GFR conditions 1–2, 5.8% of high-risk adoptees and 2.9% of low-risk adoptees were in the spectrum, but only one of 66 in the high-risk group versus none in low-risk

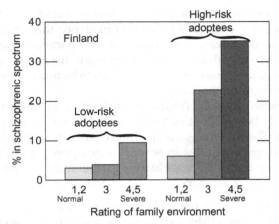

FIG. 17.3 Percentage of adopted persons in Finland diagnosed in the schizophrenic spectrum in relation to social environment in the adopting family and the presence of schizophrenia (high risk) or its absence (low risk) in a biological parent. *Reproduced with permission from Wynne, L. C. T., Niemenen, P. Sorri, A., Lahti, I., Moring, J., Naarala, M., ... Meittunen, J. (2006) Genotype-environment interaction in the schizophrenia spectrum: genetic liability and global family ratings in the Finnish adoption study. Family Process, 45, 419–434, published by John Wiley and Sons.*

was actually diagnosed as schizophrenic. The difference between high-risk and low-risk conditions was striking only when they were reared in very adverse home environments. This is very similar to the pattern seen in the recent adoption study in Sweden that assessed only broad socioeconomic environmental factors (Wicks et al., 2010).

These studies of adoption and adverse environments provide clear evidence for the importance not only of early childhood environment but also of *heredity-environment interaction* in schizophrenia. The large interaction effects contradict one of the key assumptions of heritability analysis (Chapter 14) that effects of heredity and environment are additive.

HEREDITY AND FAMILY

Things that are transmitted across generations in a family are termed not only familial but also sometimes hereditary. Genes are hereditary, and so is wealth. If something is familial, this fact tells very little about whether it involves genes or environment. Nevertheless, the study of heredity often begins with an examination of resemblance of relatives in a family. If that resemblance is very weak, then genetics probably has little to do with the particular trait. On the other hand, if parent-offspring resemblance is quite strong, this could be a sign of genetic involvement.

A Family Study

A family study usually begins by identifying *probands* or *index cases* that have some trait of interest, perhaps schizophrenia. Then, researchers try to contact all living

relatives and assess them for the same trait. Of primary interest are first-degree relatives such as parents, siblings, and children of the proband. The research design often involves a *control group* of cases that do not have the trait in question. Their relatives can be assessed, and the frequencies of the trait among relatives of index cases and controls can be compared.

Diagnosis in a family study deserves special attention because it may bias the outcome of the research and create a false impression about the role of heredity. This is apparent in the work of Franz Kallmann (1938) done in Germany in the 1930s. He characterized schizophrenia as a disease that produces "an unceasing source of maladjusted cranks, asocial eccentrics, and the lowest type of criminal offenders." In Kallmann's study, the person deciding on the diagnosis knew which person was a relative of which index case, whereas modern methodology requires "blind" *diagnosis* in which the person making the diagnosis does not know the family history or background of the patient. Kallmann firmly believed schizophrenia was hereditary before he conducted his study. He thought that a person might have a "homozygotic predisposition to schizophrenia" and yet show only mild symptoms or perhaps none at all. Likewise, he thought a person might show many symptoms of SZ and yet not have the disease, hence the need for each case to be "correctly interpreted and evaluated" by himself. How should this be done? Kallmann spelled out one criterion very clearly: "Cases which present a schizophrenic picture clinically but lack the hereditary predisposition must be excluded from the disease group of 'genuine' schizophrenia and differentiated as 'schizoform' psychoses of exogenous origin." That is, if there is no family history, the person must not have "genuine" SZ. For this and other reasons, his data are no longer taken seriously in evaluations of heredity except as negative examples from which lessons can be learned.

Family studies that span generations often run afoul of changing diagnostic criteria. If parents are judged according to DSM-1 or DSM-2 while their sons and daughters are evaluated with the stricter DSM-3 criteria, any influence of heredity may appear rather weak because the frequency of the trait is so much less among children than parents. Ideally, parents and children should be rated using the same diagnostic criteria. This was done after the publication of DSM-3 in studies wherein psychiatrists interviewed both kinds of relatives using the same criteria as part of a research project.

One of the first to be done with the new criteria delivered a surprise. Pope, Jonas, Cohen, and Lipinski (1982) located 39 probands with definite schizophrenia and collected extensive data on 199 first-degree relatives from patient charts, so that the relatives could be also be rated with DSM-3 criteria. Not even one of the 199 relatives qualified as definite schizophrenia. This created a stir and spawned further family studies. The next year,

Abrams and Taylor (1983) rated 30 schizophrenic probands and 181 first-degree relatives, finding just two of them (1.6%) to be schizophrenic.

Two much large studies soon reported higher family incidence. Kendler, Gruenberg, and Tsuang (1985) identified 253 schizophrenic probands in Iowa and 668 first-degree relatives as well as 261 controls that were awaiting surgery in the same hospital along with their relatives. There were 26 instances of schizophrenia (3.7%) among relatives of index cases but only 0.2% for control relatives. Another study done in New York City involved 90 schizophrenic probands and 376 first-degree relatives as well as a control group of similar size (Baron et al., 1985). There were 19 (5.8%) cases of schizophrenia among relatives of index cases and 0.6% among controls.

Other family studies done since then have obtained similar results. Frequency of schizophrenia is usually higher in relatives of schizophrenic probands than controls, but the frequency is generally not very high. The median incidence of schizophrenia in 18 studies of first-degree relatives was about 5% in a review by Shih, Belmonte, and Zandi (2004).

Whether the parent-offspring correlation arises from heredity or environments or both, the correlation is not very high. If somehow the separate contributions of genes and environments could be ascertained, it is likely that the correlation attributable to just one factor would be even less than the parent-offspring resemblance seen in the typical family study. This point needs to be kept in mind when examining other kinds of evidence of genetic involvement.

Adoption Studies

As discussed above, adoption places a child from one background into a new environment. The child's genes are the same before and after adoption, whereas the environment will have changed. Thus, adoption can provide information about the importance of environment, as shown in the Swedish and Finnish adoption studies. It can also provide clues about possible contributions of genotype, provided there are two comparison groups, one having some kind of familial risk for schizophrenia and the other lacking it.

Two widely cited adoption studies were done in Denmark where government agencies maintain extensive records. The work was financed by the National Institute of Mental Health (NIMH) in the United States and led by the American psychiatrists Kety and Rosenthal in collaboration with their Danish counterparts. They opted for Denmark because American "adoption agencies and the courts have been zealous in their desire to protect [the privacy of] all parties to the adoption," so the researchers "were obliged to go abroad" (Rosenthal, 1971). The studies became not only widely known but

also controversial because of doubts about just how "blind" were the diagnoses and evidence that different diagnostic criteria were used for two parallel series of cases in a way that made the influence of heredity appear larger. More recent evidence appears to be much stronger in terms of methodology.

The work led by Tienari and colleagues in Finland has already been described briefly in the section on environment. The study was a major improvement over the Danish adoption studies in several ways, partly because it was begun in 1977 after learning some unhappy lessons from the earlier work.

- A much larger, nationwide sample was obtained. There were 145 biological mothers who had given up a child for adoption and 158 controls with no diagnosis.
- Diagnosis was done using standardized interviews and stricter criteria in DSM-3R. Results were described for *both* narrow, strictly defined schizophrenia and spectrum.
- Detailed assessments were done of the adopting family environments. Standard interviews were conducted in the homes of all families and recorded on audio tape.
- Assessments of adoptees were done first at an average age of 23 years and then 21 years later in follow-up interviews when many more had developed psychiatric symptoms.
- The theoretical orientation of the study incorporated heredity-environment interaction, whereby those at genetic risk were deemed more vulnerable to adverse environments.

When data were examined without regard to family environment, the frequencies of narrowly defined schizophrenia were 11 of 145 for the index cases and 3 of 158 for controls, a clearly significant difference ($P = .009$). The difference for spectrum cases (32 in index cases and 8 in controls) was definitely significant ($P < .0001$). As shown in Fig. 17.3, the elevation of risk for adoptees from a schizophrenic mother who were reared in good home environments was quite small, but the risk was greatly magnified by rearing in unfavorable homes. A coherent picture of risk for schizophrenia and milder mental disorders was obtained only by considering *both* heredity and environment.

TWINS

In order to claim that schizophrenia or any other mental characteristic has a genetic cause, we must identify groups of people who differ in their genetic makeups but have no significant differences in their environments. Several approaches to this problem have been examined, including monozygotic (MZ) twins reared apart and MZ versus DZ (dizygotic) twins reared together.

MZ Twins Reared Apart

MZ twins reared apart seem to provide ideal material for research. If a disorder is indeed genetic, then genetically identical twins should both show it, even if they are reared in different environments. The logic seems irrefutable. If two MZ cotwins both show schizophrenia despite differences in upbringing, this would support the genetic hypothesis. But would concordance of MZ twin phenotype prove the hypothesis? No, it would not. Just because two people are reared apart does not mean they have no experiences in common. This is especially true for MZ twins who tend to look alike and are always the same sex. Consequently, if some MZ twins reared apart were concordant for schizophrenia or any other mental attribute, this finding would be interesting but inconclusive. It could reflect the action of similar experiences. The same trouble does not beset discordance of MZ twins. If one shows schizophrenia and the other does not, this most likely would not occur for genetic reasons because the genomes of MZ cotwins, while not absolutely identical, are highly similar. A study of MZ twins reared apart is an excellent way to assess the plasticity of behavior, not its heritability. Genotype is the same, whereas experiences differ.

In a study of separated MZ twins and schizophrenia, there is also a severe problem finding enough cases. Less than 1% of a population ever develops schizophrenia, less than 1% of all births are MZ twins, and MZ twins are almost never reared apart. The probability of finding even one case of schizophrenia in an adopted MZ twin is extremely small, so small that details of fewer than 30 cases have been published in the past 90 years in the entire world literature of psychiatry. There easily could have been publication bias, such that cases of separated MZ twins concordant for schizophrenia were considered worthy of publication, whereas discordant pairs were not reported or not even detected by researchers.

Some published reports of MZ twins reared apart were not quite as advertised. For example, Kallmann (1938) reported the cases of twin sisters, Kaete and Lisa, which he claimed "definitely refute the theory that schizophrenic psychosis of twins is principally conditioned by…similar environmental influence…" His argument was that the sisters were reared apart in different environments, and therefore, the appearance of schizophrenia in both of them could only be the result of identical genes. Both sisters had been admitted to the same Herzberge Hospital in Germany, Kaete in 1928 at the age of 16 and Lisa 2 years later. Although Lisa's mental state was considerably better than Kaete's, there evidently were reasonable grounds to diagnose both of them as psychotic.

How different were their experiences? The mother of Kaete and Lisa was an unmarried domestic servant who abandoned the twins after birth in hospital. Their mother's own mother had been a severe alcoholic, as was one of her four brothers. Kallmann described the other three brothers (uncles of the twins) as "eccentric borderline cases." Kaete was assigned to be reared by one uncle, and Lisa was sent to another uncle in a different town. They knew each other, met frequently after the age of 10, and came to dislike each other intensely. Kallmann himself wrote that "the twins had a similar development in point of character and social experience." If so, how could anybody possibly conclude that only their common genes could have produced mental problems? Clearly, heredity and experiences were confounded, despite rearing in different homes, and no conclusion about causes of hospitalization can be made from this pair of MZ twins.

Another pair of MZ twins, Florence and Edith, was reported from England by Craike and Slater (1945). They were separated 9 months after birth when their mother died. Edith was kept by her father, while Florence was sent to her aunt. Their childhood experiences were quite different, and they did not meet until the age of 24. Edith's father drank so heavily and was so abusive that she was sent to a children's home at the age of 8 1/2 and lived there for 11 years, whereas Florence had a reasonably happy home situation. Florence, who worked as a servant, had no serious mental problems until the age of 51, after she had known Edith for 27 years. When she was admitted to hospital, the admission record described her as "mentally very depressed, weeps and sobs copiously as she tells of all the undercurrent against her." Six months later she was described as "improved, still some depression but not so worried by delusional ideas." By current criteria, there is no way Florence would be considered schizophrenic; rather, her symptoms suggest some form of depression. As for Edith, she was never hospitalized and never diagnosed by any psychiatrist as having a mental disorder, at least not before she was sought out by Craike and Slater. Her record as a domestic servant and factory worker was very good. When Edith was interviewed for the study, one report stated, "… there was no evidence of any shallowness of affect or any bizarre quality, nor of any schizophrenic features …" and another interview 3 weeks later found "no characteristic schizophrenic symptoms." Nevertheless, Craike and Slater reviewed her life history and concluded Edith was the victim of "a chronic insidiously progressive paranoia" and that the two sisters had "a basically schizophrenic illness." Apparently, the boundaries of schizophrenia were stretched just far enough to include both sisters and call them concordant.

According to modern standards of science and psychiatric diagnosis, almost every report of MZ twins reared apart has serious flaws (Farber, 1981). A recent review by Joseph (2015) provided synopses of almost all such twins reported in the literature, and it is clear that many represented in the published literature as reared apart in fact were reared in branches of the same family and spent much time together in similar circumstances. This is not at all surprising in view of how adoptions are often arranged. Struggling families and child welfare agencies do not aim to make things easier for behavior geneticists seeking the origins of individual differences. We must make do with what society provides, even when the material falls far short of what science needs.

DZ Twins Reared Together Versus Siblings

DZ twins reared together versus siblings are much more common than separated twins, and the frequency of schizophrenia in twins is not very different from the prevalence in the whole population. There have been a fair number of pairs wherein one or both were diagnosed schizophrenic. Genetically, DZ twins are just as different as siblings born at different times. Because they are twins, their environments will tend to be more similar than for ordinary siblings. Siblings born several years apart have a somewhat different prenatal environment because their mother and her environment changed over the years, and their experiences after birth also differ. One may have an older sibling, whereas the other may not because he or she was born first. Child-rearing practices of the parents may have changed to some extent as a result of their experience with the first child. Teachers, schools, and school curricula may have changed. The mother's marital or employment status may have changed. But DZ twins are conceived at the same time, develop in the same womb, and usually grow up in relatively similar home and school environments. Thus, comparing DZ twins and their ordinary siblings provides an indication of the role of different experiences associated with age cohort and rearing conditions. Several studies have reported that DZ twins tend to be more alike with respect to psychiatric diagnosis than are nontwin siblings (Lewontin, Rose, & Kamin, 1984). Three older studies found very similar patterns. Gottesman and Shields (1966) observed concordance rates of 9.1% for DZ twins and 4.7% for nontwin siblings, whereas Kringlen (1967) found the rates to be 8.1% versus 3.0%, and Fischer (1973) reported concordance rates of 26.7% for DZ twins and 10.1% for nontwins sibs. The twin versus sib difference supports a substantial role for environment.

MZ Versus DZ Twins—Environments

It is also informative to compare the occurrence of schizophrenia in the two types of twins. MZ twins are genetically the same, whereas DZ twins are genetically different. If defective genes cause schizophrenia, then when one MZ twin is schizophrenic, the other should be too, whereas the rate of concordance in DZ twins should be much lower. At first glance, the comparison of MZ and DZ twins appears to be an ideal experiment to assess the heritability of schizophrenia. Both groups are twins and hence are conceived at the same time in the same womb, have the same older siblings, attend the same school at the same time, etc. On the other hand, they differ greatly in the degree of genetic similarity. For a twin study, it is critically important that there be no substantial differences between the environments of the MZ and DZ twins. If MZ twins have more similar experiences than DZ twins, then we expect them to have a higher rate of concordance for schizophrenia to some extent because of environmental factors. If in fact MZ twins do show higher concordance, we cannot then know for sure what yielded that result—their genes, their experiences, or some unknown combination of the two—and the study will be inconclusive.

The extent of the problem is generally well known and was revealed long ago, as exemplified by a study done in California (Wilson, 1934). Twins averaging 12 years old were asked many questions about their lives and habits. The results confirmed that the MZ pairs with the same heredities had more similar experiences. They also revealed that for most experiences, gender is very important, because M-F pairs of DZ twins are less similar than M-M or F-F DZ twins. Several other surveys of MZ and same-sex DZ twins have consistently found environments to be more similar for the MZ pairs.

When Rosenthal (1971) discussed why he and Kety decided in 1961 to investigate heredity and schizophrenia with adoption studies instead of twin studies, the question of environment was of central importance: "psychological factors unique to monozygotic twins, especially that of shared identity, have been described vividly by several investigators who maintain therefore that the equal environment assumption is ill-founded." Publications since those comments have given further reasons to doubt that assumption. Adult MZ twins tend to reside closer together (Tambs, Sundet, & Berg, 1985) and are more likely to have frequent reunions and eat similar foods (Fabsitz, Garrison, Feinleib, & Hjortland, 1978). Furthermore, those MZ twins with more extensive social contact are likely to have more similar personalities (Rose & Kaprio, 1988).

More recently, an extensive review of 11 studies compared the kinds of experiences of MZ and DZ twins that were shown in the above section on Environment to have some relevance to schizophrenia (Fosse, Joseph, & Richardson, 2015). Data from 9119 twin pairs were compiled for 24 MZ-DZ comparisons and classified into five groupings shown in Fig. 17.4. In no instance was the correlation higher for the DZ pairs, but many of the MZ correlations were substantially and significantly higher than

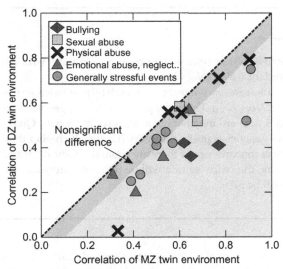

FIG. 17.4 Plot of the correlations of several environmental factors experienced by monozygotic (MZ) twin pairs versus the correlations of factors experienced by dizygotic (DZ) twin pairs. The kind of environmental factor is represented by *color* and shape of the data point. *Dashed line* represents equally similar environments for the two types of twins. *First gray zone* encompasses instances where the MZ correlation was higher than for DZ twins but the difference was not statistically significant. The second, much *larger gray zone (yellow)* includes includes instances where the MZ environmental correlation was significantly higher that the DZ correlation. *Adapted from Fosse, R., Joseph, J., & Richardson, K. (2015). A critical assessment of the equal-environment assumption of the twin method for schizophrenia. Frontiers in Psychiatry, 6, 62.*

the DZ correlations. Those extensive data contradict the commonplace assumption that MZ and DZ twins have equally similar or dissimilar experiences. MZ twins tend to have more similar experiences, and MZ-DZ comparisons of schizophrenia are confounded by differences in experience. The twin method cannot cleanly separate the effects of heredity and experience or upbringing.

MZ Versus DZ Twins: Concordance and Correlation

Despite these flaws in the twin method, it has become the most popular means of assessing heritability in behavioral and psychiatric research. Twin resemblance is described in the literature in several ways. In all older studies (Table 17.3), pairwise concordance was preferred. It is easily understood and is useful for comparing rates in MZ and DZ twin pairs using a significance test on the difference of two proportions. In the late 1970s and afterward, twin researchers began to report their results as probandwise concordance. The study by Kendler and Robinette (1983) was especially helpful because it reported both ratios (see Box 17.2). There were 30 SS twin pairs and 134 SO pairs, so pairwise rate was 30/164 = 18%. Among those pairs, there were 60 + 134 = 194 individuals diagnosed schizophrenic, and 60 of them were members of

TABLE 17.3 Published Studies of Twins and Schizophrenia

Study	Concordance method	MZ pairs (*C/N*)	MZ %C	DZ pairs (*C/N*)	DZ %C
Luxenburger (1928) Germany	Pairwise	7/14	50	0/13	0
Rosanoff, Handy, Plesset, and Brush (1934) the United States	Pairwise	18/41	44	5/53	9
Kringlen (1966) Norway	Pairwise	14/50	28	6/94	6
Gottesman and Shields (1966) the United Kingdom	Pairwise	10/24	42	3/33	9
Fischer, Harvald, and Hauge (1969) Denmark	Pairwise	5/21	24	4/41	10
Pollin, Allen, Hoffer, Stabenau, and Hrubec (1969) the United States	Pairwise	11/80	14	6/146	4
Kendler and Robinette (1983) the United States	Pairwise	30/164	18	9/268	3
Kendler and Robinette (1983) the United States	Probandwise	60/194	31	18/277	6
Shih et al. (2004) Review of eight studies	Probandwise	Median of eight studies	46.5	Median of eight studies	10.0

C, concordant pairs; *N*, no. of pairs.

SS pairs, so probandwise rate was 60/194 = 31%, much higher than pairwise. A cursory glance at twin studies from different historical periods suggests concordance rates have been going up, but this largely reflects a change in how concordance is calculated.

Many twin researchers have one overarching goal in their work—to estimate heritability of a phenotype. A massive review of virtually all published twin studies from the last 50 years (Polderman et al., 2015) did not even tabulate concordance rates. It gave only twin correlations and heritabilities of "liability."

BOX 17.2

Twin concordance. Let us symbolize someone diagnosed schizophrenic as S and not so diagnosed as O. A twin pair for which both are diagnosed schizophrenic is SS, and the number of pairs is N_{SS}, while the number of discordant pairs is N_{SO}. Thus,

Number of twin pairs in the study $= N_{SS} + N_{SO}$
Number of individuals $= 2N_{SS} + 2N_{SO}$
Number of affected individuals $= 2N_{SS} + N_{SO}$
Number of persons not affected $= N_{SO}$.

Pairwise concordance is the proportion of pairs who are concordant $= N_{SS}/(N_{SS} + N_{SO})$.

Probandwise concordance counts each SS pair twice, so it is $2N_{SS}/(2N_{SS} + N_{SO})$. This is the proportion of all people diagnosed S who are members of a concordant SS pair. Quantitative geneticists prefer this ratio. It is usually higher than pairwise concordance. If there are equal numbers of SS and SO pairs, the pairwise rate will be 50%, but probandwise rate will be 67%.

The statistical methods used in twin studies ensure that the reported correlation of MZ twins will be much higher than their concordance rate. Psychiatric diagnosis is usually all or none, a dichotomy, whereas the correlation coefficient expresses association of two continuous measures. Falconer (1965) devised a method to convert concordance to correlation, and this was adapted by Smith (1974) for use with twins. The basic idea proposed by Falconer is that a group of people varies in a quantity termed liability that is distributed like a bell-shaped curve. There is no definition for liability, and it cannot be measured; it is a mathematical device. The theory asserts that there is a fixed threshold for liability, such that anyone exceeding that threshold expresses a disorder, whereas those with lower liability scores do not. The threshold is hypothesized to have the same value for everyone, regardless of genotype or experiences. This supposition cannot be evaluated because the threshold itself cannot be measured. Through some rather involved mathematics, formulas were devised to convert the proportion of affected relatives into a correlation of their hypothetical liabilities. As Smith (1974) observed, "Nongenetic familial effects can be eliminated, [by] assuming they are the same for MZ and DZ pairs." The result is that for a probandwise concordance rate of 30% when the population frequency is 1%, the twin correlation of liabilities will be about $r = 0.73$. Thus, the correlation of hypothetical liabilities is far higher than the probandwise concordance rate. The heritability for twins is then estimated to be twice the difference between MZ and DZ twin correlations, as discussed in Chapter 14. This is how the heritability can seem so high when MZ pairwise concordance for schizophrenia is considerably less than 50%. Many of those correlations in the massive Polderman et al. (2015) review are between ethereal liabilities of twins, quantities that were not actually measured.

Thus, the mathematics of a twin study involve several major steps beyond the data that were actually observed. First, the observed pairwise concordance rates are converted to probandwise rates, which can almost double the rate. Then, the probandwise rate is converted to correlation of liabilities, which more than doubles the probandwise rate. Then, the correlations of liabilities are converted to heritability of liability.

Examples of MZ Versus DZ Twins

Table 17.3 provides data on several twin studies of schizophrenia dating back to 1928. Median pairwise concordance rates for MZ twins were about 28% compared with about 9% for DZ twins. Median probandwise concordance rates for MZ twins were about 46% in the review by Shih et al. (2004). A large study of more than 29,000 twin pairs (Prescott, Kuhn, & Pedersen, 2007) provided only correlations and heritabilities of liabilities. For liability computed from psychosis, age-adjusted heritability was about 0.35. In the Polderman review of 12 studies of schizophrenia, heritabilities of liability were about 0.45 for males and 0.65 for females. The heritability of liability of 80% that is often cited in recent articles is based on a few studies with small samples that observed unusually low frequencies for DZ twins. Studies with much larger samples yield estimates closer to 50%. All of those estimates using inflated correlations are also biased upward by the greater similarity of MZ twin environments.

The debate about the "true" magnitude of heritability could go on and on forever, but this pursuit seems increasingly pointless because it really does not matter whether the so-called heritability is 0.3, 0.5, or 0.7. The next step in the exploration of genetics and schizophrenia would proceed in the same way.

SINGLE GENES

Large resources are now being devoted to the hunt for genes that contribute to schizophrenia. The progress of this journey has unfolded in much the same way as happened for intelligence and autism. First, there were linkage studies implicating just one gene in a single report, most of which generated much excitement but were not replicated. Then, there were studies of large chromosome deletions or duplications, each involving many genes. Next came the GWAS studies that examined millions of SNPs across the entire genome, initially with little success but then uncovering several promising SNPs when sample sizes became large enough. Most recently, whole gene exomes and even the entire genome of individuals are being sequenced to detect a wide range of abnormalities, anomalies, and atypical arrangements.

Linkage Studies

A promising link was detected in Vancouver when a young man and his mother's brother were both found to have schizophrenia and an extra portion of chromosome 5 (the 11.2–13.3 segment of the q arm), whereas six other family members had neither defect (Bassett, Jones, McGillivray, & Pantzar, 1988). An association in only one small family could easily occur by chance alone. Nevertheless, the research team took this as evidence suggesting a gene for schizophrenia was located in that chromosome region, and the news was broadcast in the mass media.

The report on that one family in the medical journal *Lancet* in April of 1988 stimulated linkage studies in much larger pedigrees. There ensued an international competition to identify the culprit gene, much as occurred in the same period for the cystic fibrosis gene. No fewer than five groups of researchers working with families in Canada, the United States, Iceland, England, Scotland, and Sweden examined linkage with genes or markers known to reside in or near the 5q11-13 region. Two groups rushed into print in the 10 November 1988 issue of *Nature*. The dead heat might have set off a bitter dispute over priority and fame, except that they obtained opposite results. One group using two marker loci found a high linkage score under one quantitative model, which led them to claim "the first concrete evidence for a genetic basis for schizophrenia" (Sherrington et al., 1988), whereas another group using seven marker loci (including the two used by their competitors) obtained even more compelling evidence that a gene causing schizophrenia is *not* located in the 5q11-13 region (Kennedy et al., 1988).

In many fields of science, one report would negate the other, at least until further facts are produced; caution would prevail. However, several geneticists took the discrepancy as evidence of *genetic heterogeneity*, claiming schizophrenia *was* caused by a dominant gene in the 5q11-13 region in some families but it arose from a different gene located elsewhere or even an environmental agent in other families (e.g., Lander, 1988). Of course, another possibility is that the association with the marker genes on chromosome 5 in the first study was merely a coincidence. Even a well-designed experiment with a substantial sample size can yield a false positive, which is why replication of a finding by other scientists is a critical test of the truth of any claim. Three additional reports were soon published, each providing strong evidence against linkage with a gene in the 5q11-13 region. Two reviewers concluded the following: "There is doubt now about chromosome 5 linkage to schizophrenia, even in the original Icelandic and English sample; many investigators think that the finding may well have been spurious" (Gelernter & Kidd, 1991).

A similar fate befell a claim of linkage of a gene for manic-depressive psychosis in the Old Order Amish of Pennsylvania to marker genes on chromosome 11. The original report in *Nature* (Egeland et al., 1987) was hailed in the press as conclusive proof of a genetic cause. One scientist commented, "For the first time, molecular genetics has entered the arena of psychiatric disorders" (Kolata, 1987). In that case, the team of American researchers who had studied the large pedigree continued collecting data after publishing their first report. The original pedigree involving 81 people, 19 with some form of major affective disorder, yielded linkage (LOD) scores of 4.1 and 2.6 for linkage to the marker genes (HRAS) and INS (insulin) on chromosome 11, respectively. To be at all persuasive, a linkage score (LOD) needs to be more than 3.0, which means the probability that two genes are linked on the same chromosome is 1000 times greater than the probability that they are not linked. Later, *two* people in the original pedigree developed mental disorders, which reduced the two LOD scores to 1.0 and 1.8; and adding six more people to one family in the pedigree reduced the LOD scores further to 0.7 and 1.3. Finally, an extension of the pedigree to 31 more relatives made the LOD scores negative for close linkage (Kelsoe et al., 1989). The manic-depressive gene had disappeared.

Case reports of promising instances of schizophrenia linked to single genetic markers have continued to be published, albeit with less fanfare. When there have been several such reports, a gene may be elevated to the status of a candidate gene that warrants further study by other researchers. The SZGene database has assembled more than 1500 reports in peer-reviewed scientific journals from 1965 to 2012 (Johnson et al., 2017). Many gave conflicting results, but several seemed to have solid backing, and 25 genes connected with schizophrenia have been the subjects of more than 20 reports each. GWAS studies discussed below provide an effective means to test the replicability of those gene effects.

Instances wherein an initial positive report was followed by failures to replicate are highlighted in this chapter in order to alert the reader to the ever-present dangers of false positive results in this field. Unlike the example of Cyril Burt in the study of intelligence, none of the examples cited here were blemished by outright scientific misconduct. Instead, there are several instances of an overly enthusiastic and uncritical embrace of initial reports that involved small samples and inadequate controls. This does not amount to misconduct, but neither is it good scientific practice. The mass media can be relied upon to celebrate the "good news" fed to them by teams of researchers, and this generates an impression of steady scientific advances when a media outlet fails to take note of the many failures to replicate.

Copy Number Variations (CNVs)

The new DNA-based hybridization chips provided a means to scan the entire genome for segments of a chromosome that had been deleted or duplicated. The International Schizophrenia Consortium or ISC (2008) assessed 3391 patients and 3181 controls, and 6753 copy number variants (CNVs) were detected. Statistical comparisons identified three chromosomal regions where a large deletion was present in many more patients than controls, and in each region, more than 10 genes were known, so there was no way to be sure which gene or genes were responsible for the psychiatric problems. A larger study (Stefansson et al., 2008) assessed 9878 parent-offspring pairs or trios that allowed researchers to distinguish between de novo mutations and those transmitted in a family, and 66 de novo events were found. Two of the sites identified by the ISC study were confirmed. Another at chromosome 15q11.2 provided good evidence of a connection with schizophrenia (present in 26 or 0.55% of SZ cases and 79 or 0.19% of controls). A review of these and several other CNV scans found that almost all CNVs associated with SZ were also associated with at least one other psychiatric or neurological disorder, including autism, bipolar disorder, intellectual disability, and attention deficit hyperactivity disorder (Sebat, Levy, & McCarthy, 2009). Because of the large size of the deletion events involving several genes, it could not be determined whether the multiple disorders arose from copy number variants in the same or different genes. A further study examined 788 parent-offspring trios that included 177 SZ patients (Malhotra et al., 2011). Nine de novo mutation events were detected in the patients, including a duplication of the 22q11.21 region that was the subject of deletions in two other studies. Altogether, the studies suggested that about 3% of cases of SZ carry a large de novo CNV that is likely a major source of the person's psychiatric problems.

GWAS

Early genome-wide scans for SNPs related to schizophrenia appearing first in 2007 suffered from small sample size and generally detected no significant association with SZ (Van Winkel et al., 2010). Three studies from 2009 each on its own found no significant SNPs, but two that combined the data sets for all three studies detected a significant cluster of SNPs in the major histocompatibility complex (MHC) in chromosome 6 band p21-22. Despite the lack of significant association, the International Schizophrenia Consortium (2009) used the top SNPs in its sample to generate a polygenic prediction equation that was then applied to a different group of SZ cases and controls. The prediction was correlated with actual SZ status in a manner that might account for about 3% of variance in diagnosis. Authors claimed this result actually "explained" 3% of variance, but that was a misnomer because they had not identified even one specific gene contributing to SZ and therefore could explain nothing. Shi et al. (2009) noted that their seven SNPs were associated with a chromosomal region 1.5 Mb wide, and they remarked, "it is unclear whether the signal is driven by one or several genes, by intergenic elements, or by susceptibility alleles in many genes." At the same time, they presented a large table naming 17 genes and their functions that were in regions that fell far short of significance. The table gave the impression to the casual reader that they had learned important things about genes related to schizophrenia when, by current standards of evidence, they had learned nothing worth reporting about any specific gene. The study by Stefansson et al. (2009) found no significant SNPs in their study sample but replicated the Shi et al. findings for band 6p21-22 when they combined samples.

Large Sample GWAS

The situation changed abruptly when the Schizophrenia Working Group (2014) reported results from a total sample of 36,989 cases of SZ and 113,075 controls that was achieved by pooling data from several independent studies. Many SNPs were detected that had a clearly significant association with SZ diagnosis, and large numbers of them occurred in clusters in the vicinity of known genes. Supplementary Figure 2 in that report displayed detailed plots of the significance level of thousands of SNPs close to many genes. Fig. 17.5A shows a cluster of SNPs distributed across chromosome 6 within the limits of the SNAP91 gene. The pattern provides strong evidence that the gene has relevance for SZ, although it does not offer conclusive proof because there was no evidence of a mutation in an exon of SNAP91 or a change in its amino acid sequence. There were 52 SNP clusters that showed close association with just one known gene. In some instances, as shown for the DRD2

FIG. 17.5 Four examples of significant SNP clusters reported by the Schizophrenia Working Group (2014). Each point shows the statistical significance of a test of allele frequency at a specific SNP versus the presence or absence of schizophrenia in the sample. The criterion for genome-wide significance is shown by the dashed line. SNPs are located along the X-axis according to their positions on the chromosome. The Y-axis is the exponent of the probability value, where 6 denotes 10^{-6}, 2 is 10^{-2}, etc. Thus, higher values of Y signify smaller probabilities that the difference in allele frequency at a specific SNP represents merely chance association. Bars below each panel indicate the location of a gene, where the vertical black bars are the exons of that gene. (A) Several significant SNPs occur within the limits of the SNAP1 gene. (B) Several SNPs occur within the DRD2 gene and adjacent portions of the DNA that are not in the gene. (C) Significant SNPs span a region that contains five different genes. It is not known which of these is specifically related to the risk of schizophrenia. (D) A cluster of significant SNPs occurs in an expanse of chromosome 2 that contains no genes. *Colors of the dots have been converted to gray scale, and details about the shapes of the dots have been omitted. More complete information is provided in the original source. Based on images in Supplementary Figure 1 of the report by the Schizophrenia Working Group. Reprinted by permission from Nature/Springer, Schizophrenia Working Group of the Psychiatric Genomics Consortium (2014). Biological insights from 108 schizophrenia-associated genetic loci. Nature, 511, 421–427, Copyright 2014.*

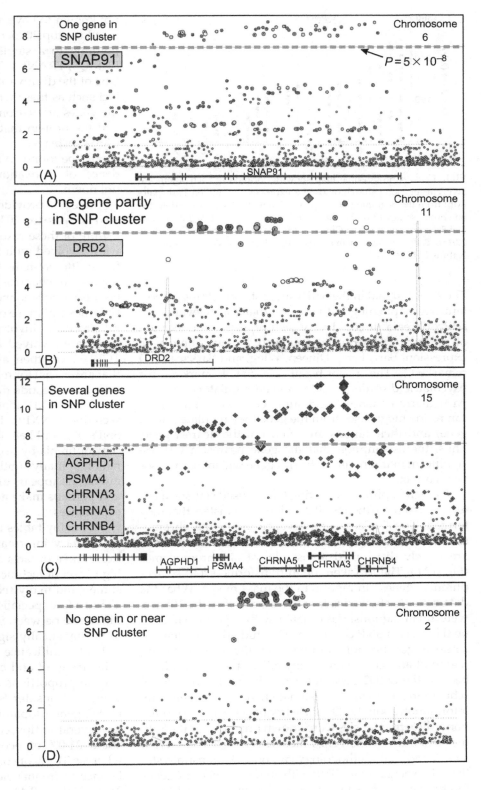

gene on chromosome 11 (Fig. 17.5B), several SNPs were located within the limits of the gene, but others were found a short distance upstream or downstream and may have involved promoter sequences. Fig. 17.5C shows another common pattern wherein the SNP cluster spanned several genes, any one or more of which might have relevance for SZ. There were more than 30 instances of multiple genes in a cluster. Fig. 17.5D shows a result for which there was no gene even close to the SNP cluster. There were 26 SNP clusters that showed such a pattern.

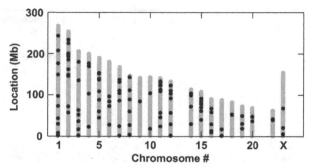

FIG. 17.6 Chromosome map showing location of each significant SNP cluster along the chromosome expressed as million base pairs (Mb) from the beginning of the p arm. No SNP clusters were detected on chromosomes 13 or 21. *Based on data in Supplementary Tables 2 and 3 of Schizophrenia Working Group of the Psychiatric Genomics Consortium (2014). Biological insights from 108 schizophrenia-associated genetic loci. Nature, 511, 421–427.*

The strong signal in the SNPs suggests something important was happening in the region that had relevance for schizophrenia. It could very well represent the location of an enhancer sequence involved in regulating gene expression. Barr, Feng, Dineen, Wigg, and Sarkar (2017) remarked, "The majority of associated markers for psychiatric disorders reside in gene regulatory regions, particularly enhancers and super-enhancers. Enhancers can reside megabases from the gene they regulate (target gene) and their targets are often not the nearest gene. Thus, the assumption that the gene nearest a GWAS-significant marker will be the risk gene will in many cases be incorrect."

The Schizophrenia Working Group used statistical criteria to assemble their SNP data into 108 clusters that they termed "loci," places of relevance along the chromosomes. It is evident that loci pertinent to schizophrenia are widely distributed across almost all chromosomes (Fig. 17.6). The loci were often transformed into schizophrenia "genes" in reports in the mass media. What the data actually showed is that less than half of the 108 loci were unambiguously associated with a particular gene on the basis of SNP clustering. The study did not demonstrate a genetic polymorphism in the protein-coding region of any gene. How many alleles there might have been at the DRD2 gene, for example, among the 36,989 schizophrenics was not estimated, nor was there any information about DRD2 allele frequencies in the population. Allele frequencies were known precisely for the SNP markers but not the genes themselves. It was a GWAS study, and that methodology can only demonstrate statistical association of a SNP with an unseen gene or genes located near it. It is just the first step in the search for a gene that contributes to the etiology of schizophrenia. In this case, it was a big step, but proving a causal connection demands more.

One of the advantages of a GWAS study is that it can detect association with a genetic locus that heretofore was not known to have any relationship with schizophrenia. Six of the top 20 loci with smallest P values among those pointing to a single gene were of this kind: IGSF9B, LRRIQ3, TRANK1, TSNARE1, ZNF804A, and ZSWIM6. One of the disadvantages of GWAS is that it can point to loci such as these where almost nothing is known about the genes apart from the location along a chromosome and the protein that is encoded. For the six loci, Gene-Cards displayed no data where knowledge of function should be found. Why they would show such clear evidence of a relation to schizophrenia is completely unknown.

Table 17.4 provides information about 14 genes associated with the top 20 loci from the large GWAS study. One feature of these results is quite striking: Different genes were involved with the function of four different neurotransmitter systems (calcium, dopamine, glutamate, and potassium channels), and this total is probably five because acetylcholine is also implicated (Fig. 17.5C). This provides a glimpse of the extraordinary diversity of neural processes involved in schizophrenia. Another two genes are involved in brain development (CNTN4 and TCF4). Five others are involved in biochemical activities that are crucial for many kinds of physiological functions such as nucleotide metabolism (DPYD), mitotic cell division (MAD1L1), zinc transport (SLC39A8), and insulin secretion (SNX19). Because these were detected using SNPs, it is unlikely that any of these genes can provoke schizophrenia by itself when mutated. Instead, a combination of many subtle effects of a wide variety of genetic mutations appears to destabilize cognitive brain function and perhaps make a person more vulnerable to adverse experiences.

The many tables in the extended report and the supplementary information of the SWG report provide a rich source of facts about the SNP method, although a high level of technical and statistical knowledge is presumed and many sections are virtually impenetrable to all but a few specialists. One pertinent fact is the size of a difference between SZ cases and controls that was held to be statistically significant at $P < 5 \times 10^{-8}$. The actual tests of significance were done on the odds ratio (OR) comparing SZ and control groups. The size of a difference in proportions can also be seen in Supplementary Table 2. Consider SNP rs2514218 that is close to the DRD2 gene (dopamine type 2 receptor that is a target for several antipsychotic drugs). Among the SZ cases, about 31.0% of people had the T allele at that SNP, whereas 31.4% of control cases had the T allele. The difference in frequencies of the T allele was a minuscule $31.0 - 31.4 = -0.4\%$. Such a small difference can be statistically significant because the sample sizes were huge—more than 36,000 SZ cases and more than 113,000 controls. One very important question in the quest for genes related to SZ is whether there are a few with substantially large effect sizes that might be

TABLE 17.4 Top 14 Genes From GWAS[a]

Symbol	Name	Location	Amino acids	Local function	Higher-level functions
CACNA1C	Calcium voltage-gated channel subunit alpha 1C	12p13.33	2221	Forms pore for calcium ion passage	Muscle contraction, neurotransmitter release, cell division
CACNA1I	Calcium voltage-gated channel subunit alpha 1 I	22q13.1	2223	Forms part of pore	Largely the same as CACNA1C
CACNB2	Calcium voltage-gated channel auxiliary subunit beta 2	10p12.33	660	Subunit of Ca channel	Increase peak calcium current
CNKSR2	Connector enhancer of kinase suppressor of Ras 2	Xp22.12	1034	Mediate MAP pathways	Widespread, sparse data
CNTN4	Contactin 4	3p26.3	1026	Axon adhesion to cells	Neurite outgrowth and neuroplasticity
DPYD	Dihydropyrimidine dehydrogenase	1p21.3	1025	Enzyme degrades uracil, thymidine	No data
DRD2	Dopamine receptor D2	11q23.2	443	Receptor coupled to G protein	Motor movement, reward, memory
GRIA1	Ionotropic glutamate receptor AMPA 1	5q33.2	906	Receptor for excitatory transmitter glu	Many, widespread in nervous system
KCNV1	Potassium voltage-gated channel modifier subfamily V member 1	8q23.2	500	Subunit of 4 part channel	Heart rate, insulin secretion, smooth muscle
LRRIQ3	Leucine-rich repeat and IQ domain-containing protein 3	1p31.1	624	No data	No data
MAD1L1	Mitotic arrest deficiency 1 (MAD1) like 1	7p22.3	718	Mitotic spindle checkpoint	Alignment of chromosomes for mitosis
SLC39A8	Solute carrier 39 member 8	4q24	460	Transport of zinc to mitochondria	Onset of inflammation, manganese uptake into brain
SNX19	Sorting nexin 19	11q25	992	Pancreas islet antigen	Insulin secretion, vesicle exocytosis
TCF4	Transcription factor 4	18q21.2	667	Recognize, bind DNA at CANNTG	Neuron differentiation

[a] Only loci associated with one gene and having sufficient information are shown.
Based on report by Schizophrenia Working Group of the Psychiatric Genomics Consortium (2014). Biological insights from 108 schizophrenia-associated genetic loci. Nature, 511, 421–427 and information in Gene Cards database.

promising aids for diagnosis or targets for therapy or perhaps there are very many genes, all with very small effects. The GWAS studies appear to have answered this question. Many genes with very small effects are implicated. Almost all of them showed a difference in allele frequency of 2% or less. This is not effect size in terms of the phenotype of interest, schizophrenia, because there is no information about alleles or allele frequencies of actual genes such as DRD2. It tells us that, if we look at the SNP alleles at rs2514218, SZ cases and controls differ slightly in the percentage with the T allele, and this pattern holds for all of the 108 loci. Another GWAS study now in progress aims to test 65,000 cases of schizophrenia and promises to expand the number of significant loci to 150 (Ripke et al., 2017) by detecting even smaller effects.

In the Wake of Large GWAS

Now that a long list of genes likely to be related to schizophrenia is available, several groups of investigators are exploring implications for other kinds of data. The SZGene database has been compiling results of linkage studies for many years and has identified genes that are regarded as historical candidate schizophrenia genes. Johnson et al. (2017) compared the top 25 historical genes with the results of the SWG GWAS study to see how many of the older findings were replicated with the more powerful SNP methods. Six of the 25 candidate genes in SZGene also met the criteria for a gene in a locus in the GWAS study, whereas 19 did not and appeared to have been false positives. Three that were supported had clear relevance to neurotransmission (DRD2, KCNN3, and GRM3).

Lencz and Malhotra (2015) noted that "a primary aim of GWAS is to identify potential targets for pharmaceutical interventions." Accordingly, they surveyed a large number of proteins that are known to be "druggable" in that they can be accessed by orally administered compounds and have good binding sites for drugs. Among those on a list of 1030 druggable gene-derived targets, 555 are targets of drugs that have already been approved for use with humans, and another 475 are now in clinical trials. The researchers then assessed how many of those genes were hits in the SWG GWAS study, because it might be possible to repurpose an existing drug to treat schizophrenia. There were 40 genes in the GWAS that seemed promising and had known functions in the nervous system, and 20 of them were targets for drugs already approved to treat a range of disorders including epilepsy, Parkinson's disease, high blood pressure, and angina. More than a dozen drugs targeting seven genes detected with GWAS are currently in clinical trials, but the results to date are not promising. The authors noted that the highly heterogeneous nature of schizophrenia with so many genetic influences of such small effect makes it unlikely that any one drug will be helpful for most patients. Nevertheless, the long development time and expense for approval of a new drug (>10 years) and large cuts to spending on drug development by large drug companies makes the pursuit of repurposing aided by GWAS something to consider.

Whole Genome Sequencing

WGS should be able to identify mutations in a family that alter protein structure and may contribute to schizophrenia. Homann et al. (2016) analyzed DNA from 63 people in nine pedigrees known to have at least one member schizophrenic. There were 25 SNPs in an exome that changed amino acid sequence in several proteins. All of the mutations turned out to be *family private*, a form of genetic heterogeneity, meaning that any particular defect occurred in only one family, so that each family harbored its own set of genetic problems. None of the 25 mutations had been detected in the SWG GWAS study, which suggests there is a large amount of genetic diversity relevant to schizophrenia that is yet to be discovered.

DRUG THERAPY

Almost everyone diagnosed schizophrenic today is prescribed one or more drugs to reduce the symptoms and make it possible to live outside the confines of an institution. None of the drugs outright cure the disorder, and all have adverse effects of some kind. There have been some helpful refinements to drug therapy since the first biochemical agent was proved effective in psychosis in 1954, but after more than 60 years, there is still no cure as such and no consensus on which drug is generally best. Many recent articles on the search for genetic bases of schizophrenia explicitly promise that genetic knowledge will at last enable psychiatrists to combat the cause, not just ameliorate a subset of symptoms.

The first effective antipsychotic agent was chlorpromazine (Fig. 17.7), a derivative of the chemical family termed phenothiazines that include the dye methylene blue (López-Muñoz et al., 2005). Early work with the phenothiazines sought a chemical substitute for quinine to use against the malaria parasite, and later, they were exploited for their antihistamine properties. One of the drugs discovered in 1946 was diphenhydramine, still in use as Benadryl. In 1949, a French surgeon working in Tunisia tested a phenothiazine for use in combatting surgical shock and found it had a remarkable calming effect *before* surgery. Those observations inspired the Rhone-Poulenc laboratories in France to synthesize various derivatives of the compound promazine, and the chlorinated form code named RP-4560 had remarkable abilities to generate relaxation and lower blood pressure. The company sent samples to several doctors for testing, and some of them found it to be a good preoperative calming agent in surgery that also lowered body temperature. A suggestion was made to psychiatrists over lunch at Hopital Val-de-Grace in France that the drug be tried on psychotic patients. In 1952, it was given to an agitated manic man who within a few hours became calm and

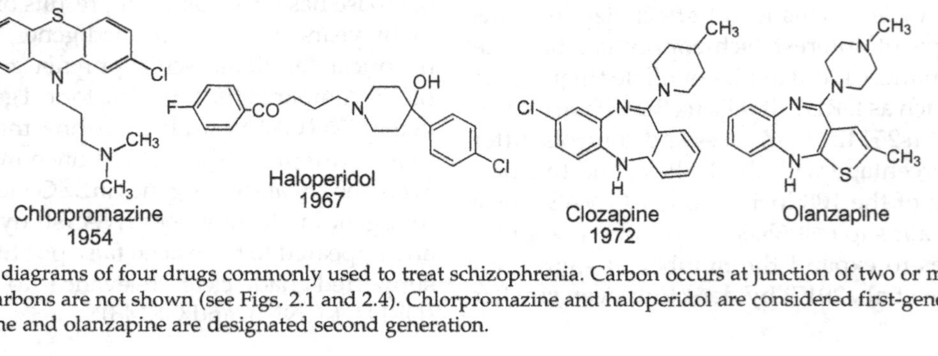

FIG. 17.7 Structural diagrams of four drugs commonly used to treat schizophrenia. Carbon occurs at junction of two or more lines. Hydrogen atoms attached to the carbons are not shown (see Figs. 2.1 and 2.4). Chlorpromazine and haloperidol are considered first-generation antipsychotic drugs, whereas clozapine and olanzapine are designated second generation.

after 3 weeks was well enough to be discharged (while remaining on the drug). The case of "Giovanni A." was also persuasive. He had a long history of disruptive behavior and was admitted to hospital for "giving improvised political speeches, getting into fights with strangers, and walking along the street with a plant pot on his head proclaiming his love of liberty." After a few days on chlorpromazine, he could converse normally and in 3 weeks was discharged.

The individual case reports were then confirmed in England in the first randomized clinical trial that used a placebo condition (Elkes & Elkes, 1954). Chlorpromazine came to North America via Canada where Lehmann obtained good results with higher doses given to manic-depressive patients (Lehmann & Hanrahan, 1954). It was resisted in the United States and licensed initially only to reduce body temperature during surgery and suppress throwing up afterward. Once it was tested on psychotic patients in large public institutions, governments noticed that this might be a way to save money by releasing patients into the community while on the drugs and closing those expensive hospitals.

Whereas chlorpromazine began life as a drug in search of a mission, its success led many large drug companies to devise their own "me too" variations on the phenothiazine structure, and the list of drugs available to treat schizophrenia grew longer. One motivation for this proliferation of drugs was the presence of adverse or "side" effects. A drug is a chemical that interacts with other chemicals and itself has no purpose. The people who synthesize it and the companies that sell it make the distinction between desirable, primary, targeted effects and unwanted adverse effects. Chlorpromazine had several side effects including drowsiness, impaired cognition, low blood pressure, and motor tremors similar to Parkinson's disease that could be intense in some people. Haloperidol proved to have strong antipsychotic effects without the sedation, but it sometimes induced severe motor tremors and, after an extended period of use, tardive dyskinesia, a form of involuntary muscle movements of the face and tongue that were irreversible. Those difficulties motivated the synthesis of the so-called second generation of "atypical" antipsychotic drugs that invoked far fewer motor abnormalities.

Chlorpromazine was eventually found to block the action of dopamine by binding to the dopamine D2 receptor. The three-dimensional features of the human D2 receptor have been elucidated (Kalani et al., 2004); it consists of seven helical structures joined into one large receptor protein molecule. The binding site for dopamine and drugs that enhance activity of the receptor (agonists) involves helices 3, 4, 5, and 6, whereas the binding site for drugs that block access of dopamine (antagonists) bind to a site formed by helices 2, 3, 4, 6, and 7.

As shown in Fig. 4.6, genes involved in dopamine synthesis, communication, and breakdown are located in many parts of the central nervous system where the neurotransmitter dopamine serves different functions. One of the shortcomings of drug therapy is that most drugs taken orally are water soluble and diffuse throughout the nervous system. The desired therapeutic effect of a drug may arise from interference with dopamine activity in one part of the brain, while the same drug is causing unwanted interference with other dopamine functions elsewhere. For example, the psychotic symptoms involving delusions and hallucinations are believed to arise from the dopamine neurons in the ventral tegmentum of the brain stem that project to the nucleus accumbens in the striatum where they signal to the next neuron in the network via D2 receptors (Stahl, 2013). Other dopamine neurons in the mesocortical system project from the brain stem to the prefrontal cortex, where they form dopamine synapses involved in cognition and motivation. Still, other dopamine neurons project from the substantia nigra in the brain stem to the striatum where they are part of the extrapyramidal motor system that utilizes dopamine as a transmitter. Blocking D2 receptors with chlorpromazine thereby reduces delusions and hallucinations while at the same time impairing cognitive and motor functions.

It is sometimes said that a psychiatric drug has a highly specific effect on just one kind of neurotransmitter receptor. However, virtually all known drugs used to treat schizophrenia have effects on several kinds of receptors (Table 17.5). Pharmacologists quantify their potency and specificity by assessing receptor-binding abilities in pieces of nervous tissue in a glass dish in the laboratory. Different doses of a drug are applied to different culture dishes, and researchers measure the percentage of a particular receptor that is bound by the drug. *Receptor affinity* is then indicated by the dosage needed to occupy 50% of the receptors. A very powerful drug would need only a small dose to achieve this, whereas some drugs would have very little binding and show an effect only at unrealistically high doses. Data provided by Solmi et al. (2017) and Correll (2010) in Table 17.5 reveal the wide differences between first- and second-generation antipsychotics and within a single category. The pattern of relative receptor binding is crucial for the desired antipsychotic benefits and adverse effects.

A thorough clinical review of the use of different drugs (Solmi et al., 2017) noted that first-generation drugs such as chlorpromazine and haloperidol usually have high affinity for the D2 receptor, whereas most second-generation drugs such as olanzapine have high affinity for the serotonin 5HT2A receptor and in some instances D2. The relative binding to D2 and 5HT2A receptors seems to be related to the kinds of adverse effects. A high ratio of D2 to 5HT2A is associated with motor side effects and tardive dyskinesia, whereas high 5HT2A relative to D2 is associated with weight gain and metabolic disturbances. Haloperidol is well known for its motor

TABLE 17.5 Receptor Affinity for Antipsychotic Drugs[a]

Drug	Type of receptor					Side effects[b]
	D2	5HT2A	5HT2C	α1-Adr	H1	
Chlorpromazine[c]	3	4.1	32	2.6	3	LBP, SED, WTG
Haloperidol[c]	2.6	61	4700	17	260	EPS, SEX, TD
Perphenazine[c]	1.4	5	132	1	8	
Aripiprazole	0.7	8.7	22	26	30	
Clozapine	210	2.6	4.8	6.8	3.1	LBP, SED, WTG
Olanzapine	20	1.5	4.1	44	0.1	WTG
Risperidone	3.8	0.1	32	2.7	5.2	
Quetiapine	770	31	3500	8.1	19	
Ziprasidone	4.8	0.4	1.3	10	4.6	

Abbreviations: D2, dopamine type 2; 5HT2A, 2C, serotonin type 2A, 2C; α1-Adr, alpha-1 adrenergic; H1, histamine type 1; EPS, extrapyramidal motor tremor; LBP, low blood pressure; SED, sedation; SEX, sexual dysfunction; TD, tardive dyskinesia; WTG, weight gain.
[a] *Source: Condensed from Solmi et al. (2017), Table 2; Correll, (2010). Data represent dosage in nanomoles of the drug required to occupy 50% of receptors in tissue culture of brain explants. Higher dosage denotes lower receptor affinity.*
[b] *Side effects listed in table were rated *** in review by Solmi et al. (2017).*
[c] *"First-generation" or conventional antipsychotic drug; others are second-generation, atypical antipsychotics.*

side effects, and olanzapine is known to produce extreme obesity and diabetes-like symptoms in some cases. Sedation is related to high affinity for histamine type-1 receptor affected by chlorpromazine; clozapine; and, to a lesser extent, olanzapine.

Reviews and meta-analyses of the research literature on antipsychotic drugs have yielded mixed results. Hartling et al. (2012) evaluated the quality of evidence in each study and found that in most instances, it was quite low. They observed that functional outcomes in terms of health-related quality of life do not differ markedly between first- and second-generation drugs and differences in global effects on psychotic symptoms were also small. They suggested that olanzapine is the best available drug at this time. They also expressed concern that 69% of the studies they reviewed were sponsored by drug companies and prior evidence had shown this could generate bias in favor of second-generation drugs (Sismondo, 2008). Zhang et al. (2013) found a small but significant effect in which sponsorship by a drug company yielded evidence in favor of second-generation drugs, whereas government sponsorship favored the first generation. They suggested risperidone was the best choice for treating

first-episode psychosis. Their meta-analysis found generally small differences between first- and second-generation drugs (effect sizes 0.1–0.2) for reduction of psychotic symptoms but larger effects of 0.4–0.65 for side effects.

The ability to synthesize "designer drugs" aimed at a specific neural or cognitive function is still an aspiration, not yet a tried and true capability. None of the drugs in common use was formulated with the aid of information about genes related to schizophrenia. This is not at all surprising in view of the highly polygenic features of the genetic part of the equation, for example, those on the list in Table 17.4. Some of the "family private" mutations with much larger effects tend to be relatively rare in the population, which discourages people and companies in search of new therapies. This is still a research frontier where investigators are probing for safe passage through a maze of dangerous obstacles.

HIGHLIGHTS

- Schizophrenia is diagnosed by a cluster of behavioral and cognitive symptoms. There is no biological or neurological indicator of the disorder.
- Diagnostic criteria have changed considerably over the past 100 years in a way that now requires more severe symptoms to qualify as schizophrenia.
- Many studies have found that the prevalence of schizophrenia is higher when there were adverse experiences in childhood and young adulthood. Those who have had several major adverse experiences show greatly increased risk of becoming schizophrenic.
- Among adopted children who had a biological parent with schizophrenia, their vulnerability to adverse experiences was greatly increased when there was more than one adverse experience.
- If a biological relative is schizophrenic, about 5% of first-degree relatives in the same family (parents, children, and siblings) are also schizophrenic.
- The experiences of monozygotic (MZ) twins tend to be more similar than those of dizygotic (DZ) twins. This renders the typical twin study that compares MZ and DZ pairs incapable of separating the influences of heredity and environment on schizophrenia.
- Genome-wide association studies utilizing SNPs have detected several dozen loci that implicate genes that influence the risk of schizophrenia. The loci are widely distributed across almost all chromosomes. The genetic aspect of schizophrenia involves a large number of genes, each with a very small effect on the phenotype.
- Genes implicated by the SNP data include some pertinent to the transmitters dopamine and glutamate, others related to calcium and potassium ion channels,

and still others pertinent to axon growth and mitotic cell division.

- Sequencing the entire genome of all family members has uncovered a few extremely rare mutations having relatively large effects that are "family private," occurring in only one pedigree.
- Most drugs used to reduce psychotic symptoms in schizophrenia block the activity of dopamine and/or serotonin transmitters by binding to the D2 and/or 5HT2A receptors, but some common antipsychotic drugs bind to more than a dozen kinds of receptors involving five different transmitters.
- Unwanted side effects of psychiatric drugs are related to the specific array of receptors blocked by a particular drug.

References

Abrams, R., & Taylor, M. A. (1983). The genetics of schizophrenia: a reassessment using modern criteria. *The American Journal of Psychiatry, 140*, 171–175.

Baron, M., Gruen, R., Rainer, J. D., Kane, J., Asnis, L., & Lord, S. (1985). A family study of schizophrenic and normal control probands: implications for the spectrum concept of schizophrenia. *The American Journal of Psychiatry, 142*, 447–455.

Barr, C., Feng, Y., Dineen, A., Wigg, K., & Sarkar, A. (2017). 8. CRISPR/Cas9 genome-editing of the RERE super-enhancer alters expression of genes in independent schizophrenia GWAS regions. *Biological Psychiatry, 81*, S4.

Bassett, A., Jones, B., McGillivray, B., & Pantzar, J. T. (1988). Partial trisomy chromosome 5 cosegregating with schizophrenia. *The Lancet, 331*, 799–801.

Correll, C. (2010). From receptor pharmacology to improved outcomes: individualising the selection, dosing, and switching of antipsychotics. *European Psychiatry, 25*, S12–S21.

Craike, W., & Slater, E. (1945). Folie à deux in uniovular twins reared apart. *Brain, 68*, 213–221.

Egeland, J. A., Gerhard, D. S., Pauls, D. L., Sussex, J. N., Kidd, K. K., Alien, C. R., ... Housman, D. E. (1987). Bipolar affective disorders linked to DNA markers on chromosome 11. *Nature, 325*, 783–787.

Elkes, J., & Elkes, C. (1954). Effect of chlorpromazine on the behaviour of chronically overactive psychotic patients. *British Medical Journal, 2*, 560–565.

Fabsitz, R., Garrison, R., Feinleib, M., & Hjortland, M. (1978). A twin analysis of dietary intake: evidence for a need to control for possible environmental differences in MZ and DZ twins. *Behavior Genetics, 8*, 15–25.

Falconer, D. S. (1965). The inheritance of liability to certain diseases, estimated from the incidence among relatives. *Annals of Human Genetics, 29*, 51–76.

Farber, S. L. (1981). *Identical twins reared apart: A reanalysis*. New York: Basic Books.

Fischer, M. (1973). Genetic and environmental factors in schizophrenia: a study of schizophrenic twins and their families. *Acta Psychiatrica Scandinavica. Supplementum, 238*, 9–142.

Fischer, M., Harvald, B., & Hauge, M. (1969). A Danish twin study of schizophrenia. *The British Journal of Psychiatry, 115*, 981–990.

Fosse, R., Joseph, J., & Richardson, K. (2015). A critical assessment of the equal-environment assumption of the twin method for schizophrenia. *Frontiers in Psychiatry, 6*, 62.

Gelernter, J., & Kidd, K. K. (1991). The current status of linkage studies in schizophrenia. *Research Publications—Association for Research in Nervous and Mental Disease, 69*, 137–152.

Gottesman, I. I., & Shields, J. (1966). Schizophrenia in twins: 16 years' consecutive admissions to a psychiatric clinic. *The British Journal of Psychiatry, 112*, 809–818.

Gruenberg, E. M. (1968). Foreword. In *Diagnostic and statistical manual of mental disorders* (2nd ed.). New York, NY: American Psychiatric Association.

Hartling, L., Abou-Setta, A. M., Dursun, S., Mousavi, S. S., Pasichnyk, D., & Newton, A. S. (2012). Antipsychotics in adults with schizophrenia: comparative effectiveness of first-generation versus second-generation medications: a systematic review and meta-analysis. *Annals of Internal Medicine, 157*, 498–511.

Homann, O. R., Misura, K., Lamas, E., Sandrock, R. W., Nelson, P., McDonough, S. I., & DeLisi, L. E. (2016). Whole-genome sequencing in multiplex families with psychoses reveals mutations in the SHANK2 and SMARCA1 genes segregating with illness. *Molecular Psychiatry, 21*, 1690–1695.

Insel, T. (2013). *Post by former NIMH director Thomas Insel: Transforming diagnosis*. National Institute of Mental Health.

Insel, T. R. (2014). The NIMH Research Domain Criteria (RDoC) project: precision medicine for psychiatry. *American Journal of Psychiatry, 171*, 395–397.

International Schizophrenia Consortium. (2008). Rare chromosomal deletions and duplications increase risk of schizophrenia. *Nature, 455*, 237–241.

International Schizophrenia Consortium. (2009). Common polygenic variation contributes to risk of schizophrenia and bipolar disorder. *Nature, 460*, 748–752.

Johnson, E. C., Border, R., Melroy-Greif, W. E., de Leeuw, C. A., Ehringer, M. A., & Keller, M. C. (2017). No evidence that schizophrenia candidate genes are more associated with schizophrenia than noncandidate genes. *Biological Psychiatry, 82*, 702–708.

Joseph, J. (2015). *The trouble with twin studies: A reassessment of twin research in the social and behavioral sciences*. New York: Routledge.

Kalani, M. Y. S., Vaidehi, N., Hall, S. E., Trabanino, R. J., Freddolino, P. L., Kalani, M. A., ... Goddard, W. A. (2004). The predicted 3D structure of the human D2 dopamine receptor and the binding site and binding affinities for agonists and antagonists. *Proceedings of the National Academy of Sciences of the United States of America, 101*, 3815–3820.

Kallmann, F. J. (1938). Heredity, reproduction and eugenic procedure in the field of schizophrenia. *The Eugenical News, 23*, 105–113.

Kelsoe, J. R., Ginns, E. I., Egeland, J. A., Gerhard, D. S., Goldstein, A. M., Bale, S. J., ... Conte, G. (1989). Re-evaluation of the linkage relationship between chromosome 11p loci and the gene for bipolar affective disorder in the Old Order Amish. *Nature, 342*, 238–243.

Kendell, R. E., Cooper, J. E., & Gourlay, A. J. (1971). Diagnostic criteria of American and British psychiatrists. *Archives of General Psychiatry, 25*, 123–130.

Kendler, K. S., Gruenberg, A. M., & Tsuang, M. T. (1985). Psychiatric illness in first-degree relatives of schizophrenic and surgical control patients: a family study using DSM-III criteria. *Archives of General Psychiatry, 42*, 770–779.

Kendler, K. S., & Robinette, C. D. (1983). Schizophrenia in the National Academy of Sciences-National Research Council Twin Registry: a 16-year update. *The American Journal of Psychiatry, 140*, 1551–1563.

Kennedy, J. L., Giuffra, L. A., Moises, H. W., Cavalli-Sforza, L., Pakstis, A. J., Kidd, J. R., ... Kidd, K. K. (1988). Evidence against linkage of schizophrenia to markers on chromosome 5 in a northern Swedish pedigree. *Nature, 336*, 167–170.

Kolata, G. (1987). Manic-depression gene tied to chromosome 11. *Science, 235*, 1139–1140.

Kraepelin, E. (1898). Diagnose und Prognose der Dementia Praecox. Heidelberger Versammlung 26/27. *Dohr Neurol Psychiatry, 254*–263.

Kringlen, E. (1966). Schizophrenia in twins: an epidemiological-clinical study. *Psychiatry, 29*, 172–184.

Kringlen, E. (1967). *Heredity and environment in the functional psychoses: An epidemiological-clinical twin study*. London: William Heinemann Medical Books Ltd.

Lander, E. S. (1988). Splitting schizophrenia. *Nature, 336*, 105–106.

Lehmann, H. E., & Hanrahan, G. E. (1954). Chlorpromazine: new inhibiting agent for psychomotor excitement and manic states. *AMA Archives of Neurology & Psychiatry, 71*, 227–237.

Lencz, T., & Malhotra, A. (2015). Targeting the schizophrenia genome: a fast track strategy from GWAS to clinic. *Molecular Psychiatry, 20*, 820–826.

Lewontin, R. C., Rose, S. P., & Kamin, L. J. (1984). *Not in our genes: Ideology and human nature*. New York: Pantheon Books.

Lipkowitz, M. H., & Idupuganti, S. (1983). Diagnosing schizophrenia in 1980: a survey of US psychiatrists. *The American Journal of Psychiatry, 140*, 52–55.

López-Muñoz, F., Alamo, C., Cuenca, E., Shen, W. W., Clervoy, P., & Rubio, G. (2005). History of the discovery and clinical introduction of chlorpromazine. *Annals of Clinical Psychiatry, 17*, 113–135.

Luxenburger, H. (1928). Vorläufiger Bericht über psychiatrische Serienuntersuchungen an Zwillingen. *Zeitschrift für die Gesamte Neurologie und Psychiatrie, 116*, 297–326.

MacDonald, N. (1960). Living with schizophrenia. *Canadian Medical Association Journal, 82*, 218.

Malhotra, D., McCarthy, S., Michaelson, J. J., Vacic, V., Burdick, K. E., Yoon, S., ... Sebat, J. (2011). High frequencies of de novo CNVs in bipolar disorder and schizophrenia. *Neuron, 72*, 951–963.

Matheson, S., Shepherd, A. M., Pinchbeck, R., Laurens, K., & Carr, V. J. (2013). Childhood adversity in schizophrenia: a systematic meta-analysis. *Psychological Medicine, 43*, 225–238.

Molloy, C., Conroy, R. M., Cotter, D. R., & Cannon, M. (2011). Is traumatic brain injury a risk factor for schizophrenia? A meta-analysis of case-controlled population-based studies. *Schizophrenia Bulletin, 37*, 1104–1110.

Neale, J. M., & Oltmanns, T. F. (1980). *Schizophrenia*. Chichester: John Wiley & Sons.

Polderman, T. J., Benyamin, B., de Leeuw, C. A., Sullivan, P. F., van, B. A., Visscher, P. M., & Posthuma, D. (2015). Meta-analysis of the heritability of human traits based on fifty years of twin studies. *Nature Genetics, 47*, 702–709.

Pollin, W., Allen, M. G., Hoffer, A., Stabenau, J. R., & Hrubec, Z. (1969). Psychopathology in 15,909 pairs of veteran twins: evidence for a genetic factor in the pathogenesis of schizophrenia and its relative absence in psychoneurosis. *American Journal of Psychiatry, 126*, 597–610.

Pope, H. G., Jonas, J. M., Cohen, B. M., & Lipinski, J. F. (1982). Failure to find evidence of schizophrenia in first-degree relatives of schizophrenic probands. *The American Journal of Psychiatry, 139*, 826–828.

Prescott, C. A., Kuhn, J. W., & Pedersen, N. L. (2007). Twin pair resemblance for psychiatric hospitalization in the Swedish Twin Registry: a 32-year follow-up study of 29,602 twin pairs. *Behavior Genetics, 37*, 547–558.

Ripke, S., Schizophrenia Working Group, & O'Donovan, M. (2017). Current status of schizophrenia Gwas. *European Neuropsychopharmacology, 27*, S415.

Rosanoff, A. J., Handy, L. M., Plesset, I. R., & Brush, S. (1934). The etiology of so-called schizophrenic psychoses: with special reference to their occurrence in twins. *American Journal of Psychiatry, 91*, 247–286.

Rose, R. J., & Kaprio, J. (1988). Frequency of social contact and intrapair resemblance of adult monozygotic cotwins—or does shared experience influence personality after all? *Behavior Genetics, 18*, 309–328.

Rosenhan, D. L. (1973). On being sane in insane places. *Science, 179*, 250–258.

Rosenthal, D. (1971). A program of research on heredity in schizophrenia. *Systems Research and Behavioral Science, 16*, 191–201.

Schizophrenia Working Group of the Psychiatric Genomics Consortium. (2014). Biological insights from 108 schizophrenia-associated genetic loci. *Nature, 511*, 421–427.

Sebat, J., Levy, D. L., & McCarthy, S. E. (2009). Rare structural variants in schizophrenia: one disorder, multiple mutations; one mutation, multiple disorders. *Trends in Genetics, 25*, 528–535.

Sherrington, R., Brynjolfsson, J., Petursson, H., Potter, M., Dudleston, K., Barraclough, B., ... Gurling, H. (1988). Localization of a susceptibility locus for schizophrenia on chromosome 5. *Nature, 336*, 164–167.

Shevlin, M., Houston, J. E., Dorahy, M. J., & Adamson, G. (2008). Cumulative traumas and psychosis: an analysis of the National Comorbidity Survey and the British Psychiatric Morbidity Survey. *Schizophrenia Bulletin, 34*, 193–199.

Shi, J., Levinson, D. F., Duan, J., Sanders, A. R., Zheng, Y., Pe'er, I., ... Gejman, P. V. (2009). Common variants on chromosome 6p22.1 are associated with schizophrenia. *Nature, 460*, 753–757.

Shih, R. A., Belmonte, P. L., & Zandi, P. P. (2004). A review of the evidence from family, twin and adoption studies for a genetic contribution to adult psychiatric disorders. *International Review of Psychiatry, 16*, 260–283.

Silverstein, M. L., Warren, R., Harrow, M., Grinker, R., & Pawelski, T. (1982). Changes in diagnosis from DSM-II to the research diagnostic criteria and DSM-III. *The American Journal of Psychiatry, 139*, 366–368.

Simmons, J. M., & Quinn, K. J. (2014). The NIMH research domain criteria (RDoC) project: implications for genetics research. *Mammalian Genome, 25*, 23–31.

Sismondo, S. (2008). Pharmaceutical company funding and its consequences: a qualitative systematic review. *Contemporary Clinical Trials, 29*, 109–113.

Smith, C. (1974). Concordance in twins: methods and interpretation. *American Journal of Human Genetics, 26*, 454–466.

Solmi, M., Murru, A., Pacchiarotti, I., Undurraga, J., Veronese, N., Fornaro, M., ... Seeman, M. V. (2017). Safety, tolerability, and risks associated with first- and second-generation antipsychotics: a state-of-the-art clinical review. *Therapeutics and Clinical Risk Management, 13*, 757–777.

Spitzer, R. L. (1976). More on pseudoscience in science and the case for psychiatric diagnosis. A critique of D. L. Rosenhan's "On Being Sane in Insane Places" and "The Contextual Nature of Psychiatric Diagnosis". *Archives of General Psychiatry, 33*, 459–470.

Spitzer, R. L., & Fleiss, J. L. (1974). A re-analysis of the reliability of psychiatric diagnosis. *British Journal of Psychiatry, 125*, 341–347.

Spitzer, R. L., & Williams, J. B. (1985). Classification of mental disorders. *Comprehensive Textbook of Psychiatry, 4*, 591–613.

Stahl, S. M. (2013). *Stahl's essential psychopharmacology: Neuroscientific basis and practical applications* (4th ed.). New York: Cambridge University Press.

Stefansson, H., Ophoff, R. A., Steinberg, S., Andreassen, O. A., Cichon, S., Rujescu, D., ... Collier, D. A. (2009). Common variants conferring risk of schizophrenia. *Nature, 460*, 744–747.

Stefansson, H., Rujescu, D., Cichon, S., Pietilainen, O. P., Ingason, A., Steinberg, S., ... Stefansson, K. (2008). Large recurrent microdeletions associated with schizophrenia. *Nature, 455*, 232–236.

Stephens, J. H., Astrup, C., Carpenter, W. T., Shaffer, J. W., Goldberg, J., et al. (1982). A comparison of nine systems to diagnose schizophrenia. *Psychiatry Research, 6*, 127–143.

Stepniak, B., Papiol, S., Hammer, C., Ramin, A., Everts, S., Hennig, L., ... Ehrenreich, H. (2014). Accumulated environmental risk determining age at schizophrenia onset: a deep phenotyping-based study. *The Lancet Psychiatry, 1*, 444–453.

Tambs, K., Sundet, J., & Berg, K. (1985). Cotwin closeness in monozygotic and dizygotic twins: a biasing factor in IQ heritability analysis? *Acta Geneticae Medicae et Gemellologiae, 34*, 33–39.

The task force on nomenclature and statistics of the American Psychiatric Association. (1978). The American Psychiatric Association [DSM-III Draft].

Tienari, P., Wynne, L. C., Sorri, A., Lahti, I., Läksy, K., Moring, J., ... Wahlberg, K.-E. (2004). Genotype-environment interaction in schizophrenia-spectrum disorder: long-term follow-up study of Finnish adoptees. *The British Journal of Psychiatry, 184,* 216–222.

van Os, J., Kenis, G., & Rutten, B. P. (2010). The environment and schizophrenia. *Nature, 468,* 203–212.

Van Winkel, R., Esquivel, G., Kenis, G., Wichers, M., Collip, D., Peerbooms, O., ... Van Os, J. (2010). Genome-wide findings in schizophrenia and the role of gene–environment interplay. *CNS Neuroscience & Therapeutics, 16,* 185–192.

Varese, F., Smeets, F., Drukker, M., Lieverse, R., Lataster, T., Viechtbauer, W., ... Bentall, R. P. (2012). Childhood adversities increase the risk of psychosis: a meta-analysis of patient-control, prospective- and cross-sectional cohort studies. *Schizophrenia Bulletin, 38,* 661–671.

Wicks, S., Hjern, A., & Dalman, C. (2010). Social risk or genetic liability for psychosis? A study of children born in Sweden and reared by adoptive parents. *American Journal of Psychiatry, 167,* 1240–1246.

Wilson, P. (1934). A study of twins with special reference to heredity as a factor determining differences in environment. *Human Biology, 6,* 324–354.

Wynne, L. C. T., Niemenen, P., Sorri, A., Lahti, I., Moring, J., Naarala, M., ... Meittunen, J. (2006). Genotype-environment interaction in the schizophrenia spectrum: genetic liability and global family ratings in the Finnish adoption study. *Family Process, 45,* 419–434.

Zhang, J.-P., Gallego, J. A., Robinson, D. G., Malhotra, A. K., Kane, J. M., & Correll, C. U. (2013). Efficacy and safety of individual second-generation vs. first-generation antipsychotics in first-episode psychosis: a systematic review and meta-analysis. *International Journal of Neuropsychopharmacology, 16,* 1205–1218.

18

Sexuality and Gender

In some sectors of society, sex and behavior used to seem so simple. Men married women; next, they had intercourse, and then, they generated a family with several boys and girls. Everybody knew how they were supposed to behave and fit into a simple and orderly society. Now, we realize things were never so simple, and dichotomies were pervasive fictions propagated by those who had strong opinions about how other people ought to behave and what they should feel. There was an abundance of moral teachings and legal strictures but very little scientific knowledge. Well-designed scientific studies of sexuality have been uncommon, partly because they were considered too controversial to receive public funding (Bailey et al., 2016). Even now, there is not a solid foundation of well-researched facts in this domain. Instead, there have been and still are many heated debates, political movements and demonstrations, court challenges, and new laws. Striking differences are apparent between countries and even among states and provinces within a country. Nevertheless, many things have been learned, and there have been major changes in some countries (Table 18.1). Today, it is obvious that sex and gender are very complex traits (Box 18.1).

VARIETIES OF TRAITS

Sex exists along several dimensions and is measured in a wide variety of ways.

Chromosomal sex can be determined from a karyotype. As discussed in the chapter on the XYY male (Table 13.3), there are many variations on the theme of XX and XY, including multiple copies of an entire chromosome (e.g., trisomy) or large deletions or duplications of just part of a chromosome (Chapter 12). Generally speaking, an embryo with a Y chromosome that carries the SRY gene will embark on a male kind of gonadal development, but when the androgen receptor (AR) gene on the X is mutated, the male-typical anatomy will not form. The development of sexual and reproductive organs depends on many genes that are not located on either the X or the Y (Chapter 10).

Anatomical sex can also take more than the two typical forms. In several conditions, the external genitalia are ambiguous (McNamara, Swartz, & Diamond, 2017; Wherrett, 2015). The clitoris of a female newborn can be enlarged to the extent that it is sometimes mistaken for a penis (clitoromegaly), and the male genitalia may be so small that the infant is tallied as female (micropenis), as often occurs in androgen insensitivity syndrome (AIS). Virtually, every degree of departure from the typical male or female anatomy has been observed. The Prader or Quigley rating scales involving five degrees of departure from typical anatomy are often used to describe the situation at birth (Wherrett, 2015). The phenotypic range in partial androgen insensitivity syndrome is remarkably wide. When in doubt, a pediatrician and the parents may decide to take a "wait and see" approach after birth, knowing that things should become more apparent at puberty. There may be a decision to embark on genital reconstruction surgery. Prior to 2006, this was often done soon after birth on the recommendation of the pediatrician, but the result often was disappointing, and further surgery was sometimes undertaken. Modern standards advocate delaying any decision about surgery until the child is old enough to have adopted a particular gender and can play some role in the decision. In some European countries, surgical alteration of the genitals without consent is now forbidden by law. The Intersex Society of North America offers guidelines and extensive resources for families and professionals seeking the best options for the child. Alice Dreger (Dreger, 1998) advocates that "'Ambiguous' genitalia do not constitute a disease. They simply constitute a failure to fit a particular (and, at present, a particularly demanding) definition of normality." Many causes of ambiguous genitalia are known (Wherrett, 2015), including 4 chromosome anomalies, 19 different causes in XY, and 17 in XX newborns, about half of which arise from a genetic defect in a specific enzyme or receptor involved in steroid hormone function

TABLE 18.1 Timeline of Same-Sex Policies

Year	Country	Policy
1952	The United States	DSM-1 declares homosexuality a mental disorder
1953	The United States	Homosexuals banned from federal government employment
1967	The United Kingdom	Homosexual acts are legalized
1969	Canada	Homosexual acts removed from Criminal Code
1973	The United States	American psychiatrists vote to remove homosexuality from DSM
1978	Canada	Homosexuals allowed to immigrate
1980	The United States	Transexualism added to list of mental disorders in DSM-3
1990	The United States	Homosexuals allowed to immigrate
1992	Canada	End of ban on gays and lesbians in military
1995	The United States	Sexual orientation included in offenses under Hate Crime Law
1996	Canada	Sexual orientation included in Human Rights Act
1996	The United States	Defense of Marriage Act bans gay marriage
2000	The United Kingdom	Gays and lesbians allowed to serve in military
2003	The United States	Homosexual behavior legalized
2005	Canada	Same-sex couples given legal right to marry
2011	The United States	End of ban on gays and lesbians in military
2013	The United States	Supreme Court extends benefits to same-sex couples
2015	The United States	Supreme Court legalizes gay marriage
2017	Canada	Government apology for past wrongs against LGBTQ2 people
2017	The United States	Presidential decree bans transgendered people from military

or sexual development (Figs. 10.2 and 10.5). These various conditions generally do not predict the sexual orientation of a child at puberty, although androgen insensitivity syndrome (Chapter 10) almost always results in a person with the XY karyotype whose external anatomy is primarily female and who is raised as a female and attracted to males.

Gender identity is the psychological sense of whether an individual experiences the self as male or female. It is assessed by self-report. This is often manifested in a child long before puberty or the emergence of sexual desire. When gender identity concurs with anatomical sex assigned at birth, this is the typical or *cis* condition. When a boy senses that she is a girl or a girl regards himself as a boy, this is the *trans* condition. In the case of a transexual child or adult, there may or may not be surgical and hormonal treatments to transform the anatomy into the opposite sex. There may be a substantial change in apparel and lifestyle to fit the gender identity of a trans person, but these surface features are not part of the definition of trans. The issue is what the person believes his or her true gender should be. *Gender fluidity* happens when a person changes gender identity or sexual orientation over a period of time, something that is particularly common among teens and young adults. It also can happen that a person does not regard the self as either male or female. Trans people sometimes reject the typical gendered pronouns *he/she* and *his/her* in favor of *they*, *their*, and *ze*.

Sexual orientation denotes the subjects of erotic attraction or desire. Various terms are used to describe this phenomenon. Today, heterosexual denotes attraction to the opposite sex, but the term was not always used in that sense. Prior to 1890, it was not used at all by theologians or psychiatrists, and its first use was to describe someone who desired both males and females (Katz, 1995). In the 1901 edition of *Dorland's Medical Dictionary*, heterosexuality was defined as "abnormal or perverted appetite toward the opposite sex" (Ambrosino, 2017). Only in the 1930s, largely through the influence of Freud and his followers, did the term come to denote an exclusively male-female attraction that psychiatrists then embraced as "normal."

Today, homosexual is sometimes used to denote attraction to the same sex, but the term is more often used to denote male-male attraction, not female-female. Some researchers prefer more precise terms such as androphilic (attracted to males) and gynophilic (attracted to females), so that a homosexual male is an androphilic man, while a lesbian is a gynophilic female (Bailey et al., 2016). Bisexual involves attraction to both sexes, whereas asexual means no attraction to either sex. Sexual orientation does not necessarily concur with sexual identity. Orientation and attraction are behaviors, whereas identity involves an enduring concept of the self. Same-sex behavior has been recognized for millennia, but only recently has psychiatry embraced the idea that some people have an inherent quality of being "homosexuals" or "heterosexuals."

Orientation is usually assessed with a simple scale of self-report, such as the Kinsey scale that ranges from 0 (always opposite sex attraction), to 3 (bisexual), to 6 (exclusively same sex). Recent surveys have employed a three-point scale (straight, bisexual, and lesbian) or a five-point scale shown in Fig. 18.1. Because people may

"Colette," age 14, lived with her father and older brother; her mother died when she was 7. She first had sex at age 12, with her best friend from Girl Scouts; 8 months later, she dated a boy from school and found she liked sex with him too. Now, she's in love with Jenny, also age 14, and they've been together most of the school year. When she decided to confide in her brother that she's bisexual, she was unprepared for his outrage. He told their father, who kicked her out of the house, and then her brother got a couple of his friends to assault Jenny. Convinced

Jenny's family would be just as hostile, the two girls ran away. To survive, Jenny panhandles, while Colette works as a stripper and has sex with their landlord in exchange for the basement room they stay in. She has come to the teen clinic 4-month pregnant.

Source: Case study taken from *Saewyc, E. M., Skay, C. L., Pettingell, S. L., & Reis, E. A. (2006). Hazards of stigma: the sexual and physical abuse of gay, lesbian, and bisexual adolescents in the United States and Canada.* Child Welfare, 85, 195–213.

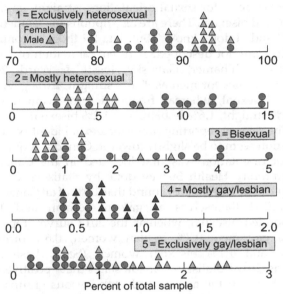

FIG. 18.1 Frequency of self-ratings of sexual orientation or attraction on a five-point scale. Each symbol represents the result of one large population survey that involved at least 1000 people. Scales on the X-axis are expanded to show the full range of values. *Reprinted from Savin-Williams, R. C., & Vrangalova, Z. (2013). Mostly heterosexual as a distinct sexual orientation group: a systematic review of the empirical evidence.* Developmental Review, 33, 58–88, Copyright 2013, with permission from Elsevier.

not wish to have their orientation known, some sex researchers have tried to get at the "true" self by measuring genital arousal with a *penile plethysmograph* or *phallometer* that senses pressure from an erection in males. Evidence from a phallometer test is sometimes used in sentencing decisions about convicted pedophile offenders (Purcell, Chandler, & Fedoroff, 2015). Other indicators such as time spent viewing erotic photos of people of different sexes and pupillary diameter have been employed in lab settings (Bailey et al., 2016).

The pupil diameter test was employed in a device developed by the Canadian security service to detect

homosexuals in the civil service (Hauen, 2017). The device later became known in the media as the fruit machine. A series of photos were shown to a person seated in a dental chair, and pupil diameter and eye movements were recorded (Kinsman, 1995). The pictures showed half-naked men from a physical culture magazine, neutral images in paintings, and even an image of Christ on the cross. Investigators believed a homosexual would be unable to take his eyes away from the genital area of Christ. The work began in 1962 and was abandoned in 1963 when sufficient volunteers to test the device could not be recruited. In practice, the Canadian security police relied on grueling interviews and surveillance of gay entertainment venues to identify homosexuals, thousands of whom were dismissed from public service in the 1950s and 1960s.

Sexual identity is the self-identified label a person uses to describe the self to other people. Common identities include hetero or straight, gay, lesbian, bisexual, transexual, asexual, and uncertain or undecided. In some instances, nonheterosexual people may prefer the all-inclusive label "queer," a former term of vulgar abuse. In other contexts, "queer" denotes a person who is not comfortable with the conventional labels gay, lesbian, or bi. The term "gay" can also be more inclusive, as in the formation of "gay-straight" alliance clubs in many schools that can include girls and "gay pride" events that include women and girls. Those celebrations are increasingly known simply as "pride" events. Among some of the First Nations of Canada, those who engage in same-sex activities are often called "two-spirited," which accounts for the 2 at the end of the acronym LGBTQ2. The existence of asexual and undecided identity may be recognized by adding + to yield the all-inclusive LGBTQ2+.

In population surveys, the precise wording of a question conveys to the respondent which sense of sex or gender is being queried. For example, in the large review by Savin-Williams and Vrangalova (2013), the wording used in each study is given in their Table 1.

A question about identity can be expressed as "which of the categories best describes yourself" or "which answer best describes how you currently think of yourself." Sexual orientation can be questioned as which option "best describes who you have ever felt sexually attracted to." Some wording may not clearly point to one specific dimension, such as "which of the following best describes your feelings." This matter can be important when a person identifies as a married heterosexual while experiencing some degree of attraction toward persons of the same sex. Scientific journals specializing in sexual behavior are now opting for the awkward term "men who have sex with men" in the titles of their articles to denote sex behavior rather than identity or attraction.

Sexual behavior can occur with the opposite sex, the same sex, or both sexes. Typically, intimate behavior takes place one-on-one in private, but many variations on the theme are known to occur. The behavior may involve intercourse in a variety of body positions, manual, oral, or anal stimulation. Many other body parts have erogenous value for some people. People may also make use of electromechanical devices and a diverse array of toys.

Legal Classifications

Marriage requires a special license from a government or religious authority, whereas common-law unions involve living together under the same roof as partners but without the license or certain legal protections afforded by marriage. Unrelated people living together as "just friends" differ from common-law unions especially when the relationship breaks apart and ownership of a home and common property is disputed. Governments also make laws about age limits for sexual behavior, availability of birth control, and many other aspects of sexual relations. Some research reports that give data on number of sex "partners" do not use the term in the legal sense of a partner but simply mean those who have participated in some kind of sexual behavior with another person. Certain categories such as adultery involve conventional forms of behavior, attraction, and identity but with a person who is married to another, and in some countries, such conduct is punished harshly.

These various traits and qualities as well as the options available for each need to be carefully considered when a researcher seeks to determine population frequency of different characteristics and possible genetic and environmental influences. The challenges are serious indeed when many of the behaviors and preferences are proscribed by law, dominant religions, or the psychiatric profession in a country. Sexual minorities are often the targets of discrimination, persecution, and sometimes even murder, which make it virtually impossible to obtain accurate and valid estimates of some behaviors in many jurisdictions. Minorities may have very good reasons to conceal desires and habits.

POPULATION PREVALENCE

Precisely, how many people in a country belong in a category other than straight heterosexual? This can be assessed with a well-constructed population survey that avoids several kinds of biases in sampling, so that the results are broadly representative. The National Health Interview in the United States uses face-to-face interviews with follow-up by telephone to fill in some of the blanks about many features of a person's health. It uses a three-category scale for sexual orientation—straight, gay/lesbian, and bisexual. There are also options for "something else" and "I don't know," but data for those uncommon choices are not described. A recent tabulation of 53,016 adults (Dahlhamer, Galinsky, Joestl, & Ward, 2017) reported rates for men as 97.6% straight, 2.0% gay, and 0.4% bisexual, whereas frequencies for women were 97.2% straight, 1.6% lesbian, and 1.2% bisexual.

The rates of reporting nonheterosexual identity or orientation seem to be slightly lower in Canada. A question about orientation was first included in the Canadian Community Health Survey done by Statistics Canada in 2003. The 2009 survey found that 1.0% of all Canadians identified themselves as gay or lesbian and 1.0% responded bisexual, whereas the 2012 survey reported 1.5% gay men, 0.7% lesbian women, 0.6% bisexual men, and 1.1% bisexual women (Scott, Lasiuk, & Norris, 2017). Statistics Canada states that "positive rapport between the agency and with various groups and individuals, coupled with assurances of anonymity, contribute to respondents feeling very comfortable with the interviewing arrangements" (Statistics Canada, 2015). Despite the rapport, more than one-third of those contacted in a recent survey chose not to register their sexual identity with the government agency.

Using a five-point scale reveals some interesting differences between men and women (Savin-Williams & Vrangalova, 2013). Fig. 18.1 shows data reported in several surveys based on different populations and age groups. Only those involving more than 1000 of each sex and questions clearly directed at sexual orientation or attraction were extracted from the larger report to construct Fig. 18.1. Several broad generalizations can be drawn: (1) A rate of exclusively heterosexual attraction exceeding 90% occurred often for males but far less for females. (2) Many more females chose the options "mostly heterosexual" or "bisexual." (3) The rate for exclusively gay males was much higher than for exclusively lesbian females. (4) There was a wide range of values within each category. Across many samples from the United States in the Savin-Williams and Vrangalova (2013) review, the frequency of males who are exclusively or mostly gay is roughly 2.5%, whereas women who are exclusively or mostly lesbian account for about 1.3% of recent samples.

Official sanctions applied against nonheterosexuality have been lifted in several countries (Table 18.1), but the extent of proximal peer and family pressures is difficult to quantify. In many societies, there continue to be strong, discriminatory social pressures or punishments directed at those who express something other than straight heterosexual qualities, despite official proclamations of equality and fairness (Chard, Finneran, Sullivan, & Stephenson, 2015).

WHO IS ABNORMAL?

A population survey can take a neutral stance on whether different categories of sexual orientation enrich a society or are lamentable exceptions. It seeks simply to describe behavior. The psychiatric profession, on the other hand, passes judgment on who is normal and who has a significant mental disorder or illness that warrants treatment. At the outset of the DSM project in 1952, homosexuality was designated a mental illness, despite the fact that many gay and lesbian persons were otherwise healthy, highly educated, and achieved noteworthy successes in business, science, and the arts. It has been noted that "cultural values play a greater role in psychiatry than they do in the rest of medicine when it comes to deciding what constitutes a 'disorder'" (Shorter, 2014; see also Gaines, 1992). Of course, in the 1950s, when there were harsh measures used against gays and lesbians by most governments, the vast majority of nonheterosexual persons concealed their sexual identities. In retrospect, it appears that much of the difficulty in life experienced by gay and lesbian persons has been a consequence of popular prejudice, repressive laws, and stigmatizing psychiatric diagnoses.

After WW2, some prominent individuals "came out" as gay or lesbian, sometimes at great personal sacrifice including loss of life because of their gender identity and sexual orientation. Sexual acts between consenting adults of the same sex were held to be crimes punishable by many years in prison. A gay rights movement gradually gathered support. The core demand was that all citizens should be afforded the same rights and protection under the law, regardless of gender identity or sexual orientation. A step in that direction was taken in 1967 when Canada's Minister of Justice Pierre Trudeau introduced amendments to the criminal code to remove homosexual acts from the list of crimes. He stated at the time, "the view we take here is that there's no place for the state in the bedrooms of the nation" (CBC News, 2015).

One noteworthy feature of the demand for equality was the desire to serve in the armed forces. This was forbidden for many years, and those whose nonheterosexual behaviors or identities were discovered while in service were unceremoniously expelled and lost all accumulated benefits. Another target of activism was the lack of access to the institutions of marriage and adoption of children. All of those official forms of discrimination were encouraged by the DSM designation of homosexuality as a psychiatric disorder outside the range of acceptable behaviors.

Pressure was exerted on the American Psychiatric Association from its members and others outside the ranks of psychiatry to remove homosexuality from the DSM. This was partially accomplished after a vote of APA members passed a resolution in 1973 (Bayer, 1981). Nevertheless, the new DSM-3 in 1980 retained one form of homosexuality on its list and added transexualism for the first time. DSM-4 changed the name of transexualism to gender identity disorder and the number from 302.5 to 302.6 but kept the definition the same. Category 302.9 could be used when someone experienced distress as a result of being gay or lesbian, a common occurrence during a period of disrespect and persecution. Finally, in 2013, the APA removed homosexuality from DSM altogether, 40 years after the resolution of members called for this in 1973. DSM-5 changed the name of transexualism to gender dysphoria. Now, there is a movement to have this category expunged as well. Some have argued that keeping it in the DSM will help to obtain professional help. Lev (2013), who favors its removal, argues: "Trans people deserve access to medical care, not because they are mentally ill and fit the criteria within a diagnostic manual, but rather precisely because they are sane and actualizing their authentic gender is their civil right."

Meanwhile, the lists of psychosexual disorders and paraphilias proscribed under the DSM have remained unchanged for more than 50 years. The list of psychiatric disorders that are given official code numbers in the DSM includes insufficient vaginal lubrication, pain during sex, and inability to orgasm during intercourse in females. In males, both premature and delayed ejaculation are considered psychiatric disorders. The list of forbidden paraphilias includes desire to have sex with animals or children, sexual arousal from watching others have sex, or having others watch you performing sex. During a period when societal mores have been radically transformed, several of the paraphilias are now practiced widely, such as sadomasochism and bondage that are featured in the widely read book *Fifty Shades of Grey*. Tight-fitting black leather regalia with metal adornments, body piercing, and heavy chains have become common fashion statements. The authors of DSM category 302 appear to be not only "societies' moral gatekeepers"

(Shorter, 2014) but also antiquated prudes. "Kinky" practices such as sadism or masochism are now performed with consent to achieve sexual excitement among many otherwise mentally normal adults. However, when inflicted on unwilling persons, they are criminal acts, not just psychiatric curiosities.

HEREDITY

Investigations of genetics in relation to nonheterosexual attractions and behaviors have followed a familiar pattern seen for intelligence, autism, and schizophrenia. First, there were twin studies to show that heredity plays a significant role. Next came widely publicized reports of linkage of a few markers on one chromosome to a gene, followed by failures to replicate. Then, new methods allowing thousands of markers to be assessed detected several weak signals amidst the noise. Continuing research now seeks to determine which specific gene might have different alleles relevant to the phenotype.

Politics and Religion

There is one major complication for studies of heredity and same-sex attraction. From the outset, people with widely differing opinions about gay rights and legalization of same-sex behavior and marriage have claimed that a predominant role of either heredity or environment is important for their political agendas (Bailey, Wallace, & Wright, 2013; Whiteway & Alexander, 2015). Many writers maintain that if same-sex attraction is genetically determined, then gays and lesbians deserve the same rights as everyone else because they really have no choice but to be who they are. Others see that homosexuality is a free choice and therefore those who follow that path may rightfully be punished for wrongdoing. If environmental factors that encourage same-sex attraction can be discovered, they argue that things can then be done to prevent people from becoming gay or lesbian.

A survey of more than 1000 American adults done by the Pew Research Center questioned respondents about their opinions on genetic determination of sexual orientation, the legality of same-sex relations, equal rights for gays and lesbians, same-sex marriage, and related topics (Haider-Markel & Joslyn, 2008). They also collected extensive data about respondents' education and age, religiosity, religious creed (protestant and born-again), political orientation (liberal and conservative), and other potentially relevant factors. There was a very strong tendency for those who believed in a genetic source of same-sex attraction to favor same-sex rights and marriage. Likewise, political liberals were much more likely to believe in genetic determination and support civil rights

for people in same-sex relationships, whereas political conservatives usually took the opposite stances. "Born-again" Christians were strongly against civil rights for same-sex persons and at the same time did not believe gay and lesbian orientations arose from heredity.

These patterns seem to be specific to the issue of same-sex attraction and are not applicable to many other phenotypes of current interest and controversy. Consider the matter of race and intelligence, a topic to be explored in the next chapter. There has been a long-standing political divide on this topic. Liberals tend to see that racial differences in achievement arise from unjust discrimination, impoverished environments, and unequal opportunities, and they tend to favor government policies that help disadvantaged minorities to improve their lives. Many conservatives, on the other hand, ascribe racial and ethnic inequality to inherited abilities and oppose government education, health, and welfare policies designed to assist the poor. When it comes to same-sex rights, on the other hand, the political camps appear to trade places. These strongly held opinions make research on a controversial topic such as same-sex attraction and behavior more difficult. When reviewing what is currently known on the topic or what people think they know, it is wise to be alert for the many ways that popular prejudice can sway those designing, funding, and interpreting research.

Family and Twins

Available evidence has consistently found that the frequency of same-sex attraction in a family is higher if a person has a nonheterosexual brother. One study reported that 13.5% of brothers of gay probands were themselves gay (Hamer, Hu, Magnuson, Hu, & Pattatucci, 1993), while another reported a 6.5% rate (Bailey & Bell, 1993), substantially higher than the population prevalence of 1%–2%. Both studies relied on subjects recruited via ads in homophile publications or AIDS/HIV clinics, which makes them liable to sampling biases. Government-sponsored surveys designed to yield representative samples usually ask about the orientation of the person who agrees to participate in the survey, but those kinds of surveys are loathe to ask a person to tell others about his brother or sister. There are few data on rates in parents versus their children. Furthermore, many nonheterosexual people do not have children. As with other phenotypes, knowing that the rate is higher in first-degree relatives does not tell anything about the origins of the correlation, given that people in the same family share both heredity and many features of the environment.

Twin research on same-sex attraction has been reported for more than 50 years, but earlier studies suffered from numerous shortcomings in their methodologies and small sample sizes. More recently, there have been two population surveys with large samples, both based on existing registers of all twins known in a country. The Swedish registry used data on lifetime number of same-sex partners for both men and women, not just attraction, and it found a surprisingly high rate of 6.9% lifetime prevalence of having one or more same-sex partners (Långström, Rahman, Carlström, & Lichtenstein, 2010). "Partner" in that study did not mean cohabitation or long-term commitment, just a participant in a sex act; some individuals reported 21 same-sex partners. The Australian study reported an overall rate of 3.7% same-sex orientation (Zietsch et al., 2012). Each study involved more than 1000 pairs of each kind of twin and a substantial number of MZ and DZ pairs wherein at least one twin was rated as nonheterosexual. As shown in Table 18.2, concordance rates were higher for MZ than DZ twins, which supports a role for heredity. For the MZ twins having the same genotypes, only about 12% of relevant pairs both showed same-sex behavior or attraction. As discussed in Chapter 17, pairwise concordances were about half the values of probandwise concordances. Both indicators suggest that nongenetic factors must play a major role.

Penetrance

Strength of genetic influence on the development of the individual can be expressed in a simple way using data on MZ twins. It is widely accepted among geneticists that many genes influence the development of a phenotype but do not code for it in an all-or-none manner. Instead, persons with an atypical genotype often show only a probability of expressing an atypical phenotype. This phenomenon is termed *incomplete penetrance* in technical parlance (Chapter 11), and the probability of showing the atypical phenotype is termed "penetrance." Penetrance can be estimated directly only if we know which persons have the atypical genotype. But in a research project in its early stages, we have no idea of which gene or genes are involved. Therefore, estimates of penetrance must be indirect. Suppose we have two MZ cotwins with the same atypical genotype. If the probability that any one person with that genotype shows the typical phenotype is P, then the probability of showing the atypical phenotype is $1 - P = Q$, the degree of penetrance. Table 18.3 provides probabilities of observing all four possible pairings of two MZ cotwins. Note that the combination of two cotwins wherein neither shows the atypical phenotype would not be included in the sample in many twin studies. A twin study often would involve only pairs where at least one twin showed the atypical phenotype, as discussed in Chapter 17. Using a bit of algebra, it is possible to express penetrance in terms

of pairwise concordance: $Q = 1 - (1 - C)/(1 + C)$. For the twin studies in Table 18.2 in which pairwise concordances are about 0.12 and 0.13, the penetrance values are about 0.22. That is, at least 78% of the people who have the atypical genotype do *not* show the atypical phenotype. Genetic factors probably have some relevance because penetrance is clearly above 0, but nongenetic factors must play major roles in the story of same-sex attraction.

The "Gay" Gene

Many people have heard about the "gay gene" that is alleged to determine male same-sex attraction. The term can be traced to discussions of a report in *Science* in 1993 that claimed to have discovered genetic linkage between several markers in band 28 of the q arm of the X chromosome of males and an unknown gene nearby that supposedly determines homosexuality (Hamer et al., 1993). The study examined 40 pairs of gay brothers who had been recruited for the research by ads in homophile magazines and at homophile picnics and local HIV clinics. It compared the frequency of pairs in which both men inherited the same allele of the marker to the frequency expected merely by chance. Five closely spaced markers at one end of the X were found to be transmitted to both men in 33 of 40 pairs, a pattern that was statistically significant. The markers spanned a region of more than 4 million base pairs (MB) that contained hundreds of different genes. Nothing in the data pointed to any specific gene. Nevertheless, in a widely read article in *Scientific American*, the authors stated that "the gene itself" had not yet been located, implying there was indeed one gene in that genetic jungle that made a man gay (LeVay & Hamer, 1994). This idea was propagated widely through the news media and became common knowledge.

The study immediately drew criticism. Fausto-Sterling and Balaban (1993) pointed out that the research needed a control group of heterosexual brothers in order to show that they would *not* have highly similar alleles at the same markers. They also noted that, in a period of highly politically charged debate, it would be prudent to confirm the preliminary results from a small sample of men before going public with such controversial claims. This advice proved to be prescient. An independent study of markers on the X chromosome was done with 52 pairs of nonheterosexual brothers and 33 control pairs (Rice, Anderson, Risch, & Ebers, 1999). There was no evidence of linkage with any of four markers in the Xq28 region.

Another report by the authors of the original study added 73 families with two gay brothers to the original sample of 40 and a subsequent sample of 33 (Hu et al., 1995), then tested linkage with a new set of 403 markers that spanned all chromosomes (Mustanski et al., 2005).

TABLE 18.2   Twin Studies of Nonheterosexuality

Study	Prevalence in total sample	Type of twin	Number of C pairs	Number of D pairs	Pairwise[a] concordance	Probandwise concordance
Långström et al. (2010) Sweden 2826 total pairs	6.9%	MZ	33	252	11.6%	22.6%
		DZ	16	177	8.3%	16.6%
Zietsch et al. (2012) Australia 3256 total pairs	3.7%	MZ	13	82	13.7%	24.0%
		DZ	8	115	6.5%	13.0%

C, concordant; D, discordant.

[a] *Pairwise data estimated from probandwise figures given in the article plus prevalence. Both male and female pairs are included.*

TABLE 18.3   MZ Twin Concordance and Penetrance (Q) for "Atypical" Phenotypes

		Twin 1					Typical (P = .8)	Atypical (Q = .2)
	Theory	Typical (P)	Atypical (Q)	Pairs with one or more atypical	Example			
Twin 2	Typical (P)	P^2	QP	$2PQ+Q^2$	Typical (P = .8)		.64	.16
	Atypical (Q)	PQ	Q^2		Atypical (Q = .2)		.16	.04
	Pairwise concordance $C=Q^2/(2PQ+Q^2)$				$C=0.04/0.36=0.11$			

For the full set of 146 brother pairs in which both were rated nonheterosexual, the new data fell far short of showing linkage with markers in region Xq28. Because the added sample of 73 families was part of the same study reported in 1993 and was not analyzed separately, it does not constitute a replication. It is just an extension of the same study by the same lead scientists with a larger sample. Thus, the failure to achieve statistical significance supplants the original Hamer et al. (1993) finding. Mustanski et al. (2005) did find a region in chromosome 7 (7q36) that showed significant linkage with same-sex attraction.

Meanwhile, genetic technology continued to advance, and a fine grained assessment of huge numbers of markers across the entire genome became possible. This was implemented with a larger sample of 409 pairs of nonheterosexual brothers using a technique that could assess 440,793 SNPs for possible linkage anywhere in the genome (Sanders et al., 2015). The study found strong evidence of linkage with a gene somewhere in band 8q12 but did not replicate the result for 7q36 that had been reported by Mustanski et al. (2005). Neither did it confirm the role of the CYP19 aromatase gene on chromosome 15 that had previously been advocated as a candidate gene for same-sex attraction by the Hamer group (DuPree, Mustanski, Bocklandt, Nievergelt, & Hamer, 2004). There was a hint of relevant markers in Xq28, but the data did not meet accepted criteria for statistical significance.

It is illuminating to reiterate the findings according to accepted standards of evidence in genetic research, where each study is counted only once. There have been three studies:

1. Hamer et al. (1993) originally reported significant linkage with a gene or genes in a wide region of Xq28, but when their sample size was increased (Mustanski et al., 2005), there was no significant linkage.
2. Rice et al. (1999) found no evidence of significant linkage to Xq28.
3. Sanders et al. (2015) found no evidence of significant linkage to Xq28.

Thus, the evidence is consistent: three studies, none of which found linkage with Xq28.

Now, let us examine how results were described by the authors and commentators.

First, Hamer in the 1993 report elevated the findings to evidence for "the Xq28 candidate locus" when it had not even been replicated by anyone. This one report of something interesting in a region containing hundreds of genes falls far short of what is needed to make something a candidate gene. In the 1994 *Scientific American* article by LeVay and Hamer, this became "the gene itself," and they described how it might function physiologically, when they actually had not even one clue as to which gene or genes might be involved. There was no scientific basis to say anything at all about how the alleged gene might work. Second, the 2005 Mustanski report referred to "interesting peaks" in the plot of LOD scores when in fact nothing was significant, which makes the location of

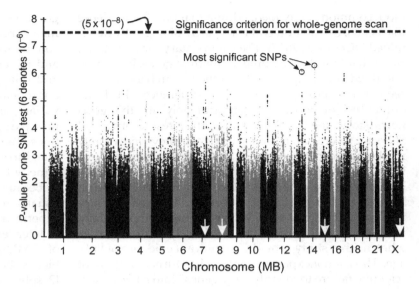

FIG. 18.2   Results of a GWAS study of 1077 homosexual men, 818 of whom were the 409 brother pairs studied previously by Sanders et al. (2015). No SNP anywhere on any chromosome showed significant linkage. Arrows indicate the locations of markers reported in previous publications to show significant linkage. None was replicated in this much larger sample. *From Sanders, A. R., Beecham, G. W., Guo, S., Dawood, K., Rieger, G., Badner, J. A., … Martin, E. R. (2017). Genome-wide association study of male sexual orientation.* Scientific Reports, 7, 16950.

"peaks" indistinguishable from noise. It invited the reader to form favorable opinions about the "findings" by just looking at a graph while ignoring the actual test statistic that did not point to significance. Third, the 2015 report by Sanders pointed to "the second strongest region" (Xq28) for linkage that had been "implicated in prior research." But it was not implicated because none of the previous data from other groups yielded significant evidence of linkage. There was nothing to replicate. The Sanders description strongly implied that there really was an important gene there, but it just did not meet strict criteria for significance. This phenomenon is termed *criterion drift*, whereby researchers are swayed by data that do not meet the widely accepted criteria for significance.

A commentary by Servick (2014) on the Sanders et al. (2015) report was headlined "New support for the 'gay gene'" that it elevated to a "replication" of Hamer's 1993 result. Servick wrote that Hamer felt "vindicated" by the new study that also "fingered" Xq28, when in fact neither Hamer's nor Sander's data showed significant linkage with genetic markers in Xq28. Servick also quoted Risch, a leading authority in the field, who pointed out that the Sanders study did not meet criteria for statistical significance. All this equivocation leaves the reader with the impression that the evidence for a "gay gene" is mixed and the concept has advocates in the ranks of scientists. It suggests that the gene really is there at one end of the X chromosome but the effect is just not large enough to convince everyone.

Whereas the articles published in peer-reviewed journals usually are circumspect and mention limitations to their findings, science journalists are not so restrained and often act as touts for genetic determinism of same-sex attraction. Their writings reach a far larger audience and are clearly responsible for the original claim of a "gay gene" that was not made in the Hamer et al. (1993) article.

Science writers and headline makers are largely responsible for the pervasive belief in the existence of a "gay gene," sometimes aided and abetted by scientists who would dearly love to find such a creature on the X and are willing to bend the rules about statistical significance in order to persuade readers to take seriously the substandard findings. The fact remains that there never was any evidence of a "gay gene" and 24 years of further research and discussion have not provided any credible evidence.

GWAS

A very recent report made the news and confirmed what was written in the above paragraph. The article by Sanders et al. (2017) was published online in *Nature* on 7 December 2017. It used GWAS to scan the entire genome with microarrays that could assess 5,642,880 SNPs. The sample of 1077 homosexual men included the 409 brother pairs studied previously by Sanders et al. (2015), a study that claimed some support for linkage to genes in 8q12 and Xq28. In the expanded sample, no SNP reached the criterion of significance of $P < 5 \times 10^{-8}$ for a whole genome scan. There were two locations where the most significant SNP reached close to $P = 5 \times 10^{-7}$, well short of the accepted criterion in the field of genetics (Fig. 18.2). One small cluster of SNPs on chromosome 13 was located *between* the genes SLITRK6 and SLITRK5. Another on chromosome 14 was located in an intron in the TSHR gene. There was no evidence of any genetic variation in a protein-coding exon. Nevertheless, the authors discussed how the genes SLITRK6 and TSHR might be related to homosexuality through influences on brain function.

A commentary on the study in *New Scientist* claimed falsely that "two gene variants have been found to be more common in gay men, adding to mounting evidence that sexual orientation is at least partly biologically determined" (Coghlan, 2017). The commentary made a strong claim that "for the first time, individual genes have been identified" or "pinpointed" and the researchers had succeeded in finding "single-letter differences in DNA sequences" of two genes. Unfortunately, for this good news story, the Sanders et al. (2017) results involved SNPs that were not located in any gene; they were merely markers on chromosomes 13 and 14. Furthermore, there is no mounting evidence of a gay gene anywhere on the human genome. The latest study did not itself confirm Xq28 as a site of genetic influence. The commentary implied that Hamer's 1993 study linked homosexuality to Xq28 and that the Sanders' 2017 study confirmed this, when neither study provided significant evidence of linkage. Thus, a nonexpert in genetics has turned a litany of scientific failure to identify "gay genes" into a tale of success and steady progress.

Sanders et al. (2017) made no mention in their article of the fact that their scan failed to provide any evidence of SNPs near genes in Xq28, 8q12, and 7q36 or on chromosome 15 (CYP19) that had been flagged in previous studies. The new report could not show that such genes were irrelevant, because the sample size of 1077 gay men and 1231 controls was far too small. Twin and family studies indicate that the genetic influence on same-sex attraction is probably real but is not very strong. When genetic variants exert only a weak influence on the development of a phenotype, very large samples of people are needed to detect the signatures of specific genes.

Congenital Adrenal Hyperplasia (CAH)

An unusual syndrome featuring ambiguous genitalia at birth in genetic XX females was shown to be caused by overproduction of androgens by the adrenal gland (Wilkins, Lewis, Klein, & Rosemberg, 1950). A very large pedigree of a Swiss family made it apparent that the syndrome arose from an autosomal recessive gene (Prader, 1962), and that gene encoded the enzyme 21-hydroxylase (now termed CYP21A2 gene) that is located at chromosome band 6p21.33. Normally, the 21-hydroxylase enzyme converts progesterone to deoxycorticosterone in the adrenal gland, which then gives rise to corticosterone, cortisol, and aldosterone (Fig. 10.2). When that pathway is blocked, the alternative route from progesterone to testosterone in the adrenal cortex becomes more active, and the excess of androgens can affect the external genitalia of the female fetus. Null mutations in CYP21A2 also produce serious deficiency of aldosterone that is important for salt-water balance, and the infant then excretes far too much salt, a trait termed salt-wasting that can be life-threatening. Certain alleles of CYP21A2 do not generate severe salt-wasting and instead are noteworthy mainly because of virilization of the female genitalia. The condition can be treated by administering glucocorticoid and mineralocorticoids to infants to restore salt balance and other drugs to suppress the production of excess androgens. About 1/15,000 newborns show this genetic variant, and it is one of the Mendelian disorders that is now part of routine newborn screening in many jurisdictions (Table 8.2). Prompt treatment of the condition soon after birth can restore hormonal balances, but it does not remedy the anatomical anomaly of the genitals.

Because of the prenatal imbalance of hormones that is usually remedied after birth, CAH provides a natural experiment about the role of prenatal androgen levels in later female sexuality. There are four kinds of alleles of CYP21A2 that are most common. The null allele blocks the corticosterone pathway altogether, while the I2 splice variant seriously impairs its functions. Those two kinds of mutations generate high androgen levels and contribute to the clinical condition known as saltwasting (SW) CAH. The lesser form of simple virilizing (SV) CAH does not involve salt-wasting and results in milder increases in prenatal androgen levels and anomalies of the external genitalia. The SV form is usually seen with the I172N mutation, whereas the milder nonclassical form (NC) is evident mainly with the V281L and P30L mutations.

Girls with CAH almost always exhibit female gender identity. For a wide range of behavioral phenotypes, however, there are often patterns of behavior more typical of boys and men. A large study of 62 adult females with CAH in Sweden found that those with the null mutation showed a higher frequency of participation in rough sports, outdoor activities, interest in automobiles, and engagement in male-dominant occupations (Frisén et al., 2009). In the realm of sexual life, differences were strongly related to the kind of mutation, which in turn was related to prenatal androgen levels. Bisexual or lesbian orientations were very common with the I2-splice and null mutations (Fig. 18.3A). The number of women with no sex partner was also greatly elevated. Patients reported considerable shame and embarrassment about the anatomy of their genitals, even after multiple attempts to make surgical repairs. The sense of shame is provoked by an environmental or cultural influence that follows from an anatomical effect that follows from a genetic mutation. A very small change in a gene is magnified into a large change in the social functioning of women with CAH. Treating that aspect of the syndrome will require adjustments in attitudes through education of people with CAH and their families and peers, perhaps even an entire society.

A study of 141 adult women with CAH and 22 controls in New York City collected detailed information on sexual history, preferences, imagery, and actual sex partners.

FIG. 18.3 Studies of adult women with congenital adrenal hyperplasia (CAH) syndrome that caused elevated prenatal levels of androgens. (A) Frequency of nonheterosexual orientation and no sexual partners among genetically normal controls and 62 women in Sweden with four different mutations of the CYP21A2 gene that contribute to CAH. Null has no functional enzyme at all. (B) Kinsey lifetime scores in 141 women in New York City with different forms of CAH versus 22 non-CAH controls. *(A) Adapted from Frisén, L., Nordenstrom, A., Falhammar, H., Filipsson, H., Holmdahl, G., Janson, P. O., ... Nordenskjold, A. (2009). Gender role behavior, sexuality, and psychosocial adaptation in women with congenital adrenal hyperplasia due to CYP21A2 deficiency. The Journal of Clinical Endocrinology & Metabolism, 94, 3432–3439, by permission of Oxford University Press. (B) Reprinted by permission from Nature/Springer, Meyer-Bahlburg, H. F., Dolezal, C., Baker, S. W., & New, M. I. (2008). Sexual orientation in women with classical or non-classical congenital adrenal hyperplasia as a function of degree of prenatal androgen excess. Archives of Sexual Behavior, 37, 85–99, Copyright 2008.*

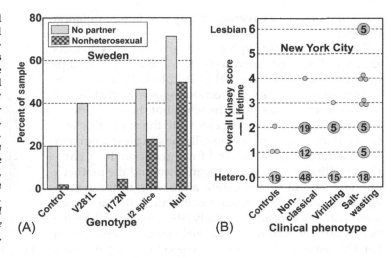

Data were analyzed according to phenotypic class (SW, salt-wasting; SV, simple virilization; and NC, nonclassical with mild phenotypes), and sexual behaviors were summarized with a lifetime Kinsey score ranging from 0 (always heterosexual) to 6 (exclusively lesbian). As shown in Fig. 18.3B, there was a high frequency of nonheterosexual orientation among those with the SW phenotype (16/39) and moderately elevated levels for the SV and NC conditions. Among the 39 SW cases, 47% experienced same-sex imagery, and 21% had one or more same-sex partners.

CAH is a rare alteration of prenatal steroid hormone metabolism in the adrenal gland that makes only a very small contribution to the population of women who experience same-sex attraction. Nevertheless, it is quite informative. The defective enzyme is part of the chemical pathway to cortisol and aldosterone, not to sex hormones, but blockage of the pathway can increase androgen levels indirectly. The CYP21A2 gene or its mutation is not simply a gene *for* same-sex attraction. It is not even expressed in the brain. It is part of some rather complex biochemical machinery (Figs. 10.2 and 10.5) that has an organizing role in fetal sexual development. In this case, knowing about those alternative pathways helps to formulate an effective treatment for the hormonal imbalance, but the balance can be restored only after birth when the unusual genital anatomy is noticed, and there are long-lasting effects on the lives of many adult women with CAH. This phenomenon provides clear evidence that prenatal hormone levels can influence later same-sex attraction (Hines, 2011). In that sense, the attraction might be seen as "inborn," and "born this way" may be a reasonable way of expressing things for CAH. The situation is considerably more complex than that slogan implies, because most women, even among those with the most severe salt-wasting form of CAH at birth, do *not* later develop same-sex attraction.

ENVIRONMENT

Evidence of an environmental influence on same-sex attraction does not accrue from failures to identify genes relevant to that phenotype. Evidence that environment plays a role must come from studies of people who experience different environments.

Older Brothers

One factor has been repeatedly associated with the frequency of male same-sex attraction: family size and birth order of male siblings. It was first reported by Blanchard and Sheridan (1992) that men with more older brothers are more likely to express same-sex attraction. Most, although not all, subsequent reports have supported this finding. The observation is not without complications because gay men also tend to have more older sisters (e.g., Kangassalo, Polkki, & Rantala, 2011). This is not surprising because larger families would involve more older sisters and older brothers. A recent review and meta-analysis by Blanchard (2018) of 26 studies published over a period of 25 years showed that the critical factor is the number of older brothers. He constructed an *older brother ratio* of the average number of older brothers to the average number of siblings other than older brothers (older sisters plus later-born brothers plus later-born sisters) and found that it was consistently higher for nonheterosexual men. In a meta-analysis, the odds ratio of being nonheterosexual versus being heterosexual for those with more older brothers was about 1.47, a clearly significant ($P < 0.00001$) but phenotypically small effect. Estimates based on older data when family size was larger indicate that about 15%–25% of nonheterosexual males arise in families with more older brothers (Blanchard & Bogaert, 2004), so there must be several other environmental factors that are relevant. This factor will be of lesser importance as average family size in a population declines.

The older brother hypothesis aligns with several facts about prenatal male nervous system and sexual development (Bogaert & Skorska, 2011). The placenta usually prevents bacteria and other infectious agents carried by the mother from crossing into the fetus, but there are often a few cells from the fetus that cross into the mother's circulation. If the fetus is male, the cells may express a number of proteins (antigens) derived from genes on the Y chromosome that are foreign to the mother, and the mother's immune system will generate antibodies to those antigens such as H-Y. The effect is cumulative; the more sons a woman has carried prenatally, the greater will be the population of antibodies to male-specific antigens in her bloodstream. Some of the Y-specific antigens, such .as those from the PCDH11Y and NLGN4Y genes, are known to be expressed in the fetal brain and could play a role in sexual differentiation of the brain. It is therefore plausible to propose that maternal antibodies to those antigens are more likely to alter prenatal brain development in male fetuses whose mothers have carried one or more sons previously. Those changes could influence later sexual orientation. At present this is an unproved hypothesis.

Childhood Sexual Abuse

Another factor that shows a consistent and strong relationship with adult sexual identity and orientation is childhood sexual abuse. In most studies, sexual abuse was defined as sexual touching, interference, or intercourse involving an adult or sibling when the child was younger than a specific age. By the very nature of such incidents, investigators must rely on the self-report of an adult who had been abused when a child. Most studies also collected information on physical abuse and other indicators of adverse home environments. Many earlier studies relied on convenience samples of gay and lesbian adults recruited from gay/lesbian picnics, publications, and parades, but three recent studies examined large population samples.

A longitudinal study in the United States involved 34,653 adults interviewed in the home as part of the National Epidemiologic Survey on Alcohol and Related Conditions (Sweet & Welles, 2012). It included questions about sexual identity, behavior, and attraction that were reduced to a five-point scale for data analysis. Sexual abuse was defined as sexual touching or actual or attempted intercourse by an adult, or unwanted sexual involvement with a person under the age of 18. The frequency of experiencing sexual abuse was far higher for those who later identified as gay, lesbian, bisexual, or heterosexual with some same-sex partners.

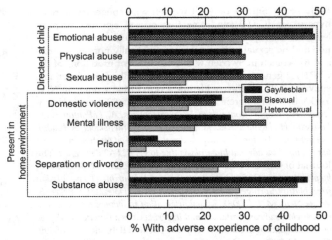

FIG. 18.4 Data from a population sample of 22,071 adults living in three states in the United States that used self-ratings of sexual identity in adults and interviews to complete the adverse childhood experience scale. Abuse directed at the child had a strong association with later sexual identity. *Adapted from Andersen, J. P., & Blosnich, J. (2013). Disparities in adverse childhood experiences among sexual minority and heterosexual adults: results from a multi-state probability-based sample. PLoS One, 8, e54691.*

A study in Australia utilized the nationwide Australian Twin Registry that provided phone interview data from 9884 adults (Zietsch et al., 2012). Sexual abuse was regarded as sexual touching with an adult family member or intercourse before the age of 14 with someone at least 5 years older. The average age of experiencing abuse was 8.7 years, and the perpetrator in more than 94% of cases was a male relative. The study also recorded features of the home environment that might increase the risk of mental disorders and, possibly, same-sex attraction. The frequencies of childhood sexual abuse and risky family environments were considerably higher in those who as adults experienced same-sex attraction. In both of the studies, the rate of sexual abuse for bisexual and lesbian adult women had been 20% or higher in childhood, more than double the rate for heterosexual women.

A third study in the United States obtained unbiased population samples totaling 22,071 people from three different states as part of the Behavioral Risk Factor Surveillance System survey and rated sexual identity with the same scale (Andersen & Blosnich, 2013). Interviewers rated eight different indicators of childhood experiences in the home environment, three of which involved abuse inflicted directly on the child. For all but the separated/divorced factor, adverse childhood experiences were significantly (P < .001) more common among adults who reported being gay/lesbian or bisexual (Fig. 18.4). The rate of abuse in the backgrounds of gay/lesbian or bisexual adults was about twice as high as in heterosexual adults, and it was higher in those who had experienced more kinds of adverse childhood events. Thus, the findings of those large surveys were consistent with the

TABLE 18.4 Prevalence (%) of Sex Abuse in Three Survey Populations for Grades 7–12 Youth[a]

Population sampled	Girls				Boys			
	Heterosexual	Mostly hetero.	Bisexual	Lesbian	Hetero sexual	Mostly hetero.	Bisexual	Gay
British Columbia	18.2	25.2	29.8	36.9	3.3	8.2	20.9	24.4
Minnesota	24.0	ND	36.1	31.6	5.6	ND	24.6	19.8
Seattle	11.6	ND	26.6	29.0	5.6	ND	30.9	24.5

[a] ND denotes no data.

Source: *Reprinted from Saewyc, E. M., Skay, C. L., Pettingell, S. L., & Reis, E. A. (2006). Hazards of stigma: the sexual and physical abuse of gay, lesbian, and bisexual adolescents in the United States and Canada. Child Welfare, 85, 195–213.*

notion that adverse childhood experiences, especially sexual abuse, cause an elevated frequency of showing same-sex identity or attraction in adults.

There is also a possibility that some of the associations between environmental variables and same-sex attraction are unrelated to effects of sexual abuse or, alternatively, that they mediate effects of sexual abuse. Several of the environmental factors shown in Fig. 18.4 are correlated with each other. A valid question is whether sexual abuse exerts a noteworthy effect that is independent of the other variables. This cannot be readily addressed in human populations in which there is no control of who experiences what conditions. There are statistical methods that may provide some helpful insights.

Roberts, Glymour, and Koenen (2013, 2014) examined three categories of variables in the National Epidemiologic Survey on Alcohol and Related Conditions data. One was the indicator of sexual abuse that was scored as none, low, medium, or high. Another was a composite of nonsexual maltreatment measures. A third was "instrumental variables" that increased the likelihood of sexual abuse in the home but had little influence on adult same-sex attraction. Those included poverty, parental alcohol abuse and mental illness, and having a stepparent before the age of five. The association of sexual abuse in childhood with adult same-sex measures was considerably stronger than the relation with nonsexual maltreatment. The authors concluded that "None of the instruments were associated with sexual orientation after adjustment for maltreatment and sexual abuse." They estimated that 9%–23% of variation in adult sexual orientation was attributable specifically to sexual abuse in childhood. This analysis was disputed by several other scholars and then rebutted by Roberts et al. (2014). The debate continues.

Abuse in Adolescence

Three large population surveys of adolescent youth in Grades 7–12 done in British Columbia, Minnesota, and Seattle in two different years asked questions about sexual orientation and experiences of abuse (Saewyc, Skay, Pettingell, & Reis, 2006). The Minnesota and Seattle surveys incorporated three types of orientation—het, bisexual, and lesbian/gay—while the BC survey also included mostly heterosexual (MH). The results for MH were between heterosexual and gay/lesbian. Table 18.4 presents prevalence values averaged over 2 years. Sexual abuse of girls was far more common than of boys, but for both genders, rates were considerably higher for nonheterosexual youth. These data do not directly tell much about the origins of nonheterosexuality among youth because their reported sexual orientation was already established when the survey was done, and the survey questions did not ask when the *first* experience of abuse occurred. Authors of the study noted that it could not be determined whether the abuse preceded a declaration of being bi, gay, or lesbian or whether "coming out" came first and then the young person was abused.

Sexual abuse of a child has multiple harmful effects on health (Roberts et al., 2013) and is not specific to later sexual attraction and behavior. Schizophrenia also appears to be associated with sexual abuse (Table 17.3), and that experience may have a more severe influence on high-risk children who have a family history of schizophrenia. What is truly shocking is the high rate of sexual abuse of children of all sexual identities and orientations, especially among girls, something that is a serious crime punishable by many years in prison but is seldom targeted by law enforcement authorities and rarely prosecuted.

ANOTHER LOOK AT "BORN THIS WAY"

The embrace of "born this way" and the "gay gene" as centerpieces of the struggle for rights of sexual minorities was a strategic political choice, not something derived from scientific knowledge. It is interesting to assess how the doctrines fare in the light of what we know now about same-sex identity and attraction.

Same-Sex Attraction is a Complex Trait

It is clear today that same-sex attraction is not a simple all-or-none trait and there is no major gene effect that can account for a large portion of variance in the multidimensional phenotypes. With the modern tools of genetic analysis now available, a genetic mutation with a large effect could be detected quite easily. Failure to detect any kind of "gay gene" does not imply that genes count for nothing and environment is all. Instead, it suggests that genes involved in this domain likely have rather small individual effects and sample sizes of 50,000 or more people with the phenotypes of interest may need to be assessed with genetic tools in order to attain significant, replicable results. There might even be a gene in the Xq28 region that contributes to same-sex attraction, but samples of just 1000 people experiencing same-sex attraction are not going to find it. It is possible that *family private* mutations with large effects may be involved in one or a few families, as sometimes happens with autism and schizophrenia, but this would likely shed light on only a few exceptional cases, as happens with CAH.

The doctrine of the "gay gene" was bad biology from the outset. There is no behavior that is simply encoded in a single gene. Genes work at the molecular level and encode sequences of amino acids in proteins. Every behavior that has been studied carefully has been found to involve the activities of numerous neurons, multiple hormones and neurotransmitters, and many hundreds or even thousands of genes (Chapters 4 and 5). In a few rare cases, an especially damaging mutation has been found to seriously disrupt neural functioning and development of behavior. Exploring how such a mutation exerts a large effect on a phenotype can teach us relevant things about biological systems. Androgen insensitivity syndrome (AR gene) and congenital adrenal hyperplasia (CAH) reveal small parts of an intricate and dynamic puzzle.

Genetic and Environmental Variance

Statistical models based on MZ and DZ twins are sometimes used to estimate sources of individual differences in a phenotype. Shortcomings of such models have already been discussed (Chapter 15). Keeping those in mind, it is interesting to review data presented in the large review of twin studies by Polderman et al. (2015) that spans a wide range of phenotypes including physical, physiological, and behavioral traits. The review identified only two twin studies of "gender identity disorders" and found that those studies did not allow an estimate of variance components. There were several other behavioral phenotypes related to family and social behaviors with more data. They estimated that approximately 30% of total phenotypic variance was attributable to genetic variance and about 35% arose from environmental factors shared by the twins, the remainder being attributable to environmental factors not shared by a twin

pair. The estimates for genetic variance are far below what one would expect for a trait that is almost entirely determined by genetic differences. Combined with evidence from two large twin studies of nonheterosexuality in Table 18.2 where pairwise concordance in MZ twins was less than 15%, they suggest that values for genetic contributions to variations in same-sex attraction are relatively small. This result argues strongly against the possibility of finding a so-called gay gene or even a set of such genes that could explain most of the variation in same-sex attraction.

Heritability and Modifiability

The doctrine of the gay gene asserts that because same-sex attraction is largely genetically determined, then nothing can be done to change it using environmental treatments. This doctrine conflicts with what is known about many other phenotypes: heritability does not determine modifiability. There are highly heritable traits, such as PKU, that turn out to be easily modified by changing the diet or other aspect of the environment. CAH is an example of a disorder that not only is readily modified after birth with hormone therapy but also has effects during prenatal life that are not easily altered. Studies of behavioral phenotypes sometimes show evidence of heredity-environment interaction (Fig. 17.3), whereby heredity appears to regulate the risk of developing an adverse phenotype when exposed to an adverse environment. It is well known that the mechanisms underlying learning and memory as well as hormonal responses to environmental changes involve large numbers of genes (Chapter 4). Behavioral plasticity itself depends on gene functions. So much depends on when and where a gene exerts its effects that broad generalizations are not warranted without first undertaking investigations in depth of specific syndromes.

A corollary of the doctrine of genetically fixed behavioral phenotypes is the idea that things that appear early in life, prior to birth, reflect primarily genetic actions whereas environment can shift the gears only after birth. Thus, if something is apparent at birth, it is inborn and therefore genetic. This is a false doctrine with many contrary examples in both lab animals and humans. Stress on the pregnant mother, for example, can alter nervous system function of the fetus if it occurs within a critical period of development. Whether those kinds of interactions are pertinent to the broad range of same-sex attraction is not currently known for sure, but they are plausible, whereas a doctrine of developmental fixity of sexual identities and attractions prior to birth is not plausible. Even when a gene is known for sure to influence sexual orientation, as in the case of CAH, it often shows incomplete penetrance, and only a minority of those with the genetic mutation show altered behavioral phenotypes (Fig. 18.3).

Eugenics

Caution about "born this way" is warranted because of the long and sordid history of eugenic measures applied against people believed to harbor a genetic defect. Eugenic policies implemented in the United States and Alberta in Canada and then later in Nazi Germany were explicitly aimed at eliminating people from the population who were held to carry untreatable, incurable genetic traits. Widespread belief that someone was "born that way" was no insurance against persecution and discrimination. To the contrary, it often landed someone in a large state institution for "defectives" and then on a surgeon's table for sterilization.

HIGHLIGHTS

- Sexuality and gender are very complex traits having several dimensions, each showing a continuum of different degrees of phenotypic expression. These include chromosomal sex, anatomical or genital sex, gender identity, sexual orientation, sexual identity, and sexual behavior.
- The words used to describe different phenotypes are culturally defined. In many instances, they differ between countries and have changed substantially from the 19th century to the present time. Authority for conventional usage has shifted from religion to psychiatry.
- Many forms of sexuality have been strongly prohibited and punished by the state and religious authorities as well as organizations of professional psychiatrists. Homosexual acts were decriminalized in several countries in the late 1960s but not removed from the Diagnostic and Statistical Manual of the American Psychiatric Association until DSM-5 was adopted in 2013.
- Available data suggest that in the United States, Canada, and several countries in Europe, the prevalence of gay men is about 2.0%–3.0%, and bisexual men is about 1%, whereas about 1.5%–2.0% of women reported being lesbian, and more than 1% are bisexual. Because of widespread discrimination, the true frequencies are probably higher.
- Research with twins indicates that the influence of heredity on same-sex attraction is relatively weak. Fewer than 15% of monozygotic (MZ) twin pairs are concordant for same-sex attraction.
- Despite an initial claim in 1995 of a "gay gene," no research study has found evidence of a major gene in

chromosome band X28 or anywhere else in the genome that is significantly related to male same-sex attraction. Whole genome scans of all chromosomes using large numbers of SNPs have not detected any significant genetic influence on male same-sex attraction.
- Congenital adrenal hyperplasia (CAH) is caused by a mutation in the CYP21A2 gene on chromosome 6 that leads to production of high levels of testosterone by the cortex of the adrenal gland. In XX females, the null allele of the gene can increase the frequency of same-sex attraction between women to a level of about 50%. The hormonal imbalance can be corrected at birth, but the prenatal hormonal environment can influence later same-sex attraction in adult women.
- One well-established environmental influence results in a higher frequency of same-sex attraction in men who had a higher number of older brothers. Why this happens is not yet known.
- Many studies have found that abuse directed at a child, especially sexual abuse, is associated with higher frequency of same-sex attraction in adult men and women.

References

Ambrosino, B. (2017). *The invention of 'heterosexuality'*. BBC News [March 16].
Andersen, J. P., & Blosnich, J. (2013). Disparities in adverse childhood experiences among sexual minority and heterosexual adults: results from a multi-state probability-based sample. *PLoS One, 8*, e54691.
Bailey, J. M., & Bell, A. P. (1993). Familiality of female and male homosexuality. *Behavior Genetics, 23*, 313–322.
Bailey, J. M., Vasey, P. L., Diamond, L. M., Breedlove, S. M., Vilain, E., & Epprecht, M. (2016). Sexual orientation, controversy, and science. *Psychological Science in the Public Interest, 17*, 45–101.
Bailey, J., Wallace, M., & Wright, B. (2013). Are gay men and lesbians discriminated against when applying for jobs? A four-city, internet-based field experiment. *Journal of Homosexuality, 60*, 873–894.
Bayer, R. (1981). *Homosexuality and American psychiatry: The politics of diagnosis*. Princeton, NJ: Princeton University Press.
Blanchard, R. (2018). Fraternal birth order, family size, and male homosexuality: meta-analysis of studies spanning 25 years. *Archives of Sexual Behavior, 47*, 1–15.
Blanchard, R., & Bogaert, A. F. (2004). Proportion of homosexual men who owe their sexual orientation to fraternal birth order: an estimate based on two national probability samples. *American Journal of Human Biology, 16*, 151–157.
Blanchard, R., & Sheridan, P. M. (1992). Sibship size, sibling sex ratio, birth order, and parental age in homosexual and nonhomosexual gender dysphorics. *Journal of Nervous and Mental Disease, 180*, 40–47.
Bogaert, A. F., & Skorska, M. (2011). Sexual orientation, fraternal birth order, and the maternal immune hypothesis: a review. *Frontiers in Neuroendocrinology, 32*, 247–254.
Chard, A. N., Finneran, C., Sullivan, P. S., & Stephenson, R. (2015). Experiences of homophobia among gay and bisexual men: results from a cross-sectional study in seven countries. *Culture, Health & Sexuality, 17*, 1174–1189.
Coghlan, A. (2017). What do the new 'gay genes' tell us about sexual orientation? *New Scientist*, (December 7).

Dahlhamer, J. M., Galinsky, A. M., Joestl, S. S., & Ward, B. W. (2017). Sexual orientation and health information technology use: a nationally representative study of US adults. *LGBT Health, 4*, 121–129.

Dreger, A. D. (1998). "Ambiguous sex"—or ambivalent medicine?: ethical issues in the treatment of intersexuality. *Hastings Center Report, 28*, 24–35.

DuPree, M. G., Mustanski, B. S., Bocklandt, S., Nievergelt, C., & Hamer, D. H. (2004). A candidate gene study of CYP19 (aromatase) and male sexual orientation. *Behavior Genetics, 34*, 243–250.

Fausto-Sterling, A., & Balaban, E. (1993). Genetics and male sexual orientation. *Science, 261*, 1257–1259.

Frisén, L., Nordenstrom, A., Falhammar, H., Filipsson, H., Holmdahl, G., Janson, P. O., ... Nordenskjold, A. (2009). Gender role behavior, sexuality, and psychosocial adaptation in women with congenital adrenal hyperplasia due to CYP21A2 deficiency. *The Journal of Clinical Endocrinology & Metabolism, 94*, 3432–3439.

Gaines, A. D. (1992). From DSM-I to III-R; voices of self, mastery and the other: a cultural constructivist reading of US psychiatric classification. *Social Science and Medicine, 35*, 3–24.

Haider-Markel, D. P., & Joslyn, M. R. (2008). Beliefs about the origins of homosexuality and support for gay rights: an empirical test of attribution theory. *Public Opinion Quarterly, 72*, 291–310.

Hamer, D. H., Hu, S., Magnuson, V. L., Hu, N., & Pattatucci, A. M. (1993). A linkage between DNA markers on the X chromosome and male sexual orientation. *Science, 261*, 321–327.

Hauen, J. (2017). Canada 'poured thousands and thousands' into 'fruit machine'—a wildly unsuccessful attempt at gaydar. *National Post*, (May 24).

Hines, M. (2011). Prenatal endocrine influences on sexual orientation and on sexually differentiated childhood behavior. *Frontiers in Neuroendocrinology, 32*, 170–182.

Hu, S., Pattatucci, A. M., Patterson, C., Li, L., Fulker, D. W., Cherny, S. S., ... Hamer, D. H. (1995). Linkage between sexual orientation and chromosome Xq28 in males but not in females. *Nature Genetics, 11*, 248–256.

Kangassalo, K., Polkki, M., & Rantala, M. J. (2011). Prenatal influences on sexual orientation: digit ratio (2D:4D) and number of older siblings. *Evolutionary Psychology, 9*, 496–508.

Katz, J. N. (1995). *The invention of heterosexuality*. New York: Plume.

Kinsman, G. (1995). "Character weaknesses" and "fruit machines": towards an analysis of the anti-homosexual security campaign in the Canadian civil service. *Labour/Le Travail, 35*, 133–161.

Långström, N., Rahman, Q., Carlström, E., & Lichtenstein, P. (2010). Genetic and environmental effects on same-sex sexual behavior: a population study of twins in Sweden. *Archives of Sexual Behavior, 39*, 75–80.

Lev, A. I. (2013). Gender dysphoria: two steps forward, one step back. *Clinical Social Work Journal, 41*, 288–296.

LeVay, S., & Hamer, D. H. (1994). Evidence for a biological influence in male homosexuality. *Scientific American, 270*, 44–49.

McNamara, E. R., Swartz, J. M., & Diamond, D. A. (2017). Initial management of disorders of sex development in newborns. *Urology, 101*, 1–8.

Mustanski, B. S., DuPree, M. G., Nievergelt, C. M., Bocklandt, S., Schork, N. J., & Hamer, D. H. (2005). A genomewide scan of male sexual orientation. *Human Genetics, 116*, 272–278.

News, C. B. C. (2015). TIMELINE | Same-sex rights in Canada. (May 25). Retrieved from: *https://www.cbc.ca/news/canada/timeline-same-sex-rights-in-canada-1.1147516*.

Polderman, T. J., Benyamin, B., de Leeuw, C. A., Sullivan, P. F., van, B. A., Visscher, P. M., & Posthuma, D. (2015). Meta-analysis of the heritability of human traits based on fifty years of twin studies. *Nature Genetics, 47*, 702–709.

Prader, A. (1962). Die Klinik der häufigsten chromosomalen Störungen. *Helvetica Medica Acta, 29*, 403.

Purcell, M. S., Chandler, J. A., & Fedoroff, J. P. (2015). The use of phallometric evidence in Canadian criminal law. *The Journal of the American Academy of Psychiatry and the Law, 43*, 141–153.

Rice, G., Anderson, C., Risch, N., & Ebers, G. (1999). Male homosexuality: absence of linkage to microsatellite markers at Xq28. *Science, 284*, 665–667.

Roberts, A. L., Glymour, M. M., & Koenen, K. C. (2013). Does maltreatment in childhood affect sexual orientation in adulthood? *Archives of Sexual Behavior, 42*, 161–171.

Roberts, A. L., Glymour, M. M., & Koenen, K. C. (2014). Considering alternative explanations for the associations among childhood adversity, childhood abuse, and adult sexual orientation: reply to Bailey and Bailey (2013) and Rind (2013). *Archives of Sexual Behavior, 43*, 191–196.

Saewyc, E. M., Skay, C. L., Pettingell, S. L., & Reis, E. A. (2006). Hazards of stigma: the sexual and physical abuse of gay, lesbian, and bisexual adolescents in the United States and Canada. *Child Welfare, 85*, 195–213.

Sanders, A. R., Beecham, G. W., Guo, S., Dawood, K., Rieger, G., Badner, J. A., ... Martin, E. R. (2017). Genome-wide association study of male sexual orientation. *Scientific Reports, 7*, 16950.

Sanders, A. R., Martin, E., Beecham, G., Guo, S., Dawood, K., Rieger, G., ... Kolundzija, A. (2015). Genome-wide scan demonstrates significant linkage for male sexual orientation. *Psychological Medicine, 45*, 1379–1388.

Savin-Williams, R. C., & Vrangalova, Z. (2013). Mostly heterosexual as a distinct sexual orientation group: a systematic review of the empirical evidence. *Developmental Review, 33*, 58–88.

Scott, R. L., Lasiuk, G., & Norris, C. M. (2017). Sexual orientation and depression in Canada. *Canadian Journal of Public Health, 107*, 545–549.

Servick, K. (2014). New support for 'gay gene'. *Science, 346*, 902.

Shorter, E. (2014). Sexual sunday school: the DSM and the gatekeeping of morality. *Virtual Mentor, 16*, 932–937.

Statistics Canada (2015). *Same-sex couples and sexual orientation... by the numbers. The Daily*. Ottawa: Statistics Canada. (June 25). Retrieved from: *https://www.statcan.gc.ca/eng/dai/smr08/2015/smr08_203_2015#a3*.

Sweet, T., & Welles, S. L. (2012). Associations of sexual identity or same-sex behaviors with history of childhood sexual abuse and HIV/STI risk in the United States. *Journal of Acquired Immune Deficiency Syndromes, 59*, 400–408.

Wherrett, D. K. (2015). Approach to the infant with a suspected disorder of sex development. *Pediatric Clinics, 62*, 983–999.

Whiteway, E., & Alexander, D. R. (2015). Understanding the causes of same sex attraction. *Science and Christian Belief, 27*, 17–40.

Wilkins, L., Lewis, R. A., Klein, R., & Rosemberg, E. (1950). The suppression of androgen secretion by cortisone in a case of congenital adrenal hyperplasia. *Bulletin of the Johns Hopkins Hospital, 86*, 249–252.

Zietsch, B. P., Verweij, K. J., Heath, A. C., Madden, P. A., Martin, N. G., Nelson, E. C., & Lynskey, M. T. (2012). Do shared etiological factors contribute to the relationship between sexual orientation and depression? *Psychological Medicine, 42*, 521–532.

19

Race

Humans comprise a young and diverse species that arose in Africa about 200,000 years ago and migrated to all parts of the globe. People living in different geographic regions today show a multitude of phenotypic differences in physical form and behaviors, but complex patterns of migration and interbreeding have intertwined our ancestries so extensively that it is now very difficult to perceive the separate contributions of genetics and environment to group differences. Some writers believe people can be partitioned into genetically distinct races, while others claim race is something arbitrary, a classification system that has done much harm.

A BRIEF HISTORY OF HUMANITY

The history of our species involves what our ancestors looked like (anatomy) compared with modern humans, where they lived (geography), tools and ceremonial objects they created (archaeology), and historical data derived from their DNA (biology). Anatomy involves bones and skeletons that are occasionally well preserved in sediments or caves. Age of things from long ago is difficult to judge, and methods to estimate age warrant a brief description.

Age of ancient events or objects is sometimes expressed as years before Christ (BC) and anno Domini (AD, year of our Lord), but science prefers a system less tightly bound to a specific religion. Years before the Common Era (BCE) and of the Common Era (CE) still use the birth of Christ as the 0h. BP is now commonly used to express years before the present time, and thousands or kilo years ago (Kya) or years ago (ya) has the same meaning, as does million years ago (mya). Kya thus looks backward from the year an article or book was published. For dating ancient events, the shifting origin for 0 years and 0 days is of little consequence. Kya is sufficient for describing the sequence and timing of major events.

Age of Rocks and Fossils

A few human fossils were buried with pieces of wood having tree rings that could reveal their age, but the method is not useful for things older than about 10,000 years. Instead, three indirect methods can yield good approximations of age.

Carbon Dating

Most carbon in nature consists of ^{12}C with six protons and six neutrons that is very stable, but intense cosmic rays from beyond our solar system bombard our atmosphere and from time to time collide with a nitrogen atom to generate ^{14}C that is radioactive and decays spontaneously to ^{14}N with a half-life of 5730 years. Half-life is the time required for half of the radioactive form of an element in a sample to decay. The ^{14}C and ^{12}C in the atmosphere form carbon dioxide (CO_2) that is then utilized by plants and trees to generate their tissues. The ratio of ^{14}C to ^{12}C in the atmosphere is very stable at 1.5:1. No new ^{14}C is created in plant tissues because the molecules are protected from cosmic rays. Thus, in very old tissues derived from plants, including tissues of animals that ate the plants, the ratio of ^{14}C to ^{12}C will be low and eventually 0 for the oldest fossils. Radiocarbon dating is useful for ages up to about 100,000 years and is therefore valuable for dating our more recent fossilized ancestors. Special methods with elaborate controls are used to obtain valid estimates with carbon dating (Grün, 2006), and different methods sometimes yield substantially different age estimates from the same sample. For example, a fossil of Neanderthal found in the Vindija Cave in Croatia initially gave an age of about 28,000 years, but a new method that first purged several contaminants from the sample revised the carbon-dating age to more than 40,000 years (Deviese et al., 2017). For older fossils embedded in ancient rocks, the decay of uranium (^{235}U) to lead (^{207}Pb) with a half-life of 710 million years or potassium (^{40}K) to argon (^{40}Ar) with a half-life of 1.25

Genes, Brain Function, and Behavior
https://doi.org/10.1016/B978-0-12-812832-9.00019-1

billion years can be informative for estimating ages of remote ancestors but not for anatomically modern humans.

Mutations: Clicks in the DNA Clock

It would be very helpful to know when the lines of descent leading to apes and humans split, but there are no fossils known for this event. Instead, we have very good information about the living descendants of the two lines—modern chimpanzees and modern humans. The complete genome sequence consisting of 3.1 billion nucleotide bases (Table 1.2) has been determined for both species. Comparing those bases throughout the two genomes, about 1.30% of them are different. This is primarily a result of accumulated mutations that cause genomes to gradually diverge. Mutations are believed to occur mostly by chance. The difference between living people and the hypothetical common ancestor would amount to about 0.65% of base pairs, and the same figure would apply to chimps, such that 0.65%+0.65%=1.3%. Thus, 0.65% of 3×10^9 base pairs would amount to about 20 million mutations since the lines diverged. Most would be SNPs in introns or the space between genes and have little or no impact on function. The figure now commonly used for the mutation rate during ancient times is estimated at about 10^{-9}-base-pair change per year. This implies that an individual accumulates about 3.1 new mutations per year until the age when he or she breeds. To accumulate 20 million mutations in a long line of ancestors would require about $20 \times 10^6 / 3.1 = 6.45 \times 10^6$ or 6.45 million years. This estimate of the time since chimps and humans diverged is based not on fossils but on DNA of living species. Thus, evolutionary time can be estimated from similarity of DNA sequences when there are no fossils to be dated.

The Haplotype Clock

Mutations in any one base in the DNA are very rare, but rearrangement of chromosomal material by crossing-over of the strands is common (Fig. 14.1). In the larger human chromosomes, there are usually one to three crossover events per meiosis that creates germ cells in every generation. The process does not change any of the alleles of a gene, but it does alter the pattern of which alleles on nearby genes are located on the same chromosome strand. If two people of different ancestry mate, the alleles on one strand will all be of one source, and the other strand will have all alleles from the other source, until crossing-over occurs. Crossing-over leaves large segments of the chromosome with the ordering of alleles unchanged. The segment with alleles having the same origin is termed a *haplotype* (haploid genotype). During the next generation, there will again be crossing-over, most likely at some other location, and the size of the haplotypes will get smaller. The more

generations that have passed since the original crossing of two ancestors, the smaller will be the length of the haplotypes. Knowing the average length of the haplotypes, the researcher can then estimate how many generations have passed since the ancestries merged or crossbred. This approach is especially useful for dating recent events such as mixing (admixture) of groups that met several centuries or a few thousand years ago following a long migration.

A Chronology of Important Events

Scientific study has yielded a coherent account of human history beginning about 6.5 million years ago when lines of descent from a common ancestor diverged for the great apes and the line leading to modern humans. Apes, humans, and orangutans comprise the *hominid* family of species. The great apes are now part of the genus *Pan*, whereas humans belong to the genus *Homo*. Lines of descent after the divergence are not well defined because so much of the evidence about our ancestors, especially the testimony of ancient DNA, has been erased over great expanses of time in hot climates. The approximate age of anatomically distinctive fossils can be estimated, but without a high-quality DNA sample from ancient remains, there is no way to be certain that a form arising earlier was indeed the direct, biological ancestor of a later form.

Apes usually walk on their knuckles, whereas the posture of upright walking that frees the hands for carrying things and making tools appeared in the fossil record in skeletons of *Australopithecus* about 3.6 million years ago (Table 19.1). Tool making appeared about 2 million years ago in assemblages of bones associated with *H. habilis*. It is significant that all of the ancient fossils of *Australopithecus* and *Homo* prior to 1.8 million years ago have been found in Africa. No fossils of any of the great apes of *Pan* or ancient human ancestors have ever been discovered in the Americas. About 1.8 million years ago, individuals having the upright anatomy now termed *H. erectus* migrated out of Africa to Europe and Asia. Their fossils have been found in present-day China, Indonesia, Spain, Turkey, France, and Hungary. The last of their kind disappeared from the fossil record somewhere in the interval from 200,000 to 350,000 years ago.

Skeletal fragments of the species *H. neanderthalensis* were first noted in sites in Europe from around the time when *H. erectus* vanished. Whether Neanderthals were direct descendants of *H. erectus* is not at all certain because a more recent fossilized form termed *H. heidelbergensis* was present in Europe, Africa, and Asia more than 500 Kya. No DNA has been extracted from *H. erectus* or *H. heidelbergensis* fossils to test those possibilities. It is clear that *H. neanderthalensis* did not evolve in

TABLE 19.1 Timeline of Human Evolution and Migrations[a]

Time (Kya)	Event
6500	Lines leading to great apes and humans diverge
3600	Upright walking (*Australopithecus*)
2300	Tool making (*H. habilis*?)
1800	*Homo erectus* migrates out of Africa
550	Lines leading to Neanderthal and humans diverge
250	*Homo neanderthalensis* appears in fossil record in Europe
150–200	Anatomically modern humans (AMHs) appear in Africa
55–60	AMH first migrations out of Africa
40–60	Interbreeding of AMH and Neanderthals in Eurasia
37–25	Extinction of Neanderthals
16–13	AMH migration to the Americas via Beringia
12–8	Beginning of agriculture in Levant and Central Africa
6–1.2	Admixture, cultural change across Europe, within Africa
3–0.8	Peopling of the remote Pacific Islands
0.5–0.2	North Africans take European slaves
0.5–0	Europeans colonize the world
0.4–0.15	Americans take African slaves

[a] *Kya = 1000 years ago.*

Africa and never lived there. They spread throughout Europe and Asia, and their remnants have been found in present-day Great Britain, Italy, Croatia, Belgium, Russia, Uzbekistan, France, and Iraq. Archaeological sites have established that they made use of fire, crafted fine spear points for hunting large mammals, built shelters, made clothing from animal skins, and used ceremonial objects. That species was remarkably modern, but they were not the principal ancestors of modern humans.

Modern humans (*H. sapiens*) first appeared in the fossil record in Africa (Ethiopia) about 190,000 years ago and spread across the continent. It cannot be said for sure that modern humans arose for the first time in Ethiopia, because fossils closest to the emergence may not have been preserved. Furthermore, the emergence of genuine *H. sapiens* may have been very gradual, involving several intermediate steps or gradations, many of which were not fossilized. A recent report dated remains of *Homo* in Morocco to 315,000 years ago and found bones with a remarkably modern jaw and face structure that lacked the large braincase of modern humans (Hublin et al., 2017), so there is some debate about whether they were truly human. Of great importance was the migration of anatomically modern humans (AMHs) out of Africa into the Middle East about 60,000 years ago (Fig. 19.1).

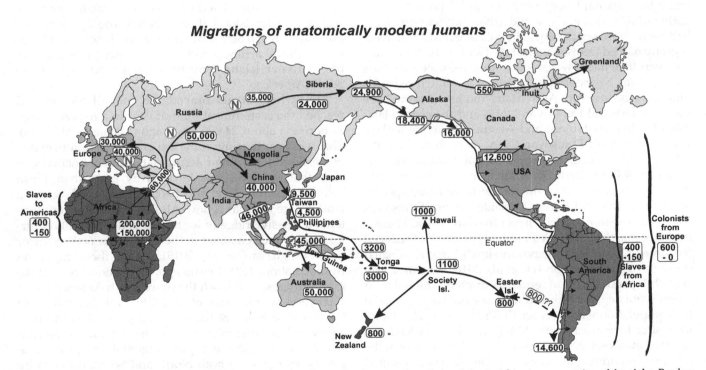

Migrations of anatomically modern humans

FIG. 19.1 Routes and timing (ya) of migrations of anatomically modern humans from their origins in Africa to every region of the globe. Borders are shown for modern countries. Colors are for convenience in viewing. Northern portions of regions in blue were heavily glaciated in the past. Based on several authorities cited in the text. N in *circle* shows a site where Neanderthal fossils were found.

Neanderthals—Our Kissing Cousins

The early migrants found themselves in a new land with many large animals to pursue. They also encountered others who were anatomically very similar to themselves and knew much about hunting and tools, a group we now call Neanderthals. Investigators extracted some samples from remains of those beings and sequenced the mitochondrial DNA, but further study revealed that the dating of the remains was not to be trusted because almost 80% of the DNA was actually contamination from modern humans, including the researchers themselves (Wang et al., 2014). The methodological issues were finally overcome by a large team that sequenced the entire genome from just the toe phalanx of a *H. neanderthalensis* specimen found in the Vindija Cave in Croatia (Prüfer et al., 2014). Carbon dating of the collagen remaining with the bone placed its age at 50,300 years ago. The DNA nucleotide base differences between the fossil specimen and present-day humans established the age of the common ancestor at about 550,000 years, at which time that ancestor would have lived in Africa and neither the Neanderthals nor anatomically modern humans would have yet emerged. Just who was that ancestor? This is not known.

The full DNA sequence confirmed what had been suspected from previous samples: there had been some interbreeding of Neanderthals and AMH. About 1.5%–2% of the DNA from present-day humans of European ancestry appears to have come from Neanderthals. The estimate provided by National Geographic's Geno 2.0 project for the author of this book is 1% Neanderthal. Furthermore, there had been a third species afield in Asia when AMH arrived. A group named Denisovans (for the Denisova Cave in Russia where their fossils were found) appeared to be a separate species from Neanderthals, and they also interbred with humans who reached Asia. Many modern Asians have traces of ancestral Denisovan DNA, and those who later migrated to New Guinea and Australia have the highest known fraction of Denisovan DNA, more than 1% of the DNA of some aboriginal people (Sankararaman, Mallick, Patterson, & Reich, 2016). Thus, the species we now identify as *H. sapiens* is actually a mixture of three *Homo* species. Modern Africans lack traces of Neanderthal and Denisovan DNA because those species never lived in Africa. There is evidence that the AMH-Neanderthal hybrids had reduced fertility and Neanderthal genes are slowly being eliminated through natural selection (Fu et al., 2016; Sankararaman et al., 2016) and that Neanderthal themselves suffered from a significant degree of inbreeding that would be expected if their population were very small (Prüfer et al., 2014). Not long after their trysts with AMH, the Neanderthals and Denisovans as species became extinct. *Homo sapiens* is the sole survivor among the species of *Homo* that once roamed Africa, Europe, and Asia. Our ancestors more or less took over the world.

Accelerating Dispersal Around the Globe

The earliest AMHs spread out and dispersed throughout mainland Europe and east to Asia via both southern and northern routes. They were not able to reach northern portions of Scandinavia and Russia because those regions were still covered by a thick sheet of ice. Those taking the southern route through Asia via what is now Thailand and Myanmar reached Borneo and New Guinea about 45,000 years ago, walking across land exposed by a sea 135 m lower than today, and Australia about 50,000 years ago (Duggan & Stoneking, 2014; Skoglund et al., 2016). They must have been able to make watercraft to traverse the 70 km of open water between some of the many islands along the way. The first wave of migration appears to have halted for more than 30,000 years at deep water separating today's Solomon Islands from the Santa Cruz Islands, an expanse of 500 km of open ocean. A second wave of AMH migration into the Pacific region commenced much later via Taiwan and the Philippines, passed through the islands north of New Guinea, and crossed 600 km of open water to reach Fiji and Tonga about 3000 years ago (Duggan & Stoneking, 2014; Skoglund et al., 2016). After a pause of almost 2000 years, migrants who had mastered the difficult art of navigation over vast expanses of the Pacific Ocean reached the Society Islands, Hawaii, New Zealand, and finally Easter Islands (Matisoo-Smith, 2015). Thus, the peopling of the Earth was completed very recently. Pacific Islanders reached Easter Island shortly before the Spanish conquistadors reach South America. There is some evidence from cultural artifacts and sweet potatoes that the people of Easter Island may have even made a foray to South America.

Meanwhile, others braved the cold of Siberia and reached the easternmost limit of Asia in a region known as Beringia about 24,900 years ago (Llamas et al., 2016). That was near the time of the last glacial maximum extent of ice. So much ice had accumulated in continental glaciers that sea level was lowered by more than 135 m below its current level and people could walk across a land bridge to what is now Alaska. There, they encountered a massive ridge of ice that impeded migration further south and remained in the vicinity for 6000 years or more (Raghavan et al., 2015). When their migration resumed about 16,000 years ago, they were able to make rapid progress and reach the tip of South America about 1500 years later (Llamas et al., 2016), probably following the coastline much of the way and giving birth to new ancestral lines that migrated into the interior of the continents to establish the populations that occupied virtually every region of both North and South America by the time when European explorers first arrived in 1492 CE (526 ya).

Admixture and Continuing Migration

No sooner had a region of the globe been populated than the various groups then began to expand and move again to new homes, where they interbred with other populations. DNA of modern humans has provided valuable tools for tracing the more recent forays of populations. According to Pickrell and Reich (2014), in some regions, the spread of agriculture was achieved not by a more recent population replacing an older one but rather by migration and admixture (interbreeding) of local populations. Some of the movements can be traced via languages and also by DNA similarities among living people.

Africa was the first continent populated by AMH, and some of the local populations lived in isolation from others for a very long time, as much as 200,000 years, resulting in marked divergences of DNA sequences. The greatest genetic differences within any continental group today occur within Africa. Nevertheless, Africa has been the site of many migrations over long distances within the continent. North Africans today show the genetic signature of merging with populations from West Africa and the Middle East. South of the Sahara desert, a major movement of people speaking Bantu languages took place beginning about 5600 ya and proceeded to occupy most of western and southern regions (Li, Schlebusch, & Jakobsson, 2014). The original Bantu speakers lived in the vicinity of present-day eastern Nigeria and western Cameroon. They had developed agriculture after 8000 ya, and their treks over thousands of kilometers spread agriculture, Bantu languages, and DNA over vast regions. A few local populations such as the San hunter-gatherers of the Namibian desert were not greatly affected, but the rainforest hunter-gatherers of West, Central, and South Africa merged extensively with the newcomers. A second wave of migration about 2000–1000 ya spread Bantu features further (Patin et al., 2017).

More recent changes involving Africa have entailed out-migration to European colonies around the world and the United States because of the slave trade that extended from before 1500 CE to the early days of the 20th century in many places. Modern African Americans show on average about 75% DNA sequences of African origin, and some present-day countries such as Barbados show more than 85% African alleles (Montinaro et al., 2015). The distribution of the African DNA segments of African Americans closely approximates the historical origins of slaves captured and exported from West Africa (13% from Senegambia, 7% from the Windward Coast, 50% from the Bight of Benin, and 30% from West-Central Africa; Patin et al., 2017). A vivid example of admixture is provided by Barack Obama, the former president of the United States (Box 19.1).

Europe was also the locus of many migrations and admixtures. A considerable number of archaeological

---

BOX 19.1

Barack Obama speech on race, 18 March 2008 (Obama, 2008)

"I am the son of a black man from Kenya and a white woman from Kansas. I was raised with the help of a white grandfather who survived the Depression to serve in Patton's Army during World War II and a white grandmother who worked on a bomber assembly line at Fort Leavenworth while he was overseas. I've gone to some of the best schools in America and lived in one of the poorest nations. I am married to a black American who carries within her the blood of slaves and slave owners - an inheritance we will pass on to our two precious daughters. I have brothers, sisters, nieces, nephews, uncles and cousins of every race and every hue, scattered across three continents... it is a story that has seared into my genetic makeup the idea that this nation is more than the sum of its parts - that out of many, we are truly one."

---

sites have yielded useful DNA samples from 45 Kya to more recent times, and the ancient DNA revealed that the DNA signatures of the older AMH in Europe are not found in present-day Europeans. Instead, all are descended from a single population present in the area from about 37 to 14 Kya, and then, a branch from the Eurasian steppe in Near Eastern Europe was added after 14 Kya (Fu et al., 2016). It was found that, after about 10 Kya, "admixture is a universal property across all groups" and almost all of West Eurasia now contains genetic material with signatures in the samples corresponding with DNA derived from several regions of Africa, Asia, and the Near East (Busby et al., 2015). The authors describe the patterns of admixture as a "mixture of mixtures." Detailed genetic histories of several migrations across Europe from east to west and west to east over the past 5000 years were documented by Brandt, Szecsenyi-Nagy, Roth, Alt, and Haak (2015).

Large population movements between continents brought dramatic changes to both the cultures and the DNA of distant regions of the Earth. The results are evident today in the DNA signatures of modern people. Prior to the advent of European colonialism, the Americas were populated by the descendants of the long migrations from Eurasia (Fig. 19.1), and the people had lived there so long (>15,000 years) that they had evolved a distinctive array of Native American DNA markers. A recent survey of DNA from several modern countries of South America (Homburger et al., 2015) documented the extensive and thorough mixing of genetic material from four distant sources (Table 19.2). The markers even allowed the European component to be traced to specific

TABLE 19.2 Average Genetic Contributions of Four World Regions to Countries of South America[a]

Country	European (%)	Native American (%)	West African (%)	East Asian (%)
Argentina	67.3	27.7	3.6	1.4
Chile	57.2	38.7	2.5	1.7
Colombia	62.5	27.4	9.2	0.9
Ecuador	40.8	50.1	6.8	2.3
Peru	26.0	68.3	3.2	2.5

[a] *Values for individuals within a country can be substantially above or below the averages shown in the table.*

Excerpted from Homburger, J. R., Moreno-Estrada, A., Gignoux, C. R., Nelson, D., Sanchez, E., Ortiz-Tello, P., … Bustamante, C. D. (2015). Genomic insights into the ancestry and demographic history of South America. PLoS Genetics, 11, e1005602 (Table 1).

regions such as the Iberian Peninsula or Italy, in agreement with the historical records.

A picture of admixture on a world scale was generated by Hellenthal et al. (2014) by analyzing DNA samples using 474,491 SNPs documented in 1490 people from 95 populations living in diverse regions around the globe. Statistical analysis detected more than 100 instances of genetic admixture dating back almost 4000 years. It was possible to group those 95 populations into 18 large clusters, each of which shared many instances of some unique genetic signatures while differing to some extent from nearby clusters. Each cluster spanned a geographic region that was assigned a unique color shown in Fig. 19.2. Then, for each population, the authors determined which regions had contributed the largest fractions of DNA sequences. The complex figure warrants

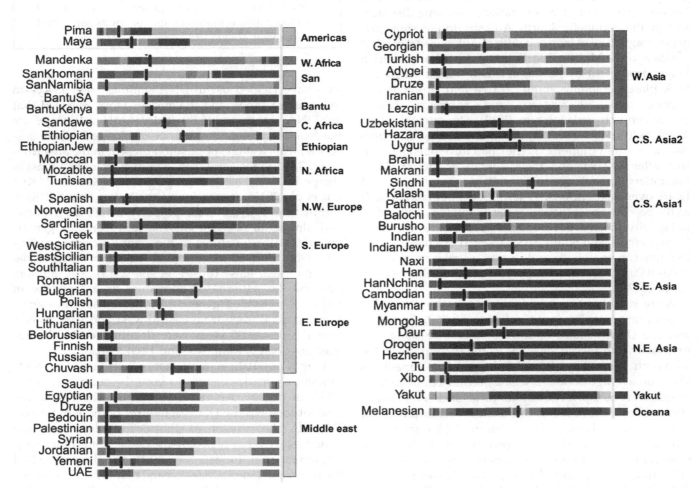

FIG. 19.2 Origins of DNA markers in the genomes of living people from several geographic regions. Colors in 18 vertical bars denote the broad region encompassing several closely related populations. Each population within a region has sets of SNPs that provide genetic signatures of ancestry. Extent of color indicates the percentage of the current genomes that are derived from different regions through admixture. Portions to the right of the black bars indicate the most common fraction of ancestry. This figure does not convey full information when viewed in shades of *gray*. The color version in the original article or online is essential for perceiving group differences. *Reprinted from Hellenthal, G., Busby, G. B. J., Band, G., Wilson, J. F., Capelli, C., Falush, D., & Myers, S. (2014). A genetic atlas of human admixture history. Science, 343, 747–751, with permission from AAAS.*

careful study. In only a few populations was a large majority of DNA sequence derived from just one region (San from Namibia, Norwegians, Lithuanians, Palestinians, and Brahui plus a few others not shown in the chart). For most of the other populations, the ancestries could be traced to at least two or three major groups outside of their own region. Some, such as the Mayans, Sardinians, Egyptians, and Melanesians, showed a complex pattern of diverse ancestors using haplotypes. It was possible to estimate how many generations had elapsed since some of the major admixture events. Those included the Bantu expansion, North African trade in European slaves, American importation of African slaves, European colonial migrations, Slavic and Turkic migrations across Eurasia, and migrations related to the Mongol and Khmer empires.

The patterns in Fig. 19.2 depend to some extent on which specific populations were originally chosen for the analysis. The authors tried to sample those that were relatively intact after the massive migrations from Europe and Africa. For example, there are hundreds of tribes and nations among the Native Americans of North and South America, but the sample included only the Pima from the United States and Maya from Mexico. The number of clusters emerging from such a complex analysis is somewhat arbitrary, and the choice of groups to include can change the apparent structure (Rosenberg et al., 2002; Tishkoff et al., 2009).

WHAT IS A RACE?

Anatomically modern humans consist of a large number of groups that have been given various names by geographers and anthropologists (Fig. 19.2), and there are several defensible systems for dividing us into these groups. Are some or all of the group races?

Geneticists' Views on Race

The answer offered by specialists in population genetics is evident in many published articles cited in the previous section on human migration and admixtures. None of them used the term "race" in their analyses or discussions of human origins. Instead, there are descriptions of populations and groups that differ in ancestry or lineage. When Raghavan et al. (2015) discussed the Athabascan and northern and southern Amerindians, they simply termed them *branches* from a single population that migrated from Siberia, and they termed that single wave of migration a "monophyletic group," meaning it was derived from a single ancestral population of Siberians. Those Siberians were themselves the descendants of several groups in the more remote past. Along this journey,

there was no period when a branch was formally designated a race. It was all one long journey involving multiple branching and admixture events that continue at the present time.

An earlier discussion of the race concept from a geneticist's perspective (Sankar & Cho, 2002) noted that how populations cluster in a data analysis depends on the populations chosen for study, criteria for establishing boundaries for groups, and resolution of the analysis that is strongly influenced by sample size. They noted that substituting the term "ethnicity" for race does not solve any of these problems. They proposed that authors be asked to define race if and whenever they use the term and state clearly whether they are using it as a proxy for genetic similarity or factors such as socioeconomic status or both.

Over the ensuing decade, the use of race in genetic research has declined, and just recently, a prominent article in *Science* argued that the concept is so "mired in confusion" that "It is time for biologists to find a better way" (Yudell, Roberts, DeSalle, & Tishkoff, 2016). The point was made emphatically by the image accompanying the article of a shredder consuming a page with the term "race" on it. The authors argued that racial categories are "genetically heterogeneous and lack clear-cut boundaries." They propose that we use terms such as "ancestry" and "population," a practice that is now widely adopted. Race as a social construct may continue to have a place in research on human behavior, but it is not a genetically based concept.

Anthropologists' Views of Race

Race has for many years been a central focus of anthropology, the study of human physical and cultural diversity. There have been calls to dispense with the term (Smedley & Smedley, 2005) along with defenses of its utility (Duster, 2005). A survey of anthropologists found that the profession was almost evenly split on the usefulness and validity of the concept of human races (Lieberman & Reynolds, 1996). Consequently, a new survey was done (Wagner et al., 2017) that asked 55 questions pertaining to the topic, including some with the exact same wording as the 1978 survey. It was directed at members of the American Anthropological Association (AAA), 3286 of whom completed the online survey in 2013. Most of the respondents were professional anthropologists. There was a major shift in opinion since the 1978 survey. To the blunt statement "No races exist now or ever did," 17% of respondents agreed in 1978, while 53% agreed in 2013. Support for a more nuanced statement ("Biological variability exists, but this variability does not conform to the discrete packages labeled races") also rose from 1978 (79%) to 2013 (89%) but was

TABLE 19.3 Opinions of Members of the American Anthropological Association in 2013[a]

Item wording	Majority opinion
"There are discrete biological boundaries among races"	93% disagree or strongly disagree
"Biological variability exists, but this variability does not conform to the discrete packages labeled races"	89% agree or strongly agree
"Genetic ancestry—inferred from genetic markers—rather than race, is a better proxy for genetic relationships among sub-Saharan Africans, Asians, Europeans, Pacific Islanders, and Native Americans"	75% agree or strongly agree
"Genetic differences between racial groups explain most behavioral differences between individuals of different races"	95% disagree or strongly disagree
"The use of the term 'race' to describe human groups should be discontinued"	71% agree or strongly agree
"The term 'race,' as used to described human groups, should be replaced by a more appropriate and precise term"	71% agree or strongly agree

[a] The entire survey involved 55 questions.
Excerpted from Table 2 in Wagner, J. K., Yu, J. H., Ifekwunigwe, J. O., Harrell, T. M., Bamshad, M. J., & Royal, C. D. (2017). Anthropologists' views on race, ancestry, and genetics. American Journal of Physical Anthropology, 162, 318–327.

markedly high on both occasions. Several items on the 2013 survey evoked a strong majority of opinion (Table 19.3) that generally indicated a lack of support for the old concept of racially based differences in behavior.

Official Government Views of Race

In many countries, the government sanctions one particular way of categorizing and labeling its citizens for conducting a census and evaluating social policies. In the United States, this is decided by the Office of Management and Budget in collaboration with the Census Bureau. The difficulties are great because of the diverse origins of the population. The Census Bureau (2015) recognizes that "the terms 'race,' 'ethnicity,' and 'origin' are confusing or misleading to many respondents, and they mean different things to different people."

Different versions of census questions are being evaluated with the 2015 National Content Test that will inform the choice of words for the 2020 census. One version replaces the terms race and ethnicity with the generic term "category." Many people in the United States today identify as Hispanic or Latino, but the census recognizes this as an ethnicity, not a racial grouping, so the survey asks if someone is of Hispanic or Latino origin. A version of the forthcoming census asks first if the person identifies as Hispanic or Latino and then in the next

question asks if he or she is white, black, American Indian, etc. One version stipulates that members of "white" include "German, Irish, English, Italian, Lebanese, and Egyptian," while another version has a separate category for Middle Eastern or North African that would allow a person from Egypt or North Africa to avoid being classified as African American, even though their country of origin certainly is part of the African continent. Thus, a possible outcome would classify some people of Latin American origin as non-Whites, while Egyptians are grouped with Whites. It is no wonder that many people are confused by the questions and concepts.

In Canada, a change in immigration policy in 1967 introduced a point system for admitting immigrants that emphasized job skills and education while removing country of origin or ancestry as a criterion. The result was a large increase in immigration from Asia, Africa, and Latin America, a trend that continues today. While the population was becoming more diverse, Statistics Canada decided to remove race as a category in the census. It stated officially that the old racial standard for reporting data "is no longer recommended for use and is not to be used" (Statistics Canada, 2017). The agency noted that race "is based primarily on genetically imparted physiognomic features" but is rendered ambiguous by mixed ancestries and changing terminologies. The Employment Equity Act of 1986 designated four groups that were often subjected to discrimination in employment: women; aboriginal people; persons with disability; and "persons who are, because of their race or color, in a visible minority." In the 1986 census, 10 origins were specified that qualified as being in a visible minority: "Black, Indo-Pakistani, Chinese, Korean, Japanese, Southeast Asian, Filipino, other Pacific Islanders, West Asian and Arab, and Latin American." That year, 6.3% of Canada's population consisted of visible minorities (Li, 2000).

The 2011 and 2016 Canadian censuses had separate questions for ethnicity and "population group." The 2016 census refrained from defining the vague term population group and instead gave a list of permissible groups that could be checked (Census of Population, 2016). It bore a strong resemblance to older definitions of race. Question 17 asked, "What were the ethnic or cultural origins of this person's ancestors?" Question 19 spoke of concern for equal opportunity and listed the following categories without asking any question about ancestors: White, South Asian, Chinese, Black, Filipino, Latin American, Arab, Southeast Asian, West Asian, Korean, and Japanese, plus a box to enter some other groups. The census form allowed multiple responses for both ethnicity and population group. Instructions cautioned that former nationality is not a good guide to ethnicity or population group: "…a person who has Canadian citizenship, speaks Punjabi and was born in the United State may report Guyanese ethnic origin" (Census of Population, 2016). Given the confusion of categories, there is no way that race can be derived from the multitude of responses. "White" and "black" are certainly not biological races.

The 2016 census revealed that 22.3% of Canada's population consisted of officially recognized visible minorities. Ethnicity was very diverse, including more than 200 recognized ancestries. For many ethnicities, multiple responses were given, largely because of intermarriage or common-law unions among groups that had lived in Canada for several generations. Thus, official statistics collected in Canada no longer use the term "race" and do not define current measures in a way that allow them to serve as a proxy for race.

Admixture in the Modern World

One glaring omission from the US Census questionnaire is a means to indicate that a person derives from more than one place or race of origin. A partial indicator exists in the questions about legal marriage, in which respondents are asked to report the race of origin of themselves or their partners. The portion of newlyweds in the United States who report being of different races or origins has increased dramatically from about 3% in 1967 to 17% in a 2015 survey (Livingston & Brown, 2017). The increase has been gradual and steady over a period of 50 years. Rate of intermarriage is particularly high among people of Asian ancestry (29%) and Hispanic origin (27%). Those figures are even higher for respondents who themselves were born in the United States (46% for Asian and 39% for Hispanic).

From the beginning of the United States as a republic and for many decades after the American Civil War of 1861–65 that ended slavery, several states in the United States, especially those in the southern region, enforced laws against "miscegenation" or crossing of races. The test of black ancestry was the "one drop of blood" rule that classified any person with even one African ancestor as black, even if the ancestor was several generations back in time (Wallenstein, 2005).

State antimiscegenation laws were finally overturned by the US Supreme Court in 1967 in the case of *Loving v. Virginia*. Muriel Jeter married Richard Loving in June of 1958 in Washington, DC, because interracial marriage was illegal in their home state Virginia. Shortly after the newlyweds returned to Virginia to live, police raided their home in the middle of the night and charged them with a felony. They appealed their conviction, and the case worked its way to the supreme court where Virginia's law was ruled contrary to the Equal Protection Clause of the 14th Amendment to the US Constitution. The court's decision noted, "The fact that Virginia prohibits only interracial marriages involving white persons demonstrates that the racial classifications must stand on their own justification, as measures designed to maintain White Supremacy."

Considering this historical background, it is no wonder that interracial marriages were reported at a low frequency in 1967. It was not only the letter of the law but also widespread social attitudes keeping those laws in force for so many years that must have discouraged many to refrain from doing what love urged.

Increases in intermarriage are partly a reflection of society-wide changes in social attitudes. Whereas in 1990, 63% of Americans expressed opposition to mixed marriage involving someone who is black; by 2016, that figure had fallen to 14% (Livingston & Brown, 2017). In the past 7 years, the portion of respondents who thought interracial marriage was good for society rose from 24% to 39%.

In the United States as a whole, the fraction of the population that identifies as not "White" is steadily increasing, and acceptance of leadership from prominent persons of color has increased to the point where Barack Obama, himself the child of an interracial marriage and born in Hawaii, was elected president two terms in a row and won respect as a world leader. Clearly, there are people who are very unhappy about these trends and would like to see a return to the old ways, but interracial mating or admixture is not biologically reversible. Once mixed, alleles distributed widely across 23 different chromosomes cannot be unmixed. Chromosomal crossing-over will continue to generate new combinations long into the future. Unlike matings of *H. sapiens* and *H. neanderthalensis* that are effectively different species and whose hybrids seem to suffer some disadvantages, matings among contemporary human ancestral groups are quite fertile.

It appears that race as a biological or genetic category has already been firmly rejected by a majority of scientists studying human group differences, and many ordinary people are now finding ancestry less of a barrier to mating. The proportion of children from diverse backgrounds is steadily increasing today, just as it did at times in past millennia. Evidence for the existence of distinct races was never very strong, and now, people are shredding the already fuzzy boundaries of old categories.

The April 2018 special issue of *National Geographic*, dedicated to issues of race and ancestry, provides compelling visual evidence of the continuum of skin colors around the world, and it celebrates the marriages of people of diverse ancestries. The cover photo of two American girls tells so much about race today. One girl has fairly light skin, whereas the other has somewhat darker brown skin. Many people look at the photo and perceive one being racially "white" and the other "black." In fact, they are dizygotic twins born at the same time from the same two parents and reared together (Edmonds, 2018). They differ in alleles of a few minor skin-color genes, as do their parents.

In another sign of the times, Prince Harry of the United Kingdom married the American actress Meghan Markle on 19 May 2018 at St. George's Chapel at Windsor, a historic site that took its current form in the 1470s. Markle's father Thomas descended from Irish and Dutch immigrants, and her mother Doria Ragland is an African American. During her childhood, Markle was encouraged to play with both white and black dolls and playacted the creation of mixed-race families. When young Meghan struggled to fill in a school form that had just two boxes, one for white and another for black, her father told her, "If that happens again, you draw your own box" (Seward, 2018).

PHENOTYPES

When people think of "race" as a category, they often focus on group differences in visible phenotypes, but a mere association of some phenotype with geographic population cannot establish the relevance of genetic variation. Fortunately, there are now methods to locate specific genes that may be involved in group differences, and solid evidence for the roles of several genes has been found. As with the examples of PKU, Huntington's disease, and androgen insensitivity (Chapters 8–10), knowing how a gene functions to alter a phenotype can be illuminating.

Skin Color

Perhaps, the most salient feature of geographic group differences is skin color, a phenotype that ranges very widely in our species. Much is now known about which genes affect pigmentation and how they work. Furthermore, there are now good explanations for why geographic populations differ so greatly in skin color. Skin color is a complex trait but not impossibly complex. Several genes involved in pigmentation differ among and within groups. Table 19.4 presents seven genes that have been found to influence skin color in at least two different studies. A review of the older research literature on single-gene effects identified seven that have since been confirmed in recent work. Taken together, they can account for a substantial portion of phenotypic variation in skin color but certainly not all of it. For this trait, there appear to be a few genes with large effects that are readily detected and a much larger number of genes that individually exert only small effects that are not readily identified. Environment matters too. The propensity to tan in sunlight has a genetic basis. A recent study of tanning

TABLE 19.4 Genes Affecting Skin Color Identified Using SNPs[a]

Study	Population	Sample size	Genes important for skin-color variation
Rees and Harding (2012)	Literature review	NA	ASIP, **HERC2/OCA2**, IRF4, MC1R, **SLC24A5**, **SLC45A2**, **TYR**
Beleza et al. (2013)	Cape Verde Islands	899	**HERC2**, SLC24A5, **SLC45A2**, **TYR**
Liu et al. (2015)	Groups within Europe	17,662	ASIP, **HERC2**, IRF4, MC1R, **SLC45A2**
Crawford et al. (2017)	Groups within Africa	1570	**HERC2/OCA2**, **SLC45A2**
Wilde et al. (2014)	Europe, ancient DNA	932	**HERC2**, **SLC45A2**, **TYR**

[a] NA, not applicable. Bold type indicates a gene identified in three or more of the five studies.

ability in Europeans (Visconti et al., 2018) identified several genes wherein allelic differences had a major effect on the tanning response to sunlight (HERC2, IRF4, MC1R, RALY, SLC45A2, and TYR). Dietary deficiencies can also make a big difference.

The Cape Verde Islands offer a fine opportunity to search for genes because admixture of African and European populations beginning several centuries ago has produced a well-intermixed population with a very wide range of skin colors and large genetic differences as well (Beleza et al., 2013). A GWAS study of 903,837 SNPs in 899 people revealed that in four genes, an allele that reduces the amount of pigmentation had a high frequency in current Europeans and a low frequency in Africans, and it showed intermediate frequency among those living in Cape Verde. The study measured skin reflectance very carefully and found that the combined effects of alleles at four loci associated with lighter skin in Cape Verde could account for about 35% of the variance in skin color. The genes with largest effects on the full skin-color range (100%) were SLC24A5 (13%), SLC45A2 (8%), TYR (8%), and OCA2 (6%), all of which were identified as being relevant to skin color in previous GWAS studies using SNPs (Table 19.4).

Current African populations differ greatly in skin color, ranging from a notably light skin among the San of Botswana in the far southern part of the continent to a very dark skin in the Nilo-Saharan people of Ethiopia. A GWAS study of 1570 Africans from diverse regions within the continent identified four genes associated with the degree of pigmentation that could account for about 29% of the variation in skin color in Africa (Crawford et al., 2017). Two (SLC24A5 and OCA2/HERC2) were also detected by the Cape Verde study. Using DNA-dating methods, they found that alleles contributing to the diversity predated the appearance of anatomically modern humans in Africa. A similar GWAS study of people from the Netherlands, the United Kingdom, and Australia (Liu et al., 2015) identified five genes involved in skin-color variation, all of which had been detected in at least one other study. It is interesting that several of the genes shown to be important for skin-color differences *between* African and European populations are also relevant to phenotypic variation *within* Africa and Europe. Clearly, genetic and phenotypic diversity are ubiquitous in modern humans.

Most of the genes listed in Table 19.4 have clear relations with skin pigmentation. Tyrosinase enzyme (TYR) mediates the first two steps in the synthesis of melanin pigment from tyrosine (Fig. 2.5). Being homozygous for a null or loss-of-function mutation can eliminate the production of the pigment altogether, resulting in albinism (Fig. 2.6). Nevertheless, there are other TYR mutations that reduce the activity of the enzyme but do not eliminate it, and one of these (Table 19.4, rs1042602) is known to lighten the skin pigmentation. When melanin has been synthesized in the melanocyte, it is packaged and transported in the melanosomes. Mutations in the HERC2/OCA2 complex can reduce the activity of the OCA2 protein that normally maintains the ion balance of the melanosomes, thereby reducing pigmentation. Two solute carrier proteins are encoded by genes that can lighten skin color; the mutant form of SLC24A5 reduces the number of melanosomes, while a defect in SLC45A2 impairs melanin precursor transport. Generally speaking, it appears that alleles that reduce melanin-related protein can function without totally obliterating them and can yield reduced pigmentation without causing overt albinism. Whereas natural selection works to reduce the frequency of alleles that cause albinism, those that merely reduce pigmentation may confer some advantage.

A study of ancient DNA has shown how dramatically the frequencies of certain alleles related to pigmentation have changed in Europe (Wilde et al., 2014) after the migration of AMH out of Africa. The authors analyzed the DNA sequences associated with three genes known to be related to pigmentation in prior studies and determined SNP genotypes (Table 19.5). Ancient DNA was available from 63 specimens of AMH dating from 6500 to 4000 ya that had been found in the steppe region of what is today the Ukraine. Allele frequencies were considerably lower than those of present-day Ukrainians but well above the frequencies in Africans. There must have been strong selective forces favoring the survival and multiplication of the alleles that yielded lower levels of pigmentation.

Humans arose in regions of Africa not far from the equator where sunlight is intense all year. Dark pigmentation of the skin serves to protect against cancer-causing mutations resulting from ultraviolet radiation (UVB) from the sun. When AMH made their gradual northward trek, they entered regions of Eurasia with much colder weather where winter nights were long and clothing was essential. This greatly reduced the amount of sunlight striking the skin, and this came at a cost. Sunlight stimulates the production of vitamin D in the human skin, and vitamin D is a dietary essential that our bodies cannot manufacture any other way. The lack of vitamin D impairs health in several ways (Yuen & Jablonski, 2010). The effect would have been compounded in Europe by the spread of agriculture that made copious amounts of vitamin-poor grains part of the staple diet. Thus, there was clear advantage in northern latitudes to having lighter skin that would be more sensitive to sunlight. Around the world today, there is a very strong correlation between average annual temperature and sunlight versus pigmentation of the skin of people native to those regions.

These observations about skin color do not fit well with notions of fixed racial characteristics or racial superiority/inferiority. A large continent such as Africa has within it high degrees of population diversity (Tishkoff et al., 2009). The diversity reflects adaptations to differing local conditions. The nature of those adaptations becomes apparent when people from one region migrate to another and then gradually acquire new characteristics that are better suited to their new homes. Africans living in tropical climes need genes that generate dark skin to protect the DNA in their skins from damage by UVB. As their descendants migrated northward, a new contingency arose that gave an advantage to lighter skin from reduced melanin. Initially, rare alleles of several genes increased in frequency, even though they *impaired* the ability to make melanin. What might ordinarily be seen as a genetic defect that greatly reduces the production of a substance that the genes evolved to manufacture (melanin) becomes an advantage under changed conditions. Gradually, over thousands of years, natural selection favored the reproduction of alleles among northern people that produce only modest amounts of melanin but enable greater synthesis of vitamin D. Thus, the allele frequencies are attuned to a new kind of life. Then, the mode of life changes again when modern clothing,

TABLE 19.5 Frequencies of Alleles of Three Genes Contributing to Lighter Skin in Three Populations[a]

Gene	Ancient Ukraine	Modern Ukraine	Modern Europe
HERC2	0.160	0.651	0.710
SLC45A2	0.432	0.927	0.970
TYR	0.043	0.367	0.368

[a] *Allele frequency associated with lighter skin was effectively 0 in most African populations and close to 0 among most Asians.*
Excerpted from Wilde, S., Timpson, A., Kirsanow, K., Kaiser, E., Kayser, M., Unterlander, M., . . . Burger, J. (2014). Direct evidence for positive selection of skin, hair, and eye pigmentation in Europeans during the last 5,000 y. Proceedings of the National Academy of Sciences of the United States of America, 111, 4832–4837 (Table 1).

sunscreens, and diets attenuate the effects of sunlight, making skin color less biologically relevant.

Eye Color

Blue eyes are a rarity in Africa but are fairly common in Scandinavia and other regions of northwestern Europe. Eye color is not a racial characteristic, but it differs in frequency among populations. Eye color is relatively stable during a person's lifetime and depends on specific changes in the DNA. Two independent studies detected the identical genetic polymorphism that was clearly related to eye color, one involving a large family in Denmark (Eiberg et al., 2008) and another involving more than 3000 people from more diverse regions of Europe (Sturm et al., 2008). They assessed genotypes at several SNPs and found one on chromosome 15 band q13.1 where two alleles were perfectly correlated with eye color. The SNP known as rs12913832 was located in intron 86 of the HERC2 gene and involved a substitution of the base C for the base T at position 467 of the intron. A portion of the DNA sequence of the ancestral allele was GAACTTGACATTTAATGCTCAAA, whereas the sequence with the new mutation (C in bold) was GAACTTGACACTTAATGCTCAAA. Further study revealed that portions of that intron served as an enhancer of transcription of the adjacent OCA2 gene that was known to affect skin color. The binding site (underlined) involved six bases where a transcription factor could bind to the enhancer of OCA2. In the mutant form, OCA2 transcription and therefore pigmentation were reduced but not eliminated altogether. The result was what we perceive as blue color in the iris. As shown in Fig. 19.4, blue occurs when both alleles possessed by an individual are base C, while having just one C often results in a greenish iris, and being TT results in mainly brown iris. The effects of SNP rs12913832 on this particular phenotype were very large, but it was evident that some other gene or genes must also play a role. One identified by Sturm et al. (2008) was the mutation R419Q in the OCA2 gene that had only a mild effect on the iris. There are probably many others with much smaller effects that make them difficult to detect.

It appears that the mutation resulting in blue eyes arose in Europe and spread widely through northern populations, but the reason for this spread is not entirely clear. Eye color and skin color are genetically distinct traits with only a few mutations in common. One HERC2 mutation that determines eye color (Fig. 19.3) also lightens the skin (Table 19.5), but most of the many mutations at other places in the genome that influence skin color do not change eye color appreciably. People with dark skin can have blue eyes. There have been speculations that blue eyes were such a novelty that it became

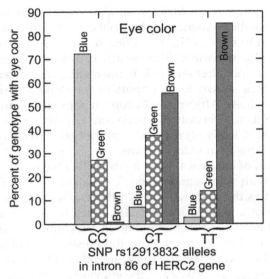

FIG. 19.3 Eye color distributions for three genotypes (CC, CT, and TT) at one position in an enhancer of the expression of the HERC2 gene on chromosome 15. *Reprinted from Sturm, R. A., Duffy, D. L., Zhao, Z. Z., Leite, F. P., Stark, M. S., Hayward, N. K., … Montgomery, G. W. (2008). A single SNP in an evolutionary conserved region within intron 86 of the HERC2 gene determines human blue-brown eye color. American Journal of Human Genetics, 82, 424–431, Copyright 2008, with permission from Elsevier.*

a desirable trait in a mate more or less for decorative reasons. Blue eyes do not seem to confer any advantage in the visual system and may actually expose the retina to more intense UV radiation. It is feasible that selection for lighter skin that helped to make more vitamin D was accompanied by more blue eyes as a more or less accidental covariate of the HERC2 mutation.

Health and Disease

It is well known that there are large disparities in almost every indicator of health and disease between Americans classified as "white" and "black" (Nestel, 2012). Several models have been proposed to explain the differences, all of which have some supporting evidence. The prevalence of high blood pressure among African Americans is considerably higher (37%) than white Americans (24%), which has led some writers to posit a genetic origin of the difference, but the prevalence of high blood pressure is even lower (16%) among West Africans (Dressler, Oths, & Gravlee, 2005), which is suggestive of environmental factors. One multifaceted variable of considerable importance is social class, which sometimes manifests as poverty involving low income and education. Although data on race are routinely collected in US studies, suitable indicators of social class are often not available (Kawachi, Daniels, & Robinson, 2005). When they are both available, researchers are often frustrated by the obvious confounding of race and class that make it difficult to separate the effects of the two.

As an example, one study collected data on health indicators and five categories of race as well as parental years of education (Chen, Martin, & Matthews, 2006). Children designated as black came from families with less education and also showed higher rates of poor health and activity and reduced school attendance because of poor health. There was a very strong relation between health and parents' years of education for white and black children but not Hispanic or Asian children, an effect the authors attributed to stronger social support networks among more recent immigrants.

Considering Canada's system of universal health-care insurance, one might expect there to be no association with factors such as social status and ancestry. Nevertheless, there are disparities among ethnic groups, especially the more recent immigrants from South Asia who have high rates of heart disease and diabetes along with less well-ingrained health behaviors such as consulting a physician early, before serious illness is manifest (Nestel, 2012). Whether any of the group differences arise from heredity is impossible to judge because data on race are no longer routinely collected in Canada.

Intelligence

From its beginnings, psychology has devoted much attention to the question of racial differences in intelligence. Francis Galton, one of the founders of the field, addressed the matter in his treatise on *Hereditary Genius* (Galton, 1869). He devised a rating scale to evaluate "the worth of different races" and assigned Africans a rating three grades below the average white man, one grade of which he surmised might be attributable to different upbringings. In fact, neither he nor any other psychologist of the time gave any kind of mental tests to native Africans, made no systematic comparisons among their relatives, and collected no measurements of their environments. Instead, he and his friends interviewed British explorers who had been to Africa. He reported that "... the white traveller almost invariably holds his own in

their presence. It is seldom that we hear of a white traveller meeting with a black chief whom he felt to be the better man." His writings on the matter reflected racial prejudice that was prominent among intellectuals in a major European colonial power, a prejudice that justified the subjugation, expropriation, and extermination of so many indigenous peoples around the world. There was nothing in Galton's writings about race that could pass muster as genuine science.

While psychology advanced with the invention of credible IQ tests (Chapters 14 and 15) and a better understanding of how to separate cause and effect with controlled experiments, renewed attention to the question of race and intelligence aroused controversy and strong opinions throughout the post-WWII period, such as the article by Jensen (1969) and the book by Herrnstein and Murray (1994). With the completion of the sequencing of the human genome in 2001, the field shifted gears and directed its energies toward the discovery of single genes having major effects on intelligence and many other behavioral phenotypes. Some scholars hoped that finding a gene or two that clearly altered intelligence would give renewed impetus to research on racial differences, but as described in Chapter 15, this did not happen. The large GWAS studies of intelligence summarized in Table 15.7 reported only a few SNPs with very small effects that were apparent only with immense samples of more than 70,000 people.

It is informative to compare GWAS studies of intelligence (Table 15.7) or intellectual disability (Table 15.8) with SNP and GWAS studies of skin color (Table 19.4). Not one of the genes clearly related to skin-color variation appears in any of the lists of genes implicated in linkage studies of intelligence or intellectual disability. Genetically, variations in intellectual functioning seem to have nothing at all to do with variations in skin color.

Race continues to appear in studies of intelligence done in the United States mainly as a control variable or covariate included in data analyses when the study is focused on some other kind of influence. For example, a study of secondhand smoke (Yolton, Dietrich, Auinger,

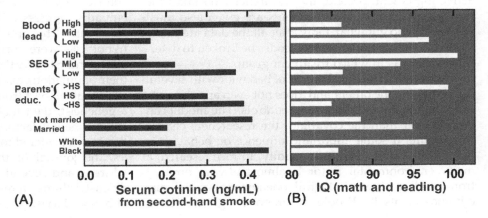

FIG. 19.4 (A) Levels of cotinine in blood derived from secondhand smoke in children from families differing in several attributes. (B) Mean IQ test scores of those same children show a similar pattern of differences in relation to covariates. *Adapted from Yolton, K., Dietrich, K., Auinger, P., Lanphear, B. P., & Hornung, R. (2005). Exposure to environmental tobacco smoke and cognitive abilities among U.S. children and adolescents.* Environmental Health Perspectives, 113, 98–103.

Lanphear, & Hornung, 2005) presented data for a number of other measures that were correlated with the amount of the smoke-derived metabolite cotinine in the bloodstream (Fig. 19.4). When blood cotinine values were high, so was blood level of the environmental pollutant lead, and those high values were evident especially in people from lower socioeconomic status families with relatively less formal education. All of those indicators were associated with poverty, as were race and marital status. Determining which measure influenced others or might be caused by them was a statistical task of major proportions. The pattern of results supported the contention that secondhand smoke had a strong negative influence on childhood intelligence above and beyond the effects of an array of covariates.

Experience of Racial Discrimination

Statistical models that incorporate several covariates may help to identify significant influences on intelligence and other phenotypes, but they cannot prove what things cause group differences. The design of population surveys falls far short of what is needed to demonstrate causation. Consider the design of a clinical trial to assess changes in behavior caused by a new drug. After volunteers are recruited, they are randomly assigned to treatment conditions—drug or placebo. This insures that there are no substantial differences between treatment groups in any important characteristic before the treatment is administered. If there happen to be variations in skin color or ethnicity among the volunteers, they should be equally distributed across the treatment groups (if sample size is large). Then, the drug is given, and measures of its possible effectiveness are collected. A double-blind study incorporates two special conditions. First, the subjects themselves must not know who is getting the drug and who receives a placebo that lacks the active drug. This is usually achieved by having code numbers on the pill or vial of solution so that even the researcher administering the treatment does not know who gets the active ingredient. Second, the researcher who takes the measurements after the treatments are given also must not know who got what. Only after all the data are safely loaded in the computer will the codes be broken to determine who was in which treatment group.

Evaluating a statistical model of behavior with several covariates, on the other hand, does not even approximate random assignment with blind controls. The model can be no better than the covariates the researchers choose to measure. If some important influence on behavior is ignored, the study will lack validity. In race research, a major environmental factor is almost always omitted from studies of psychological phenomena—racial discrimination itself. People in minority groups have reported time and time again that cruel taunts and daily aggressions because of skin color and other characteristics are common experiences throughout childhood that impair self-esteem and disrupt concentration on the task at hand. There are ways to measure the experiences of discrimination. Were they employed routinely in psychological research, any tendency to attribute group differences to biology would be invalidated by even a cursory examination of the data.

When groups differ in ancestry in a way that gives rise to differences in skin color and other "visible" features, the machinations of modern societies wherein racial discrimination is prevalent will automatically generate noteworthy differences in experience that are correlated with skin color and other features. There is no way to render the research subjects themselves or their compatriots and teachers "blind" to this reality. Survey research that includes race as a variable will always have aspects of biological ancestry and recent experience confounded to a substantial degree and cannot separate cause from consequence.

BEYOND SCIENCE

Every few years, a scientist working in genetics makes controversial claims about racial differences in behavior, especially intelligence, that provoke furious replies and condemnation that in turn are dismissed as products of political bias. An example is provided by the 25th annual meeting of the Behavior Genetics Association in Richmond, Virginia, 2 June 1995. Glayde Whitney of Florida State University had been elected president of the BGA and was entitled to present a presidential address. The title of his address was innocuous: "25 years of behavior genetics." Whitney in his prepared remarks first accused many of his colleagues of being guilty of environmentalism. This seemed like an odd thing to say at a meeting of professional geneticists. Eventually, he got to the core of his message: "… some, perhaps much, of the race difference in murder rate is caused by genetic differences in contributory variables such as low intelligence, lack of empathy, aggressive acting out, and impulsive lack of foresight." The only data he cited in support of his hypothesis were correlations between murder rates in states and cities in the United States and the fraction of their populations that were black males, the kind of data one would expect from a criminologist or sociologist, not a geneticist, data that could plausibly be explained by environmental factors. No data on genotypes of murderers were offered to support his claims. *Behavior Genetics*, the journal of the BGA, declined to publish the address, and several people resigned from the BGA in disgust. Whitney remained defiant and accused his critics of being Marxists.

BOX 19.2

Watson embroiled in controversy involving race in 2007

October 17. *Times Online*. Watson announced he was "inherently gloomy about the prospect of Africa" because "all our social policies are based on the fact that intelligence is the same as ours - whereas all the testing says not really." One bit of evidence he cited was his own experience that "people who have to deal with black employees find this [equality] not true." He predicted that genes for intelligence would be found within a decade.

October 18. *Fox News*. Watson's remarks "created a racial firestorm." He also started a DNA learning center near Harlem and wanted to have more blacks at his lab, "but there's no one to recruit," so he said. The implication was clear that bad genes were the cause. He had earlier called for using gene therapy to eliminate low intelligence.

October 19. *Times Online*. The board of the Cold Spring Harbor Laboratory where Watson had been director for many years announced that it had "decided to suspend the administrative responsibilities of Chancellor James D. Watson, Ph.D."

October 19. *NY Times*. Watson tried to backtrack by issuing a statement that "I cannot understand how I could have said what I am quoted as having said. There is no scientific basis for such a belief."

October 26. *Washington Post*. Watson retired from his position, saying, "The passing on of my remaining vestiges of leadership is more than overdue."

A more prominent scientist made a bigger fool of himself 12 years later. James Watson, the winner of the Nobel Prize as codiscoverer of the structure of DNA, made an unfortunate foray in 2007 into the behavior genetics of race and intelligence that evoked swift condemnation and precipitated his resignation as the director of the prestigious Cold Spring Harbor Laboratory. The rapid-fire sequence of events is given in Box 19.2. His claim that policies are based on the presumption that intelligence is the same in everyone fails to recognize that intelligence is defined and measured in ways that recognize ubiquitous phenotypic differences among people. There is nothing that requires any two groups of people to have the same mean IQ. If two groups have different mean IQ scores, this tells us nothing about the origins of the differences. Watson was a molecular biologist and made important contributions to our understanding of things that transpire at the molecular level, but in the realm of behavior, he valued his own narrow experiences above the carefully considered conclusions of population biologists and behavior geneticists.

There is something about the concept of race that can engender strong beliefs in the absence of compelling reason. A recent book by a journalist about race took a stand in favor of the existence of five distinct races of humans (Wade, 2014), despite abundant recent writings to the contrary by population geneticists. The book also claimed evolution had generated race differences in intelligence and other valuable behaviors, despite the lack of clear evidence about specific genes relevant to those behaviors (e.g., Chapter 15). It was criticized in a letter to the *NY Times Book Review* by 147 population geneticists who objected to the way their published work was misappropriated and misinterpreted by Wade (Coop et al., 2014).

Several of the signatories had written much about various ways of grouping humans using DNA variants, including Tishkoff et al. (2009) who found that the data were fit best by a model involving 14 groups worldwide. A brief review of Wade's book by Balter (2014) in *Science/AAAS News* was followed by a blog involving thrust and parry from more than 56 commentaries that featured *ad hominem* attacks interspersed among valiant but futile attempts to focus the discussion on established facts.

Does it really matter whether humanity is divided into 14, 6, 5, or 4 main groups on the basis of genetic variation? What use would be made of this knowledge? Whatever figure is adopted by an advocate of racial grouping, how would the abundant evidence of historical and continuing admixture be incorporated into the scheme? It is because of the indistinct and shifting boundaries between groups that many geneticists have lost confidence in the method of grouping today's humans into races. Those with strongly held racial prejudices do not care about the exact number of *bona fide* races, because their main claim is superiority of "whites" over all the others.

HIGHLIGHTS

- Anatomically modern humans (AMHs) arose in Africa more than 150,000 years ago and migrated out of Africa into the Middle East, Europe, and Asia from 60,000 to 40,000 years ago. Migration continued until all regions of the globe were populated about 800–500 years ago.
- In Europe and Asia, AMH interbred with Neanderthal and Denisovan branches of the genus *Homo*, and many

people now carry remnants of the genomes of those two species.

- As the entire globe was becoming populated, many groups migrated locally and interbred with neighboring groups. The resulting admixture, a mosaic of ancestral genomes, is apparent in the DNA of all living people.
- Statistical algorithms can be used to arrange the many groups of modern humans into clusters according to similarity of their DNA, but several patterns of clustering are consistent with the data, and the choice of the number of clusters must be made by the researchers and is somewhat arbitrary. One credible choice results in 14 clusters worldwide.
- Clusters of populations are generally not regarded as genuine races because the borders between clusters are indistinct and overlapping to a substantial degree.
- Large-scale immigration and intermarriage of people with different ancestries has further blurred the boundaries between clusters, and this trend is accelerating.
- A large majority of population geneticists and anthropologists no longer regard race as a meaningful way to describe human groups.
- Governments of different countries define race in different ways or do not use the term at all. When it is used, it serves as a socially defined construct, not a biologically precise means to separate people into groups.
- Skin color, a phenotype associated with race in the minds of many people, is influenced by several genes that are involved in the synthesis and transport of melanin, and certain alleles of these genes can lighten skin color. Genes that appear to contribute to group differences in skin color are also responsible for skin-color variations within groups.
- Lighter skin evolved slowly as the ancestors of modern Europeans migrated northward to colder regions exposed to far less sunlight, and lighter skin appears to offer some advantages for health in the far north.
- Genes known to be relevant to skin color show no association with genes thought to be relevant to intelligence or intellectual disability on the basis of genome-wide scans (see Chapter 15).

References

Balter, M. (2014). Geneticists decry book on race and evolution. *Science News*,(August 8).

Beleza, S., Johnson, N. A., Candille, S. I., Absher, D. M., Coram, M. A., Lopes, J., ... Tang, H. (2013). Genetic architecture of skin and eye color in an African-European admixed population. *PLoS Genetics*, 9, e1003372.

Brandt, G., Szecsenyi-Nagy, A., Roth, C., Alt, K. W., & Haak, W. (2015). Human paleogenetics of Europe—the known knowns and the known unknowns. *Journal of Human Evolution*, 79, 73–92.

Busby, G. B. J., Hellenthal, G., Montinaro, F., Tofanelli, S., Bulayeva, K., Rudan, I., ... Capelli, C. (2015). The role of recent admixture in forming the contemporary West Eurasian genomic landscape. *Current Biology*, 25, 2878.

Census of Population. (2016). *Ethnic origin reference guide*. Ottawa: Statistics Canada.

Chen, E., Martin, A. D., & Matthews, K. A. (2006). Understanding health disparities: the role of race and socioeconomic status in children's health. *American Journal of Public Health*, 96, 702–708.

Coop, G., Eisen, M. B., Nielsen, R., Przeworski, M., Rosenberg, N., et al. (2014). *Letters: 'A troublesome inheritance'*. New York Times Book Review. (August 8). Retrieved from: (2014). *http://www.nytimes.com/2014/08/10/books/review/letters-a-troublesome-inheritance.html?_r=0*.

Crawford, N. G., Kelly, D. E., Hansen, M. E. B., Beltrame, M. H., Fan, S., Bowman, S. L., ... Tishkoff, S. A. (2017). Loci associated with skin pigmentation identified in African populations. *Science*, 358, eaan8433.

Deviese, T., Karavanic, I., Comeskey, D., Kubiak, C., Korlevic, P., Hajdinjak, M., ... Higham, T. (2017). Direct dating of Neanderthal remains from the site of Vindija Cave and implications for the Middle to Upper Paleolithic transition. *Proceedings of the National Academy of Sciences of the United States of America*, 114, 10606–10611.

Dressler, W. W., Oths, K. S., & Gravlee, C. C. (2005). Race and ethnicity in public health research: models to explain health disparities. *Annual Review of Anthropology*, 34, 231–252.

Duggan, A. T., & Stoneking, M. (2014). Recent developments in the genetic history of East Asia and Oceania. *Current Opinion in Genetics & Development*, 29, 9–14.

Duster, T. (2005). Race and reification in science. *Science*, 307, 1050–1051.

Edmonds, P. (2018). People are made how they are. *National Geographic*, 233(April, 12).

Eiberg, H., Troelsen, J., Nielsen, M., Mikkelsen, A., Mengel-From, J., Kjaer, K. W., & Hansen, L. (2008). Blue eye color in humans may be caused by a perfectly associated founder mutation in a regulatory element located within the HERC2 gene inhibiting OCA2 expression. *Human Genetics*, 123, 177–187.

Fu, Q., Posth, C., Hajdinjak, M., Petr, M., Mallick, S., Fernandes, D., ... Reich, D. (2016). The genetic history of Ice Age Europe. *Nature*, 534, 200–205.

Galton, F. (1869). *Hereditary genius: An inquiry into its laws and consequences*. London: Macmillan and co.

Grün, R. (2006). Direct dating of human fossils. *American Journal of Physical Anthropology*, (Suppl. 43), 2–48.

Hellenthal, G., Busby, G. B. J., Band, G., Wilson, J. F., Capelli, C., Falush, D., & Myers, S. (2014). A genetic atlas of human admixture history. *Science*, 343, 747–751.

Herrnstein, R. J., & Murray, C. (1994). *The bell curve: Intelligence and class structure in American life*. New York: Free Press.

Homburger, J. R., Moreno-Estrada, A., Gignoux, C. R., Nelson, D., Sanchez, E., Ortiz-Tello, P., ... Bustamante, C. D. (2015). Genomic insights into the ancestry and demographic history of South America. *PLoS Genetics*, 11, e1005602.

Hublin, J. J., Ben-Ncer, A., Bailey, S. E., Freidline, S. E., Neubauer, S., Skinner, M. M., ... Gunz, P. (2017). New fossils from Jebel Irhoud, Morocco and the pan-African origin of *Homo sapiens*. *Nature*, 546, 289–292.

Jensen, A. (1969). How much can we boost IQ and scholastic achievement? *Harvard Educational Review*, 1–123 [Reprint series no. 2].

Kawachi, I., Daniels, N., & Robinson, D. E. (2005). Health disparities by race and class: why both matter. *Health Affairs*, 24, 343–352.

Li, P. S. (2000). *Cultural diversity in Canada*. Ottawa, Canada: Minister of Justice.

Li, S., Schlebusch, C., & Jakobsson, M. (2014). Genetic variation reveals large-scale population expansion and migration during the expansion of Bantu-speaking peoples. *Proceedings of the Royal Society B: Biological Sciences*, 281, 1–9.

Lieberman, L., & Reynolds, L. T. (1996). Race: the deconstruction of a scientific concept. In L. Lieberman, & L. T. Reynolds (Eds.), *Race and other misadventures: Essays in honor of Ashley Montagu in his ninetieth year* (pp. 142–173). Dix Hills, NY: General Hall.

Liu, F., Visser, M., Duffy, D. L., Hysi, P. G., Jacobs, L. C., Lao, O., ... Kayser, M. (2015). Genetics of skin color variation in Europeans: genome-wide association studies with functional follow-up. *Human Genetics, 134*, 823–835.

Livingston, G., & Brown, A. (2017). Intermarriage in the U.S. 50 years after Loving v. Virginia. *Pew Social Trends, 8*.

Llamas, B., Fehren-Schmitz, L., Valverde, G., Soubrier, J., Mallick, S., Rohland, N., ... Haak, W. (2016). Ancient mitochondrial DNA provides high-resolution time scale of the peopling of the Americas. *Science Advisor, 2*, e1501385.

Matisoo-Smith, E. (2015). Ancient DNA and the human settlement of the Pacific: a review. *Journal of Human Evolution, 79*, 93–104.

Montinaro, F., Busby, G. B., Pascali, V. L., Myers, S., Hellenthal, G., & Capelli, C. (2015). Unravelling the hidden ancestry of American admixed populations. *Nature Communications, 6*, 6596.

Nestel, S. (2012). *Color coded health care: The impact of race and racism on Canadians' health.* Toronto: Wellesley Institute.

Obama, B. (2008). Barack Obama's speech on race. *The New York Times,* (March 16). Retrieved from:(2008). https://www.nytimes.com/2008/03/18/us/politics/18text-obama.html.

Patin, E., Lopez, M., Grollemund, R., Verdu, P., Harmant, C., Quach, H., ... Quintana-Murci, L. (2017). Dispersals and genetic adaptation of Bantu-speaking populations in Africa and North America. *Science, 356*, 543–546.

Pickrell, J. K., & Reich, D. (2014). Toward a new history and geography of human genes informed by ancient DNA. *Trends in Genetics, 30*, 377–389.

Prüfer, K., Racimo, F., Patterson, N., Jay, F., Sankararaman, S., Sawyer, S., ... Paabo, S. (2014). The complete genome sequence of a Neanderthal from the Altai Mountains. *Nature, 505*, 43–49.

Raghavan, M., Steinrucken, M., Harris, K., Schiffels, S., Rasmussen, S., DeGiorgio, M., ... Willerslev, E. (2015). Genomic evidence for the Pleistocene and recent population history of Native Americans. *Science, 349*, aab3884.

Rees, J. L., & Harding, R. M. (2012). Understanding the evolution of human pigmentation: recent contributions from population genetics. *Journal of Investigative Dermatology, 132*, 846–853.

Rosenberg, N. A., Pritchard, J. K., Weber, J. L., Cann, H. M., Kidd, K. K., Zhivotovsky, L. A., & Feldman, M. W. (2002). Genetic structure of human populations. *Science, 298*, 2381–2385.

Sankar, P., & Cho, M. K. (2002). Toward a new vocabulary of human genetic variation. *Science, 298*, 1337–1338.

Sankararaman, S., Mallick, S., Patterson, N., & Reich, D. (2016). The combined landscape of denisovan and Neanderthal ancestry in present-day humans. *Current Biology, 26*, 1241–1247.

Seward, I. (2018). Different lives. *Majesty, 39*, 42.

Skoglund, P., Posth, C., Sirak, K., Spriggs, M., Valentin, F., Bedford, S., ... Reich, D. (2016). Genomic insights into the peopling of the Southwest Pacific. *Nature, 538*, 510–513.

Smedley, A., & Smedley, B. D. (2005). Race as biology is fiction, racism as a social problem is real: anthropological and historical perspectives on the social construction of race. *American Psychologist, 60*, 16–26.

Statistics Canada. (2017). *Previous standard—Race.* Ottawa: Statistics Canada. Retrieved from:(2017). https://www.statcan.gc.ca/eng/concepts/definitiond/pevious/preethnicity.

Sturm, R. A., Duffy, D. L., Zhao, Z. Z., Leite, F. P., Stark, M. S., Hayward, N. K., ... Montgomery, G. W. (2008). A single SNP in an evolutionary conserved region within intron 86 of the HERC2 gene determines human blue-brown eye color. *American Journal of Human Genetics, 82*, 424–431.

Tishkoff, S. A., Reed, F. A., Friedlaender, F. R., Ehret, C., Ranciaro, A., Froment, A., ... Williams, S. M. (2009). The genetic structure and history of Africans and African Americans. *Science, 324*, 1035–1044.

Visconti, A., Duffy, D. L., Liu, F., Zhu, G., Wu, W., Chen, Y., ... Falchi, M. (2018). Genome-wide association study in 176,678 Europeans reveals genetic loci for tanning response to sun exposure. *Nature Communications, 9*, 1684.

Wade, N. (2014). *A troublesome inheritance: Genes, race and human history.* New York: Penguin Publishing Group.

Wagner, J. K., Yu, J. H., Ifekwunigwe, J. O., Harrell, T. M., Bamshad, M. J., & Royal, C. D. (2017). Anthropologists' views on race, ancestry, and genetics. *American Journal of Physical Anthropology, 162*, 318–327.

Wallenstein, P. (2005). Reconstruction, segregation, and miscegenation: interracial marriage and the law in the Lower South, 1865–1900. *American Nineteenth Century History, 6*, 57–76.

Wang, F., Smith, N. R., Tran, B. A., Kang, S., Voorhees, J. J., & Fisher, G. J. (2014). Dermal damage promoted by repeated low-level UV-A1 exposure despite tanning response in human skin. *JAMA Dermatology, 150*, 401–406.

Wilde, S., Timpson, A., Kirsanow, K., Kaiser, E., Kayser, M., Unterlander, M., ... Burger, J. (2014). Direct evidence for positive selection of skin, hair, and eye pigmentation in Europeans during the last 5,000 y. *Proceedings of the National Academy of Sciences of the United States of America, 111*, 4832–4837.

Yolton, K., Dietrich, K., Auinger, P., Lanphear, B. P., & Hornung, R. (2005). Exposure to environmental tobacco smoke and cognitive abilities among U.S. children and adolescents. *Environmental Health Perspectives, 113*, 98–103.

Yudell, M., Roberts, D., DeSalle, R., & Tishkoff, S. (2016). Taking race out of human genetics. *Science, 351*, 564–565.

Yuen, A. W., & Jablonski, N. G. (2010). Vitamin D: in the evolution of human skin colour. *Medical Hypotheses, 74*, 39–44.

Further Reading

Jones, N. A. (2015). Update on the U.S. Census Bureau's race and ethnic research for the 2020 census. *Survey News. U.S. Census Bureau, 3*, 4.

20

The Future

WHAT GENES DO

One of the foremost goals of this book has been to explain what real genes actually do. This is presented as an antidote to the simplistic but popular notion that a gene codes for a specific behavior or mental disorder. A gene is a very large molecule that codes for the structure of mRNA and proteins that are also very large molecules. But many other molecules and processes such as methylation play important roles in determining where and when any particular gene is expressed in mRNA and protein. Things that occur outside the cell can have a major influence on how the exons from that gene are spliced in different ways to create different proteins. The gene does not have an executive function; it is not the conductor of the symphony. It is controlled by many factors inside and outside a cell. It is part of a system.

Single Gene Perspective

A specific gene is huge in the world of molecules but very small in the world of cells. Its location and function can often be well enough understood in the context of a small organelle such as a synapse, a sarcomere, or the growth cone at the tip of a pioneering axon. Its role in a series of steps in synthesis of important neuroactive chemicals can often be portrayed well in a fairly simple path diagram showing products of several gene-encoded enzymes. Invariably, the picture reveals that any particular gene has a definite place in the system of diverse parts, but it is nonetheless just one of many essential parts.

Zoom out to the level of a cell, perhaps a neuron, and it becomes very difficult to see exactly what any one gene is doing. Its activity can be detected by antibodies or mRNA expression arrays, but at that level, it is just one of thousands of molecules that are active in the same cell. Zoom out a little further to the level of networks of neurons active in different anatomical parts of the brain, and it becomes apparent that a specific molecule such as a neurotransmitter can serve many different functions, depending on where it occurs.

The neurotransmitter acetylcholine triggers muscle contraction, but it also is involved in diverse activities of the parasympathetic nervous system and the functioning of many glands. The Ach molecule is fairly small, but its presence is detected by a great diversity of immense cholinergic receptor molecules that are widely distributed throughout the nervous system. Dopamine too has many different roles in the drama of behavior. Even at the level of sensory receptors with their exquisite sensitivities to specific chemicals or colors of light, it is difficult to ascribe a simple function to just one gene-encoded receptor because the nervous system senses the identity of environmental stimuli according to the combinations of different kinds of receptors that are activated. Thus, any notion of one gene-one neural function is quite unrealistic.

Behaviors are products of networks of neurons and many other cells, and those neurons and cells engage many kinds of proteins and genes to make things work the right way. One sign of this is the occasional mutation in one gene that has a major impact on many behaviors. As knowledge of PKU, AIS, or CAH teaches, behavior depends on complex systems of molecules and cells.

It is clear that the role of any one gene is highly specific and cannot be anticipated by just looking at its DNA sequence. Extensive research is needed to learn how any one gene fits into the larger picture of neural and behavioral function. The research often uncovers real surprises and opens our eyes to new possibilities. The picture is clearest when the mutation in question has a very large effect.

Complex Traits

Complex traits influenced by large number of genes with small effects offer no way to know what any specific gene or even an assembly of genes is doing. We presume that the variety of gene effects seen in single-gene studies also underlies multiple gene influences on complex traits, but there is no way to verify this. The original objective of

Genes, Brain Function, and Behavior
https://doi.org/10.1016/B978-0-12-812832-9.00020-8

269

the genetic analysis of complex traits was to dissect the complex into a number of simpler components. The hope was that complex traits arise from a relatively small number of genes with moderate-sized effects and that larger samples of people would make it possible to find those genes. This has not been achieved because the gene contributions are just too small individually and too numerous. The statistical methods used to study complex traits reveal almost nothing about what genes actually do.

The current approach of quantitative geneticists to complex traits is to use even larger samples of people to localize even more relevant genes. A recent analysis of human height gives a foretaste of things that likely will be detected for human intelligence and schizophrenia. (Boyle, Li, & Pritchard, 2017) synthesized the findings of the massive GIANT study that combined large samples into one large data set using meta-analysis (Wood et al., 2014). The total sample of 253,288 people examined with GWAS detected 697 loci significant at $P = 5 \times 10^{-8}$ that accounted for about 16% of phenotypic variance in height. They estimated that there were at least 100,000 SNPs dispersed widely across the genome that had a median effect size of 0.14 mm on height and that "a substantial fraction of all genes contribute to variation." They concluded that "there is an extremely large number of causal variants with tiny effect sizes on height." Their review of other phenotypes, some related to behavior and psychiatric phenotypes, suggested this pattern was widespread. In such a large forest of small trees, it will not be possible to ascertain the role of any single-gene polymorphism. Instead, vast numbers of genes appear to be organized in networks with multiple functions, giving rise to a pattern of "network pleiotropy."

ANTICIPATING FUTURE TRENDS

Discovery of Genetic Effects

Research on genes, the brain, and behavior seems to be still in the early stage of finding genes that are relevant to differences between people. Although several mutations of large effect have been known for decades, the ability to scan the entire genome for lesser or rarer variants is quite recent, and many initial claims are still in need of replication. Many genes implicated by data from SNPs still need to be verified.

Polygenic effects have been confirmed for several complex traits such as intelligence in the normal range. GWAS studies have indicated there are considerable numbers of genetic variants with very small effects on behavior and an almost complete absence of effects with moderate magnitudes. Very small effects were not detectable until sample sizes exceeded 50,000 people

(Table 15.6). Some researchers have proposed that even larger samples are needed (Plomin & von Stumm, 2018).

It is important to remind ourselves that GWAS studies do not provide evidence of genetic effects on any phenotype. They demonstrate statistical association between SNP marker alleles and some phenotype. The difficult challenge of making a credible connection between a SNP and a real gene in the vicinity cannot be surmounted by studying larger samples. Larger samples can only add to the compendium of SNPs whose relations to real genes are unknown.

Statistical theory leads us to expect that larger samples will inevitably show evidence of many more SNPs with even smaller effect sizes than the 0.7 IQ point difference found in recent research. Fig. 20.1 shows the critical size of a difference in IQ for two genotypes at a single SNP locus needed to reject the null hypothesis when a study involves different sample sizes. Expanding the sample to 200,000 or even 1 million people is virtually certain to yield a richer harvest of apparently significant SNP effects. Some unknown fraction of positive "hits" in a large GWAS study will survive replication. Increasing the sample size will not enable researchers to uncover genes with larger effects. Those kinds of genes would have been plainly evident with much smaller samples.

De novo mutations of large effect have proved to be readily detectable with whole exome sequencing in parent-offspring trios when the offspring chosen for study show substantial behavioral deficits. Prime examples involve genes and copy number variations pertinent to intellectual disability, autism, and schizophrenia. The difference between child and parent IQ in the ID studies is often 50 IQ points or more, a huge effect by anyone's standards. The general pattern shows a considerable

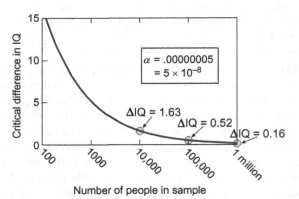

FIG. 20.1 Approximate number of people needed in a sample to detect an IQ difference between two genotypes of an SNP marker. The larger the sample size, the smaller the phenotypic difference that can be detected. Alpha level in the inset is the conventional genome-wide significance criterion for a GWAS study. With a very large sample, a very small difference in IQ test score contributed by just one locus can be detected.

number of relevant mutations, each present in only a few or even one family. The study by de Ligt et al. (2012) reported major mutations in 29 genes among 100 patients, and Rauch et al. (2012) reported mutations in 24 genes among 51 patients. One noteworthy feature of the two studies is that they detected only one gene in common (SCN2A).

When a new study with a larger sample is done, there most likely will be even more "hits," including a few seen in previous studies. It seems likely that more studies of this kind will uncover more plausible gene effects, but eventually, the studies will deliver diminishing returns of new and informative genes with large effects. What the total count might eventually be is difficult to estimate, but 100 genes of major effect on ID seem like a realistic number. For some of the ID mutations already well known, the frequency is sometimes 1/10,000 births. If there are about 100 such genes in a large population, we would expect about 1/100 or 1% of children to show ID, which is a realistic figure.

This estimate applies only to de novo mutations that show dominant expression, where just one copy of the gene that yields a defective protein is sufficient to impair a behavioral phenotype. How many there might be with recessive expression would require testing far larger samples. Exome sequencing could easily detect individuals who carry two aberrant copies of a gene, even if they are different alleles, and both parents should be carriers. To date, those kinds of recessive genes seem to be quite rare in ID and autism spectrum disorder.

A few mutations of moderate effect can combine to shift some phenotypes by substantial amounts. This pattern has not yet been found for behavioral phenotypes or psychiatric disorders, but it does occur for biomedical maladies such as diabetes. An example is shown in Chapter 19 for skin color, where about 35% of phenotypic variation in Cape Verde can be accounted for by knowing someone's alleles at just four relevant genes (SLC24A5, SLC45A2, TYR, and HERC2/OCA2; see Table 19.4). Genes with such large phenotypic effects would be easily detected by a GWAS or an exome sequencing study. Perhaps, this pattern will appear for behavioral phenotypes not yet evaluated.

Discovery of Treatments, Therapies

The ultimate objective of most genetic research with humans is to learn things eventually that will have practical consequences, hopefully beneficial ones. Simply, compiling catalogs of genes that appeared as "hits" in one or more studies seems pointless. It is fair to ask what are the practical consequences of recent genetic studies for understanding and altering nervous system and behavioral or psychiatric traits.

Polygenic effects involving a few dozen or even hundreds of genes are inconsequential in any practical sense when the specific genes are unknown. This pessimistic assessment is magnified by the observation that a prediction equation based on all significant SNPs in a study typically is not associated with even 5% of phenotypic variance (Table 15.6). Genes work in highly specific ways, and without knowledge of those ways, there is no insight into treatments that will ameliorate adverse genetic effects. The entire enterprise of statistical genetics since its origins with Galton and Pearson made no effort to discover environmental or medical treatments. Many of the devotees of that approach argued forcefully that attempts to improve outcomes by better environments are futile. New studies with larger samples will inevitably boost the percent of variance associated with a phenotype of interest, but those studies will do this by accumulating more and more ever smaller SNP effects that offer steadily diminishing prospects of beneficial therapies.

De novo mutations of large effect open doorways to treatments. Given what we already know about the challenges in treating single-gene disorders such as PKU, HD, AIS, and LHON, a concentrated effort by several teams of investigators will be needed to unlock the secrets of just one of those genes pertinent to ID or autism. The quest is complicated by the relative scarcity of children suffering from a particular kind of mutation. Some of those mutations are extremely rare, even "family private," whereas others will be seen more often. The more frequent ones should be good targets for new research efforts. This will be a slow process, and we can anticipate that for some kinds of mutations, the conclusion will be that nothing can be done to change the course of development. But without making a serious effort to change something, its plasticity will remain entirely unknown.

Mutations of moderate effect, if they can be identified, will be less promising for further research than those of large effects detected through whole exome scans. An end run around this difficulty may be possible if some of the major mutations detected and then explored in depth turn out to have alleles of lesser but moderate effect. For example, some genes that result in albinism when there is a null mutation have alleles that exert a moderate effect on the amount of melanin expressed in skin color within the normal range (Chapter 19). The same thing could happen for genes pertinent to nervous system function.

Genetic Screening

Screening, whether done shortly after conception, at the fetal stage or at birth, is not something done merely out of curiosity. Parents and medical staff generally have this done only if there might be some outcomes that would be "actionable." As described in Fig. 9.4, removing just one cell of an eight-cell embryo can identify a wide range of genetic mutations or chromosome copy number variants. In vitro diagnosis is almost always done on several viable embryos, and those with evident genetic defects are then discarded. At the fetal stage, amniocentesis to withdraw cells of the fetus or a sample of the mother's blood to detect fetal DNA for genotyping is feasible. It could detect a wide range of defects, and pregnancy termination would often be the result when a serious defect is found. Screening newborns, on the other hand, is likely to benefit some children and their family by identifying deficits that can be readily treated.

Polygenic traits could potentially be assessed in one cell of an eight-cell embryo. This appears to be a motivation for the emphasis on polygenic prediction that otherwise has no clear utility. The recent review by Plomin and von Stumm (2018) suggests that with enough data, it might be possible to make a reasonably accurate prediction of a child's future intelligence. The technology to do this kind of embryo selection exists now. At the present time, the accuracy of the prediction will not be very high. It seems unlikely that a surge of customers will soon be lining up for this service. Nevertheless, private companies are already promoting this kind of application to generate new business. Apart from possible use in embryo selection, there does not appear to be any other practical application of polygenic prediction.

MIRACLE CURES?

In this age of biotechnology, many new methods are claimed by some to have the potential to revolutionize treatment of genetic disorders. Much of this is hype driven by profit motives, but there are indeed new approaches being evaluated now.

One of these is termed clustered regularly interspaced short palindromic repeats CRISPR/Cas9 system, a technique to snip out small segments of DNA that may encode a mutation and then insert the normal DNA sequence into the gap. It seems to work with most species of plants and animals, especially when done in the lab with tissue culture. It is being applied now with mice, and there are preliminary studies with humans as well (Bassuk, Zheng, Li, Tsang, & Mahajan, 2016). Lessons are being learned. One of the big technical hurdles is getting the molecular probe that performs the magic into the cell. Large biomolecules do not readily pass through a living cell membrane, unlike small drugs that are water soluble and end up everywhere in the body and nervous system. Molecular probes can sometimes do the trick in human cells in tissue culture. The mouse is an ideal subject for testing ways to deliver the package in a whole animal to the address where it is needed.

In the course of those mouse studies, researchers recently uncovered an unpleasant surprise. CRISPR is promoted as a fast and incredibly precise way to alter the DNA by targeting specific kinds of small repeated DNA sequences near a gene. Fast it is, but a closer look at the results raises concerns about off-target mutations. A team of researchers decided to check the precision of the change by using whole-genome sequencing on the mice that had the CRISPR alteration. They identified more than 1500 single-base mutations scattered widely across the genome in mice treated with CRISPR-Cas9 and more than 100 deletions or insertions of multiple nucleotide bases here and there (Schaefer et al., 2017), some of which were in exons and likely to change a protein structure. The study used a biochemical approach and did not search for phenotypic alterations in a live animal caused by all those mutations, but the risk could be substantial. This kind of genetic "side effect" could negate the therapeutic benefits.

Shortly after the article was published, doubts about its conclusion were expressed by the editor of the journal where it was published, *Nature Methods*. There were concerns that the mice compared for the study might have differed at many loci even without the CRISPR treatment. The paper was then retracted, but four of its six authors stated they did not agree with the retraction. This can happen with a relatively new technology—high hopes followed by doubts and complications. More work is needed on this topic.

Meanwhile, a battle is taking place over patents on parts of the CRISPR method. The Broad Institute at Harvard and Massachusetts Institute of Technology holds 13 patents but is being challenged by the University of California, and the French drug company Cellectis claims to have won a new patent that is broader than the American ones (Cohen, 2016).

Whether in the future CRISPR will be hailed as a miracle method is difficult to foresee. It can be applied with tissue transplantation where cells from an individual are repaired and then used to replace cells with a mutation. It is not apparent that this approach can change the germ cells and generate multigenerational repairs.

CYSTIC FIBROSIS—DIZZYING COMPLEXITY AND COST

The challenges to deriving health benefits from genetic knowledge are well illustrated by the example of cystic fibrosis. It is one of the most common recessive genetic disorders that imperils health, having a prevalence of about 1/3600 in European populations, which means that about one person in 60 is a carrier of the mutant gene. It seriously impairs lung function and other organ systems because of excess accumulation of mucus. A concerted research effort has been devoted to diagnosing, treating, and perhaps someday curing the malady that significantly shortens life. The gene was discovered in 1989, and almost 30 years later, there still is no cure. A chronology shows the main steps taken in this quest (Orenstein, O'Sullivan, & Quinton, 2015; Pearson, 2009):

1982: Salty sweat of CF patients provides clue to biochemical defect in ion transport.

1985: The gene is mapped to chromosome 7.

1989: Gene is sequenced and named cystic fibrosis transmembrane conductance regulator (CFTR).

1993: Clinical trial of gene therapy ends in failure.

1994: The drug dornase alfa helps to break up thick mucus in the lungs.

1995–2010: 1500 different mutations in CFTR are found to cause CF.

2007: The drug ivacaftor (VX-770), a potentiator of CFTR, used to treat the G551D mutation.

2007: Newborn screening programs begin.

2009: Second gene therapy trial begins; eventually abandoned.

2012: The drug ivacaftor (Kalydeco; VX-770) approved to treat CF-G551D in the United States.

2014: Cystic Fibrosis Foundation sells royalty rights to VX-770 for $3.3 billion.

2017: The drug combination lumacaftor-ivacaftor is marketed to aid those with the F508del mutation but is not approved for funding in Canada.

2018: A new consortium is formed to develop an improved gene therapy for CF.

According to Pearson (2009), "From the beginning, the goal was gene therapy." The general idea was to insert the normal human gene into a bacterium that could then be transfected to human lungs where it would make a normal CFTR protein. The human immune system blocked that approach. A number of new drugs were tested for their ability to potentiate the action of the defective CFTR protein, and a few of them appeared to help those with specific CFTR alleles. This was not gene therapy, and it did not cure the malady, but it did reduce CF symptoms in about 5% of CF patients. Researchers compiled evidence that there might be other genes that modify the actions of CFTR. Pearson (2009) remarked that the data suggest CF "will spiral into a new realm of dizzying complexity."

Meanwhile, the most effective treatment for symptoms is behavioral. Skillful pounding on the chest followed by deep coughing can clear large amounts of the mucus from the lungs. The patient is still vulnerable to chest infections but, with the aid of drugs to break up the mucus, can lead a reasonably active life. Treating just the symptoms and providing general medical care has extended the median survival time from 15 years in 1975 to 41 years in 2013 (Orenstein et al., 2015). None of this prolongation of life is attributable to the knowledge of the CFTR gene.

A formidable challenge arose recently not from the difficult science but from the practicalities of sales and profits from an "orphan" drug with so few beneficiaries. Medical researchers and cystic fibrosis patients and their families were outraged by the $311,000 annual cost of ivacaftor and the proposed $259,000 per year for the combo drug lumacaftor-ivacaftor (Orenstein et al., 2015). The public health-care system in British Columbia, Canada, has not yet approved funding for either drug (BC Ministry of Health, 2018). Part of the reason for the high prices, according to drug companies, is that development costs are high and there are few customers with any one allele. So, the price must be high in order to realize a healthy profit.

The example of costs associated with CFTR raises questions about the whole enterprise of precision medicine. Most of the genetic mutations pertinent to behavioral or psychiatric disorders are quite rare. In the case of de novo mutations, there are many different genes affected but only a very small number of patients with any particular mutation. If it becomes possible after a large amount of research to discover a way to suppress symptoms of a disorder with gene-based knowledge, what will the miracle treatment cost? It is highly likely that the price will exceed that of ivacaftor. Just how many genetic mutations can a society afford to treat at these rates? The publicity at the launch of the precision medicine initiatives was silent on this matter. The matter will lie dormant in the realm of behavioral and psychiatric disorders until there are finally a few breakthroughs that bring miracle drugs to market.

RENEWED EUGENICS?

The old eugenics movement promised a rapid decline in people with hereditary defects as a result of small-scale sterilization policies. Of course, the political advocates never did the simple calculations needed to show that this would not work. They also failed to require evidence that a particular inmate in a public institution did indeed harbor a defect in a gene that would be transmitted to future generations. The old policies were based on bad science mixed with social prejudice against the poor and racial minorities.

Given the new and powerful tools to detect genetic variants that are available today, it is interesting to explore whether a new kind of eugenics based on good science is feasible. The objective of eugenics is to reduce the burden on society of hereditary disease by preventing procreation of people who possess a mistake in the DNA code, a mutation that can cause disease in future generations. A treatment that enhances the quality of life and health of a person suffering the effects of a genetic mutation is not a eugenic measure if all it does is change a phenotype. It must prevent the mutation from being reproduced in offspring. Several scenarios are pertinent.

Embryo Selection

Prospective parents might have reason for concern that their child will possess a mutation that is likely to impair development. This could happen when one of the parents is a known carrier, as in the case of Huntington disease, or both are carriers, as happens with cystic fibrosis. It can also happen when there is a de novo mutation or chromosome copy number variant that is present in neither parent. A DNA test targeted at a specific mutation known to occur in a family might be informative, or a couple might decide to have whole-genome sequencing done for each fertilized egg in order to check for new mutations (Fig. 9.4). Any embryo found to carry a clearly harmful mutation would not be implanted back into the mother's womb. This approach would prevent a genetic disorder from being expressed in the current generation or being passed to future generations.

Screening Fetal Cells and DNA in Maternal Blood

The same kind of genetic screening could be done on fetal cells of fragments of fetal DNA found in the mother's blood. The technology is much easier than embryo selection that requires growing embryos in a glass dish outside the mother's body. The downside is that selection would require pregnancy termination of a fetus several months old. The fetal cells for genetic analysis could also be obtained via amniocentesis or chorion biopsy.

Newborn Screening

Blood from a newborn is easily obtained from the umbilical cord. Once there is a live birth, the objective of a genetic test is to find children in need to treatment to prevent them developing symptoms. Terminating a live-born infant is not an option in most jurisdictions. Newborn screening is thus aimed at the phenotype and is not a eugenic measure.

Sterilization of a Prepubertal Young Person

Sterilization would have no impact on the phenotype of the individual but would prevent a genetic mutation from being transmitted.

Selective Chemical Ablation of a Disease-Related Allele

Presuming the CRISPR or some similar technique is eventually perfected, it might be possible to remove the mutation from a child's body cells, which would alter the phenotype but probably not prevent transmission. A variant of the method aimed precisely at germ cells or eight-cell embryos might reduce the likelihood of transmitting the mutation to future generations. Only the latter would have eugenic implications.

Deciding Not to Have Children

A couple who decides not to have children because of concern that a mutation might be passed to future generations would be implementing a eugenic choice. It would actually amount to eugenics only if the risk of passing the mutation is well founded.

Certain of these scenarios entail the parents' wish to avoid the burden of rearing a child with a genetic mutation. Those cases qualify as eugenic measures if they also prevent transmission of a mutation. These options differ from the old eugenics in that they would hopefully be based on good science, true facts, and voluntary participation by well-informed parents. Under this scenario, there is no longer a political movement called eugenics that strives to sway families and governments to wage war against harmful mutations. Instead, there is a widely dispersed, case-by-case consideration of options and actions that will have implications for future populations. Sometimes, a decision about reproduction will utilize guidance from a professional genetic counselor. Under contemporary ethical guidelines of genetic counselors (e.g., the National Society of Genetic Counselors), the best practice is not to recommend a specific course of action but to outline several possibilities and the likely consequences of each, so that the prospective parents can decide what is best for them and their future child, if there will be one.

If new treatments to relieve the symptoms of a genetic disorder enable a person to have children who previously could not, the result will be the direct opposite of a eugenic measure. More of that kind of mutation will be passed to the next generation than happened previously. The consequences of this change would be ameliorated, however, by the existence of an effective treatment. The gene would be passed on, but the formerly impaired phenotype would no longer be seen. What was formerly regarded as a genetic disease might become just another genetic variant that requires specific kinds of environments in order to thrive. It could become a trait like nearsightedness that is very common but is treated readily with corrective lenses or a minor corneal surgery.

ETHICAL STANDARDS

Only 100 years ago, many people suffering from some kind of genetic or chromosomal anomaly were regarded as not fully human and undeserving of full rights in society. They were often confined in institutions and treated badly, even used as guinea pigs in medical experiments. Now, we see many instances where people considered intellectually disabled, autistic, or schizophrenic are demanding full rights in mainstream society. Patient and family activism is having a major impact on government policies in some countries, even influencing the criteria for diagnosis of a disorder or preferred treatments (Berryessa & Cho, 2013).

Genetic knowledge now contributes to the way we perceive many disabled people. It has become difficult to see how a change in just one nucleotide base of a gene would make someone less than human. We all differ from each other at thousands of genetic loci. Everyone has a chunk of a chromosome missing or added here and there. There is no "normal" genotype that can serve as a "gold standard" for ideal humanity. Some mutations have unfortunate consequences for development of the brain or other organs, whereas others are phenotypically silent or neutral. Genetic variation is part of being human and being alive. In every generation, new alleles of most genes are coming into existence by chance, while others are passing away into history. The mutations that are most harmful for development are rare because they make it very difficult for someone to have children and pass on a malformed gene. The process tends to be self-limiting.

The terms used to refer to people have changed considerably. Instead of saying someone is abnormal or a mutant, we now say the person has special needs and may be the carrier of a mutation. Even the word "anomaly" can stigmatize, whereas "variant" is less pejorative. Someone who is diagnosed as expressing schizophrenia or autism need not carry a lifelong label as being a "schizophrenic" or "autistic," as though they are interlopers from some special strain of human. Schizophrenia and autism are phenotypes, and phenotypes can change, while the genotype remains the same. Sometimes, the person experiences a long episode of schizophrenic thought and behavior, but then, the symptoms recede, and the storm is over. It is likely that the person will remain unusually vulnerable to a recurrence of schizophrenic symptoms, but when the symptoms are absent, so is the disease in situations where the disease is defined by a phenotype. Having a mutant form of the enzyme phenylalanine hydroxylase does not give someone PKU for life. PKU is a phenotype, and it is expressed only in certain environments. The same is true for schizophrenia (Fig. 17.3).

Prenatal Screening

There is a distinction between good ethical standards of professional conduct and legal versus illegal acts.

Among many countries there is a wide range of practices allowable under the law that may not be seen as ethical by everyone in the medical profession. This is especially true for the period from conception to birth. In many countries, it is now acceptable to terminate a pregnancy for social reasons. This choice may be made by the mother who must carry that child throughout pregnancy and then rear it for many years, whereas for married couples, it is often a joint decision.

In the 1920s and 1930s, laws in several jurisdictions authorized compulsory sterilization of someone with Huntington disease. Those laws have all been repealed, and the future of someone carrying the HTT mutation that can lead to HD is now decided by the parents or even the person who carries the HTT mutation. Embryo selection to prevent HD (Fig. 9.4) is legal in many places. The scope of things that could be the basis for selection is wide. A couple might decide they want a child with blue eyes. Many would disagree with such a choice and the entire concept of designer babies, but the current law would stand aside. At a public medical facility, the procedure might not be covered in cases of mere cosmetic choice, but there would be private clinics willing to do the procedures for a fee.

This can be quite controversial in the case of sex selection. In some cultures, the birth of a son to continue the family name and wealth on the male side is of great importance. This tendency can persist even after immigration to a country such as the United States or Canada that generally does not show any marked preference for male births. Analysis of birth and abortion records for women of South Asia ancestry observed a striking bias for the third birth when the first two children had both been girls (Wanigaratne et al., 2018). The ratio of boys to girls on the third birth was 2.4:1 following two female births, compared with 0.6:1 following two male births, and the rate of abortions following two female births was far higher (28/100) than following two male births (11/100). These data indicate a much stronger tendency to abort a fetus after having two girls, with the result that many more sons were third born. The sex of the aborted fetus was not recorded in official data, but the pattern of data for live-born infants strongly supports the hypothesis of pervasive preference for having at least one son.

It is legal in many jurisdictions to have whole-genome sequencing done for several embryos or a fetus and not continue a pregnancy when some major mutation or chromosome copy number variation is found. For a copy number variant such as Down syndrome, this would have no eugenic consequences because the trisomy would not be transmissible to future offspring. Considering how few of the known genetic mutations are currently treatable, it seems likely that knowledge of most genetic mutations would be used to opt for termination of pregnancy when a prenatal test returns evidence of a genetic mutation. There has been very little discussion of this aspect of precision medicine and personal genome assessment, and data on the practice will be very difficult to assemble. Nevertheless, there are grounds for concern that the major application of the new genetic knowledge may be eugenic, even though it is advertised as ameliorative and aimed to cure. The major motive in launching the DNA sequencing efforts may very well have been to discover cures, but this now appears to have been a case of wishful thinking. Cures for most genetic disorders involving behavior or mental traits are very far away, whereas embryo selection is doable here and now. In making this statement, there is no intention to suggest embryo selection is a good idea or something to be discouraged. It is something deeply personal, and few governments will want to intrude at the prenatal phase of reproduction.

Infancy

Once a child with a serious genetic mutation is born, the issue becomes treatment, not continuation of life. Treatment will depend strongly on the health-care system that prevails in a country. If there is universal health care provided by a government, then most infants affected by a particular medical disorder would probably receive some kind of special care. Lacking universal health insurance, infants from wealthy and well-educated parents might receive the best available care, while those born into poverty and ignorance receive little or nothing, depending on the country. When the government is the source of funds for treatment, an issue may arise, as with cystic fibrosis, concerning how much money can be devoted to treating rare genetic disorders over a lifetime. Could the system be bankrupted by discovery of many new treatments that are outrageously expensive?

Adulthood

Once a person carrying a serious genetic mutation becomes an adult, a number of issues may arise that would require the Wisdom of Solomon to resolve. Suppose a man 50 years old has five children, two of whom have already married and given him four grandchildren. His own father died in an auto accident at the age of 27, and little is known of ancestors further back who all lived in countries in Eastern Europe during the Cold War. He then develops neurological symptoms and is eventually diagnosed with Huntington disease. What should he do? Knowing that nothing can be done to prevent the onset of symptoms in carriers of the HTT mutation, should he tell his five children? If he tells them, will any or all of them decide to take the test for the HD allele?

Should his adult children tell their young children they might be at risk? If an adult child of the man takes the test and is a carrier, should she tell any prospective marriage partner of her situation? If a person is a confirmed carrier, should he inform his insurance company of a preexisting condition? Eventually, when one of the man's children develops symptoms and is diagnosed of HD, probably everyone in the family will find out about HD and then want to know why they were not told. The father may have decided, with very good reasons, that it would be better for his children who are carriers to live a good life as long as possible without the cloud of eventual HD hanging over them.

Privacy

Who is entitled to know the results of someone's genetic test? If it is ordered through a family physician or a medical geneticist, it ought to be confidential, but what about results of a spit-in-the-tube genetic testing service? What if the data banks of that service get hacked via the Internet, as has happened for some of the big banks and even police agencies?

Many deep and difficult ethical issues arise from the occurrence of genetic mutations, questions that are beyond the scope of this book. Everything would be so much easier if precision medicine could quickly and accurately prescribe a curative treatment, but this does not seem to be likely in the near future.

References

Bassuk, A. G., Zheng, A., Li, Y., Tsang, S. H., & Mahajan, V. B. (2016). Precision medicine: genetic repair of retinitis pigmentosa in patient-derived stem cells. *Scientific Reports, 6,* 19969.

BC Ministry of Health (2018). *BC PharmaCare Drug Information-Lumacaftor-ivacaftor (Orkambi). (2018). https://www2.gov.bc.ca/assets/gov/health/health-drug-coverage/pharmacare/lumacaftor-ivacaftor-3589-info.pdf.*

Berryessa, C. M., & Cho, M. K. (2013). Ethical, legal, social, and policy implications of behavioral genetics. *Annual Review of Genomics and Human Genetics, 14,* 515–534.

Boyle, E. A., Li, Y. I., & Pritchard, J. K. (2017). An expanded view of complex traits: from polygenic to omnigenic. *Cell, 169,* 1177–1186.

Cohen, J. (2016). Dramatic twists could upend patent battle over CRISPR genome-editing. *Science News,* https://doi.org/10.1126/science.aah7381.

de Ligt, J., Willemsen, M. H., van Bon, B. W., Kleefstra, T., Yntema, H. G., Kroes, T., ... Vissers, L. E. (2012). Diagnostic exome sequencing in persons with severe intellectual disability. *New England Journal of Medicine, 367,* 1921–1929.

Orenstein, D. M., O'Sullivan, B. P., & Quinton, P. M. (2015). Cystic fibrosis: breakthrough drugs at break-the-bank prices. *Global Advances in Health and Medicine, 4,* 8–57.

Pearson, H. (2009). Human genetics: one gene, twenty years. *Nature, 460,* 164–169.

Plomin, R., & von Stumm, S. (2018). The new genetics of intelligence. *Nature Reviews: Genetics, 19*, 148–159.

Rauch, A., Wieczorek, D., Graf, E., Wieland, T., Endele, S., Schwarzmayr, T., Di Donato, N., (2012). Range of genetic mutations associated with severe non-syndromic sporadic intellectual disability: an exome sequencing study. *The Lancet, 380*, 1674–1682.

Schaefer, K. A., Wu, W. H., Colgan, D. F., Tsang, S. H., Bassuk, A. G., & Mahajan, V. B. (2017). Unexpected mutations after CRISPR-Cas9 editing in vivo. *Nature Methods, 14*, 547–548.

Wanigaratne, S., Uppal, P., Bhangoo, M., Januwalla, A., Singal, D., & Urquia, M. L. (2018). Sex ratios at birth among second-generation mothers of South Asian ethnicity in Ontario, Canada: a retrospective population-based cohort study. *Journal of Epidemiology and Community Health,* https://doi.org/10.1136/jech-2018-210622. epub ahead of print.

Wood, A. R., Esko, T., Yang, J., Vedantam, S., Pers, T. H., Gustafsson, S., ... Frayling, T. M. (2014). Defining the role of common variation in the genomic and biological architecture of adult human height. *Nature Genetics, 46*, 1173–1186.

Index

Note: Page numbers followed by *f* indicate figures, *t* indicate tables, and *b* indicate boxes.